A Stratigraphical Basis for the Anthropocene

Geological Society books refereeing procedures

The Society makes every effort to ensure that the scientific and production quality of its books matches that of its journals. Since 1997, all book proposals have been refereed by specialist reviewers as well as by the Society's Books Editorial Committee. If the referees identify weaknesses in the proposal, these must be addressed before the proposal is accepted.

Once the book is accepted, the Society Book Editors ensure that the volume editors follow strict guidelines on refereeing and quality control. We insist that individual papers can only be accepted after satisfactory review by two independent referees. The questions on the review forms are similar to those for *Journal of the Geological Society*. The referees' forms and comments must be available to the Society's Book Editors on request.

Although many of the books result from meetings, the editors are expected to commission papers that were not presented at the meeting to ensure that the book provides a balanced coverage of the subject. Being accepted for presentation at the meeting does not guarantee inclusion in the book.

More information about submitting a proposal and producing a book for the Society can be found on its website: www.geolsoc.org.uk.

It is recommended that reference to all or part of this book should be made in one of the following ways:

Waters, C. N., Zalasiewicz, J. A., Williams, M. Ellis, M. A. & Snelling, A. M. (eds) 2014. *A Stratigraphical Basis for the Anthropocene*. Geological Society, London, Special Publications, **395**.

Snowball, I., Hounslow, M. W. & Nilsson, A. 2014. Geomagnetic and mineral magnetic characterization of the Anthropocene. *In:* Waters, C. N. Zalasiewicz, J. A., Williams, M., Ellis, M. A. & Snelling, A. M. (eds) 2104. *A Stratigraphical Basis for the Anthropocene*. Geological Society, London, Special Publications, **395**, 119–141. First published online November 19, 2013, http://dx.doi.org/10.1144/SP395.13

GEOLOGICAL SOCIETY SPECIAL PUBLICATION NO. 395

A Stratigraphical Basis for the Anthropocene

EDITED BY

C. N. WATERS
British Geological Survey, UK

J. A. ZALASIEWICZ
University of Leicester, UK

M. WILLIAMS
University of Leicester, UK

M. ELLIS
British Geological Survey, UK

and

A. M. SNELLING
British Geological Survey, UK

2014
Published by
The Geological Society
London

THE GEOLOGICAL SOCIETY

The Geological Society of London (GSL) was founded in 1807. It is the oldest national geological society in the world and the largest in Europe. It was incorporated under Royal Charter in 1825 and is Registered Charity 210161.

The Society is the UK national learned and professional society for geology with a worldwide Fellowship (FGS) of over 10 000. The Society has the power to confer Chartered status on suitably qualified Fellows, and about 2000 of the Fellowship carry the title (CGeol). Chartered Geologists may also obtain the equivalent European title, European Geologist (EurGeol). One fifth of the Society's fellowship resides outside the UK. To find out more about the Society, log on to www.geolsoc.org.uk.

The Geological Society Publishing House (Bath, UK) produces the Society's international journals and books, and acts as European distributor for selected publications of the American Association of Petroleum Geologists (AAPG), the Indonesian Petroleum Association (IPA), the Geological Society of America (GSA), the Society for Sedimentary Geology (SEPM) and the Geologists' Association (GA). Joint marketing agreements ensure that GSL Fellows may purchase these societies' publications at a discount. The Society's online bookshop (accessible from www.geolsoc.org.uk) offers secure book purchasing with your credit or debit card.

To find out about joining the Society and benefiting from substantial discounts on publications of GSL and other societies worldwide, consult www.geolsoc.org.uk, or contact the Fellowship Department at: The Geological Society, Burlington House, Piccadilly, London W1J 0BG: Tel. +44 (0)20 7434 9944; Fax +44 (0)20 7439 8975; E-mail: enquiries@geolsoc.org.uk.

For information about the Society's meetings, consult *Events* on www.geolsoc.org.uk. To find out more about the Society's Corporate Affiliates Scheme, write to enquiries@geolsoc.org.uk.

Published by The Geological Society from:
The Geological Society Publishing House, Unit 7, Brassmill Enterprise Centre, Brassmill Lane, Bath BA1 3JN, UK

The Lyell Collection: www.lyellcollection.org
Online bookshop: www.geolsoc.org.uk/bookshop
Orders: Tel. +44 (0)1225 445046, Fax +44 (0)1225 442836

British Library Cataloguing in Publication Data

A catalogue record for this book is available from the British Library.
ISBN 978-1-86239-628-9
ISSN 0305-8719

Distributors
For details of international agents and distributors see:
www.geolsoc.org.uk/agentsdistributors

Typeset by Techset Composition India (P) Ltd., Bangalore and Chennai, India.
Printed by Berforts Information Press Ltd, Oxford, UK

Contents

The technosphere concept

Dedication

We take pleasure as editors in dedicating this volume to Paul Crutzen of The Max Planck Institute for Chemistry, Germany. Professor Crutzen played a pivotal role in the creation of the Anthropocene concept in its modern sense, publishing initially with Eugene Stoermer in the *IGBP Newsletter* of 2000, and then in a brief, but vivid paper in *Nature* in 2002. These papers provided the springboard for the subsequent extraordinary spread of this concept through both the sciences and the humanities. Regardless of whether this term will eventually be formalized to become part of the Geological Time Scale (with the papers in this volume being devoted to examining the stratigraphic evidence) the Anthropocene has already proved extraordinarily fruitful in placing both human history and recent environmental changes within the deep time context of Earth history.

Paul Crutzen is among the most respected of living scientists: he was the most cited scientist in the 1990s, received a Nobel Prize in 1995 for his work on the mechanisms behind ozone destruction in the stratosphere, and was also a lead figure in the 'nuclear winter' hypothesis. Having essentially conceived, and driven, the Anthropocene concept in its present form, he remains deeply interested in, and supportive of, the analysis of the Anthropocene in geological terms. Without him, it is unlikely that this new scientific field would exist in anything like its current form, and it certainly would not have the scientific and public profile that, not least, have taken the whole of the science of stratigraphy to significantly greater levels of public recognition and significance. As he recently wrote, among the key attributes of the Anthropocene are its 'interdisciplinary, intensity and internationality': qualities that we think will prove to be of durable value.

Colin Waters
Jan Zalasiewicz
Mark Williams
Michael Ellis
Andrea Snelling
March 2014

A stratigraphical basis for the Anthropocene?

COLIN N. WATERS[1]*, JAN A. ZALASIEWICZ[2], MARK WILLIAMS[2], MICHAEL A. ELLIS[1] & ANDREA M. SNELLING[3]

[1]*Environmental Science Centre, British Geological Survey, Keyworth, Nottingham NG12 5GG, UK*

[2]*Department of Geology, University of Leicester, Leicester LE 1 7RH, UK*

[3]*NERC Isotope Geosciences Laboratory, British Geological Survey, Keyworth, Nottingham NG12 5GG, UK*

**Corresponding author (e-mail: cnw@bgs.ac.uk)*

Abstract: Recognition of intimate feedback mechanisms linking changes across the atmosphere, biosphere, geosphere and hydrosphere demonstrates the pervasive nature of humankind's influence, perhaps to the point that we have fashioned a new geological epoch, the Anthropocene. To what extent will these changes be evident as long-lasting signatures in the geological record?

To establish the Anthropocene as a formal chronostratigraphical unit it is necessary to consider a spectrum of indicators of anthropogenically induced environmental change, and to determine how these show as stratigraphic signals that can be used to characterize an Anthropocene unit and to recognize its base. It is important to consider these signals against a context of Holocene and earlier stratigraphic patterns. Here we review the parameters used by stratigraphers to identify chronostratigraphical units and how these could apply to the definition of the Anthropocene. The onset of the range of signatures is diachronous, although many show maximum signatures which post-date1945, leading to the suggestion that this date may be a suitable age for the start of the Anthropocene.

The 'Anthropocene' is in many respects a novel potential geological unit. Stratigraphy, which deals with the classification of geological time (geochronology) and material time-rock units (chronostratigraphy), has historically defined geological units based upon significant, but temporally distant, events. These events are typically, although not exclusively, associated with major changes in the fossil contents of rocks below and above a particular horizon, and therefore with the temporal distribution of life forms. It was only following such observations that new stratigraphical units were proposed and ultimately defined. For example, J. Phillips used the major mass extinction at the end of the Permian in 1840 to recognize the beginning of both the Triassic Period and of the Mesozoic Era. The ultimate definition of the base of the Triassic, however, was accomplished only in 2001, when the Global Stratotype Section and Point (GSSP) was taken at the base of a specific bed in a section in Meishan, China, coinciding with the lowest occurrence of the primary marker, the conodont *Hindeodus parvus* (Yin *et al.* 2001). In contrast, the Anthropocene was proposed as a term (Crutzen & Stoermer 2000) before any consideration of the nature of the signature of this new stratigraphical unit was given. For the first time in geological history, humanity has been able to observe and be part of the processes that potentially may signal such a change from the preceding to succeeding epoch.

What are the key 'events' over the last decades to millennia that have the potential to leave a recognizable record in sediments/ice that could be used to define the base of the Anthropocene? The options cover a diverse range of geoscientific fields and need not be restricted to the biostratigraphical tools typically used throughout much of the geological column to define chronostratigraphical units. Potential stratigraphical tools and techniques that may be used to define the base of the Anthropocene include the following (Fig. 1):

(1) Appearance and increased abundance of anthropogenic deposits:
  - artificial anthropogenic deposits;
  - anthropogenic soils (anthrosols);
  - novel minerals and mineraloids;
  - anthropogenic subsurface structures ('trace fossils');
  - anthropogenic modification of terrestrial and marine sedimentary systems.

(2) Biotic turnover:
  - megafauna;
  - reef ecosystems;
  - microflora;
  - microfauna.

From: WATERS, C. N., ZALASIEWICZ, J. A., WILLIAMS, M., ELLIS, M. A. & SNELLING, A. M. (eds) 2014. *A Stratigraphical Basis for the Anthropocene*. Geological Society, London, Special Publications, **395**, 1–21.
First published online March 24, 2014, http://dx.doi.org/10.1144/SP395.18

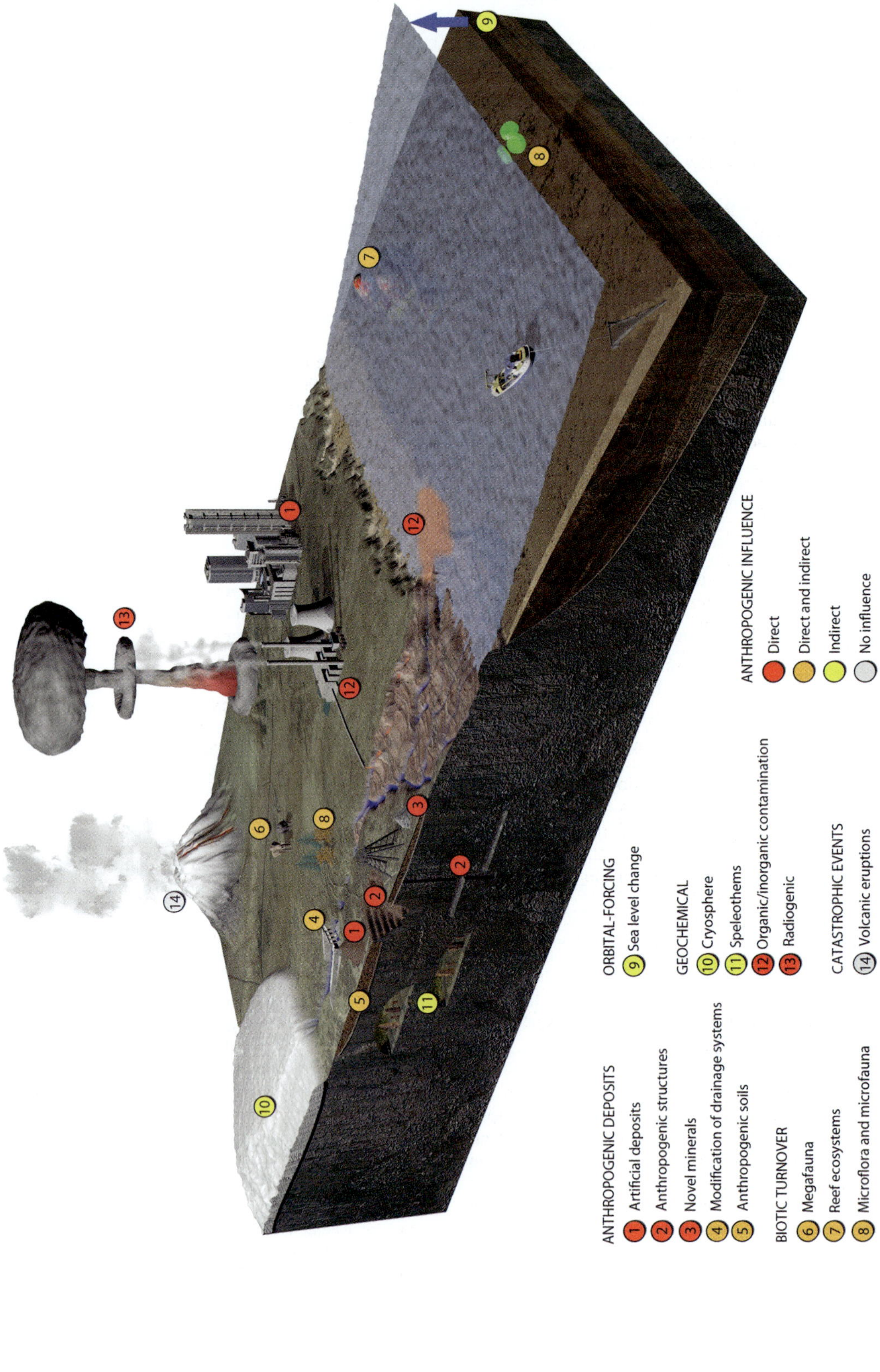

Fig. 1. Examples of key 'events' that could produce stratigraphical signatures that could be used to define the base of the Anthropocene.

(3) Geochemical:
- evidence preserved in the cryosphere;
- records in speleothems;
- organic and inorganic contributions to sediments.

(4) Climate change:
- ocean geochemistry;
- oceanic biodiversity;
- continental to ocean sediment flux;
- sea-level change.

(5) Catastrophic events:
- radiogenic spikes from nuclear bomb tests/accidents;
- volcanic eruption;
- meteorite/asteroid (bolide) impact.

The 17 contributions to this Special Publication, *A Stratigraphical Basis for the Anthropocene*, mainly cover those events that have been directly the result of humanity's growing influence on the Earth – techniques (1)–(3) above – and it is most likely that one or more of these signatures could be used to define the basal boundary of the Anthropocene. In addition, the practical use of tephrochronology, dating historical events through volcanic ash deposits, clearly provides an important stratigraphical tool for quantifying Anthropocene events.

In this paper, we begin by presenting a description of the process by which the Anthropocene is being considered for ratification. We consider the hierarchical stratigraphical level to which the Anthropocene might be applied, or remain a popular but entirely informal unit which exists outside the formal Geological Time Scale. We outline some of the techniques for dating sediments/ice, detail the three main suggestions forwarded as potential ages for the start of the Anthropocene: pre-Industrial Revolution; 1800 and the start of the Industrial Revolution in parts of the planet; and 1950, the 'Great Acceleration' in global economic activity following World War II (Steffen *et al.* 2007). Potential future ages are also considered. A glossary of commonly used terms is also provided.

Process of ratification of the proposed Anthropocene Epoch

The Anthropocene Working Group (AWG) of the Subcommission on Quaternary Stratigraphy (SQS) was established in 2009 to consider the informal proposal that we no longer live in the Holocene Epoch, but in a time period which should be referred to as the Anthropocene. The AWG is tasked to assess evidence that there are environmental signatures preserved in sedimentary or cryospheric successions that can be attributed uniquely to the Anthropocene. If accepted, the AWG would need to define a Global Stratigraphic Section and Point (GSSP or 'golden spike') in a type locality, or to define a Global Standard Stratigraphic Age (GSSA or numerical age), that defines the Holocene–Anthropocene boundary.

The process by which a new epoch can be ratified is described by **Finney (2013)**, who also raises a series of pertinent questions that he feels need to be addressed by the AWG, although many of these questions are unique to the Anthropocene. **Zalasiewicz *et al.* (2014*a*)** describe some of the problems related to the short timescales inherent in the definition of the Anthropocene, such as bioturbation and pedogenesis. The ability to locate a boundary through counting varves in sediments or layers in ice core to the nearest year, or at least decade, would provide a scale of rigour not previously faced during the definition of older chronostratigraphical boundaries, where potential diachroneity of many thousands of years cannot be resolved by current dating techniques.

Status as epoch or age

Given the hierarchical nature of chronostratigraphy (Salvador 1994), the higher the rank of the Anthropocene, the greater the change has occurred between it and the previous stratigraphical unit (**Gibbard & Walker 2013**). The term proposed, even if by accident (Steffen *et al.* 2004), implies by use of the ending 'cene' to be of Epoch status. Stages/ages typically end in 'ian' and as such if the new division was considered to be of this rank would need to be named as Anthroposian, or similar. To warrant Epoch status the scale of changes in key criteria (biostratigraphical, sedimentological and geochemical) need to be of comparable magnitude to those used as evidence for earlier Epoch boundaries, such as that between the Pleistocene and Holocene (Gibbard & Walker 2013). Hence, consideration as a potential Epoch has the scientific benefit of overtly testing the implicit hypothesis in Crutzen (2002): that the Holocene, defined by fundamental aspects of the Earth system, has terminated.

The Holocene is being considered to be divided into three Stages/Ages along the lines of 'Early', 'Mid' and 'Late' Holocene, with internal boundaries at 8.2 and approximately 4.2 ka (Walker *et al.* 2012). This does not leave open the option of the Anthropocene to be considered a Late Holocene Stage/Age.

The base of the Quaternary Period is formally defined at a GSSP (Gibbard *et al.* 2010), although a concept associated with this definition is that it also reflects the onset of the major northern hemisphere glaciation. **Wolff (2013)** faces the possibility that the end of the sequence of northern hemisphere glaciations should signal the end of the Quaternary, but he suggests that current evidence does not preclude glacial inception in the future (timescales of 10–100 ka). Levels of atmospheric CO_2 (Lüthi *et al.* 2008), CH_4 (Loulergue *et al.* 2008) and N_2O (Schilt *et al.* 2010) in ice cores are at levels higher than observed for the last 800 ka (Wolff 2013), and, in the case of CO_2, at levels unprecedented since the warmer Pliocene Epoch (see Haywood *et al.* 2011). Such signatures would distinguish the Anthropocene from the Holocene and part, if not all, of the Pleistocene.

Between the 1500s and 1700s, the number of species extinctions (plants and animals) ran at less than 50 per century, with extinctions rising to 125 in the 1800s and 500 in the 1900s (Barnosky 2013). **Barnosky (2013)** concludes that, although extinction rates are elevated at 3–12 times normal background rates, less than 1% of species have become extinct. For vascular plants, at least 5% of native species appear to have been lost across half of the terrestrial biosphere, but in many cases native plant species are able to maintain viable populations even in heavily managed anthropogenic biomes (Ellis *et al.* 2012). Therefore, at present, we are not experiencing something equivalent to the Big Five mass extinctions, where an estimated 75–96% of known species became extinct, or, as regards large terrestrial vertebrates, the Late Quaternary Megafauna Extinction near the Pleistocene–Holocene boundary (Barnosky *et al.* 2011; Barnosky 2013). This suggests that, as Gibbard & Walker (2013) contend, the Anthropocene does not provide a biostratigraphical signature equivalent to the epoch status defined for the Holocene. However, this extinction threshold would be exceeded in the near future, and in excess of 75% species loss can be predicted within 300–500 years at current extinction rates, unless conservation methods become markedly more effective (Barnosky *et al.* 2011; Barnosky 2013). This would produce a bio-horizon on a scale of the Big Five mass extinctions and, if this is to become reality, the Anthropocene would arguably be of Period/System scale. Extinctions are not the only indicator of biostratigraphy, though, as the changes to assemblages through species invasions (Barnosky 2013) are now considerable, globally expressed and effectively permanent.

Absolute and relative dating techniques

Climatostratigraphy, or use of contrasting climatic conditions to characterize stratigraphical units, is of primary importance for correlation within Quaternary successions (Gibbard & Walker 2013). The Quaternary is subdivided into Marine Isotope Stages (MIS 1–104), reflecting orbitally forced cooling (glacials) and warming (interglacials) of the Earth's climate, the ages of which have been accurately constrained (Lisiecki & Raymo 2005). This is evident through the $\delta^{18}O$ signature of marine biogenic calcite, which reflects the increased incorporation of the light ^{16}O into expanding ice sheets (Shackleton & Opdyke 1973). MIS 1 ranges from the present to 11.7 ka, coinciding with the Holocene Epoch, the current interglacial. The Anthropocene does not fit within such a definition and clearly MIS stages are insufficient when it comes to dating anthropogenic deposits. It is the disruption of such quasi-periodic signals that makes the Anthropocene distinctive, potentially to the point that we no longer exist within a regime of orbitally dominated climate change. Alternative means need to be found of characterizing and defining the Anthropocene, as discussed below.

Radiometric dating has become an increasingly precise tool for determining the absolute age of chronostratigraphical boundaries; for example, the base of the Triassic is bracketed by two dated volcanic-ash clays and constrained at 252.16 ± 0.2 Ma (Shen *et al.* 2010), an error of only 0.001% of the total age. A number of radiometric techniques used to determine Quaternary chronology are here considered for their suitability for dating the Anthropocene. Radiocarbon (^{14}C), although routinely used by the archaeological community to date organic remains, has insufficient resolution. It has an error of several decades, which is unsuitable if the beginning of the Anthropocene is chosen to have occurred during the last 200 years (Table 1). Radioisotopes such as ^{137}Cs and ^{90}Sr are useful time markers that can be potentially linked to specific and temporally constrained emissions, are laterally extensive and have a short half-life (Table 1). However, in areas of low fallout these radionuclides may already be approaching the limits of detection (**Hancock *et al.* 2014**), and ice-core β-radioactivity on isotopes (Dibb *et al.* 1990) is unsuitable for dating signatures even for the start of the Industrial Revolution (Wolff 2013). Lamina-counting techniques used in conjunction with ^{210}Pb–^{226}Ra or ^{234}U–^{230}Th radiometric dating (Table 1) is potentially of importance in the context of dating speleothems (**Fairchild & Frisia 2013**). ^{210}Pb may also be useful for dating microfauna and microflora (**Wilkinson *et al.* 2014**), marine or lacustrine clay sediments and peats. In the more distant future, dating techniques may rely upon more long-lived isotopes, such as ^{239}Pu and ^{240}Pu (Table 1), which also bind strongly to soil and sediment particles (Hancock *et al.* 2014). The longer-lived nature and greater abundance of ^{239}Pu makes it the preferred chronometer, and

Table 1. *Commonly used radiometric dating techniques and their applicability to dating Anthropocene deposits/artefacts*

Isotope	Half-life (years)	Acceptable range	Accuracy	Suitability
^{14}C*	5568/5730	200–60 kyr	Decades–centuries	Peat, wood, charcoal, bone, shells, soil, ice core, coral, etc. (*Pre-Industrial Revolution*)
^{137}Cs†	30.17 ± 0.03	AD 1954–present	Annual (if linked to known emissions) – decades	Terrestrial – marine sediments (*mid-twentieth century*)
^{90}Sr‡	28.79	AD 1950s–present	Annual (if linked to known emissions) – decades	Terrestrial – marine sediments (*mid-twentieth century*)
^{210}Pb-^{226}Ra§	22.3 (^{210}Pb)	<150 years	Decades	Carbonates, speleothems, microflora, microfauna (*mid-twentieth century*)
^{234}U-^{230}Th§	245 560	<500 kyr	Centuries	Carbonates, speleothems, bone, teeth (*Pre-Industrial*)
^{239}Pu‖	24110	<100 kyr	Centuries; annual if linked to known emissions)	Soil, sediment (*mid-twentieth century*)
^{240}Pu‖	6563	<30 kyr	Centuries; annual if linked to known emissions	Soil, sediment (*mid-twentieth century*)

Text in italics indicates which of the three main options of the age of the Anthropocene could be most usefully dated using the respective isotopes.
*Stuiver & Polach (1977); †Unterweger (2013); ‡Browne (1997); §Elert (2013); ‖cf. Hancock *et al.* (2014).

in many regions the signal is likely to be detectable in sediments for 100 kyr or longer (Hancock *et al.* 2014).

Radiogenic methods such as Luminescence and Electron Spin Resonance (ESR) are relatively new techniques becoming increasingly used by archaeologists and Quaternary geoscientists. The Luminescence method dates the last time an object was heated (particularly useful for pottery) or exposed to sunlight (potentially useful to delimit burial of artificial deposits). It can provide dates that range from 10 years up to 1 Myr, but has comparatively low accuracy, with errors of typically 5–10% (Duller 2008). ESR dates, mainly used on corals, speleothems, teeth and bone, range from a few thousand years to 300 kyr and so may be of little practical use for dating the Anthropocene if it is to fall within the last two centuries.

Cosmogenic Radionuclides (CRNs) dating relies upon the accumulation of ^{3}He, ^{10}Be, ^{21}Ne, ^{26}Al and ^{36}Cl in response to the duration of exposure of the upper 1–2 m of sedimentary deposits or ice to cosmic rays (Gosse & Phillips 2001). The technique has the ability to date the timing of surface exposure through excavation using CRN production (range 100 years to 5 Ma) or the date of burial through decay of CRNs (range of *c.* 0.1–5 Ma) (Akçar *et al.* 2008).

Radiometric dating of volcanic ash deposits has become an intrinsic part of the characterization of GSSPs. For example, the base of the Triassic Period at the Meishan GSSP is bracketed by dated volcanic-ash clays 18 cm below and 8 cm above the base of the Triassic (Shen *et al.* 2010). Such regionally extensive deposits could be used as marker bands to demarcate the base of the Anthropocene. Each eruption can be characterized by a distinctive geochemical 'fingerprint', and a combination of radiometric dating and the historical documentation of events can lead to age constraints at annual resolution (Smith 2013). **Smith (2013)** identifies a number of useful marker tephra deposits but suggests, in agreement with Zalasiewicz *et al.* (2008), that the 1815 CE eruption of Tambora, Indonesia – the largest eruption in recorded history – would be most suitable of such markers, particularly as it aligns with the early phase of the Industrial Revolution. Although the ash deposits were spatially restricted and constrained by wind direction, the effects are evident globally with development of associated sulphate peaks within ice cores and temporary climatic events evident in tree rings (Delmas 1992; Briffa *et al.* 1998; Smith 2013).

Prior to the introduction of radiometric dating techniques in the twentieth century, the relative age of deposits was constrained through biostratigraphy, which has formed the basis for defining most pre-Quaternary chronostratigraphical units (e.g. Gradstein *et al.* 2012). Assemblage and abundance biostratigraphical zones, based upon mixes of native and non-native species in both terrestrial and marine settings and lineage zones, based on the evolution of crop plants, are likely to be most useful in defining the Anthropocene (Barnosky 2013). Interval zones based upon extinctions over recent centuries are of limited use, as most extinct species were formerly not widespread and/or unlikely to leave a fossil record (Ager 1993; Barnosky 2013). Biostratigraphical zones used to recognize chronostratigraphical boundaries are diachronous to some degree as new taxa take time to extend their distribution from a single source origination (Barnosky 2013). Furthermore, there appears to be a time-lag between the onset of anthropogenic activity and the resultant influence upon microbiota (Wilkinson *et al.* 2014). With a deep-time perspective, these diachroneity and time-lag effects fall within the range of error of most radiometric and biostratigraphic dating techniques, and are not considered significant. For definition of the base of the Anthropocene, which is likely to be resolved at annual or decadal accuracy, such diachroneity severely limits the use of biostratigraphy in our current proximal view of events, but it is likely to become negligible in the future use of biostratigraphy as a tool for recognizing the Anthropocene.

Human artefacts, routinely used as an indicator of age in archaeological investigations, could be used as an equivalent of the geological 'type-fossils', with potentially greater resolution than biostratigraphical fossils (Barnosky 2013; Edgeworth 2013; Ford *et al.* 2014). The evolution of these artefacts, which may be considered human-produced trace fossils (Barnosky 2013; Williams *et al.* 2013) or technofossils (Zalasiewicz *et al.* 2014*b*), is a function of cultural dynamics rather than natural selection (**Edgeworth 2013**). These artefacts are prone in recent decades, certainly since the 1950s, to evolve from invention (equivalent to the biostratigraphical First Appearance Datum or FAD) to global distribution (equivalent to biostratigraphical acme) and then to obsolescence (equivalent to biostratigraphical rarity) within comparatively few years, as a function of the globalization of trade. Also, the lithological composition of wastes in landfills is equivalent to the biostratigraphical assemblage zone and can be indicative of age, as illustrated by **Ford *et al.* (2014)**. Such artefacts and anthropogenic facies variations provide a very high-resolution (potentially annual to decadal) tool for dating deposits (Zalasiewicz *et al.* 2014*b*). However, the long-term preservation potential of such artefacts and anthropogenic sediments will be variable (Price *et al.* 2011; Ford *et al.* 2014), such that only part of today's wide range of artefacts will be recognizable in the distant future.

Annual layer-counting techniques can produce very high precision dating, potentially to annual resolution. Potential techniques include dendrochronology, coral laminations, seasonally layered sediments in glacially influenced lakes, speleothem layers and ice cores. Details of the various techniques are summarized in Bradley (1999). Dendrochronology not only has anchored chronologies extending throughout most of the Holocene; the pattern of rings is indicative of local climatic conditions within temperate zones and can also potentially be used to determine wood provenance.

There is no single global palaeomagnetic spike that could be used to define the base of the Anthropocene (Snowball *et al.* 2013). However, **Snowball *et al.* (2013)** note that there is a global event, most strongly developed in mid to high latitudes coincident with a low in dipole latitude and peak in dipole moment at 2.55 ka cal. BP (the European 'f-event'), which may be a potential chronostratigraphic marker. A new archaeological dating technique uses high-frequency secular variation of the geomagnetic field. This permits annual to decadal age resolution for Fe-oxide-bearing materials, including artefacts such as fired ceramics, formed in the last few thousand years (Snowball *et al.* 2013).

The significance of the history of excavation or 'cut' in archaeology in helping to determine the history and timing of events (Edgeworth 2013) has analogues in the use of geological unconformities to constrain the timing of events through allostratigraphy (Ford *et al.* 2014). It is clear that the complexity of such 'cut' surfaces, although of value at the local scale, makes regional-scale correlation of erosional/non-depositional surfaces almost impossible. The only unconformity that can be correlated with any certainty is the bounding surface between the lowermost artificial deposits from underlying natural deposits that predate human modification of the immediate landscape. This bounding surface is highly diachronous overall, although times of marked expansion of cities (e.g. post the mid-twentieth century) represent traceable stratigraphic 'plateaux'.

Definition of a boundary stratotype or numerical age

The International Stratigraphic Guide (Hedberg 1976; Salvador 1994) requires that all major chronostratigraphical subdivisions are defined with reference to boundary stratotype localities in sedimentary reference sequences, designated as GSSPs. Definition of the Holocene differed in that the GSSP was defined in ice core rather than a sedimentary deposit (Walker *et al.* 2009), but essentially followed principles outlined in the International Stratigraphic Guide. Zalasiewicz *et al.* (2014*a*) review how an Anthropocene signature may be recognized in a range of terrestrial and marine settings. This is helpful when considering potential environments to seek the location of a GSSP, if a traditional route to defining the base is to be chosen.

It has recently been argued by Smith *et al.* (in press) that the precision in radiometric techniques in the determination of the age of chronostratigraphical boundaries is such that definition of a GSSP in a single section based upon the evolution of a specific indicator faunal/floral species should be replaced by a GSSA (or numerical age). With the definition of the base of the Anthropocene possibly at a time of tens to hundreds of years before present, the resolution of dating techniques is at least decadal if not annual, and definition of a GSSA at a specific year is feasible and arguably preferable to using a proxy indicator (Zalasiewicz *et al.* 2011). Smith *et al.* (2014) propose that, in general, GSSAs should be decided based upon a spectrum of signatures. In this section we consider four distinct options for the potential placement of the base of the Anthropocene: (1) pre-Industrial Revolution age; (2) Industrial Revolution age; (3) mid-twentieth century age; and (4) the future.

Evidence for a pre-Industrial Revolution age

Gibbard & Walker (2013) characterize the Holocene Epoch as a time in which there has been a progressive increase in the prominence of humans as an agent influencing natural environments and processes. They argue that the anthropogenic signature is a hallmark of the current Holocene interglacial and this is distinct from previous interglacials that occurred during the Pleistocene. They contend that it is not then possible to further use the activities of humans to define a post-Holocene Epoch.

If the Anthropocene is to be considered the Epoch that humanity has created, it is evident that human influence on the planet in the form of directly deposited terrestrial anthropogenic deposits are markedly diachronous in their nature, are laterally impersistent, may include numerous disconformities, may be reworked by continued human landscape modification and, ultimately, have relatively low propensity for preservation in the geological record (Ford *et al.* 2014). The earliest signatures approximate to the onset of the Holocene with Edgeworth (2013) describing a significant and long-lived urban development which commenced some 11 ka BP (Fig. 2). It may be misleading, though, to think of the Anthropocene just as the 'human epoch'. The key factor is the level of geologically significant global change, with humans currently happening to be the primary drivers: future, potentially yet more pronounced change (cf. Wolff 2013) may be primarily driven by Earth system feedbacks such as methane release, and yet would still clearly be part of the same phenomenon.

Anthropogenic influence is not necessarily first seen through urban development. More often it is evident through the initiation of agricultural practices, with forest clearances increasing atmospheric CO_2 levels from 8 ka BP, and cultivation and irrigation techniques increasing atmospheric CH_4 levels from about 5 ka BP (Ruddiman 2003, 2005) (Fig. 2). Prior to 1700 CE, the deforestation was almost exclusively of temperate forests (FAO 2010). However, CO_2 and CH_4 concentrations, trends and rates of change fall within the range recorded in ice core over the 800 kyr prior to 1800 CE, suggesting that there is no strong evidence that humanity has driven these cycles outside of their natural range prior to the Industrial Revolution (Wolff 2013). Also, it has been argued that the rise in CH_4 levels over the last 5 kyr does not need to be linked to changes in agriculture, but could be the product of natural changes in the Earth's orbit associated with precession-induced modification of seasonal rainfall in the southern hemisphere tropics (Singarayer *et al.* 2011).

Human indirect influence upon rivers provides a recognizable signature in the fluvial system, including coastal deltas. This is associated with increasing sediment loading in response to erosion due to deforestation, animal grazing and changing agricultural practices, mill development, transport networks, and the influence upon global climate systems including effects such as increased precipitation intensity or desertification and sea-level rise resulting in coastal inundation (Merritts *et al.* 2011; Syvitski & Kettner 2011). In particular, the impact of introduction of intensive agricultural practices is noted as causing a widespread stratigraphical marker across many continents associated with a transition from basal gravels with organic channel fills to a thick capping of sandy silt (Brown *et al.* 2013). In two nearby river systems in the UK, this boundary is dated at 3600–4400 years cal BP and 1300–220 years cal BP, showing that this boundary is significantly time-transgressive and makes it difficult to consider as a sedimentary boundary for the start of the Anthropocene (Brown *et al.* 2013).

Mineral magnetic studies in lake sediments, which are a strong indicator of deforestation events and soil erosion, suggest a complex and diachronous history of clearance (Snowball *et al.* 2013). The largest mineral magnetic signatures associated with catchment disturbance during expansion of agriculture in Europe began around 1100 CE ± 100 years (Fig. 2) with similar signatures evident in China and Mexico at broadly the same time, although they are dependent on cultural and not geological controls, and are not isochronous (Snowball *et al.* 2013). Anthropogenic disturbance of soil horizons is also clearly recorded in speleothems and is also notably diachronous (Fairchild & Frisia 2013). Although initiation of forest clearances can be discounted as an adequate signature for recognizing the base of the Anthropocene, it is clear that the expansion and intensification of

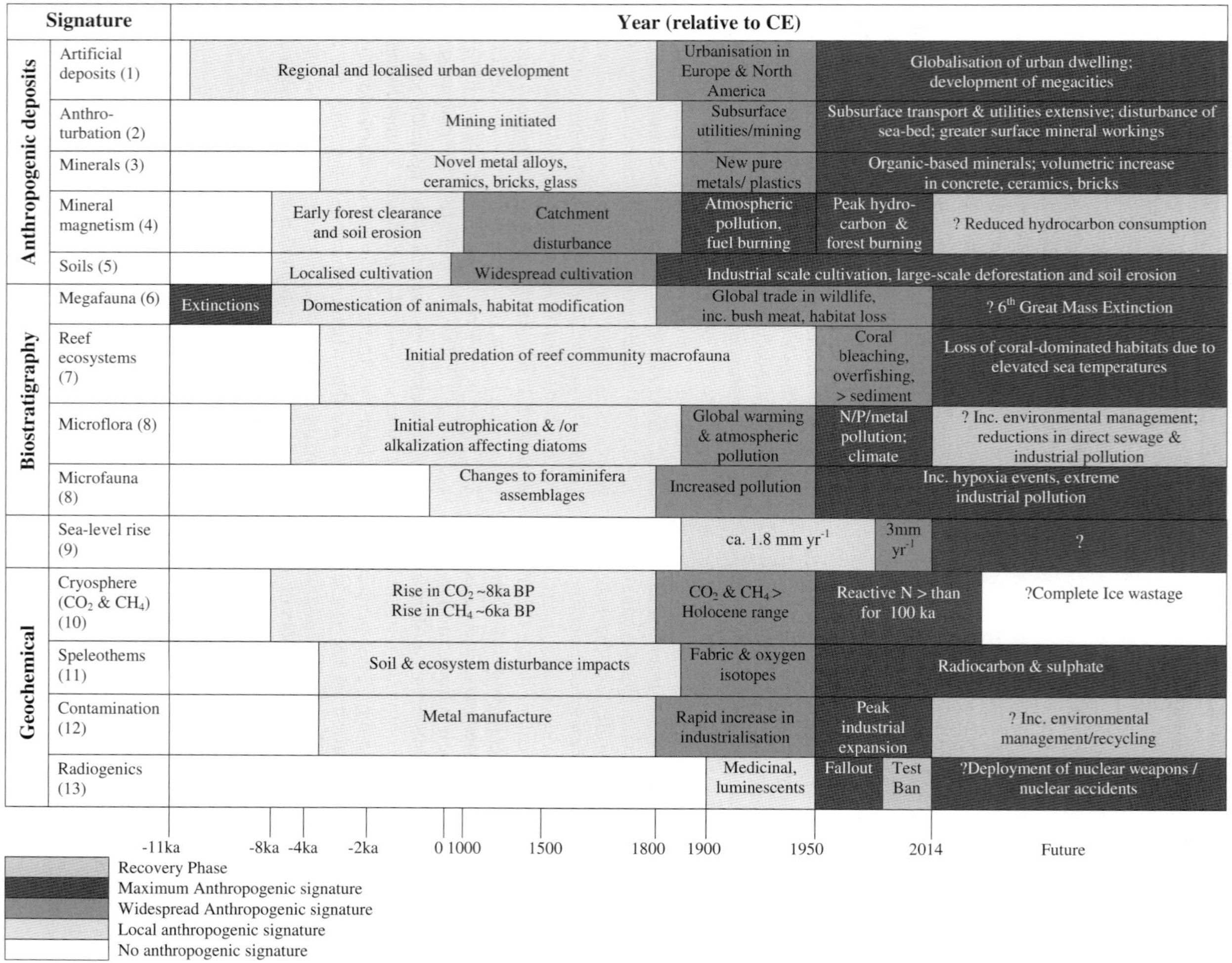

Fig. 2. Relative significance of anthropogenic signatures with time. Note the non-linear timescale. (1) Edgeworth (2013) and Ford *et al.* (2014); (2), Williams *et al.* (2013) and Ford *et al.* (2014); (3) Zalasiewicz *et al.* (2013); (4) Snowball *et al.* (2013); (5) Certini & Scalenghe (2011) and Edgeworth (2013); (6) Barnosky (2013); (7) Hoegh-Guldberg (2014); (8) Wilkinson *et al.* (2014); (9) Church & White (2011); (10) Wolff (2013); (11) Fairchild & Frisia 2013; (12) Gałuszka *et al.* (2013); and (13) Hancock *et al.* (2014).

agricultural land use has resulted in extensive clearances of native vegetation and megafauna, and replacement with domesticates in excess of 3 ka ago (Ellis *et al.* 2013). The onset of these agricultural practices also resulted in significant modifications of fluvial systems, especially the rapid siltation and increase in sedimentation rate (Dearing & Jones 2003; Poirier *et al.* 2011).

The influence of humanity on the generation of soils is vast. Anthropogenic influence includes increasing atmospheric CO_2 leading to acidification, addition of lime or fertilizers, management of biota through insecticides and herbicides, physical mixing and movement of soils through ploughing, and accelerating soil-forming processes (Richter 2007). It has been suggested that the base of such an extensive anthropogenic soil horizon could make a suitable 'golden spike' at approximately 2 ka BP (Certini & Scalenghe 2011). However, as for anthropogenic deposits, the age of onset of significant development of anthrosols is highly diachronous. For example, the charcoal-enriched 'terra preta' of the Amazon Basin is somewhat younger, potentially up to 500 BCE (Woods 2008). Much of Europe includes evidence for the development of plaggen soils, potentially up to 4 ka BCE in age (Simpson 1997), but mainly the product of a type of farming cultivation during the medieval period and post-medieval times (Edgeworth 2013). Soils have low preservation potential and represent an open system prone to modification, and are probably the product of numerous events or phases of modification, which are still ongoing. Consequently, Gale & Hoare (2012) and Zalasiewicz *et al.* (2014*a*) argue that the resolution of the age of the base of gradational soil horizon is not suitable to define the Anthropocene.

Human impacts on diatom assemblages in lakes, the product of eutrophication and/or alkalization linked to deforestation and introduction of agriculture, extend back at least 5 kyr (Wilkinson *et al.* 2014) (Fig. 2). Similarly, changes to land use and land cover, and the resultant increase in soil erosion and transport of sediment into the near-shore setting, result in changes to foraminiferal assemblages considerably earlier than other environmental drivers (Wilkinson *et al.* 2014). The impact of humans on coral reefs was minimal during the early Holocene, with first evidence of decreasing ecological diversity of the large marine herbivores and carnivores beginning around 3.5 ka BP (**Hoegh-Guldberg 2014**) (Fig. 2).

It is clear that there are major anthropogenic signatures evident during pre-Industrial Revolution times. However, the range of signatures, their magnitude and their spatial extent are typically less than that evident during later times. The timing of these impacts overall is more markedly diachronous across the Earth at the scale of our perspective, and a single isochronous marker is not apparent prior to the Industrial Revolution.

Evidence for an Industrial Revolution age

Early descriptions of the Anthropocene argued in favour of it starting coincident with the initiation of the Industrial Revolution in western Europe (Crutzen 2002; Zalasiewicz *et al.* 2008). Gibbard & Walker (2013) consider the clearest marker horizon is a rise in atmospheric CO_2 levels above any previous Holocene level from around 1750 CE, coincident with the start of an upward rise in CH_4 and N_2O (Fig. 2), although it is important to recognize that this is not directly observed in the rock record. In the ice record, the termination of the 'Little Ice Age', a time of modest cooling mainly of the northern hemisphere from about 1350 to 1850 CE (Solomon *et al.* 2007), may be a response to that change in atmospheric composition. Changes in the chemical and physical properties of speleothems (Fairchild & Frisia 2013) can be linked to the start of this climate amelioration (Fig. 2).

Gibbard & Walker (2013) argue that although the CO_2 signature is recognized globally, including within polar ice cores (Lüthi *et al.* 2008; Wolff 2013), the cause of the signature reflects industrialization in only a small part of the Earth, mainly western Europe and eastern North America. This is, however, not an argument raised against the definition of the Cretaceous–Tertiary (K–T) boundary marking the base of the Cenozoic. Here, the crater associated with bolide impact is only 180 km across, but the signature of this impact through tektites, the iridum-rich clay layer, climatic change and biotic extinctions was global. Barnosky (2013) contends that when all traces of humanity are considered, it forms a boundary layer more widespread than the iridium layer used to recognize the K–T boundary.

The onset of the Industrial Revolution (Fig. 3) resulted in a marked change in the characteristics of anthropogenic deposits (Price *et al.* 2011; Ford *et al.* 2014). These include: increased use of building and construction materials; increased exploitation of subsurface deposits; widespread inclusion of processed metals and associated manufactured goods; and increased human activities at depth, either for mineral exploitation or subsurface infrastructure. However, the onset of the Industrial Revolution is diachronous, not reaching many developing countries until the middle of the twentieth century (Fig. 3) and, consequently, Gibbard & Walker (2013) argue it should not be used as a criterion for defining the Anthropocene.

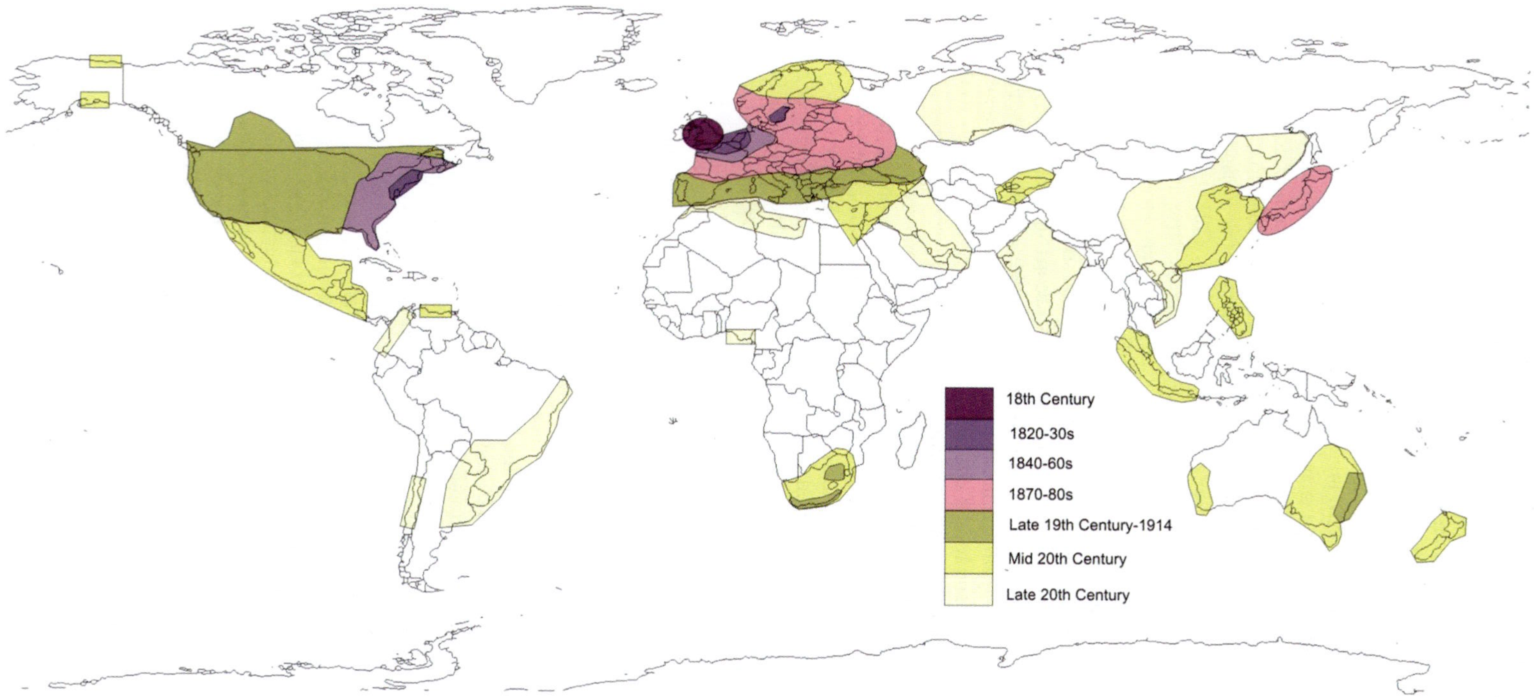

Fig. 3. Map showing the approximate age for the beginning of the Industrial Revolution and subsequent industrialization across the planet. This is a subjective event, here interpreted as the widespread growth of mechanization in respect to manufacturing, transport and innovation.

Williams *et al.* (2013) propose that the base of the Anthropocene should coincide not with the start of the Industrial Revolution, but with the radical evolution of the urban environment in the mid-eighteenth century. The increased size of conurbations resulted in the need to evolve subsurface transport and sewerage systems in order to keep them functioning. Such subsurface developments have greater long-term preservation potential than surface urban deposits, but the cross-cutting, non-stratiform nature of these subsurface structures precludes their use in recognition of a traditional GSSP. Williams *et al.* (2013) use a particular event, the inception in London of the first Metro system in 1863, as the criterion for defining the start of the Anthropocene. Williams *et al.* (2013) compare the increased complexity of the urban environment to be analogous to the increasing complexity of the trace fossils used to define the base of the Cambrian System. Such comparisons to an extent diminish some of the arguments made against definition of the base of the Anthropocene. The base of the Cambrian was recognized initially through a concept of increasing biological complexity and, ultimately, one ichnospecies was chosen to represent this changing complexity. This first appearance has ultimately proved to be diachronous over hundreds of thousands of years and sections where this transition can be observed are few. In contrast, the complex urban environment has taken only decades to promulgate globally and now covers about 1% of the Earth's surface. This suggests that while the location of a worldwide, precisely synchronous boundary for the Anthropocene is challenging, it is no less so in consideration of existing chronostratigraphical units.

In European lakes, diatom assemblages show significant changes in response to human-induced acidification between 1800 and 1850, with the first evidence of eutrophication in these lakes between 1850 and 1900 (Battarbee *et al.* 2011; Wilkinson *et al.* 2014). This signature is also widespread in Arctic, northern European and North American lakes, with prominent changes to diatom assemblages since about 1850 inferred to be a response to global warming and atmospheric pollution (Fig. 2), but that the timing varied between regions and lakes (Wilkinson *et al.* 2014). Whereas, in the oceans, increased nitrogen fixation and elevated concentrations of soluble iron due to increased deposition of iron-rich desert dust since 1870 caused intensified growth of phytoplankton (**Gałuszka *et al.* 2013**).

In terrestrial environments there was an initial introduction of new plants and domestic animals from the 1500s marking the early age of global exploration and trade, although introduced species in Australasia began mainly in the 1800s (Barnosky 2013). There were significant introductions of alien plant species from the 1800s in many continents (Barnosky 2013), coinciding with scientific investigations during the Enlightenment, the notable British attempts to develop plantations of important exotic commercial plants within their colonies and the increased interest in horticulture. It is a feature of regional plant species richness that the losses of native species are more than offset by the increases in exotic species (Ellis *et al.* 2012). Globally, the percentage abundance of humans and domestic animals increased relative to wild megafauna in the 1750s, with a second acceleration in the mid-twentieth century (Barnosky 2013).

The relationship of black magnetic spherules and atmospheric pollution through fossil fuel burning results in increased magnetic mineral abundance in sediments associated with the Industrial Revolution (Snowball *et al.* 2013). This is expressed by magnetic susceptibility or isothermal remanent magnetization, and is particularly preserved in peat bogs, soils, lakes, coastal and offshore sediments. They often occur in association with increased heavy metal concentrations. These particles first become abundant in England and the eastern seaboard of North America around 1800, and a spread of industrial sources during the nineteenth century. The largest number of sites shows initial increases of magnetic pollution particles forming an 'AD 1900-event', representing an expression of major fuel burning in these industrialized areas (Locke & Bertine 1986; Snowball *et al.* 2013) (Fig. 1). However, in other parts of the world, signatures appear later – for example, the 1950s in eastern Asia (Snowball *et al.* 2013) – noting also that there is an approximately 100 year lag in the appearance of these magnetic pollution signatures in lake sediments.

Evidence for a mid-twentieth century age

This time interval coincides with the 'Great Acceleration' in global economic activity following World War II (Steffen *et al.* 2007). The extraordinary growth of cities and megacities and major infrastructure projects (Fig. 4), and their associated deposits, may be considered a distinctive feature of the Anthropocene (Williams *et al.* 2013; Zalasiewicz *et al.* 2014*a*). This is perhaps the most apparent signature of anthropogenic impact in that today some 52.4% of the global population live in urban areas (UNDESA 2012). This represents an increase from around 7% in 1800. Despite their focus for human habitation, urban areas cover only about 1% of the ice-free land surface (Klein Goldewijk *et al.* 2010) (Fig. 4). This coverage increases to about 91.9% in densely populated countries such as Japan (UNDESA 2012).

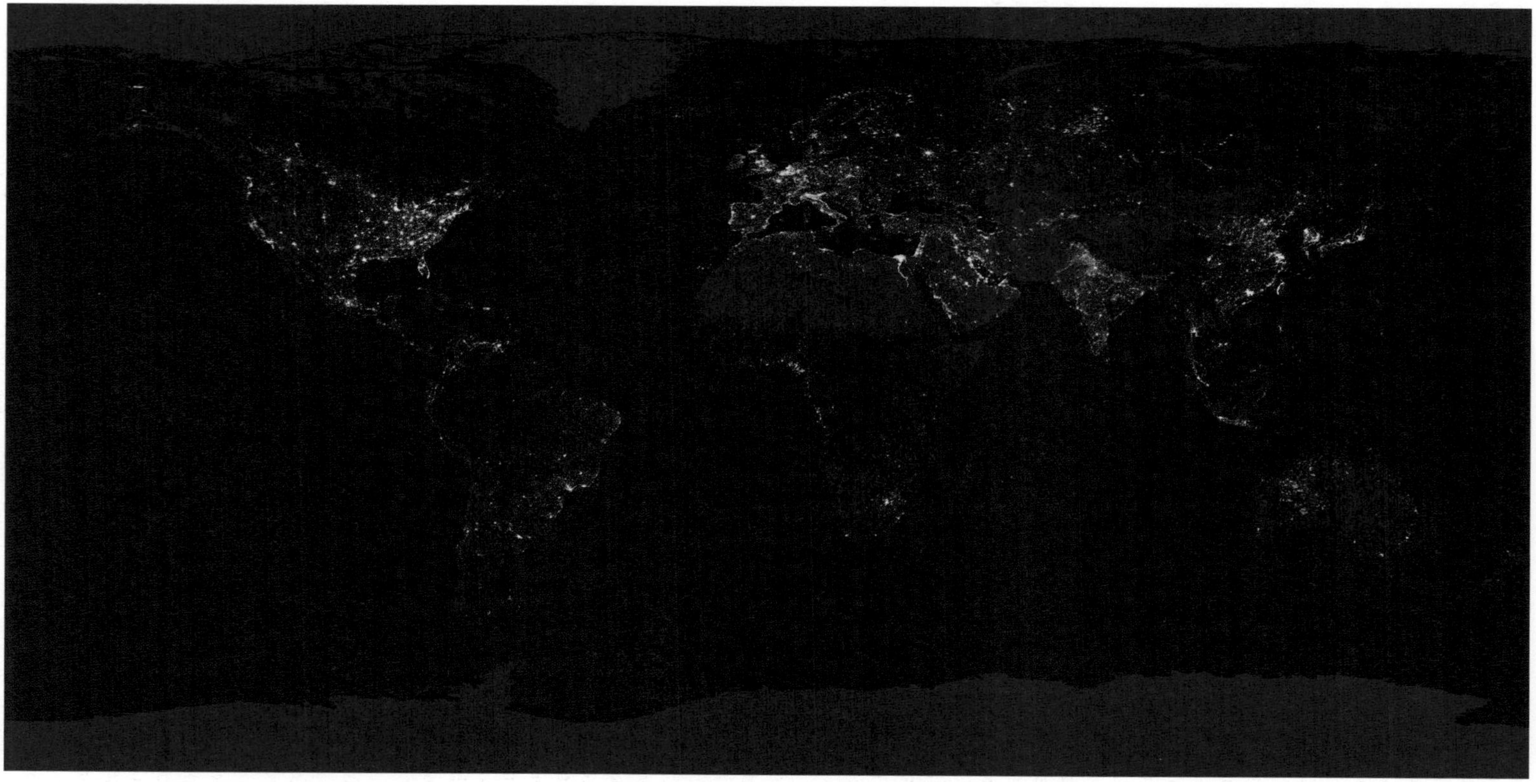

Fig. 4. View of the Earth at night, 2012, showing the distribution of urban conurbations, but not necessarily the most populated areas, through the presence of city lights. The image shows the domination of city construction in coastal areas, particularly in South America and Africa, and along major transport networks in North America, Europe, Russia, India and China. From Earth Observatory, NASA (http://earthobservatory.nasa.gov/Features/NightLights/page3.php).

There has been a significant change in the nature and volume of physical artificial anthropogenic deposits from 1945 onwards (Fig. 2). The natural gradient of sediment transfer from high to low topographical areas has been overtaken by the anthropogenic flux of materials from resources extracted mainly from rural areas to deposition in urban areas in the form of construction schemes (Hooke 2000; Wilkinson 2005). There has been a dramatic rise in overburden and spoil ratios associated with mineral extraction, and the volumes of material worked and used for construction (Ford *et al.* 2014). This is a response to increased demand through population growth and resource consumption, and to technical innovations such as the rapid spread in use of bulldozers from the 1950s. Ford *et al.* (2014) argue that this time interval is characterized by electronic equipment, extensive concrete manufacture, deep mining and generation of vast amounts of waste. These stratigraphical signals are both sharp (to decadal level) and globally widespread.

Minerals such as mullite (present in fired brick and ceramics), ettringite, hillebrandite and portlandite (found in cement and concrete) are present in archaeological times, but have become significantly more common since the mid-twentieth century (Fig. 2) and are sufficiently stable to provide a lasting signature (**Zalasiewicz *et al.* 2013**). New metal alloys, mineraloid glasses, semiconductors, synthetic 'minerals' and emerging nanomaterials may be uniquely indicative of the Anthropocene (Zalasiewicz *et al.* 2013). Plastics have appeared in the environment in significant volumes since the mid-1940s, but by 2008 an estimated 260×10^6 t of plastic was produced, increasing annually by 9% (Thompson *et al.* 2009). Much of this output finds its way to landfill or the sea, with microplastics becoming an abundant trace fossil within marine sediments since the 1950s (Barnosky 2013). However, uncertainty exists as to their persistence following burial or within the marine environment, although their decomposition would be associated with the release of toxic compounds which, in themselves will result in a geochemical signature.

Forests cover about 31% of the Earth's land surface (FAO 2010) but as human populations increase, so too does the rate of deforestation. Since the 1950s, the scale of deforestation has increased by about 44%, compared with the average for the past 5 kyr, with the net loss of 5.2×10^6 ha of forest over the first decade of the new millennium (FAO 2010). However, it is too early to see resultant increased mineral magnetic signatures as a consequence of deforestation since the mid-twentieth century (Snowball *et al.* 2013). The areal extent of deforestation is now predominantly in tropical forests, in part due to demand for the timber, but also the clearance of forests for agricultural development and fuel supply (FAO 2010). Erosion of soils have undoubtedly increased as a result of this deforestation, (e.g. Dearing & Jones 2003 and references therein) and has resulted in increased sediment input to fluvial systems. In contrast, the role of agriculture on the unintentional erosion of soils is considered to have declined dramatically since the 1950s in response to modern soil conservation practices in developed countries (Hooke 2000). Since the 1940s there has been an increase in direct management of rivers, such as the construction of dams, channel maintenance and the urbanization of floodplains (Merritts *et al.* 2011; Syvitski & Kettner 2011). The proliferation in the building of major dam schemes across the globe (Syvitski & Kettner 2011) has caused about 20% of sediment load to be retained by reservoirs (Syvitski *et al.* 2005). As a result, global river systems typically show a peak flux of sediments to the oceans by the early twentieth century (Syvitski & Kettner 2011). Many coastal deltas have seen net subsidence since the 1930s, partly in response to this reduced sediment influx from the rivers, but also the impact of water and hydrocarbon extraction from deltas (Syvitski & Kettner 2011). Increased development of urban areas on deltas results in greater loading and compaction. Sediment flux is further reduced as a result of the construction of flood-prevention schemes designed to prevent sediment recharge of the inhabited parts of the delta top.

Traditional biostratigraphical signatures, used elsewhere in the stratigraphical column, may continue to have applicability to the definition of the Anthropocene. Although the introduction of exotic species are first documented following the onset of the Industrial Revolution in parts of the globe (Fig. 3), there has been accelerated introductions during increased global transportation during World War II and, subsequently, in the 1970s with the introduction of supertankers (Barnosky 2013) (Fig. 2). The release of ballast waters is now the main route of transporting invasive species, particularly with colonization within estuaries near to port facilities (Roberts 2012). It is, perhaps, in the controlled evolution of crops that most precise biostratigraphical tools may be found. Lineage zones based on maize hybrids may be suitable as the crop is geographically widespread, can only reproduce through human cultivation and may be preserved in sediments for thousands of years (Barnosky 2013). Morphologically distinct and widespread hybrids have been developed in the 1840s and 1930s, although, if molecular biology techniques are considered, mutations from the 1950s and genetically modified variants marketed since 1998 may also be recognized (Barnosky 2013).

Microfaunal and microfloral signatures within the marine environment are the most widely used biostratigraphical tool in the Phanerozoic, and signatures can also be recognized which may help resolve the definition of the Anthropocene. Potential drivers include increasing nutrient loading (N and P) and eutrophication, acidification, the presence of inorganic pollutants, alkalization, and climate change. In the Arctic and alpine lakes, there is evidence of diatom assemblage responses indicative of eutrophication coinciding with increasing atmospheric N deposition since about 1950–1970 and after 1980 (Wolfe *et al.* 2013; Wilkinson *et al.* 2014). Benthic foraminifera are sensitive to eutrophication, heavy metal and organic pollutants, changes in water management practices, introduction of non-indigenous species, and land-use changes. Foraminiferal records show a dramatic increase in the frequency and intensity of bottom-water hypoxia events since the mid-twentieth century, coinciding with the increased use of N- and P-based fertilizers (Blackwelder *et al.* 1996; Wilkinson *et al.* 2014). Ostracod abundance and diversity is related to eutrophication in freshwater and marginal marine settings, industrial pollution, sewage effluents, oil pollution, fish farming and salinity variations, and a marked reduction in diversity has become more widespread and profound during the mid-twentieth century (Wilkinson *et al.* 2014).

During the latter half of the twentieth century, fish stocks have fallen dramatically. This is, in part, the consequence of overfishing caused by the increasing use of factory fleets using improved technologies such as echo location and satellite data, and increasingly larger ships, drift nets and longlines (Roberts 2012). But, increasing artisanal fishing in coastal areas is also an important factor. The consequence is likely to be evident in sediments as reduced numbers and diversity of fish remains. In coastal and marine shelf/slope settings down to about 1 km depth, this biostratigraphical signature would coincide with extensive anthropogenic modification of surface sediments (Puig *et al.* 2012; Roberts 2012), with bottom trawling and dredging affecting $15 \times 10^6\ km^2\ a^{-1}$, about half the area of global continental shelves (Watling & Norse 1998). Contrast this with the early nineteenth century when only about 1% of the oceans were exploited (Roberts 2012). The effect may be most apparent in deep oceans where fish have low reproductive rates and stocks are largely unprotected from overfishing. Other pressures, such as the increased introduction of fish farming in estuaries, may also influence fish stocks through transmission of disease, and the introduction of nutrients and pollution from antibiotics, pesticides and fungicides (Roberts 2012). The massive growth of jellyfish populations, formerly controlled, in part, by a healthy fish population, can also threaten fish numbers through competition and consumption of fish eggs (Roberts 2012).

Coral reefs extend over only 0.1–0.2% of the oceans and are less extensive than the urbanized zones are on land. Despite being of limited areal extent, modern reefs account for a significant component of marine biotic diversity. Pollution, warming and acidification of ocean waters, eutrophication, and reduction in light levels due to increased sediment flux as coastal forests were removed have seriously stressed coral reefs (Hoegh-Guldberg 2014). The frequency and severity of mass coral bleaching episodes, beginning in 1979, have increased with time (Hoegh-Guldberg 2014). There has been a 50% reduction in the abundance of reef-building corals over the past 40–50 years (Fig. 2), with rates of change several orders of magnitude higher than for much of the last million years (Hoegh-Guldberg 2014). The effects of oceanic acidification, however, is still uncertain. Upper ocean pH across open oceans to coastal waters at various latitudes is highly site-dependent, with pH values that range markedly (0.024–1.43 units) on a monthly basis (Hofmann *et al.* 2011). Even within the comparatively stable open oceans, episodic variations in pH are greater than the annual rate of acidification, meaning that the influence of ocean acidification on oceanic biota is unlikely to be simple to predict in detail (Hofmann *et al.* 2011) and some organisms will be more sensitive to pH change than others. However, as the entire range of pH variation will be shifted to lower values by anthropogenic CO_2 emissions, significant consequences are likely.

Anthropogenic production of carbon, nitrogen and phosphorus from activities such as fossil fuel combustion, agriculture and fertilizer production increased (Fig. 2), both in volume of output and spatial reach, during the mid-twentieth century (Gałuszka *et al.* 2013). One of the more significant signatures over the past century has been a doubling of reactive nitrogen at the Earth's surface, particularly in response to the invention and implementation of the Haber process from 1913 (Zalasiewicz *et al.* 2011). Influx of excess reactive nitrogen to the ocean has resulted in increased eutrophication and increasing algal blooms, which, in turn, causes oxygen deficiency that changes the redox potential, with this process intensifying over the last 30 years (Gałuszka *et al.* 2013). Remote northern hemisphere lakes show depletion in $\delta^{15}N$ values (Holtgrieve *et al.* 2011; Wolfe *et al.* 2013) that started in 1895 CE $\pm$ 10 years, but has been accelerating over the past 50 years (Fischer *et al.* 1998). In Greenland ice, the main phase of increase was 1950–1980, culminating in levels higher than those observed for the previous 100 kyr (Wolff 2013) (Fig. 2), representing a marker that is distinct from the Holocene background.

Sulphate concentrations in Greenland ice rose by a factor of 4 over Pre-Industrial levels, with the main increases between 1900–1920 CE and 1940–1980 CE (Wolff 2013). Atmospheric disturbance of the sulphur cycle is also evident in both speleothems and trees (Fairchild & Frisia 2013). However, the increases in sulphate concentration fall within ranges possible from both large volcanic eruptions (Smith *et al.* 2013) and associated with the last glacial maximum, suggesting this is not a suitable primary marker for the Anthropocene.

Industrially produced metal pollutants, including Pb, Cd, Cu and Zn, can undergo long-range atmospheric transport, and commonly occur at levels above natural background across many depositional environments that are likely to persist in the future geological record (Gałuszka *et al.* 2013). Stable lead isotopes are particularly important for recognizing the global pollutant signature associated with alkylead additives used in gasoline as an anti-knocking agent from 1940 to 1980 (Gałuszka *et al.* 2013). Lead concentrations in Greenland snow in 1960 were a factor of 200 above the Holocene background level (Boutron *et al.* 1991; Wolff 2013). Emerging pollutants that are uniquely associated with modern technological advances may represent an important signature for the Anthropocene. Rare earth elements (REEs), used in modern high-technology industries and medicine, are now appearing in the environment, and are very persistent and non-biodegradable (Gałuszka *et al.* 2013). Persistent organic pollutants (POP) also provide potential signatures because of their long residence time in different environments and their resistance to degradation, but these would still not represent long-term signatures when viewing the start of the Anthropocene several thousands of years hence.

Temporal trends in accumulation of pollutants in sediments will differ regionally, dependent upon the diachronous expansion of industrialization (Fig. 3). Signatures may also be affected by changes in pH and redox potential of sediments, which may result in remobilization of substances (Gałuszka *et al.* 2013). Ultimately, the diachroneity in many geochemical anthropogenic signals may limit their use for defining the base of the Anthropocene. However, the most dramatic isochronous contamination signature of the mid-twentieth century is the beginning of the nuclear age and the global spread of artificial radionuclides. Global-scale enrichment in artificial radioisotopes has resulted from atmospheric nuclear weapon testing, mainly from 1945 to 1980 (Fig. 2), with more localized, although still widespread, signatures associated with discharges from nuclear reactors (Gałuszka *et al.* 2013; Hancock *et al.* 2014). There is a ^{137}Cs fallout peak of 1963–1964 (mainly in northern hemisphere sediments) and a more globally extensive peak in 1964 for ^{239}Pu. Ice cores show jumps in beta-radioactivity in 1954 and 1964, with a peak in 1966 that is a factor of 100 above background levels (Wolff 2013). Speleothems record a widespread and unambiguous radiocarbon signal that commenced in 1955 and peaked in 1962, relating to atmospheric nuclear testing. Following the test-ban treaty, levels of ^{14}C in the atmosphere have declined exponentially (Fairchild & Frisia 2013). This signature has been recorded in corals and salt marshes. However, it is probably the initial post-1945 rise in concentrations that would be used to mark a putative base of the Anthropocene, rather than the peak signature.

Given the weight of evidence, including some of the issues described above, Zalasiewicz *et al.* (2014*a*) argue the case for an approximately 1950 CE date for the onset of the Anthropocene.

Future perspective

There is a strong argument, forwarded by Wolff (2013), that the characteristics of the fully developed Anthropocene are still uncertain and that we may be living through a transition towards a new epoch, rather than being fully within it. With the exception of the definition of the Holocene, decisions made to ratify chronostratigraphical units have been made with the understanding that the events or signatures characteristic of that time period have finished. This is not true for the Anthropocene, but the erection of a new Anthropocene Epoch can only be made on the basis of material evidence of elapsed events. Projections of future trends are simply predictions: some are more robustly founded than others and they can provide a sense of perspective when considering recent patterns. Ultimately, if there is a consensus that the main environmental changes lie ahead of us, it might be concluded that it is too early to judge the position of the base of the Anthropocene, even if there is sufficient material evidence (including that detailed in this Special Publication) that the stratigraphic change to date is significant.

Each of the various events that have been proposed or discussed here as starting points for an Anthropocene Epoch is diachronous and spatially heterogeneous. But, the level of diachroneity varies from several millennia (e.g. urbanization) to a very few years (e.g. artificial radionuclide deposition). Virtually all stratigraphic boundaries are diachronous and spatially heterogenous to an extent that would make any of the potential Anthropocene bounding events seem effectively instantaneous, in a far-future perspective. The key question here is whether the range of evidence currently existing can enable contemporary Earth scientists to *effectively* and *usefully* demarcate and correlate the Anthropocene as a stratigraphic unit.

There is also concern, rightly so, for the potential for preservation of an Anthropocene bounding event (e.g. Ford *et al.* 2014). The traces of existing bounding events for deep-time stratigraphic boundaries are, of course, not universally preserved. Preservation depends on many factors, but it can be reasonably predicted, say, that cities sited on subsiding deltas are much more likely to enter the stratigraphic record in some form than those sited in mountainous terrain. Ice core from Greenland has been used to define a GSSP for the Pleistocene–Holocene boundary (Walker *et al.* 2009) and has the potential to also be used for the Anthropocene (e.g. Smith 2013; Wolff 2013). However, with extensive wasting of ice within a realm of increasing global temperatures, with greatest increases in polar regions, the likely preservation of ice formed a little over half a century ago is uncertain.

Despite the imposition of anthropogenically induced environmental stresses on global flora and fauna over recent centuries and decades, there is presently no justification for associating a mass extinction horizon with this time interval (Barnosky *et al.* 2011; Barnosky 2013). However, if currently elevated extinction rates continue, the sixth mass extinction (75% species loss) would occur within three to five centuries and that an extinction threshold exceeding the late Quaternary Megafaunal Extinction could occur even sooner (Barnosky 2013). Probably the single most significant extinction event of a single species would be that of mankind itself, but could not be used to justify the introduction of a term such as the Anthropocene.

The extent to which current demands for environmental and biotic conservation can be effective in the future are difficult to predict. However, locally, there is evidence that measures to clean up once heavily polluted environments are having an effect. For example, changes in local dinoflagellate cyst assemblages in response to human-induced eutrophication initially in the mid- to late-nineteenth century and, particularly, during the early to mid-twentieth century have shown trends of recovery in the 1980s–1990s in response to improvements in sewage treatment works (Wilkinson *et al.* 2014). A future reduction of input of humanity's wastes into the oceans may at least allow recoveries of other micro- and macro-faunal communities in the future (Fig. 2), although perhaps not into the same patterns as those that existed prior to human perturbation.

Ocean temperatures have increased by 0.1 °C over the past century, although seasonal and diurnal changes are greater. In the northern hemisphere there is already evidence of the northwards 'march to the poles' of fish and plankton species in response to this warming (Roberts 2012). There is a potential, with increased ocean temperatures in the future, for a more marked affect on ostracod assemblages (Wilkinson *et al.* 2014). Perhaps most susceptible to temperature increases are biota in high-latitude regions. High northern latitudes are likely to experience the greatest temperature increases over this century (see Haywood *et al.* 2011), and fauna in this region may have no alternative environment for retreat. With projected oceanic temperature rises, it is realistic to envisage the loss of coral-dominated reef habitats by the middle of this century (Hoegh-Guldberg 2014) (Fig. 2). The late Paleocene–early Eocene thermal maximum event (PETM), with associated spike in CO_2 and ocean acidification, may be a close analogue to our current climate trajectory. This event saw a stepwise transition of platform reef assemblages (Scheibner & Speijer 2008). During the Paleocene, coralgal reef associations with diverse coral types dominated in low and mid latitudes. A transitional late Paleocene stage is marked by persistent or coralgal reefs in mid latitudes, but large foraminifers dominate in low latitudes. By the start of the PETM, larger foraminifers and encrusting foraminifers form large reefs with coralgal build-ups generally absent (Scheibner & Speijer 2008). Such transitions appear to have beeen recorded in modern reefs by Leinfelder *et al.* (2012) with the evolution of low-diversity reefs in Almirante Bay, Panama. This reef appears to be thriving in an environment of increasing terrigenous run-off and reduced salinities, and may reflect a pattern for evolution of Anthropocene reefs.

Global estimates of sea-level rise in response to thermal expansion and melting of land-based ice for the twentieth and twenty-first centuries are approximately 1.8 mm a^{-1} (Church & White 2011), considered an acceleration on previous centuries. Precise determinations using satellite altimetry indicate rates of sea-level rise of 3.2 ± 0.4 mm a^{-1} from 1993 to 2003 (Church & White 2011), which suggests a continuation of this acceleration. This should be compared with the >40 mm a^{-1} rate during the last deglaciation at approximately 14 ka BP (Fairbanks 1989). Modelling limits predictions to 2100 CE, and estimates for subsequent increased rates in sea-level elevation are difficult to quantify, but the extreme estimate would be melting of all ice sheets, leading to a sea-level rise of 80 m (Williams & Hall 1993). By comparison with the Cretaceous and Eocene 'greenhouse intervals', it is expected that with a doubling of CO_2 from Pre-Industrial levels there will be an increase in the precipitation rate (Haywood *et al.*

2011), which would be expected to cause both increased soil erosion and increased discharges in fluvial systems.

Increased output of anthropogenic CO_2 may result in future acidification of the oceans and significant undersaturation of $CaCO_3$. This may cause a shallowing in the average carbonate compensation depth and production of a prolonged carbonate gap (for several thousand years) in deep marine deposits, such that paler coccolith and foraminifer oozes will become rare, and darker clay and silicic deposits will dominate (Tyrrell 2011).

Haff (2013) suggests an alternative view of the Anthropocene as an age of technology, with increasing domination of our environment by an emergent technosphere, of which humans are components. Haff (2013) suggests that the technosphere has evolved into a dynamic system, but as a juvenile system that has not reached equilibrium, being a poor recycler of critical resources. Appropriation of energy by the technosphere has resulted in disruption to the lithosphere, atmosphere, hydrosphere and biosphere. Time will tell if this event is like the Great Oxidation Event of about 2.4 billion years ago that resulted in a shift in the global state. Or, if unsustainable, the evolution and demise of the technosphere represents a brief episode, comparable to the K–T impact event.

Summary and conclusions

In summary, it is recognized that in order to define the Anthropocene as a formal chronostratigraphical unit, it is necessary to apply the same rigorous evidence-based approach to recognizing key signatures as has been used for the definition of older units. However, there should be concerns if special criteria are being imposed to justify the definition that could not be met by these older units.

Ultimately, there is a requirement to identify a critical change to a new regime in which anthropogenic influence is a dominant controlling factor upon aspects as diverse as, for example, biotic abundance and variability, sediment flux and sediment composition, geochemical and radiogenic signatures, climate change, sea-level rise, and ice-cover loss. Ideally, the definition of the Anthropocene should be based on a single, globally expressed signature. This could be, for example, the appearance of radiogenic fallout, although there remains the questions as to whether the initial post-1945 rise or the peak signature some two decades later should be used. However, a definition drawing upon a spectrum of signatures would enable characterization of the unit to reflect a profound change across many environmental indicators. As demonstrated in this paper, the onset of the broad range of signatures is diachronous, spanning almost 11 kyr or more (Fig. 2). Many, although not all, of the indicators covered in this Special Publication show maximum signatures that post-date 1945, leading to the suggestion that this date may be a suitable age for the commencement of the Anthropocene should it prove useful and necessary to define it (Fig. 2). What cannot be quantified is the extent that the acme of many of these signatures lies ahead in the future, indicating that we lack the full perspective of geological time to review the total impact of humans on Earth. It is important to recognize that human decision-making has the potential to shape the future geological record. For the present, we must continue to work with a developing narrative, even as it unfolds.

The authors wish to acknowledge the helpful comments from S. Price and V. Banks, and constructive feedback from D. Schmidt is also gratefully acknowledged. Figure 1 was designed by C. Wardle. This article is published with the permission of the Executive Director of the British Geological Survey (NERC).

Appendix

Glossary of key terms

- *Anthropocene* (derived from *anthrōpos 'human being'* and *kainos 'new'*) was first proposed as an epoch by Crutzen & Stoermer (2000) to denote the present time interval, in which human activities have profoundly altered the global environment. The term is currently undefined and is used only informally.
- *Anthropozoic* (derived from *anthrōpos 'human being'* and *zōion 'animal' or* zoic *'life'*) was proposed as an era by Antonio Stoppani (1873) in the 1870s in recognition of the increasing power and effect of humanity on the Earth's systems. The epithet of '-zoic' is used to name units of era ranking (e.g. Palaeozoic, Mesozoic, Cenozoic; that is, the rank above that of period, in turn an order above epoch.
- *Anthrocene* is a term proposed by Revkin (1992), which had essentially the same meaning as Anthropocene.
- *Anthropocene deposits* refer to those sediments and contained materials of various sources (e.g. plastics, metals, glass) created by processes that reflect either human or natural agents, or a combination of the two, that accumulated during the time interval known as the Anthropocene.
- *Anthropogenic deposits* refer to those sediments that have been created either directly or indirectly by human activities, but in which there is a dominant proportion of redeposited or novel material (Price *et al.* 2011; Ford *et al.* 2014). Such deposits may include *artificial deposits/artificial ground (direct anthropogenic deposits)*. If natural processes are present, such as erosion and deposition within river systems, these

may be considered to be *indirect anthropogenic deposits*, where human interaction, such as agriculture, deforestation, modification of river systems, influences the location and rates of such natural processes (Ford *et al.* 2014). The above terms are purely descriptive and none of have any time connotations and do not indicate whether they relate to Anthropocene or pre-Anthropocene time. Similarly, the Anthropocene will include 'natural deposits', such as desert dune deposits, with no perceptible human influence.

- *Artificial deposits* reflecting those sediments deposited directly and purposely by human activity and which may be associated with *artificial ground*, in which the ground surface has been modified either through deposition or excavation, or a combination of the two (Price *et al.* 2011; Ford *et al.* 2014). Edgeworth (2013) distinguishes the dominance of a cultural agency as the primary force in the production of artificial ground.
- *Made ground* and *worked ground* represent physical extents of, respectively, artificial deposits accumulated above the natural ground surface and excavations into this natural ground (Price *et al.* 2011; Ford *et al.* 2014). These terms are used as part of a morphostratigraphical scheme used by the British Geological Survey to classify artificial deposits. Ford *et al.* (2014) consider the potential of developing truly lithostratigraphical schemes to classify artificial deposits.

References

AGER, D. V. 1993. *The Nature of the Stratigraphical Record*, 3rd edn. John Wiley, Chichester.

AKÇAR, N., IVY-OCHS, S. & SCHLÜCHTER, C. 2008. Application of in-situ produced terrestrial cosmogenic nuclides to archaeology: a schematic review. *Eiszeitalter und Gegenwart Quaternary Science Journal*, **57**, 226–238.

BARNOSKY, A. D. 2013. Palaeontological evidence for defining the Anthropocene. *In*: WATERS, C. N., ZALASIEWICZ, J., WILLIAMS, M., ELLIS, M. A. & SNELLING, A. (eds) *A Stratigraphical Basis for the Anthropocene*. Geological Society, London, Special Publications, **395**, first published online October 24, 2013, http://dx.doi.org/10.1144/SP395.6

BARNOSKY, A. D., MATZKE, N. ET AL. 2011. Has the Earth/'s sixth mass extinction already arrived? *Nature*, **471**, 51–57.

BATTARBEE, R. W., MORLEY, D., BENNION, H., SIMPSON, G. L., HUGHES, M. & BAUERE, V. 2011. A palaeolimnological meta-database for assessing the ecological status if lakes. *Journal of Paleolimnology*, **45**, 405–414.

BLACKWELDER, P., HOOD, T., ALVAREZ-ZARIKIAN, C., NELSEN, T. A. & MCKEE, B. 1996. Benthic Foraminifera from the NECOP study area impacted by the Mississippi River plume and seasonal hypoxia. *Quaternary International*, **31**, 19–36.

BOUTRON, C. F., GÖRLACH, U., CANDELONE, J.-P., BOLSHOV, M. A. & DELMAS, R. J. 1991. Decrease in anthropogenic lead, cadmium and zinc in Greenland snows since the late 1960s. *Nature*, **353**, 153–156.

BRADLEY, R. S. 1999. *Paleoclimatology: Reconstructing Climates of the Quaternary*. Harcourt Academic Press, San Diego, CA.

BRIFFA, K. R., JONES, P. D., SCHWEINGRUBER, F. H. & OSBORN, T. J. 1998. Influence of volcanic eruptions on Northern Hemisphere summer temperature over the past 600 years. *Nature*, **393**, 450–455.

BROWN, A., TOMS, P., CAREY, C. & RHODES, E. 2013. Geomorphology of the Anthropocene: Time-transgressive discontinuities of human-induced alluviation. *Anthropocene*, **1**, 3–13, http://dx.doi.org/http://dx.doi.org/10.1016/j.ancene.2013.06.002

BROWNE, E. 1997. Nuclear data sheets for A = 90. *Nuclear Data Sheets*, **82**, 379–546.

CERTINI, G. & SCALENGHE, R. 2011. Anthropogenic soils are the golden spikes for the Anthropocene. *The Holocene*, **21**, 1269–1274.

CHURCH, J. A. & WHITE, N. J. 2011. Sea-level rise from the late 19th to the early 21st century. *Surveys in Geophysics*, **32**, 585–602, http://dx.doi.org/10.1007/s10712-011-9119-1

CRUTZEN, P. J. 2002. Geology of mankind. *Nature*, **415**, 23.

CRUTZEN, P. J. & STOERMER, E. F. 2000. The 'Anthropocene'. *Global Change Newsletter*, **41**, 17–18.

DEARING, J. A. & JONES, R. T. 2003. Coupling temporal and spatial dimensions of global sediment flux through lake and marine sediment records. *Global and Planetary Change*, **39**, 147–168.

DELMAS, R. J. 1992. Environmental information from ice cores. *Reviews in Geophysics*, **30**, 1–21.

DIBB, J., MAYEWSKI, P. A., BUCK, C. F. & DRUMMEY, S. M. 1990. Beta radiation from snow. *Nature*, **344**(6270), 25.

DULLER, G. A. T. 2008. *Luminescence Dating: Guidelines on Using Luminescence Dating in Archaeology*. English Heritage, Swindon.

EDGEWORTH, M. 2013. Archaeological stratigraphy and artificial ground. *In*: WATERS, C. N., ZALASIEWICZ, J., WILLIAMS, M., ELLIS, M. A. & SNELLING, A. (eds) *A Stratigraphical Basis for the Anthropocene*. Geological Society, London, Special Publications, **395**, first published online October 25, 2013, http://dx.doi.org/10.1144/SP395.3

ELERT, G. 2013. *The Physics Hypertextbook*. http://physics.info/half-life/ (accessed 18 September 2013).

ELLIS, E. C., ANTILL, E. C. & KREFT, H. 2012. All is not loss: Plant biodiversity in the Anthropocene. *PLoS ONE*, **7**, e30535, http://dx.doi.org/10.1371/journal.pone.0030535

ELLIS, E. C., KAPLAN, J. O., FULLER, D. Q., VAVRUS, S., KLEIN GOLDEWIJK, K. & VERBURG, P. H. 2013. Used planet: A global history. *Proceedings of the National Academy of Sciences of the United States of America*, **110**, 7978–7985, http://dx.doi.org/10.1073/pnas.1217241110

FAIRBANKS, R. G. 1989. A 17,000-year glacio-eustatic sea-level record: Influence of glacial melting rates on the Younger Dryas event and deep-ocean circulation. *Nature*, **342**, 637–642.

FAIRCHILD, I. J. & FRISIA, S. 2013. Definition of the Anthropocene: a view from the underworld. *In*: WATERS, C. N.,

ZALASIEWICZ, J., WILLIAMS, M., ELLIS, M. A. & SNELLING, A. (eds) *A Stratigraphical Basis for the Anthropocene*. Geological Society, London, Special Publications, **395**, first published online October 24, 2013, http://dx.doi.org/10.1144/SP395.7

FAO. 2010. *Global Forest Resources Assessment 2010 – Main Report*. FAO Forestry Paper, 163. Food and Agriculture Organization of the United Nations, Rome. www.fao.org/docrep/013/i1757e/i1757e00.htm

FINNEY, S. C. 2013. The 'Anthropocene' as a ratified unit in the ICS International Chronostratigraphic Chart: fundamental issues that must be addressed by the Task Group. *In*: WATERS, C. N., ZALASIEWICZ, J., WILLIAMS, M., ELLIS, M. A. & SNELLING, A. (eds) *A Stratigraphical Basis for the Anthropocene*. Geological Society, London, Special Publications, **395**, first published online October 24, 2013, http://dx.doi.org/10.1144/SP395.9

FISCHER, H., WAGENBACH, D. & KIPFSTUHL, J. 1998. Sulfate and nitrate firn concentrations on the Greenland ice sheet. 2. Temporal anthropogenic deposition changes. *Journal of Geophysical Research*, **103**, 21,935–21,942.

FORD, J. R., PRICE, S. J., COOPER, A. H. & WATERS, C. N. 2014. An assessment of lithostratigraphy for anthropogenic deposits. *In*: WATERS, C. N., ZALASIEWICZ, J., WILLIAMS, M., ELLIS, M. A. & SNELLING, A. (eds) *A Stratigraphical Basis for the Anthropocene*. Geological Society, London, Special Publications, **395**, first published online January 23, 2014, http://dx.doi.org/10.1144/SP395.12

GALE, S. J. & HOARE, P. G. 2012. The stratigraphic status of the Anthropocene. *The Holocene*, **22**, 1478–1481.

GAŁUSZKA, A., MIGASZEWSKI, Z. M. & ZALASIEWICZ, J. 2013. Assessing the Anthropocene with geochemical methods. *In*: WATERS, C. N., ZALASIEWICZ, J., WILLIAMS, M., ELLIS, M. A. & SNELLING, A. (eds) *A Stratigraphical Basis for the Anthropocene*. Geological Society, London, Special Publications, **395**, first published online October 24, 2013, http://dx.doi.org/10.1144/SP395.5.

GIBBARD, P. L. & WALKER, M. J. C. 2013. The term 'Anthropocene' in the context of formal geological classification. *In*: WATERS, C. N., ZALASIEWICZ, J., WILLIAMS, M., ELLIS, M. A. & SNELLING, A. (eds) *A Stratigraphical Basis for the Anthropocene*. Geological Society, London, Special Publications, **395**, first published online October 25, 2013, http://dx.doi.org/10.1144/SP395.1

GIBBARD, P. L., HEAD, M. J. & WALKER, M. J. C. 2010. Formal ratification of the Quaternary System/Period and the Pleistocene Series/Epoch with a base at 2.58 Ma. *Journal of Quaternary Science*, **25**, 96–102.

GOSSE, J. C. & PHILLIPS, F. M. 2001. Terrestrial in situ cosmogenic nuclides: Theory and application. *Quaternary Science Reviews*, **20**, 1475–1560.

GRADSTEIN, F. M., OGG, J. G., SCHMITZ, M. & OGG, G. 2012. *The Geological Time Scale 2012*. Elsevier, Amsterdam.

HAFF, P. 2013. Technology as a geological phenomenon: Implications for human well-being. *In*: WATERS, C. N., ZALASIEWICZ, J., WILLIAMS, M., ELLIS, M. A. & SNELLING, A. (eds) *A Stratigraphical Basis for the Anthropocene*. Geological Society, London, Special Publications, **395**, First published online October 24, 2013, http://dx.doi.org/10.1144/SP395.4

HANCOCK, G. J., TIMS, S. G., FIFIELD, L. K. & WEBSTER, I. T. 2014. The release and persistence of radioactive Anthropogenic nuclides. *In*: WATERS, C. N., ZALASIEWICZ, J., WILLIAMS, M., ELLIS, M. A. & SNELLING, A. (eds) *A Stratigraphical Basis for the Anthropocene*. Geological Society, London, Special Publications, **395**, first published online February 28, 2014, http://dx.doi.org/10.1144/SP395.15

HAYWOOD, A. M., RIDGEWELL, A. ET AL. 2011. Are there pre-Quaternary geological analogues for a future greenhouse warming? *Philosophical Transactions of the Royal Society, London, Series A*, **369**, 933–956.

HEDBERG, H. D. (ed.) 1976. *International Stratigraphic Guide – A Guide to Stratigraphic Classification, Terminology, and Procedure*. Wiley, New York.

HOEGH-GULDBERG, O. 2014. Coral reefs in the Anthropocene: persistence or the end of the line? *In*: WATERS, C. N., ZALASIEWICZ, J., WILLIAMS, M., ELLIS, M. A. & SNELLING, A. (eds) *A Stratigraphical Basis for the Anthropocene*. Geological Society, London, Special Publications, **395**, first published online March 20, 2014, http://dx.doi.org/10.1144/SP395.17

HOFMANN, G. E., SMITH, J. E. ET AL. 2011. High-Frequency Dynamics of Ocean pH: A Multi-Ecosystem Comparison. *PLoS ONE*, **6**, e28983.

HOLTGRIEVE, G. W., SCHINDLER, D. E. ET AL. 2011. A Coherent Signature of Anthropogenic Nitrogen Deposition to Remote Watersheds of the Northern Hemisphere. *Science*, **334**, 1545–1548.

HOOKE, R. L. 2000. On the history of humans as geomorphic agents. *Geology*, **28**, 843–846.

KLEIN GOLDEWIJK, K., BEUSEN, A. & JANSSEN, P. 2010. Long-term dynamic modeling of global population and built-up area in a spatially explicit way: HYDE 3.1. *Holocene*, **20**, 565–573.

LEINFELDER, R. R., SEEMANN, J., HEISS, G. A. & STRUCK, U. 2012. Could 'Ecosystem Atavisms' help reefs to adapt to the Anthropocene? *In*: *Proceedings of the 12th International Coral Reef Symposium, Cairns, Australia, 9–13 July 2012*. Online publication http://www.reefbase.org/resource_center/publication/icrs.aspx

LISIECKI, L. E. & RAYMO, M. E. 2005. A Pliocene-Pleistocene stack of 57 globally distributed benthic $\delta^{18}O$ records. *Paleoceanography*, **20**, PA1003. http://dx.doi.org/10.1029/2004PA001071

LOCKE, G. & BERTINE, K. K. 1986. Magnetic sediments as an indicator of coal combustion. *Applied Geochemistry*, **1**, 345–356.

LOULERGUE, L., SCHILT, A. ET AL. 2008. Orbital and millennial-scale features of atmospheric CH_4 over the last 800,000 years. *Nature*, **453**, 383–386.

LÜTHI, D., LE FLOCH, M. ET AL. 2008. High-resolution carbon dioxide concentration record 650,000-800,000 years before present. *Nature*, **453**, 379–382.

MERRITTS, D., WALTER, R. ET AL. 2011. Anthropocene streams and base-level controls from historic dams in the unglaciated mid-Atlantic region, USA. *Philosophical Transactions of the Royal Society, Series A*, **369**, 976–1009.

POIRIER, C., CHAUMILLON, E. & ARNAUD, F. 2011. Siltation of river-influenced coastal environments:

Respective impact of late Holocene land use and high-frequency climate changes. *Marine Geology*, **290**, 51–62, http://dx.doi.org/10.1016/j.margeo.2011.10.008

PRICE, S. J., FORD, J. R., COOPER, A. H. & NEAL, C. 2011. Humans as major geological and geomorphological agents in the Anthropocene: the significance of artificial ground in Great Britain. *Philosophical Transactions of the Royal Society*, **369**, 1056–1084. http://dx.doi.org/10.1098/rsta.2010.0296

PUIG, P., CANALS, M. *ET AL.* 2012. Ploughing the deep sea floor. *Nature*, **489**, 286–289.

REVKIN, A. 1992. *Global Warming: Understanding the Forecast.* American Museum of Natural History, Environmental Defense Fund, Abbeville Press, New York.

RICHTER, D. DEB. 2007. Humanity's transformation of Earth's soil: pedology's new frontier. *Soil Science*, **172**, 957–967.

ROBERTS, C. 2012. *Ocean of Life: How our Seas are Changing.* Allen Lane, London.

RUDDIMAN, W. F. 2003. The Anthropogenic greenhouse era began thousands of years ago. *Climatic Change*, **61**, 261–293.

RUDDIMAN, W. F. 2005. *Plows, Plagues and Petroleum.* Princeton University Press, Princeton, NJ.

SALVADOR, A. 1994. *International Stratigraphic Guide. A Guide to Stratigraphic Classification, Terminology, and Procedure*, 2nd edn. International Union of Geological Sciences and The Geological Society of America, Boulder, CO.

SCHEIBNER, C. & SPEIJER, R. P. 2008. Decline of coral reefs during late Paleocene to early Eocene global warming. *eEarth*, **3**, 19–26.

SCHILT, A., BAUMGARTNER, M., BLUNIER, T., SCHWANDER, J., SPAHNI, R., FISCHER, H. & STOCKER, T. F. 2010. Glacial-interglacial and millennial-scale variations in the atmospheric nitrous oxide concentration during the last 800,000 years. *Quaternary Science Reviews*, **29**, 182–192, http://dx.doi.org/10.1016/j.quascirev.2009.03.011

SHACKLETON, N. J. & OPDYKE, N. D. 1973. Oxygen isotope and palaeomagnetic stratigraphy of the equatorial Pacific core V28-238: Oxygen isotope temperatures and ice volumes on a 105 and 106 year scale. *Quaternary Research*, **3**, 39–55, http://dx.doi.org/10.1016/0033-5894(73)90052-5

SHEN, S.-Z., HENDERSON, C. M. *ET AL.* 2010. High-resolution Lopingian (Late Permian) timescale of South China. *Geological Journal*, **45**, 122–134.

SIMPSON, I. A. 1997. Relict properties of anthropogenic deep top soils as indicators of infield management in Marwick, West Mainland, Orkney. *Journal of Archaeological Science*, **24**, 365–380.

SINGARAYER, J. S., VALDES, P. J., FRIEDLINGSTEIN, P., NELSON, S. & BEERLING, D. J. 2011. Late Holocene methane rise caused by orbitally controlled increase in tropical sources. *Nature*, **470**, 82–86.

SMITH, V. C. 2013. Defining the onset of the Anthropocene using tephrochronology. *In*: WATERS, C. N., ZALASIEWICZ, J., WILLIAMS, M., ELLIS, M. A. & SNELLING, A. (eds) *A Stratigraphical Basis for the Anthropocene.* Geological Society, London, Special Publications, **395**, first published online November 21, 2013, http://dx.doi.org/10.1144/SP395.11

SMITH, A. G., BARRY, T. *ET AL.* 2014. GSSPs, global stratigraphy and correlation. *In*: SMITH, D. (ed.) *Strata and Time: Probing the Gaps in Our Understanding.* Geological Society, London, Special Publications, **404**, http://dx.doi.org/10.1144/SP404.8

SNOWBALL, I., HOUNSLOW, M. W. & NILSSON, A. 2013. Geomagnetic and mineral magnetic characterization of the Anthropocene. *In*: WATERS, C. N., ZALASIEWICZ, J., WILLIAMS, M., ELLIS, M. A. & SNELLING, A. (eds) *A Stratigraphical Basis for the Anthropocene.* Geological Society, London, Special Publications, **395**, first published online December 19, 2013, http://dx.doi.org/10.1144/SP395.13

SOLOMON, S., QIN, D. *ET AL.* (eds) 2007. *Climate Change 2007: The Physical Science Basis. Contribution of Working Group I to the Fourth Assessment Report of the Intergovernmental Panel on Climate Change, 2007.* Cambridge University Press, Cambridge.

STEFFEN, W., SANDERSON, A. *ET AL.* 2004. *Global Change and the Earth System: A Planet under Pressure.* Springer, Berlin.

STEFFEN, W., CRUTZEN, P. J. & MCNEILL, J. R. 2007. The Anthropocene: are humans now overwhelming the great forces of Nature? *Ambio*, **36**, 614–621.

STOPPANI, A. 1873. *Corsa di Geologia.* Bernardoni & Brigola, Milan.

STUIVER, M. & POLACH, H. A. 1977. Discussion: Reporting of ^{14}C data. *Radiocarbon*, **19**, 355–363.

SYVITSKI, J. P. M. & KETTNER, A. 2011. Sediment flux and the Anthropocene. *Philosophical Transactions of the Royal Society*, **A369**, 957–975.

SYVITSKI, J. P. M., VÖRÖSMARTY, C. J., KETTNER, A. J. & GREEN, P. 2005. Impact of Humans on the Flux of Terrestrial Sediment to the Global Coastal Ocean. *Science*, **308**, 376–380.

THOMPSON, R. C., MOORE, C. J., VOM SAAL, F. S. & SWAN, S. H. 2009. Plastics, the environment and human health: current consensus and future trends. *Philosophical Transactions of the Royal Society*, **B364**, 2153–2166, http://dx.doi.org/10.1098/rstb.2009.0053

TYRRELL, T. 2011. Anthropogenic modification of the oceans. *Philosophical Transactions of the Royal Society*, **A369**, 887–908.

UNDESA. 2012. *World Urbanisation Prospects, 2011 Revisions.* United Nations Department of Economic and Social Affairs, New York. http://hdrstats.undp.org/en/tables/displayByRow.cfm

UNTERWEGER, M. P. 2003. *Radionuclide Half-Life Measurements at the National Institute of Standards and Technology (version 3.0).* [Online] Available: http://physics.nist.gov/Halflife. Retrieved 18 September 2013. National Institute of Standards and Technology, Gaithersburg, MD.

WALKER, M., JOHNSEN, S. *ET AL.* 2009. Formal definition and dating of the GSSP (Global Stratotype Section and Point) for the base of the Holocene using the Greenland NGRIP ice core, and selected auxiliary records. *Journal of Quaternary Science*, **24**, 3–17.

WALKER, M. J. C., BERKELHAMMER, M. *ET AL.* 2012. Formal subdivision of the Holocene Series/Epoch: a Discussion Paper by a Working Group of INTIMATE (Integration of ice-core, marine and terrestrial records)

and the Subcommission on Quaternary Stratigraphy (International Commission on Stratigraphy). *Journal of Quaternary Science*, **27**, 649–659, http://dx.doi.org/10.1002/jqs.2565

Watling, L. & Norse, E. A. 1998. Disturbance of the Seabed by Mobile Fishing Gear: A Comparison to Forest Clearcutting. *Conservation Biology*, **12**, 1180–1197, http://dx.doi.org/10.1046/j.1523-1739.1998.0120061180.x

Wilkinson, B. H. 2005. Humans as geologic agents: A deep-time perspective. *Geology*, **33**, 161–164.

Wilkinson, I. P., Poirier, C., Head, M. J., Sayer, C. D. & Tibby, J. 2014. Micropalaeontological signatures of the Anthropocene. *In*: Waters, C. N., Zalasiewicz, J., Williams, M., Ellis, M. A. & Snelling, A. (eds) *A Stratigraphical Basis for the Anthropocene*. Geological Society, London, Special Publications, **395**, first published online January 31, 2014, http://dx.doi.org/10.1144/SP395.14

Williams, M., Zalasiewicz, J., Waters, C. N. & Landing, E. 2013. Is the fossil record of complex animal behaviour a stratigraphical analogue for the Anthropocene? *In*: Waters, C. N., Zalasiewicz, J., Williams, M., Ellis, M. A. & Snelling, A. (eds) *A Stratigraphical Basis for the Anthropocene*. Geological Society, London, Special Publications, **395**, first published online October 25, 2013, http://dx.doi.org/10.1144/SP395.8

Williams, R. S. & Hall, D. K. 1993. Glaciers. *In*: Gurney, R. J., Foster, J. L. & Parkinson, C. L. (eds) *Atlas of Satellite Observations Related to Global Change*. Cambridge University Press, Cambridge, 401–422.

Wolfe, A. P., Hobbs, W. O. *et al.* 2013. Stratigraphic expressions of the Holocene–Anthropocene transition revealed in sediments from remote lakes. *Earth-Science Reviews*, **116**, 17–34.

Wolff, E. W. 2013. Ice Sheets and the Anthropocene. *In*: Waters, C. N., Zalasiewicz, J., Williams, M., Ellis, M. A. & Snelling, A. (eds) *A Stratigraphical Basis for the Anthropocene*. Geological Society, London, Special Publications, **395**, first published online October 25, 2013, http://dx.doi.org/10.1144/SP395.10

Woods, W. I. 2008. *Amazonian Dark Earths: Wim Sombroek's Vision*. Springer, New York.

Yin, H., Zhang, K., Tong, J., Yang, Z. & Wu, S. 2001. The Global Stratotype Section and Point (GSSP) of the Permian-Triassic boundary. *Episodes*, **24**, 102–114.

Zalasiewicz, J., Williams, M. *et al.* 2008. Are we now living in the Anthropocene? *Geological Society of America Today*, **18**, 4–8.

Zalaseiwicz, J., Williams, M. *et al.* 2011. Stratigraphy of the Anthropocene. *Philosophical Transactions of the Royal Society, Series A*, **369**, 1036–1055.

Zalasiewicz, J., Kryza, R. & Williams, M. 2013. The mineral signature of the Anthropocene. *In*: Waters, C. N., Zalasiewicz, J., Williams, M., Ellis, M. A. & Snelling, A. (eds) *A Stratigraphical Basis for the Anthropocene*. Geological Society, London, Special Publications, **395**, first published online December 19, 2013, http://dx.doi.org/10.1144/SP395.2

Zalasiewicz, J., Williams, M. & Waters, C. N. 2014*a*. Can an Anthropocene Series be defined and recognized? *In*: Waters, C. N., Zalasiewicz, J., Williams, M., Ellis, M. A. & Snelling, A. (eds) *A Stratigraphical Basis for the Anthropocene*. Geological Society, London, Special Publications, **395**, first published online March 11, 2014, http://dx.doi.org/10.1144/SP395.16

Zalasiewicz, J., Williams, M., Waters, C. N., Barnosky, A. D. & Haff, P. 2014*b*. The technofossil record of humans. *Anthropocene Review*, **1**, first published January 7, 2014, http://dx.doi.org/10.1177/2053019613514953

The ‘Anthropocene’ as a ratified unit in the ICS International Chronostratigraphic Chart: fundamental issues that must be addressed by the Task Group

S. C. FINNEY

Department of Geological Sciences, International Commission on Stratigraphy, California State University at Long Beach, Long Beach, California 90840, USA (e-mail: Stan.Finney@csulb.edu)

Abstract: The proposal that the ‘Anthropocene’ should be ratified as a unit of the International Chronostratigraphic Chart/Geological Time Scale deserves serious consideration by the International Commission on Stratigraphy (ICS). The ‘Anthropocene’ task group within the ICS Subcommission on Quaternary Stratigraphy is responsible for producing a recommendation to be evaluated and considered for approval at high levels in the ICS organization. It must consider the rank and extent of the unit as well as a GSSP or GSSA that defines its lower boundary or beginning. Furthermore, the task group must address several questions related to the proposed unit and the nature of the ICS chart/timescale that are unique to it. Is the ‘Anthropocene’ a concept in search of a distinct stratigraphic record? Should it be defined on a stratigraphic signal or on a date in human history? Is the ‘Anthropocene’ a unit of Earth history or human history or primarily a future projection? Is the ‘Anthropocene’ useful as a chronostratigraphic unit on geological maps? Has the perceived change to a human-dominated Earth system overwhelmed the natural Earth system? What is the conceptual basis of the International Chronostratigraphic Chart/Geologic Time Scale? Will the term ‘Anthropocene’ have value even if not ratified as a formal unit?

For ratified units of the International Chronostratigraphic Chart (ICS)/Geological Time Scale of the International Commission on Stratigraphy to have validity and authority as global geostandards, they must be based on established stratigraphic concepts and criteria and be approved following specified policies and procedures (Hedberg 1976; Cowie *et al.* 1986; Salvador 1994; Remané *et al.* 1996). Decisions are made by supermajority votes first by voting members of a subcommission from which a formal recommendation arises and then by the voting members of the ICS (three executive officers and chairs of the 16 subcommissions). Subcommission and ICS voting members have the responsibility to become fully informed about a matter upon which they vote, yet recommendations are typically developed in a small task group within a subcommission. Having not served on the task group that develops a proposal, most voting members of a subcommission and of ICS become informed by reading a formal proposal that presents a recommendation, as well as comments both for and against. Accordingly, the task group proposal must not only explain and justify its recommendation, but also address all questions that could be raised against it. Furthermore, when a task group's proposal is considered by the voting members of the subcommission, there must be extended opportunity for discussion in open meetings, via the internet, or both, and opinions from geoscientists outside of the subcommission should also be welcome and given careful consideration. It is necessary that all points are presented and all points challenged and all questions raised and all positions stated in these discussions. In such a fashion, voting members become fully informed. The same procedure occurs again when a subcommission recommendation is approved and forwarded to the ICS voting members. Final ratification by the IUGS Executive Commission of a recommendation approved by the ICS is based primarily on the assurance that ICS policies and procedures were followed. Most decisions made by the ICS are on matters receiving near unanimous agreement, but selection of boundaries for some chronostratigraphic/geochronological units, redefinition of traditional units, and replacement of traditional units by new units, particularly those at higher levels (series/epoch and system/period), have been long and contentious; some proposals have been rejected, and the recent ICS decision on the formal definition of the Quaternary System and redefinition of the Pleistocene Series followed decades of acrimonious debate. Contentious issues, novel proposals, proposals with vocal proponents, and proposals that receive international media attention are best addressed by the ICS by following these well-established procedures.

The proposal on the ‘Anthropocene’ is novel, could be contentious, has vocal proponents, and

From: Waters, C. N., Zalasiewicz, J. A., Williams, M., Ellis, M. A. & Snelling, A. M. (eds) 2014. *A Stratigraphical Basis for the Anthropocene*. Geological Society, London, Special Publications, **395**, 23–28.
First published online October 24, 2013, http://dx.doi.org/10.1144/SP395.9

has received international media attention. It is natural that media attention has focused on recognition of the immense, global impact of humans on the Earth system. However, the sudden appearance of these articles and what has appeared to make them newsworthy at this time appears to be the proposal to ratify the 'Anthropocene' as a formal unit of the ICS International Chronostratigraphic Chart, which is the basis for the Geological Time Scale (Zalasiewicz *et al.* 2008). Would this issue have been newsworthy if there was not a proposal for the 'Anthropocene' to be considered for formal recognition by the ICS? Whether or not that is the case, the proposal itself gives recognition to the ICS as the arbiter of internationally recognized geological standards, and it must give careful consideration to all proposals on stratigraphic units and procedures submitted by the greater geoscience community. It is natural that the members of the task group within the ICS Subcommission on Quaternary Stratigraphy that is considering the 'Anthropocene' are drawn largely from proponents of the initiative, and their enthusiastic contributions are greatly valued. The concern of the ICS executive, however, is that the task group must give careful consideration to all positions and arguments related to this proposal. The task group will evaluate the potential redefinition of the Holocene necessitated by definition of the 'Anthropocene' as discussed in a companion article (Gibbard & Walker 2013). However, with regard to the 'Anthropocene' being considered for ratification as a formal unit of the ICS International Chronostratigraphic Chart, several other, more fundamental issues must be considered by the task group.

Issues to be addressed

Is the 'Anthropocene' a concept in search of a distinct stratigraphic record? This question was raised by Autin & Holbrook (2012*a*, *b*) in Comments and Replies to Zalasiewicz *et al.* (2008, 2012). The issue that was raised by these authors as well as by Gibbard & Walker (2013) regards identification of a globally distinct stratigraphic marker on which to define the 'Anthropocene' as a formal chronostratigraphic/geochronological unit, suggesting that such a marker is the only basis on which to define the unit, and the only basis necessary. Consideration of a stratigraphic marker arises from the process initiated with the establishment of ICS in the late 1960s to develop a standard set of global chronostratigraphic units with the lower boundary of each unit defined by a Global Stratotype Section and Point (GSSP). These units and their boundaries also serve to define corresponding geochronological units. However, there were formal chronostratigraphic units before the concept of GSSPs. The classic units, largely defined in the early and mid-nineteenth century, were based on stratigraphic successions mostly in Europe and assembled into a temporal sequence through their superposition. They served as the basis for temporally correlating stratified Phanerozoic rocks worldwide, primarily on their palaeontological content. However, rarely were boundaries between successive units precisely defined. With the study of stratigraphic successions away from the type sections (or areas), overlaps of and gaps between many successive chronostratigraphic units were discovered. Because of natural limits to the palaeoecological and palaeogeographical distributions of palaeontological content on which the units were recognized, and because of the lack of specific boundaries, there were different interpretations of the stratigraphic extent accorded to the same unit from one region to another, and for many systems myriad sets of regional series and stages were established. It was to resolve these deficiencies and complexities that the concept of GSSPs was developed, and the goal of a single set of global units with precisely defined boundaries that could be correlated as widely as possible was established. Nevertheless, it was the stratigraphy within the units, particularly biostratigraphy, on which Phanerozoic strata were classified and temporally correlated worldwide. This then raises the question of the content of the 'Anthropocene'. When the Subcommission on Ordovician Stratigraphy began its task of establishing a single set of global series and stages and considering candidate stratigraphic signals and stratotype sections for defining boundaries, it first produced correlation charts for stratigraphic successions in each major region of the world; for example, Barnes *et al.* (1981), Webby *et al.* (1981), Ross *et al.* (1982), Hamman *et al.* (1982), Chen *et al.* (1995) and Fortey *et al.* (2000). It was through evaluation of this compilation of stratigraphic content worldwide that decisions were made on the stratigraphic extent of global chronostratigraphic units to be established and the stratigraphic signals and sections that would be used to define their boundaries. For the Holocene, which presently extends to the present, Walker *et al.* (2009) presented a well-documented summary of the stratigraphy of the Holocene and particularly of the boundary interval of the proposed GSSP in a full range of depositional settings. In this light, is there a well-documented and significant stratigraphic record for the 'Anthropocene' on which to base its recognition worldwide and to define a boundary with underlying Holocene strata? It is a question that the task group must address.

Should the base of the 'Anthropocene' be defined on a stratigraphic signal or instead should its beginning be defined on a date in human history? Zalasiewicz *et al.* (2008, 2012) considered possible definitions for defining the base of the 'Anthropocene' based upon a number of different stratigraphic signals. Such potential stratigraphic signals would be those identified as distinct, as globally widespread as possible, preferably isochronous, and resulting from human activities, specifically those of global impact on the Earth system, given its conceptual basis. The criteria for evaluating GSSP proposals are many and varied (Hedberg 1976; Salvador 1994). The most important of these is that the boundary at the candidate stratotype is defined at the level of a single stratigraphic signal within an interval of multiple, varied stratigraphic signals that allow for reliable, high-resolution correlation across the greatest possible palaeogeographical range of palaeoenvironmental settings. Definition of a unit by a GSSP is useful only if its boundary can be correlated reliably and widely with high resolution. Chronostratigraphic correlation (chronocorrelation), that is, evaluating temporal relationships between geographically widely separated stratigraphic successions, is an interpretative process that requires the evaluation of multiple, varied stratigraphic signals within a boundary interval rather than relying solely on a single signal, such as the lowest occurrence of a specific taxon, a palaeomagnetic reversal, an isotopic excursion, or a eustatically-induced vertical facies change. The matter the task group must consider should it move towards a GSSP-based definition is whether there are potential stratigraphic successions that meet this criterion.

Boundary stratotypes should be in marine successions without major vertical changes in lithofacies or biofacies in order that the boundary interval represents essentially continuous sedimentation (Hedberg 1976; Salvador 1994). However, a potential base-'Anthropocene' boundary in stratigraphic successions in the marine realm may be marked in the future by a widespread disconformity, reflecting human-induced acidification of the oceans. Furthermore, the time lag between a human-induced perturbation of the Earth system and it becoming permanently recorded in a stratigraphic succession may be great, as well as markedly variable among various depositional settings given the precision to which human history can be resolved. On the other hand, certain terrestrial successions (lacustrine, glacial ice) may offer boundary intervals with continuous deposition. The Holocene GSSP serves as an example (Walker *et al.* 2009).

Would it not be more advantageous to define the beginning of the 'Anthropocene' geochronological unit on a Global Standard Stratigraphic Age (GSSA; Remané *et al.* 1996)? A GSSA would be very accurate in contrast to the numerical age calibrated for a GSSP. For the global spread of radioactive isotopes created by the explosion of the first atomic bomb, for example, the age could be as precise as not only the year, but also the day, hour and even the millisecond. For the rise of atmospheric CO_2 levels resulting from the combustion of coal at the beginning of the Industrial Revolution in Western Europe, records of coal production could be used to assess this buildup and to select a decade or year for the GSSA.

Is the 'Anthropocene' a unit of Earth history or human history, or is it more a projection into the future? Not only would the beginning of the 'Anthropocene' geochronological unit (its GSSA) be most precisely and accurately defined and referred to in human time (century, decade, year, month, day), but so too would events that occurred or will occur within the time interval it represents. These are available from human records that extend back to the 3rd millennium BC, and thus include most of the human-induced events under consideration for defining the beginning of the 'Anthropocene'. For the last millennium, records of human activities are much more complete and precise with regard to the passage of time and with the extent of observations of the Earth system, and today the surface of the Earth, including its atmosphere and oceans, is more or less completely and continuously monitored. Thus, it is precisely recorded human observations that can be and will be used to study human impact on the Earth system, especially because the stratigraphic record is minimal and very incomplete relative to the events recorded by human observations. In contrast to the stratigraphic record being the evidence on which to interpret Earth history, events of the 'Anthropocene' are already known from human observations, and searches can then be made to find if they are recorded in the stratigraphic record. Accordingly, and in light of this fundamental difference, is not the concept of the 'Anthropocene' an aspect of human history rather than geological history for which the ICS International Chronostratigraphic Chart is being developed?

Furthermore, many of the material bodies resulting from human activities, ranging from heap-leach pads produced by mining activities to ancient cities and highways to human modified soils and even to lead pollution, are best characterized in terms of time by the civilization that produced them (e.g. Neolithic, Babylonian, Etruscan, Roman, Mayan, Industrial Revolution in Europe, gold mining in Carlin Trend of Nevada, testing of nuclear weapons). Why would one use the term

'Anthropocene' when far more specific terms are readily available and convey much more information in terms of the time they were produced and the human activities and societies that produced them? That the geological and archaeological records are diachronous, having spread geographically over time, is not a problem because their dispersal is well dated by human records, just as human-induced impacts of the 'Anthropocene' are largely diachronous. Furthermore, dispersal is an important aspect of these human activities and impacts to communicate. Should not the 'Anthropocene', however it is defined, really be considered instead as a unit of an archaeological timescale or of recorded human history?

Implicit in proposals for formal recognition of the 'Anthropocene' is its projection into the future. Certainly, human-induced effects to the Earth system are beginning to be recorded in unconsolidated sediment, but what is the certainty that a significant stratigraphic record will accumulate. Should projections into future millennia be the basis for ratifying a unit of the ICS International Chronostratigraphic Chart that up until now has been based on material, largely stratigraphic, records already well established and well documented? And are we so certain of what future millennia will bring?

What is the usefulness of the 'Anthropocene' as a material unit to be represented on geological maps? Chronostratigraphic units are material units that exist in the present. Geochronological units refer to time in the past. Lithostratigraphic units are mapped and generally shown on large-scale maps, but rock units on medium- and small-scale maps, particularly stratified rocks, are commonly shown as chronostratigraphic units. This raises the question of how human-induced material bodies of an 'Anthropocene' will be shown on medium- and small-scale maps, if at all. Yet, human-induced material bodies are well shown on large-scale topographic maps, and there exists a great range of standard symbols to show their specific nature ranging from orchards to vineyards, from highways and buildings to power lines and quarries, and including human structures of all types and ages. Furthermore, topographic maps are regularly updated and thus provide a very detailed history of human-induced changes to the surface of continents, whereas the spatial distributions of chronostratigraphic units on geological maps remain more or less stable. What use would it be to identify these material bodies as 'Anthropocene' when much more specific and thus much more meaningful designations of them already exist? Following procedures used in geological mapping, would 'Anthropocene' be used for both human-induced features on the surface as well as the soil that surrounds them, in which case both would be included in the same unit basically covering the entire map? Or would the human-induced features be shown differently than the surrounding soil, in which case the 'Anthropocene' would not be used as a chronostratigraphic unit?

Has the change to a human-dominated Earth system overwhelmed the natural Earth system, or might geological processes both internal and external still overwhelm the human influence? The case for the 'Anthropocene' as a ratified unit is based on the supposition that human activity now dominates the global environment and that human impact is significant on a global timescale. But is this true? Yes, human impact on land is widespread today, but it is restricted to shallow depths even when including the deepest mines and drill holes. In terms of geological time measured in hundreds of thousands, millions, hundreds of millions and even billions of years, is the present-day perspective relevant? And would not another extended interval of voluminous flood basalts or another large asteroid impact overwhelm any sedimentological/stratigraphic record of human activities on the Earth system? Considering that projection into the future is implicit in the concept of the 'Anthropocene', potential geological events originating within the Earth system or from outside it should also be considered.

What is, after all, the conceptual basis of the International Chronostratigraphic Chart/Geologic Time Scale? In almost all the publications that address the formalization of the 'Anthropocene' as a unit on the ICS International Chronostratigraphic Chart, whether in the scientific literature or the public media, never is there a full discussion of the nature of and basis for the ICS Chart and the Geologic Time Scale. All units presently on it are based on rock records, generally stratigraphic, of events that occurred in the past, that is in deep time. For stratigraphic successions of very young age (e.g. Neogene and Quaternary), recently developed numerical dating techniques allow for impressively high-resolution dating of units, their boundaries, and stratigraphic levels within units in those stratigraphic sections to which those techniques can be applied. For the Holocene, a wide range of 'depositional archives' can be resolved at the decadal scale or less (Walker *et al.* 2012), and the proposal by Walker *et al.* (2012) that the Holocene be subdivided into units with boundaries defined by GSSAs is under discussion. Clearly there is a transition interval from the time resolution of the ICS International Chronostratigraphic Chart to that of human calendars and clocks. However, the most fundamental question that must be addressed is whether or not the International Chronostratigraphic Chart/Geologic Time Scale is appropriate for application to chronometers of recorded and future human history. Should the 'Anthropocene' be ratified as a formal unit of the ICS Chart/Geologic Time Scale given that much of the 'Anthropocene' relative to Earth's history is really in the present and future and that events in the Earth system can be continuously recorded and best recorded and dated by human observations and chronometers whether or not they are recorded in the stratigraphic record?

Will the term 'Anthropocene' have value even if not ratified as a formal chronostratigraphic/geochronological unit? A number of very significant, long-term geological events recorded by extensive rock records and reflecting major upheavals in the Earth system are not represented by units in the ICS International Chronostratigraphic Chart/Geologic Time Scale, yet they are widely used by the geological community to convey the history of the Earth. These include Grenvillian, Variscan, Alpine-Himalayan and Snowball Earth, among others; Van Kranendonk & Kirkland (2013) provide an example with the Grenvillian of such a long-term event. Given this, is it necessary to have the 'Anthropocene' ratified as a formal unit of the ICS International Chronostratigraphic Chart? Why is it necessary or desired? Would it not have significance or relevance if it is not ratified?

The argument that formalization recognizes the immense impact of humans on the Earth system and that Earth processes are forever changed by this human impact ignores the fact that other organisms have had far greater, longer-term impacts than humans on the Earth system. The evolution of vascular land plants and their spread across continents from the late Devonian to the early Permian left a significant stratigraphic record and dramatically altered CO_2 and O_2 concentrations in the atmosphere and oceans (Berner 1998; Berner & Canfield 1989). The great Middle Ordovician diversification event (Webby *et al.* 2004) greatly altered the nature of marine sediments and biota. Yet neither is represented by a ratified chronostratigraphic/geochronological unit. Instead, their stratigraphic records and timing are expressed in terms of existing units. Why should human impact on the Earth system be different? Might the desire to establish the 'Anthropocene' as a formal unit be anthropocentric?

Conclusions

The 'Anthropocene' task group must provide a serious evaluation of each of these fundamental questions as well as of the arguments of Gibbard & Walker (2013) in any proposal that it forwards to the ICS Subcommission on Quaternary Stratigraphy for consideration. They should not have to be raised for the first time at higher levels in the approval process. Much has been made in the public media as to controversy regarding the proposal that the 'Anthropocene' be considered as a ratified chronostratigraphic/geochronological unit. Certainly, there will be debate, but first the task group must complete an evaluation of the issues presented here. The documentation of the iridium anomaly in the section at Gubbio by Alvarez *et al.* (1980) was a scientific landmark in its conclusion that a large bolide impact was the cause, but it is also an excellent example of a scientific investigation in which all possible causes of the anomaly were given serious consideration. Was this consideration of competing hypotheses a debate and controversial, or was this instead the essence of the scientific process? Although the process of ratifying a global standard chronostratigraphic/geochronological unit is partly administrative and bureaucratic, it is also clearly scientific, and it is the responsibility of the task group to give these fundamental questions serious consideration in developing a recommendation on the 'Anthropocene'.

References

ALVAREZ, L. W., ALVAREZ, W., ASARO, F. & MICHEL, H. V. 1980. Extraterrestrial cause for the Cretaceous–Tertiary Extinction, experimental results and theoretical interpretation. *Science*, **208**, 1095–1108.

AUTIN, W. J. & HOLBROOK, J. M. 2012*a*. Is the Anthropocene an issue of stratigraphy or pop culture? *GSA Today*, **22**(7), 60–61. http://dx.doi.org/10.1130/G153GW.1

AUTIN, W. J. & HOLBROOK, J. M. 2012*b*. Reply to Jan Zalasiewicz *et al.* on response to Autin and Holbrook on 'Is the Anthropocene an issue of stratigraphy or pop culture?' *GSA Today*, **22**, http://www.geosociety.org/gsatoday/comment-reply/pdf/i1052-5173-22-10-e23.pdf

BARNES, C. R., NORFORD, B. S. & SKEVINGTON, D. 1981. *The Ordovician System in Canada – Correlation Chart and Explanatory Notes*. International Union of Geological Sciences Publication, Paris, France, **8**.

BERNER, R. A. 1998. The carbon cycle and CO_2 over Phanerozoic time: the role of land plants. *Philosophical Transactions of the Royal Society of London B*, **353**, 75–82.

BERNER, R. A. & CANFIELD, D. E. 1989. A new model of atmospheric oxygen over Phanerozoic time. *American Journal of Science*, **289**, 333–361.

CHEN, X., RONG, J., WANG, X., WANG, Z., ZHANG, Y. & ZHAN, R. 1995. *Correlation of the Ordovician Rocks of China – Charts and Explanatory Notes*. International Union of Geological Sciences Publication, Trondheim, Norway, **31**.

Cowie, J. W., Ziegler, W., Boucot, A. J., Bassett, M. G. & Remané, J. 1986. Guidelines and statutes of the international commission on stratigraphy (ICS). *Courier Forschungsinstitut Senckenberg*, **83**, 1–14.

Fortey, R. A., Harper, D. A. T., Ingham, J. K., Owen, A. W., Parkes, M. A., Rushton, A. W. A. & Woodcock, N. H. 2000. *A Revised Correlation of Ordovician Rocks in the British Isles*. Geological Society, London, Special Reports, **24**.

Gibbard, P. L. & Walker, M. J. C. 2013. The term 'Anthropocene' in the context of formal geological classification. *In*: Waters, C. N., Zalasiewicz, J. A., Williams, M., Ellis, M. A. & Snelling, A. M. (eds) *A Stratigraphical Basis for the Anthropocene*. Geological Society, London, Special Publications, **395**. http://dx.doi.org/10.1144/SP395.1

Hamman, W., Robardet, M. & Romano, M. 1982. *The Ordovician System in Southwestern Europe (France, Spain, and Portugal) – Correlation Chart and Explanatory Notes*. International Union of Geological Sciences Publication, Paris, France, **11**.

Hedberg, H. D. 1976. *International Stratigraphic Guide – A Guide to Stratigraphic Classification, Terminology, and Procedure*. Wiley, New York.

Remané, J., Bassett, M. G., Cowie, J. W., Gohrbandt, K. H., Lane, H. R., Michelsen, O. & Wang, N. 1996. Revised guidelines for the establishment of global chronostratigraphic standards by the International Commission on Stratigraphy (ICS). *Episodes*, **19**, 77–81.

Ross, R. J., Jr., Adler, F. J. *et al.* 1982. *The Ordovician System in the United States – Correlation Chart and Explanatory Notes*. International Union of Geological Sciences Publication, Paris, France, **12**.

Salvador, A. 1994. *International Stratigraphic Guide – A Guide to Stratigraphic Classification, Terminology, and Procedure*, 2nd edn. The International Union of Geological Sciences. The Geological Society of America, Boulder.

Van Kranendonk, M. J. & Kirkland, C. L. 2013. Orogenic climax of Earth: the 1.2–1.1 Ga Grenvillian superevent. *Geology*, **41**, 735–739. http://dx.doi.org/10.1130/G34243.1

Walker, M., Johnsen, S. *et al.* 2009. Formal definition and dating of the GSSP (Global Stratotype Section and Point) for the base of the Holocene using the Greenland NGRIP ice core, and selected auxiliary records. *Journal of Quaternary Science*, **24**, 3–17. http://dx.doi.org/10.1002/jqs.1227

Walker, M., Björck, S. *et al.* 2012. Formal subdivision of the Holocene Series/Epoch: a discussion paper by a Working Group of INTIMATE (Integration of ice-core, marine and terrestrial records) and the Subcommission on Quaternary Stratigraphy (International Commission on Stratigraphy). *Journal of Quaternary Science*, **27**, 649–659. http://dx.doi.org/10.1002/jqs.2565

Webby, B. D., VandenBerg, A. H. M. *et al.* 1981. *The Ordovician System in Australia, New Zealand, and Antarctica – Correlation Chart and Explanatory Notes*. International Union of Geological Sciences Publication, New York, **6**.

Webby, B. D., Paris, F., Droser, M. L. & Percival, I. G. 2004. *The Great Ordovician Biodiversification Event*, Critical Moments and Perspectives in Earth History and Paleobiology. Columbia University Press, New York.

Zalasiewicz, J., Williams, M. *et al.* 2008. Are we now living in the Anthropocene? *GSA Today*, **18**(2), 4–8. http://dx.doi.org/10.1130/GSAT01802A.1

Zalasiewicz, J., Cearreta, A. *et al.* 2012. Response to Autin and Holbrook on 'Is the Anthropocene an issue of stratigraphy or pop culture?' *GSA Today*, **22**. http://www.geosociety.org/gsatoday/comment-reply/pdf/i1052-5173-22-10-e21.pdf

The term 'Anthropocene' in the context of formal geological classification

P. L. GIBBARD[1]* & M. J. C. WALKER[2]

[1]*Department of Geography, Cambridge Quaternary, University of Cambridge, Cambridge CB2 3EN, UK*

[2]*School of Archaeology, History and Anthropology, Trinity Saint David, University of Wales, Lampeter SA48 7ED, UK*

**Corresponding author (e-mail: plg1@cam.ac.uk)*

Abstract: In recent years, 'Anthropocene' has been proposed as an informal stratigraphic term to denote the current interval of anthropogenic global environmental change. A case has also been made to formalize it as a series/epoch, based on the recognition of a suitable marker event, such as the start of the Industrial Revolution in northern Europe. For the Anthropocene to merit formal definition, a global signature distinct from that of the Holocene is required that is marked by novel biotic, sedimentary and geochemical change. Although there is clear evidence of anthropogenic effects in geological sequences, it is uncertain whether these trends are sufficiently distinct, consistent and dated for the proposal for a Holocene/Anthropocene boundary to be substantiated. The current view of the Earth-Science community is that it should remain informal. For formal definition a Global Stratigraphic Section and Point (GSSP) is required. Adoption of the term 'Anthropocene' will ultimately depend on recognition of a global event horizon. Without this, there is no justification for decoupling the Anthropocene from the Holocene. If the Anthropocene is deemed to have utility, it should be as an informal historical designation rather than a formally defined stratigraphic unit (of whatever status) within the geological timescale.

The 'Anthropocene' (meaning *anthrōpos* 'human being' and *kainos* 'new') is a term that has become increasingly widely used since it was first proposed by Paul Crutzen & Eugene Stoermer (2000) to denote the present time interval, in which many geologically significant processes and conditions have been, and continue to be, profoundly altered by human activities. Processes include changes in erosion and sediment transport resulting from agriculture and urbanization, and variations in the chemical composition of the atmosphere, oceans and soils, with notable perturbations in element cycling, such as carbon, nitrogen, phosphorus and various metals. Changes in environmental conditions generated by these various processes include global warming, ocean acidification and coastal eutrophication and ensuing anoxia; habitat loss, species range shifts and extinctions; and globally significant physical and chemical changes, including cryospheric loss, ozone depletion and accelerating eustatic sea-level rise (e.g. Orr *et al.* 2005; Nicholls *et al.* 2007; Chen *et al.* 2011; Urban *et al.* 2012).

The increasing prominence of humans during the present (Holocene Series/Epoch) interglacial has been marked by a progressive increase in their effect on the natural environment and its processes, and this has long been recognized in palaeoenvironmental reconstructions. These anthropogenic effects, which have reached significant levels during the last few centuries, and particularly since the Industrial Revolution in northern Europe, have resulted in marked changes to the Earth's surface. Indeed, almost every aspect of the contemporary global environment is now modified, or at least influenced, by human activity (Crossland 2005; Zalasiewicz *et al.* 2010; Steffen *et al.* 2011). There is a body of opinion within the geological community that this accelerated human impact may be reflected in the recent stratigraphic record, and that the anthropogenic effect can be distinguished from natural 'background' conditions (Zalasiewicz *et al.* 2008). Although the term Anthropocene was initially invoked informally to describe this period of expanding human influence on the global environment (Crutzen 2002), and was apparently suggested by chance (Steffen *et al.* 2004), it is increasingly being applied to the recent geological record, where the term is viewed, by some, as being equivalent to those used as series-status subdivisions of the Cenozoic Erathem ('*recent life*'), notably the Pleistocene and Holocene.

An important consequence of the recognition of the increasing intensity of human impact on the natural environment, and its possible manifestation in the geological record, has been the initiation of

From: Waters, C. N., Zalasiewicz, J. A., Williams, M., Ellis, M. A. & Snelling, A. M. (eds) 2014. *A Stratigraphical Basis for the Anthropocene*. Geological Society, London, Special Publications, **395**, 29–37.
First published online October 25, 2013, http://dx.doi.org/10.1144/SP395.1

discussions to determine whether the stratigraphic signature of the Anthropocene is now sufficiently clearly defined to warrant its formal definition as a new period of geological time (Zalasiewicz *et al.* 2011*a*, *b*). This is currently being considered by a Working Group of the Subcommission on Quaternary Stratigraphy (SQS). In the meantime, a separate Working Group of the SQS, jointly with INTIMATE (integration of ice-core, marine and terrestrial records), has been examining the case for a formal subdivision of the Holocene Series/Epoch (M. J. C. Walker *et al.* 2012). In the light of a possible conflict between the two groups, the Holocene working group has not considered the Anthropocene question. Nevertheless, they acknowledge that although there is a clear distinction between these two initiatives, the Holocene subdivision being based on natural climatic/environmental events and the concept of the Anthropocene focusing on the human impact on the envi-ronment, there may indeed be areas of overlap. We return to the subdivision of the Holocene below.

In this paper we consider the designation of the Anthropocene as a formal time-stratigraphic unit against the background of these ongoing discussions. We examine the term Anthropocene in the context of the formal definition of geological timescale units, particularly in relation to the requirement for relating such units to unequivocal Global Stratotype Section and Point (GSSP, 'golden spike') localities, and we also evaluate the potential value and utility of defining the episode of recent human activity as a new stratigraphic unit within the geological timescale (GTS).

Stratigraphic procedures and the geological timescale

The stratigraphy of geological sequences is the foundation upon which the discipline of historical geology, and therefore the accurate reconstruction of Earth history, depends. It is concerned with all the elements of sedimentary successions, including their form, distribution, lithology, fossil content, depositional environment, vertical sequence and age. Stratigraphy is therefore a synthetic subject that examines the origin, and temporal and spatial distribution, of sediment strata and other rocks.

Chronostratigraphy ('time-rock stratigraphy') is the branch of stratigraphy that involves the application of time to rock successions, the goal of which is the establishment of a globally applicable standard timescale. The chronostratigraphical scale originated as a relative one that was calibrated primarily through the application of biostratigraphy. Reliable numerical ages have been added gradually to form a parallel, numerical timescale based on years. Chronostratigraphical divisions are 'time/rock' units, that is they refer to the sequence of rocks deposited during a particular interval of time. Geochronological divisions are the corresponding intervals of (continuous) geological time. Thus, it can be said that rocks of the Quaternary System were deposited during a time interval called the Quaternary Period.

In conventional stratigraphic practice, it is generally accepted that chronstratigraphical and geochronological boundaries should, as far as possible, be defined in continuous sedimentary sequences. Ideally, such sequences preserve multiple lines of evidence (multi-proxy records) and, in particular, continuous sedimentation spanning the boundary itself, that is, excluding time gaps. This continuity offers the opportunity for identifying, *inter alia*, palaeontological, isotopic, chemical and palaeomagnetic changes, and therefore enables the boundary to be identified as precisely as possible within the limits of the evidence contained within the sequence. Throughout the Phanerozoic, this approach provides not only a basis for accurately delimiting a boundary, but also a means by which that horizon can be extended beyond the stratotype locality for purposes of correlation.

The recommendations of the International Stratigraphic Guide (Hedberg 1976; Salvador 1994) require that all major chronostratigraphical subdivisions are defined with reference to boundary stratotype localities in sediment reference sequences. These are designated GSSPs, or 'golden spikes', and form the basis for the division of geological time in the internationally sanctioned GTS. This is the responsibility of the International Commission on Stratigraphy (ICS), which is a constituent group of the International Union of Geological Sciences (IUGS), the body that directs all aspects of global geoscience. The role of the ICS is to formulate the subdivision, classification and determination of geological time, that is, chronostratigraphy and geochronology. The ICS operates through subcommissions, each associated with a particular time period, the Subcommission on Quaternary Stratigraphy (SQS), for example, being responsible for the Quaternary (which includes the present day). Each of the subcommissions defines the time divisions for their respective periods, and the GSSPs that provide the basis for global correlation. Proposals are submitted to the IUGS, via the ICS, for formal ratification. Any changes that arise from the proposal are incorporated into the GTS (www.stratigraphy.org). Once a GSSP has been ratified by the IUGS, a 10-year moratorium on any change then applies. An alternative basis for defining a global stratotype is age and is referred to as a Global Standard Stratigraphic Age (GSSA).

This is a chronological reference point in the world's rock record that is used to define a boundary, and has been employed most widely in the Precambrian where the lack of diagnostic fossils precludes the identification of recognizable GSSPs. As with a GSSP, boundaries defined in this way must also be ratified by the ICS to be internationally valid (Remane *et al.* 1996).

Subdividing the Quaternary record

Although stratigraphical procedures are of paramount importance to geologists working in all the time periods, from the Pre-Cambrian to the Cenozoic, applications of conventional stratigraphic practice are frequently far from straightforward in the Quaternary. This is because Quaternary geologists are attempting to subdivide what might appear to be very short periods of time across a range of different environments from the tops of mountains to the depths of the oceans. The extraordinarily limited time intervals that are recognized in the Quaternary require a critical understanding of the implications of stratigraphical analysis and procedures, and the essential distinctions between the various subdivisions of rock strata that allow for the effective division, correlation and reconstruction of history from rock sequences. In this sense, the fundamental need to identify and define reference sequences (GSSPs) that can be used to equate isochronous surfaces and events around the world becomes critical. This is a requirement not only for large-scale units, such as Stages or Series, but also for smaller-scales subdivisions.

The conventional approach to the subdivision of the Quaternary stratigraphic record is to employ evidence for contrasting climatic conditions to characterize individual stratigraphic units (geologic–climatic units). This follows recommendations by the American Commission on Stratigraphic Nomenclature (1961, 1970), which recognized a geologic–climatic unit as indicating 'an inferred widespread climatic episode defined from a subdivision of Quaternary rocks' (ACSN 1970, p. 31). Although climatostratigraphy does not always find favour with geologists working in older parts of the record, in the Quaternary, where climatic change is the hallmark of the past two million years, it is difficult to envisage a scheme of stratigraphic subdivision that does not specifically acknowledge this fact (Lowe & Walker 1997). Indeed, it is now apparent that the principal driver of recent global climatic fluctuations, namely variations in Earth's orbit and axis (Milankovitch forcing), is reflected not only in key stratigraphic horizons within the Quaternary record, but can also be detected in pre-Quaternary sequences at least as far back as the late Paleocene (Lourens *et al.* 2005). 'Astronomical pacing' provides both an index of the major climatic shifts and a basis for estimating the age of these events. As a consequence, major intervals of geological time within the Neogene and the Quaternary, which are evident in the climatic record, can be further defined as chronostratigraphic units within the GTS (Zalasiewicz *et al.* 2004).

The GTS is based on a hierarchical system of classification in which time-rock sequences (chronostratigraphy) and their corresponding intervals of time (geochronology) are represented by units of progressively lower rank. The Quaternary has the rank of system (a chronostratigraphical unit of high rank) and period (the equivalent geochronological unit) within the Cenozoic Erathem/Era (Salvador 2006*a*, *b*). The Pleistocene is traditionally placed within the Quaternary at the next-lower rank of series (chronostratigraphy) or epoch (geochronology). The lower boundary of the Quaternary Series/Pleistocene Epoch is located at the GSSP at the base of the Gelasian Stage/Age in the Monte San Nicola section, Sicily, and dated at 2.58 Ma. Both the lower boundary of the Quaternary, and its status as a unit of series/stage rank, were ratified by the IUGS in 2010 (Gibbard *et al.* 2010).

The position of the Holocene

As currently defined, Holocene (meaning 'wholly modern') is the name for the most recent interval of Earth history and includes the present day. It is the second series or epoch of the Quaternary System/Period, and the boundary stratotype (GSSP) is located in the Greenland NGRIP ice core and dated to 11.7 ka (Walker *et al.* 2009).

The origin of the term Holocene is inextricably linked to the development of the nomenclature of ice-age time divisions. Two terms arose independently in the early to mid-nineteenth century to encompass near-surface, often unconsolidated, deposits: the Quaternary (Desnoyers 1829; Reboul 1833; von Morlot 1854) and the Pleistocene (Lyell 1839), although the former considerably predates this usage, having been proposed by Giovanni Arduino in 1759 for the fourth stage or 'order', which he recognized as the alluvial sediments of the Po plain, in northern Italy (Schneer 1969; Rodolico 1970). Both Quaternary and Pleistocene were originally established for marine deposits, in the Paris basin and in eastern England, respectively. However, with the recognition that the extensive 'Drift' or 'Diluvial' deposits represented greatly extended glaciation in recent Earth history, both soon became synonymous with the Ice Age, and latterly with the rise of humans. The Quaternary differed in encompassing Lyell's (1837) 'Recent' or Forbes' (1846) 'Postglacial', a period termed the Holocene by Gervais (1867–69), with the latter formally adopted by the 1885 International Geological Congress (IGC).

The terms 'Recent', or less formally 'Postglacial', as alternatives to the Holocene, although still commonly used by geologists, are invalid (cf. Gibbard & van Kolfschoten 2005). A further term, 'Flandrian', derived from marine transgression sediments on the Flanders coast of Belgium (Heinzelin & Tavernier 1957), has often been used as a synonym for Holocene (e.g. Nilsson 1983, p. 23). It has been adopted by authors who consider that the last 10 ka should have the same stage status as previous interglacial events and thus be included in the Pleistocene. In this case, the latter would thus extend to the present day (cf. West 1968, 1977, 1979; Woldstedt 1969; Hyvärinen 1978). However, this usage has been continually criticized (e.g. Morrison 1969) and has been losing ground over the last three decades (Flint 1971, p. 324; Lowe & Walker 1997, p. 16). Accordingly, the term Holocene Series/Epoch is now more widely adopted, partly in recognition of the modern usage but also, more importantly, because of the importance in the present interglacial of the evolution of the human environment. Indeed, it is this anthropogenic signature that is the hallmark of the Holocene, setting it apart from previous interglacials, and taken as a fundamental justification for its status as a time-stratigraphic unit of Series/Epoch rank as it is currently defined (Walker *et al.* 2009).

What status for the Anthropocene?

As noted above, the status of the Anthropocene is currently being examined by a Working Group of the SQS. Both within that group and outside, there is an increasing body of opinion in support of the view that the informal nature of the term 'Anthropocene' should be replaced by a formal definition (e.g. Williams *et al.* 2011). If so, the Anthropocene would become the most recent time-stratigraphic subdivision of the GTS, and would constitute a formal unit of the Quaternary System/Period. Whether such a formalization is justified (and, if so, at what rank within the geochronological/chronostratigraphical hierarchy) is now considered.

The status of a potential formal Anthropocene division is one of considerable philosophical importance. This is because the higher the term's rank, the greater the difference between conditions preserved and represented within the stratigraphic unit, and those of the immediately previous division. In the case of the term 'Anthropocene', the name selected implies, either intentionally or otherwise, that it should be seen as equivalent to Pleistocene and Holocene; that is, the term would have Series or Epoch status. If this was accepted, the corollary is that the Holocene has ended and we have entered a new geological interval, and such a transition would be equivalent to the changes between the Eocene and the Oligocene, or the Miocene and the Pliocene. In our view, a change of this magnitude is not supported by the geological evidence, the Anthropocene episode being too limited in duration when viewed from the perspective of the present, and with too variable a global stratigraphic signature to justify its elevation to a time-stratigraphic unit of Series or Epoch rank within the internationally agreed geological timescale.

The current position is that we are living in the Holocene Series/Epoch, and one of the key justifications for defining a Holocene Series, as a separate entity from the Pleistocene, is that humans (*Homo sapiens*) reached critical numbers and began influencing natural systems from the beginning of this time period onwards. As noted above, without this unique record of human impact there would be no justification for the Holocene being anything other than an interglacial, in common with all others in the Pleistocene. If the human dimension is accepted as a reasonable basis for a separate Holocene Series, as distinct from the Pleistocene, the activities of humans cannot then be used again in support of a discrete Anthropocene division. Given the extensive archaeological record of human/environment interaction throughout the Holocene (and even in earlier times), many workers find it difficult to accept the fact that 'the Holocene interglacial has ended'. Indeed, when seen in the context of the long-term climatic and environmental changes that have occurred during the Quaternary, there would seem to be little justification for this view.

We referred above to the Working Group that is considering a subdivision of the Holocene. The proposals of that Working Group are for a tripartite subdivision of the Holocene Series/Epoch (Early, Middle and Late Holocene Sub-Series/Sub-Epochs), with boundaries at 8.2 and 4.2 ka respectively. The GSSP for the Early–Middle Holocene boundary is found in a stable isotope record from the archived Greenland NGRIP ice core, and the GSSP for the Middle–Late Holocene boundary is also in a stable isotope record, this time from an archived stalagmite from Mawlmuh Cave, NE India (M. J. C. Walker *et al.* 2012). These boundaries are based on significant natural events that register globally: the 8.2 ka event was an abrupt cooling episode triggered by catastrophic meltwater release into the North Atlantic from ice-dammed lakes at the northern margins of the Laurentide ice sheet, the signal of which can be found in both hemispheres, while at *c.* 4.2 ka there was a worldwide aridification episode that was most pronounced in low- and mid-latitude regions, but which is also recorded in various proxy records from high-latitude regions. The point here is that these events are reflected in stratigraphic records from different parts of the world as broadly time-parallel marker horizons, so the GSSPs are of global significance. As such, the proposal for a subdivision of the Holocene is fully in line with the stratigraphic principles and practices described above. If a Holocene/Anthropocene boundary is to have the same credibility, then it must be underpinned by anthropogenic events that are as globally significant as the natural events that form the basis for the proposed Holocene GSSPs. In other words, the definition of the base of the Anthropocene must conform to the same stratigraphic principles.

Three questions follow from these discussions. (1) Is there a globally significant, human-induced and broadly time-parallel event that could form a basis for a stratigraphic boundary between the Holocene and Anthropocene? (2) Are we now living in a geological period that is significantly different from the preceding Holocene? (In other words, can the Holocene be considered to have ended?) (3) Would the recognition of the Anthropocene as a new geological period be of value to the wider Earth Science community?

In relation to the first of these questions, it has been suggested that the onset of the Anthropocene could be defined by the rapid industrialization and urbanization of western Europe and eastern North America in the late eighteenth century (Crutzen & Stoermer 2000), and numerous references have been made to the fact that evidence of these anthropogenic activities is preserved within the recent geological record (Steffen *et al.* 2007; Zalasiewicz *et al.* 2008; Williams *et al.* 2011). For example, the development of urban societies over the past two centuries has created 'urban strata' (e.g. the 'artificial strata' of Rosenbaum *et al.* 2003), while expanded town and city sites have influenced a range of geomorphological (e.g. fluvial) and pedological processes, and will have left their mark in stratigraphic sequences. Indeed, it has been suggested that human-affected soils ('anthropic soils') are potential 'golden spikes' for the Anthropocene (Certini & Scalenghe 2011). However, these markers, together with their geomorphological counterparts, are essentially local or, at best, regional in character, and are invariably time-transgressive (Gale & Hoare 2012). Moreover, with respect to the growth of urban centres, this is not a new phenomenon. Throughout the Middle and Late Holocene, city states from the Punjab (the Harrapan civilization), through the Fertile Crescent (the Akkadian civilization) and Egypt, to the Mayan, Aztec and Inca cultures of Central and South America, have expanded, flourished and collapsed (Diamond 2005). Each has left an imprint in the archaeological and geological record, and any of these could have been designated a marker for increased human influence, yet the limited spatial impact of these cultures makes it unlikely that their stratigraphic signal could be used to define the onset of a period of global human dominance. The rapid industrialization of Western Europe and eastern North America during the eighteenth and nineteenth centuries was, by contrast, an event of much wider significance, but both industrialization and urban development was spatially variable and diachronous. It has therefore been suggested that an independent marker horizon, such as the increase in CO_2 in polar ice cores (Crutzen 2002), could be used to designate the boundary between the Holocene and Anthropocene. A rise in CO_2 levels from around AD 1750 is evident in both Arctic and Antarctic records (Barnola *et al.* 1995; Etheridge *et al.* 1996), is accompanied by the beginnings of an upward trend in the curve for CH_4 and N_2O (MacFarling Meure *et al.* 2006; Sowers 2010), and reflects increased levels of atmospheric trace-gas loading. As Zalasiewicz *et al.* (2011*b*) have pointed out, however, although seemingly abrupt on centennial timescales, these changes tend to be too gradual to provide useful stratigraphic markers at an annual or even decadal level, and this chronological imprecision is further exacerbated by the delay in sealing of the air bubbles during snow accumulation on the ice-sheet surface, which results in a temporal offset between the annual ice-core timescale and the atmospheric CO_2 record (Bender *et al.* 1997). Stratigraphic resolution to the annual level, which is the aim of a GSSP for the Anthropocene (Zalasiewicz *et al.* 2011*b*), is therefore unlikely to be obtained from the ice-core record.

An additional difficulty with using the history of trace gases in ice cores to define the onset of the Anthropocene is that, although this is perhaps the clearest and most unequivocal indicator of human impact on the global atmospheric environment in the recent geological past, unfortunately it has little relevance to the vast areas of the planet that were not experiencing industrialization at that time. In other words, it does not coincide with a critical growth of human influence globally. Indeed, parts of the world with some of the largest concentrations of humanity, such as India and China, did not undergo their own industrial revolutions until the mid-twentieth century, some 200 years after the AD 1750 trace-gas event in the ice cores. This raises further doubts about the validity of the polar ice-core record as a potential GSSP for the base of the Anthropocene for, as we have already seen, any GSSP boundary should be a global one and, if it is to be ratified by the IUGS, it must have worldwide applicability.

In acknowledging these difficulties in the selection of an appropriate GSSP for the Anthropocene, Zalasiewicz *et al.* (2011*b*) have suggested that an alternative strategy might be to define the base of the Anthropocene numerically with a GSSA (Global Standard Stratigraphic Age). At the level of resolution sought (annual), and at this temporal distance, one possibility might be to select a numerical age, such as the beginning of 1800 in the Christian Gregorian calendar. This would allow simple and unambiguous correlation of the stratigraphical and historical records, as well as providing consistent utility and meaning to the currently used informal term. Should this proposal be deemed to be of value to the Geoscience community, the date could be proposed as a GSSA to the ICS and could then be ratified and given formal status by the IUGS. However, this compromise, although avoiding confusion regarding the onset of the Anthropocene interval, would suffer the same basic drawback as a GSSP, in that the basis for the selection of age is a perceived acceleration of human impact on the natural environment, which, as we have already seen, is globally diachronous. In other words, selecting an age such as 1800 is essentially a western perspective on world events, and the GSSA would have little meaning (or relevance) to those regions of Africa or Asia where an increase in the pace of economic development did not occur until at least a century later.

In terms of the second question posed above, the palaeoenvironmental evidence suggests that humans have been shaping the global environment to various degrees more or less throughout the course of the Holocene and, to some extent, even before (e.g. Miller *et al.* 2005; Archibald *et al.* 2012). This is clearly of considerable concern for the archaeological community, who see a continuum of human/landscape interactions at a range of temporal and spatial scales extending through the Holocene to the present day. Indeed, it has been suggested that 'by defining the Anthropocene as a geological epoch beginning only 200 years ago, Crutzen and Stoermer truncate thousands of years of human interaction with the global environment' (Periman 2006, p. 558). Clearly, even younger events that have been suggested as baselines for the Anthropocene suffer from the same difficulty. These include the 'Great Acceleration' of the second half of the twentieth century during which many human activities reached take-off points in terms of their environmental impacts (Steffen *et al.* 2011; Wolfe *et al.* 2012), and the short-lived episode of nuclear bomb testing of the mid mid-twentieth century, which is reflected in a radiogenic nuclide peak (in ^{14}C and ^{137}Cs) at around AD 1963 in sedimentary records (e.g. Goodsite *et al.* 2001; Kemp *et al.* 2009). Both events have been proposed as potential global markers for defining the onset of the Anthropocene (e.g. Zalasiewicz *et al.* 2008). However, although these are undoubtedly clear manifestations of human impact on the terrestrial and climatic environment, they simply mark one of many stages in human technological development. They certainly do not coincide with the onset of environmental changes resulting from human activity, and therefore have little value in defining the initiation of human impact (Gale & Hoare 2012). In passing, it is also worth noting that radiogenic nuclide fallout was significantly lower in the Southern Hemisphere than in the Northern Hemisphere, so the distinctive weapons' fallout peak is often difficult to discern in sediment records from the Southern Hemisphere (e.g. Collins *et al.* 2001; Humphries *et al.* 2010).

Furthermore, anthropogenic effects have not only been manifest in the terrestrial sphere throughout the Holocene, but may also have impacted on the atmospheric environment. For example, the polar ice-core records show the first increase in CO_2 concentrations *c.* 8000 years ago, while CH_4 concentrations began a similar rise *c.* 5000 years ago. The former has been related to forest clearance by human groups in China, India and Europe, while the latter has been associated with expansion of rice farming and increasing livestock numbers (Ruddiman 2003, 2005; Ruddiman *et al.* 2011). A further rise in CH_4 concentrations in ice cores from the historical period have been equated with long-term increases in agricultural emissions and biomass burning during the period of the Roman empire, the Han dynasty and the Medieval Climatic Anomaly, all of these records indicating that human activity contributed to variations in both CO_2 and CH_4 emissions to the atmosphere in pre-industrial times (Sapart *et al.* 2012). Indeed, the ice cores also contain evidence of atmospheric heavy metal pollution (As, Pb, Bi, Cu and Zn), reflecting smelting of lead and copper ores extending back to the Greek–Phoenician period (Krachler *et al.* 2009). Similar records of atmospheric Pb pollution, reflecting late prehistoric mining and smelting, have been obtained from peat sequences (Shotyk *et al.* 1998; Le Roux *et al.* 2004). These different lines of evidence are manifestations of a process of increasing human influence on the natural environment throughout the Holocene. Although no-one disputes the fact that there was a substantial amplification of human impact with the onset of the European Industrial Revolution, it is difficult to isolate a single horizon within the geological record that marks the global 'tipping point' where natural processes have been overtaken by human-induced environmental changes.

The third question concerns the utility of defining a separate geological time unit, the Anthropocene. When considering a possible subdivision of the Holocene, the SQS/INTIMATE Working Group were concerned with putting forward a scheme that the geological community would find useful and with which they would be comfortable to operate. An extensive literature review showed quite clearly that the terms 'Early', 'Middle (or Mid-)' and 'Late' Holocene were widely employed in Holocene research, so by formalizing these terms, the Working Group was effectively legitimizing what was already custom and practice. There is no doubt that the term 'Anthropocene' has caught the popular imagination, as witnessed by the number of publications that have appeared in recent years that incorporate the term; indeed, a new journal entitled *Anthropocene* is to be launched by Elsevier in 2013. However, the question remains as to whether or not the utility of the term will be as an informal designation for the period of recent enhanced human activity, or whether it can be defined in a geological sense as a formal time-stratigraphic unit of the GTS. At present, despite the implicit geological connotation of the suffix '-cene' implying a rank of Series/Epoch status, the usage of the term has been predominantly historical rather than geological in its application and definition. Because the criteria initially used to justify a Holocene Series still obtain, and while fully accepting the profound and increasing influence of humanity on the natural environment over recent centuries, we remain firmly of the view that the designation of the Anthropocene as a time unit equivalent to Series or Epoch divisions cannot yet be justified on stratigraphical grounds. At best, the Anthropocene might become another name for the later Holocene (with a usage similar, perhaps, to 'Dark Ages' or 'Middle Ages'), but in a geological context the term would remain informal. Indeed, this view has now been reinforced by the Geological Society of America, who have rejected the term from their 2012 Geological Time Scale because it has no 'internationally sanctioned standing' (J. D. Walker *et al.* 2012, p. 7). In many ways, our concerns are admirably expressed in a recent paper by Autin & Holbrook (2012), in which they write

> Before we amend our stratigraphy and end the Holocene, it would be best to settle the question of where in the stratigraphic record to drive the golden spike that defines when humanity became one of the preeminent forces of nature. Even so, will finding this layer lead to a globally relevant correlation? As stratigraphers, we require criteria to map the Anthropocene with relevant and consistent meaning. Presently, we are left to map a unit conceptually rather than conceptualizing a mappable stratigraphic unit. (Autin & Holbrook 2012, p. 61)

Conclusions

In order for the Anthropocene to be defined as a formal stratigraphical unit, a global event, distinct from that of the Holocene or of previous interglacial phases of the Pleistocene, must be identified on the basis of novel biotic (i.e. biostratigraphical), sedimentary and geochemical change(s). Although there is unequivocal evidence in recent geological records of rapidly increasing anthropogenic impacts on natural environments and systems, it is uncertain whether the geological signature of these trends is sufficiently distinct, and adequately dated at the global scale, for a Holocene/Anthropocene boundary to be substantiated on stratigraphic grounds. That being the case, if the term is to be retained for the purposes of geological time division, it must remain informal.

As currently defined, we are living in the Holocene Series or Epoch, the last 11.7 ka of Earth history. Although, on strict geological grounds, this period can be considered to be the latest of a sequence of interglacials that have characterized the Quaternary, and hence could be regarded as a unit of the Pleistocene Series/Epoch, the rise of humans (*Homo sapiens*) to critical numbers and the increasing influence they have exerted on natural systems provides the critical justification for defining this period as a separate unit of Series/Epoch status within the GTS. As such, the activities of humans cannot then be used again to justify the definition of a discrete Anthropocene division. This is the view of many Holocene geoscientists and archaeologists who do not accept that the Holocene has ended.

Although there is a substantial popular movement towards the recognition of a new period that acknowledges, and may indeed be characterized by, the overwhelming influence of humans on much of the natural environment, we continue to live within the same climatically and biologically defined time interval for which the term 'Holocene' was proposed and formally ratified by the IUGS. In the absence of a globally identifiable anthropogenic marker event that can be clearly recognized in a range of depositional contexts, that can be mapped in a conventional geological sense, and that can provide a consistent stratigraphic basis for detaching the Anthropocene from the preceding Holocene, there is, in our view, no justification for formally defining the Anthropocene as a new stratigraphic unit (of whatever status) within the established geological timescale.

We are grateful to Prof. J. Lewin and Dr A. Smith for their helpful and insightful comments on an earlier version of the manuscript, and to one anonymous referee for a constructive critical review. We also thank the editors of this volume for their invitation to us to present this paper on behalf of the Subcommission on Quaternary Stratigraphy, and for their encouragement and assistance in preparing the final draft for publication.

References

AMERICAN COMMISSION ON STRATIGRAPHIC NOMENCLATURE 1961. Code of stratigraphic nomenclature. Bulletin of the American Association of Petroleum Geologists, **45**, 645–660.

AMERICAN COMMISSION ON STRATIGRAPHIC NOMENCLATURE 1970. *Code of stratigraphic nomenclature*, 2nd edn. American Association of Petroleum Geologists Bulletin, **60**, 1–45.

ARCHIBALD, S., STAVER, C. & LEVIN, S. A. 2012. Evolution of human-driven fire regimes in Africa. *Proceedings of the National Academy of Science*, **109**, 847–852.

AUTIN, W. J. & HOLBROOK, J. 2012. Is the Anthropocene an issue of stratigraphy or pop culture? *GSA Today*, **22**, 60–61.

BARNOLA, J. M., ANKLIN, M. ET AL. 1995. CO_2 evolution during the last millennium as recorded by Antarctic and Greenland ice. *Tellus*, **47B**, 264–272.

BENDER, M., SOWERS, T. & BROOK, E. 1997. Gases in ice cores. *Proceedings of the National Academy of Sciences*, **94**, 8343–8349.

CERTINI, G. & SCALENGHE, R. 2011. Anthropogenic soils are the golden spikes for the Anthropocene. *The Holocene*, **21**, 1269–1274.

CHEN, I.-C., HILL, J. K. ET AL. 2011. Rapid range shifts of species associated with high levels of climate warming. *Science*, **333**, 1024–1026.

COLLINS, A. L., WALLING, D. E. ET AL. 2001. Using ^{137}Cs measurements to quantify soil erosion and redistribution rates for areas under different land use in the upper Kaleya River basin, southern Zambia. *Geoderma*, **104**, 299–323.

CROSSLAND, C. J. (ed.) 2005. *Coastal Fluxes in the Anthropocene*. Springer, Berlin.

CRUTZEN, P. J. 2002. Geology of mankind. *Nature*, **415**, 23.

CRUTZEN, P. J. & STOERMER, E. F. 2000. The Anthropocene. *Global Change Newsletter*, **41**, 17–18.

DESNOYERS, J. 1829. Observations sur un ensemble de dépôts marins. *Annales des Sciences Naturelles (Paris)*, **171–214**, 402–491.

DIAMOND, J. 2005. *Collapse*. Penguin, London & New York.

ETHERIDGE, D. M., STEELE, L. P. ET AL. 1996. Natural and anthropogenic changes in atmospheric CO_2 over the last 1000 years from air in Arctic ice and firn. *Journal of Geophysical Research*, **101**, 4115–4128.

FLINT, R. F. 1971. *Glacial and Quaternary Geology*. John Wiley, London & New York.

FORBES, E. 1846. On the connexion between the distribution of the existing fauna and flora of the British Isles, and the geological changes which have affected their area, especially during the epoch of the Northern Drift. *Memoir of the Geological Survey of Great Britain*, **1**, 336–432.

GALE, S. J. & HOARE, P. G. 2012. The stratigraphic status of the Anthropocene. *The Holocene*, **22**, 1491–1494, http://dx.doi.org/10.1177/0959683612449764

GERVAIS, P. 1867–69. *Zoologie et paleontology générales: Nouvelles recherches sur les animaux vertébrés et fossiles*. Paris.

GIBBARD, P. L. & KOLFSCHOTEN, VAN TH. 2005. Pleistocene and Holocene series. *In*: GRADSTEIN, F., OGG, J. & SMITH, A. G. (eds) *Geological Time Scale 2004*. Cambridge University Press, Cambridge, 441–452.

GIBBARD, P. L., HEAD, M. J. ET AL. 2010. Formal ratification of the Quaternary system/period and the Pleistocene series/Epoch with a base at 2.58 Ma. *Journal of Quaternary Science*, **25**, 96–102.

GOODSITE, M. E., ROM, W. ET AL. 2001. High-resolution AMS ^{14}C dating of post-bomb peat archives of atmospheric pollutants. *Radiocarbon*, **43**, 495–515.

HEDBERG, H. D. (ed.) 1976. *International Stratigraphic Guide*. John Wiley, London & New York.

HEINZELIN, J. DE & TAVERNIER, R. 1957. Flandrien. *In*: PRUVOST, P. (ed.) *Lexique Stratigraphique International*. Paris, Centre National de la Recherche Scientifique, **1**, 32.

HUMPHRIES, M. S., KINDNESS, A. ET AL. 2010. ^{137}Cs and ^{210}Pb derived sediment accumulation rates and their role in the long-term development of the Mkuze River floodplain, South Africa. *Geomorphology*, **119**, 88–96.

HYVÄRINEN, H. 1978. Use and definition of the term Flandrian. *Boreas*, **7**, 182.

KEMP, A. C., HORTON, B. P. ET AL. 2009. Timing and magnitude of recent accelerated sea-level rise (North Carolina, United States). *Geology*, **37**, 1035–1038.

KRACHLER, M., ZHENG, J. ET AL. 2009. Global atmospheric As and Bi contamination preserved in 3000 year old Arctic ice. *Global Biogeochemical Cycles*, **23**, GB3011, http://dx.doi.org/10.1029/2009GB003471

LE ROUX, G., WEISS, D. ET AL. 2004. Identifying the sources and timing of ancient and medieval atmospheric lead pollution in England using a peat profile from Lindow bog, Manchester. *Journal of Environmental Monitoring*, **6**, 502–510.

LOURENS, L., HILGEN, F. ET AL. 2005. The Neogene period. *In*: GRADSTEIN, F. M., OGG, J. G. & SMITH, A. G. (eds) *A Geologic Time Scale 2004*. Cambridge University Press, Cambridge, 409–440.

LOWE, J. J. & WALKER, M. J. C. 1997. *Reconstructing Quaternary Environments*. Longman, London.

LYELL, C. 1837. *Principles of Geology*, 5th edn. John Murray, London.

LYELL, C. 1839. *Nouveaux Éléments de Géologie*. Pitois-Levrault, Paris.

MACFARLING MEURE, C. D., ETHERIDGE, C. ET AL. 2006. Law Dome CO_2, CH_4 and N_2O ice core records extended to 2000 years BP. *Geophysical Research Letters*, **33**, L14810, http://dx.doi.org/10.1029/2006GL026152

MILLER, G. H., FOGEL, M. L. ET AL. 2005. Ecosystem collapse in Pleistocene Australia and human role in megafaunal extinction. *Science*, **309**, 287–290.

MORLOT, A. VON 1854. Über die quaternären Gebilkde des Thonegebeits. *Verhandlingen der Schweizerisches Gesellschaftes geschichten Naturwetenschappen*, **39**, 161–164.

MORRISON, R. B. 1969. The Pleistocene–Holocene boundary. *Geologie en Mijnbouw*, **48**, 363–372.

NICHOLLS, R. J., WONG, P. P. ET AL. 2007. Coastal systems and low-lying areas. *In*: PARRY, M. L., CANZIANI, O. F. ET AL. (eds) *Climate Change 2007: Impacts, Adaptation and Vulnerability*. Contribution of Working Group II to the Fourth Assessment Report of the Intergovernmental Panel on Climate Change. Cambridge University Press, Cambridge, 315–356.

NILSSON, T. 1983. *The Pleistocene*. Reidel, Dordrecht.

ORR, J. C., FABRY, V. J. & AUMONT, O. 2005. Anthropogenic ocean acidification over the twenty-first century and its impact on calcifying organisms. *Nature*, **473**, 681–686.

PERIMAN, R. D. 2006. Visualising the Anthropocene: human land use history and environmental management. *In*: AGUIRE-BRAVO, C., PELLICANE, P. J. ET AL. (eds) *Monitoring Science and Technology Symposium: Unifying Knowledge for Sustainability in the Western Hemisphere Proceedings RMRS-P-42CD*. US Department of Agriculture, Forest Service, Rocky Mountain Research Station, Fort Collins, CO, 558–564.

REBOUL, H. 1833. *Géologie de la Période Quaternaire et Introduction a l'Histoire Ancienne*. F.G. Leverault, Paris.

REMANE, J., BASSETT, M. G. ET AL. 1996. Revised guidelines for the establishment of global chronostratigraphic standards by the International Commission on Stratigraphy (ICS). *Episodes*, **19**, 77–81.

RODILOCO, F. 1970. Arduino (or Ardini), Giovanni. *In*: GILLESPIE, C. C. (ed.) *Dictionary of Scientific Bibliography*. Charles Scribner's and Sons, New York, 1.

ROSENBAUM, M. S., MCMILLAN, A. A. ET AL. 2003. Classification of artificial (man-made) ground. *Engineering Geology*, **69**, 399–409.

RUDDIMAN, W. F. 2003. The anthropogenic greenhouse era began thousands of years ago. *Climatic Change*, **61**, 261–293.

RUDDIMAN, W. F. 2005. *Plows, Plagues and Petroleum*. Princeton University Press, Princeton, NJ.

RUDDIMAN, W. F., KUTZBACH, J. E. & VAVRUS, S. J. 2011. Can natural or anthropogenic CO_2 and CH_4 increases be falsified? *The Holocene*, **21**, 865–879.

SALVADOR, A. 1994. *International Stratigraphic Guide*, 2nd edn. International Union of Geological Sciences and Geological Society of America, Boulder, CO.

SALVADOR, A. 2006*a*. The Tertiary and the Quaternary are here to stay. *Association of American Petroleum Geologists Bulletin*, **90**, 21–30.

SALVADOR, A. 2006*b*. A stable Cenozoic geologic time scale is indispensable. *Episodes*, **29**, 43–45.

SAPART, C. J., MONTEIL, G. ET AL. 2012. Natural and anthropogenic variations in methane sources during the past two millennia. *Nature*, **490**, 85–88, http://dx.doi.org/10.1038/nature11461

SCHNEER, C. J. 1969. Introduction. *In*: SCHNEER, C. J. (ed.) *Towards a History of Geology*. The Massachusetts Institute of Technology Press, Cambridge, MA & London, 1–18.

SHOTYK, W., WEISS, D. ET AL. 1998. History of atmospheric lead deposition since 12 370 ^{14}C yr BP from a Peat Bog, Jura Mountains, Switzerland. *Science*, **281**, 1635–1640.

SOWERS, T. 2010. Atmospheric methane isotope records covering the Holocene period. *Quaternary Science Reviews*, **29**, 213–221.

STEFFEN, W. A., SANDERSON, P. D. ET AL. 2004. *Global Change and the Earth System: A Planet Under Pressure*. Springer-Verlag, New York.

STEFFEN, W., CRUTZEN, P. J. & MCNEIL, J. R. 2007. The Anthropocene: are humans now overwhelming the great forces of Nature? *Ambio*, **236**, 614–621.

STEFFEN, W., GRINEVALD, J. ET AL. 2011. The Anthropocene: conceptual and historical perspectives. *Philosophical Transactions of the Royal Society, London*, **A369**, 842–867.

URBAN, M. C., TEWSBURY, J. J. & SHELDON, K. S. 2012. On a collision course: competition and dispersal differences create no-analogue communities and cause extinctions during climate change. *Proceedings of the Royal Society of London*, **B279**, 2072–2080.

WALKER, M., JOHNSEN, S. ET AL. 2009. Formal definition and dating of the GSSP (Global Stratotype Section and Point) for the base of the Holocene using the Greenland NGRIP ice core, and selected auxiliary records. *Journal of Quaternary Science*, **24**, 3–17.

WALKER, J. D., GEISSMAN, J. W. ET AL. 2012. The Geological Society of America geologic time scale. *Geological Society of America Bulletin*, **125**, 259, http://dx.doi.org/10.1130/B30712.1

WALKER, M. J. C., BERKELHAMMER, M. ET AL. 2012. Formal subdivision of the Holocene Series/Epoch: a Discussion Paper by a Working Group of INTIMATE (Integration of ice-core, marine and terrestrial records) and the Subcommission on Quaternary Stratigraphy (International Commission on Stratigraphy). *Journal of Quaternary Science*, **24**, 3–17.

WEST, R. G. 1968. *Pleistocene Geology and Biology*, 1st edn. Longman, London.

WEST, R. G. 1977. *Pleistocene Geology and Biology*, 2nd edn. Longman, London.

WEST, R. G. 1979. Further on the Flandrian. *Boreas*, **8**, 126.

WILLIAMS, M., ZALASIEWICZ, J. ET AL. 2011. The Anthropocene: a new epoch of geological time? *Philosophical Transactions of the Royal Sociey, London*, **369A**, 833–1112.

WOLFE, A. P., HOBBS, W. O. ET AL. 2012. Stratigraphic expressions of the Holocene–Anthropocene transition revealed in sediments from remote lakes. *Earth Science Reviews*, **116**, 17–34.

WOLDSTEDT, P. 1969. *Quartär*. F. Enke, Stuttgart.

ZALASIEWICZ, J., SMITH, A. ET AL. 2004. Simplifying the stratigraphy of time. *Geology*, **32**, 1–4.

ZALASIEWICZ, J., SMITH, A. ET AL. 2008. Are we now living in the Anthropocene. *GSA Today*, **18**, 4–8.

ZALASIEWICZ, J., WILLIAMS, M. ET AL. 2010. The new world of the Anthropocene. *Environmental Science and Technology*, **44**, 2228–2231.

ZALASIEWICZ, J., WILLIAMS, M. ET AL. 2011*a*. The Anthropocene: a new epoch of geological time. *Philosophical Transactions of the Royal Society, London*, **A369**, 835–841.

ZALASIEWICZ, J., WILLIAMS, M. ET AL. 2011*b*. Stratigraphy of the Anthropocene. *Philosophical Transactions of the Royal Society, London*, **A369**, 1036–1055.

Can an Anthropocene Series be defined and recognized?

JAN ZALASIEWICZ[1]*, MARK WILLIAMS[1] & COLIN N. WATERS[2]

[1]*Department of Geology, University of Leicester, University Road, Leicester LE1 7RH, UK*

[2]*British Geological Survey, Keyworth, Nottingham, NG12 5DP, UK*

**Corresponding author (e-mail: jaz1@leicester.ac.uk)*

Abstract: We consider the Anthropocene as a physical, chronostratigraphic unit across terrestrial and marine sedimentary facies, from both a present and a far future perspective, provisionally using an approximately 1950 CE base that approximates with the 'Great Acceleration', worldwide sedimentary incorporation of A-bomb-derived radionuclides and light nitrogen isotopes linked to the growth in fertilizer use, and other markers. More or less effective recognition of such a unit today (with annual/decadal resolution) is facies-dependent and variably compromised by the disturbance of stratigraphic superposition that commonly occurs at geologically brief temporal scales, and that particularly affects soils, deep marine deposits and the pre-1950 parts of current urban areas. The Anthropocene, thus, more than any other geological time unit, is locally affected by such blurring of its chronostratigraphic boundary with Holocene strata. Nevertheless, clearly separable representatives of an Anthropocene Series may be found in lakes, land ice, certain river/delta systems, in the widespread dredged parts of shallow-marine systems on continental shelves and slopes, and in those parts of deep-water systems where human-rafted debris is common. From a far future perspective, the boundary is likely to appear geologically instantaneous and stratigraphically significant.

The concept that we may be living in an Anthropocene geological time interval has attracted considerable interest and scrutiny since its latest restatement by Crutzen & Stoermer (2000) and Crutzen (2002) (see also Revkin 1992). These authors effectively regarded the Holocene as having terminated because of the scale and significance of human impact upon the Earth system. In this view, a new and distinct phase of Earth history has already begun, and Crutzen (2002) regarded the beginning of the Industrial Revolution as marking the beginning of profound global change.

Formalizing this concept within the Geological Time Scale (Zalasiewicz *et al.* 2008, 2011, 2012) would result in the creation of an Anthropocene Epoch. Higher levels (e.g. Period, Era) might be considered because of the lack of precedent in Earth history for some of the component stratigraphical signals, such as the lithostratigraphical signal in urban regions (Price *et al.* 2011; Ford *et al.* 2014), and the scale and character of the biotic change (Barnosky 2008, 2013; Barnosky *et al.* 2011, 2012). Lower hierarchical levels are possible too (e.g. an Anthropocene Age as subdivision of the Holocene Epoch), and this would result in less modification of the Geological Time Scale. However, we continue to discuss the Anthropocene in terms of the hierarchical level of Epoch, not least because it brings clear focus on the important scientific question of whether or not the Earth system now lies outside of the 'Holocene envelope' of stratigraphically significant environmental conditions (cf. Rockström *et al.* 2009; Steffen *et al.* 2011).

Anthropocene boundary level

To carry out the analysis below, we must provisionally select a start date for the Anthropocene. Potential dates for the beginning of this phenomenon have fallen into three categories. First, dates a few to several millennia back within the Holocene (Certini & Scalenghe 2011; Ruddiman 2013) have been suggested, reflecting the growing evidence for widespread, low-intensity human modification of the terrestrial environment (Ellis 2011; Kaplan *et al.* 2011) and, more controversially (Ruddiman 2003, 2013; cf. Elsig *et al.* 2009), resultant release of sufficient greenhouse gases to maintain the Holocene within stable conditions of climate and sea level. Secondly, the beginning of the Anthropocene at around 1800 CE, as originally suggested by Crutzen (2002): that is, around the beginning of the Industrial Revolution when the rapid increase in human numbers, energy use and atmospheric carbon dioxide levels began (Zalasiewicz *et al.* 2008, fig. 1). Thirdly, approximately 1950 CE, the beginning of the post-war 'Great Acceleration' of economic activity (Steffen *et al.* 2007).

We regard the latter two as the more suitable candidates because of the clear break between

From: Waters, C. N., Zalasiewicz, J. A., Williams, M., Ellis, M. A. & Snelling, A. M. (eds) 2014. *A Stratigraphical Basis for the Anthropocene*. Geological Society, London, Special Publications, **395**, 39–53.
First published online March 10, 2014, http://dx.doi.org/10.1144/SP395.16

Holocene global stability (or the very slow change), and the more rapid and geologically striking changes of the last two centuries (e.g. Zalasiewicz *et al.* 2008, fig. 1; Steffen *et al.* 2011, fig. 1). The Anthropocene does not represent the detectable incoming of human influence (which in any case is clearly diachronous: e.g. Kaplan *et al.* 2011) but major change to the Earth system that happens to be currently driven by human forcing, and which may geologically soon be more significantly controlled by a number of secondary positive feedbacks, such as methane release from permafrost and ice-albedo changes (e.g. Hay 2013, pp. 897–939).

For the purposes of this exercise we choose the later, approximately 1950 CE date. This level coincides with changes to lacustrine dynamics and sedimentation worldwide (expressly linked to a potential Holocene–Anthropocene boundary by Wolfe *et al.* 2013, and partly reflecting a worldwide shift in nitrogen isotopes associated with the increase in global fertilizer use: Holtgrieve *et al.* 2011). It also coincides with the beginning of the nuclear age and the spread of artificial radionuclides into contemporary sediments worldwide, and both biotic (Barnosky 2013; Wilkinson *et al.* 2014) and physical (Ford *et al.* 2014) stratigraphical signals that seem to be stratigraphically sharp (to *c.* decadal level) and globally widespread. These changes are traceable by scientists living today, and not just by hypothetical 'far-future' geologists. They represent a significant and permanent shift in the Earth system, although probably not the greatest changes, that will almost certainly take place in the coming centuries and millennia (Barnosky *et al.* 2011, 2012; New *et al.* 2011).

Thus, while it is still too early to make a formal recommendation, the approximately 1950 level currently seems to provide sharper stratigraphic definition than the relatively more diffuse and diachronous signals associated with the Industrial Revolution (e.g. the shift in carbon isotopes from the increase in fossil fuel burning: Al-Rousan *et al.* 2004).

We do not here examine the question of whether the boundary should be defined by a Global Standard Stratigraphic Age (GSSA or more simply a numerical age) or Global Stratigraphic Section and Point (GSSP = 'golden spike'). For practical purposes in current use, we consider that either would be effective. By whichever means defined, this approximately 1950 CE level might be regarded as stratigraphically challenging, in encompassing (to date) the geologically almost infinitesimally brief interval of about 65 years: over three orders of magnitude shorter than the Holocene and over five orders of magnitude shorter than the average epoch in the Cenozoic (Fig. 1).

The Anthropocene in geochronology and chronostratigraphy

Given current stratigraphic practice, we must consider the Anthropocene as a potential formal stratigraphic unit in not one but two meanings.

First, it is a potential *geochronological* unit – that is, one of geological time – over which a variety of events have taken place on Earth. An Anthropocene Epoch, as an Earth-based time unit, would (as with the Holocene Epoch and all other geochronological units) hence be used as temporal reference for events in the Earth's deep interior as much as those at the surface.

Separate geological timescales have been set up for other bodies, such as the Moon and Mars (Tanaka & Hartmann 2012), and so an Anthropocene Epoch would be limited to this planet, as holds currently true for other terrestrial geochronological units. The limits on Earth extend from the core to the atmosphere and arguably to the region of space immediately dominated by the Earth's gravitational field, although excluding the Moon, which has a separate stratigraphic scheme (Tanaka & Hartmann 2012). However, we note that it is now beginning to be possible to correlate the Anthropocene across space, in what might be regarded as the first interplanetary stratigraphic marker since the products of the Late Heavy Bombardment of the late Archaean. Infinitesimally smaller in bulk, although very much more synchronously distributed, human-projected spacecraft and associated debris have now left physical traces on and around several planets and moons of this solar system.

For all past geological units, with the exception of the later part of the Holocene Epoch (that we still, formally, live in), all of our knowledge of the history of the Earth is derived from the rock record. From the beginning of a human written record, this proxy record began to be augmented by human observations of terrestrial events. This human observation has developed, today, to the extent that many terrestrial processes are now routinely monitored, recorded and analysed; this means that geological proxy data of the Anthropocene, being captured within rock currently forming, can now be directly compared with the geological events themselves.

The Anthropocene, in tandem with other geological units, should also be considered as a material rock unit of chronostratigraphy (commonly referred to as 'time-rock'). Chronostratigraphic units are commonly regarded as the material 'rock' record of geological time, and thus the physical embodiment of (and evidence for) the passage of time. Thus, the Jurassic System comprises all of the strata formed during the Jurassic Period, while the

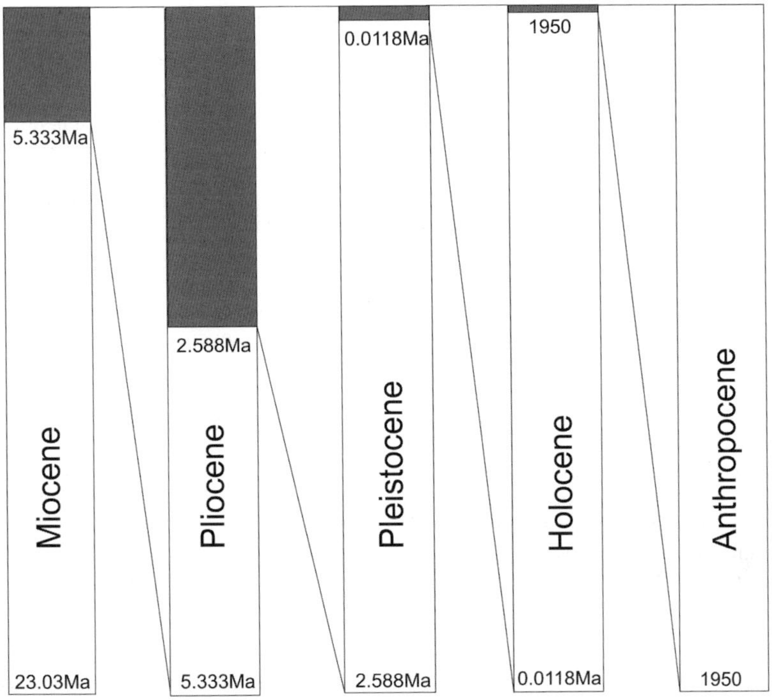

Fig. 1. Comparison of lengths of epochs from the mid-Cenozoic to the present, showing progressive shortening in time span. Dates from Gradstein *et al.* (2012).

Pleistocene Series is the equivalent rock record of the Pleistocene Epoch. There is hence a hierarchical system of chronostratigraphical terms, exactly parallel to those of geochronology. The Anthropocene, if considered as an Epoch, should also be considered as a Series.

Chronostratigraphy, scale-dependence and the Anthropocene

Not all geologists consider chronostratigraphy to be a necessary and fundamental part of the Geological Time Scale (e.g. Zalasiewicz *et al.* 2004, 2007; Carter 2007). In such an interpretation there need not be both a Pleistocene Epoch and a Pleistocene Series, but simply an Epoch, to which the material record is referred descriptively (thus: strata formed during the Pleistocene Epoch, or more simply 'Pleistocene strata'). Currently, though, most stratigraphers, as represented by voting members of the International Commission on Stratigraphy, prefer to use the dual hierarchy of geochronology + chronostratigraphy (Zalasiewicz *et al.* 2013*a*), and so we here regard consideration of an Anthropocene Series as an integral part of the analysis of the Anthropocene concept.

Chronostratigraphy in practice only effectively applies to stratified rocks, where superposition applies, and hence 'lower' equals 'older' and 'upper' equals 'younger' (Zalasiewicz *et al.* 2013*a*). Single hand specimens of igneous and (especially) metamorphic rocks commonly include a number of intermeshing fabrics of distinctly different ages (that can be dated and placed within a geochronological framework), and so 'upper' and 'lower' have no meaning, and the rock itself cannot be regarded as having 'formed' at a particular moment in time. Thus, a putative Anthropocene Series encompasses only stratified deposits currently accumulating and not (say) mineral assemblages now crystallizing (i.e. during the Anthropocene Epoch) in the roots of current mountain belts.

Chronostratigraphy is also scale-dependent (Zalasiewicz *et al.* 2007). That is, on short timescales, the superpositional fabrics of sedimentary stratification may be disrupted by such processes as bioturbation (in marine deposits especially: Anderson 2001) or by soil-forming processes (Bacon *et al.* 2012), giving disrupted sedimentary fabrics in which temporal information has been mixed or homogenized. This process commonly affects time units of durations of some thousands of years (Anderson 2001) but it can also act over timescales of millions of years (e.g. Bacon *et al.* 2012) and length scales of kilometres in the case of subsurface sedimentary diapirism (e.g. Shoulders & Cartwright 2004).

For most stratigraphic units in the deep time record, this scale-dependence effect may be neglected, given that currently achievable levels of time resolution are typically measured in fractions of millions of years. However, the duration of epochs, both actual (Holocene) and potential (Anthropocene), becomes much shorter towards the present day (Fig. 1). Thus, for Pleistocene and (especially) for Holocene strata, the scale dependence effect becomes significant, and for the Anthropocene (where decadal time resolution may reasonably be sought) it becomes an important factor in chronostratigraphic definition.

Components of an Anthropocene Series

Despite the complications noted above, an attempt to define an Anthropocene Series is both part of formal stratigraphic analysis and, independently of this, is useful in helping to understand the Anthropocene phenomenon (formal or informal) as a part of Earth history.

What might an Anthropocene Series, and its various material stratigraphic components, comprise? We consider the strata that accumulate in a range of geographical settings, from terrestrial (in the sense of 'land-based') to deep marine, and discuss how they might be recognized and characterized. We reiterate that the Anthropocene here is a time boundary, and not a boundary between anthropogenic 'artificial' and 'natural' sedimentary facies. Hence, an Anthropocene Series (and, indeed, pre-Anthropocene deposits) will include both of these facies, the boundary between them being diachronous. Nevertheless, the extent of facies diachroneity will vary, both geographically and between different types of stratigraphic signal, and this might offer the possibility of effective discrimination of an Anthropocene Series.

These strata include a number of proxies for time – not least fossils, a form of evidence that remains key to the subdivision of Phanerozoic strata (Gradstein *et al.* 2012) and that has the potential to help characterize an Anthropocene interval (Barnosky 2013; Wilkinson *et al.* 2014) when used in combination with other stratigraphic indicators (Waters *et al.* 2014).

Terrestrial settings

Geologically, the terrestrial realm may be divided into areas of erosion, particularly of older rock, and areas of deposition. The former in stratigraphy may be considered as unconformity surfaces, only to be preserved at the transition between phases of erosion and subsequent sedimentation. Although such erosion surfaces may be studied by techniques such as Terrestrial Cosmogenic Nuclide (TCN) dating (e.g. Gosse & Phillips 2001), we will not consider them further here, except via the indirect record they leave via the sedimentary deposits eroded from them. These may be broadly categorized as the following.

Soils

Soils are, perhaps, the most widespread terrestrial sedimentary facies, forming on both erosional and depositional surfaces, and having deep time equivalents, palaeosols, when preserved on depositional surfaces.

The alteration of soils by anthropogenic activities is widespread, striking and increasingly well documented (Richter 2007). However, the spread of anthropogenic soils has been strongly diachronous through the Holocene, and reflects the spread of agriculture across the globe (Ellis *et al.* 2012). At present, therefore: which soils are Holocene and which are Anthropocene?

One approach here has been to take a major phase of soil expansion 2000 years ago across northern Europe (Certini & Scalenghe 2011) and suggest that the base of that may be taken as a 'golden spike' to mark the base of the Anthropocene. This is an intriguing and imaginative suggestion, but is not without problems (Gale & Hoare 2012). First, the base of a soil upon older regolith is gradational and cannot capture a boundary with the resolution required for the Anthropocene. Secondly, and more generally, soils exemplify the 'scale-dependence' phenomenon noted above, being continually reworked by both natural and anthropogenic processes as long as they are at the Earth's surface. Hence, it may in some ways be more appropriate to place *all* surface soils in the Anthropocene because they are continually being modified, even though many of them have fabrics and components that range back for thousands and, in some cases (Bacon *et al.* 2012), for millions of years. This ongoing modification is arguably greatest for agricultural soils because of the intensive nature of human reworking. Owing to the breakdown of superposition, soils are generally problematic to classify chronostratigraphically at the very high levels of temporal resolution required for the Anthropocene. Thirdly, the criteria for definition of a 'golden spike' recommends that a section be used in which there is a continuous succession, where observed gaps in deposition are absent or at a minimum. In existing chronostratographical units, palaeosols are considered to represent time gaps and would be avoided as a basis on which to define a chronostratigraphical boundary (Remane *et al.* 1996).

Lacustrine deposits

Lake deposits are, perhaps, the most straightforward to deal with stratigraphically. Their deposits commonly form ordered strata, which – especially in those lakes with low-oxygen bottom waters – tend not to be seriously disrupted by bioturbation. The resulting high-resolution stratigraphic archives can show a clear signal of the environmental changes that may potentially characterize an approximately 1950 CE Anthropocene Series base, such as widespread, marked N isotope (Holtgrieve *et al.* 2011) and palaeontological (Wolfe *et al.* 2013) signals in northern lakes far from urban centres, while the incoming of A-bomb test-related radionuclides provides another marker (Yan *et al.* 2002; Appleby 2008; Hancock *et al.* 2011, 2014). If it was decided to define the Anthropocene boundary via a physical reference level or GSSP ('golden spike') rather than a designated numerical date GSSA (see the Discussion later), then lake deposits will figure strongly as settings for candidate stratotypes. Lacustrine sediments, though, include anthropogenic signals of other ages too, some markedly diachronous, such as sediment influxes associated with land-use changes (Edwards & Whittington 2001).

Fluvial deposits

The human management of rivers, and consequent alteration of their patterns of sedimentation and erosion, has a long history, and the consequent spread of indirect anthropogenic deposits has been marked (e.g. Syvitski & Kettner 2011; Merritts *et al.* 2011; Brown *et al.* 2013), multi-faceted (e.g. the nineteenth century modification of fluvial sedimentation in north America, as numbers of beavers – and, hence, beaver dams – fell sharply as a result of hunting: Kramer *et al.* 2011) and strikingly diachronous across the world, and even in part on a regional scale within the UK (Lewin 2012). Indeed, the difficulty of consistently recognizing an Anthropocene boundary in modern fluvial deposits was regarded by Autin & Holbrook (2012) as one reason to reject the concept of a formalized Anthropocene.

However, globally, the rate of fluvial transformation saw significant rises that coincided with the two main inflections in human economic activity at about 1800 and 1950 (Syvitski & Kettner 2011), both of which are candidate dates for the beginning of the Anthropocene. To what extent these may be generally 'traceable' within the sedimentary record seems still to be an open question. Locally, at least, major, distinct Anthropocene bodies of sediment are building up behind the major dams that in recent decades have been constructed on nearly all major rivers of the world (Syvitski & Kettner 2011), with rates of sediment supply commonly increased by deforestation and related processes (Wilkinson 2005). For instance, most sediment that used to be transported down to the Nile Delta is now trapped behind the Aswan Dam (producing a substantial, and rapidly growing Anthropocene sediment body) or held within artificially multiplied (for irrigation) distributaries within a system that has been completely altered by human activity (Stanley 1996).

Significant future rise in sea level would be expected to result in development of transgressive estuarine–marine deposits in the distal parts of river systems. However, the interplay of associated changes in precipitation, vegetation and human forcing would certainly be complex, making patterns of sedimentation hard to predict.

Aeolian deposits

Windblown deposits occur both within the major sand seas of the world, such as the ergs of the Sahara desert, as more localized dune fields, such as those associated with coastal areas, and also as far-travelled loess and related deposits. All are sensitive to local climate and to vegetation cover, and to human activity, in particular through overgrazing, overcultivation, unsustainable irrigation techniques and deforestation, which has strongly influenced the generation of loess through desertification, and whose effects include increases in dust flux (Goudie 2009). There is evidence of an increase of a factor of 2 in background dust loads over the Antlantic since the mid-1960s, the likely product of desertification caused by the doubling of the population in the Sahel region over the past 40 years (Moulin & Chiapello 2006). The extent to which these might translate into an Anthropocene Series boundary is uncertain. It seems likely that within contemporary large, long-lived dune fields, at least, the shifting sands will render a boundary difficult to locate and trace precisely – although in this the Anthropocene is not alone in facing difficulties of chronostratigraphic classification (see below).

Glacial deposits

Glacial deposits are sensitive recorders of changes in ice volume and extent, and many present-day glacial valleys in Europe include terminal moraines reflecting the greater extent of ice during the Little Ice Age of the sixteenth–mid-ninteenth centuries (Mann 2002). Similarly, the shrinking of most mountain glaciers since the 1850s, with regional variations in both retreat and advance during the mid-twentieth century and large-scale retreats since the 1980s (IPCC 2001, fig. 2.18), linked to global temperature increases, has exposed morainic deposits that may be clearly identified and mapped as of Anthropocene age, particularly where detailed cartographic and photographic records occur of glacier extents earlier in the twentieth century (e.g. Kulkarni *et al.* 2007). Associated deposits include those laid down catastrophically by dam-bursts, as increased volumes of meltwater have accumulated behind and destabilized morainic dams and wasting morainic ice cores (Nayar 2009).

Ice

This is also a terrestrial sedimentary deposit that is found on all the continents (except in Australia, and probably not for much longer in Africa, where it is represented only by rapidly thawing Kilimanjaro). Ice sheets record snow layers extending back many thousands of years, and encapsulating (in the Arctic and Antarctic) the entire interval of human history, including levels that can be identified for 1800 and 1950, and which provide data on rising CO_2 intervals. Snow layers record human pollutants from the atmosphere back to classical times (e.g. lead aerosols derived from Roman smelting). Following this, there is a succession of recorded events that might provide geochemical criteria to identify either an approximate 1800 CE or 1950 CE level. This includes the CO_2 levels preserved in air pockets (although this is compromised by the 'lock-in' time for air post-dating the deposition of the snow). However, events such as the appearance of nitrogen derived from the Haber–Bosch process (cf. Holtgrieve *et al.* 2011), the change in lead isotopes reflecting the use and then abandonment of lead additives in petrol (Bollhöfer & Rosman 2000), and the incorporation of artificial radionuclides provide useful global stratigraphic markers. The range of palaeoenvironmental proxies recorded in this medium and the annual resolution make selection of a GSSP within a snow/ice core a potential option, as for the Pleistocene–Holocene boundary (cf. Walker *et al.* 2009).

Artificial deposits

The transformation of primary raw materials (sand and gravel, limestone, mudrock, metal ores) into the fabric of urban areas represents the creation of a novel and substantial type of stratum in which the buildings themselves and the associated landscape changes (the latter mapped as various types of Artificial Ground on British Geological Survey maps, for instance: Price *et al.* 2011; Ford *et al.* 2014) provide something that combines features of a lithostratigraphic unit and of an extraordinarily large trace fossil system.

The resulting deposit is clearly anthropogenic but, because towns and cities have been a feature of human civilization since the Epi-palaeolithic (Mesolithic) at about 9000 BC (see Edgeworth 2013), also clearly diachronous. We may discuss two features of relevance here.

First, the extraordinary post-war growth of cities and megacities allows, by simply mapping the historical growth of urban areas, a distinction between post-1950 CE artificial deposits and those that predate them (Fig. 2). Prior to the 1950s, large cities tended to be located close to natural resources or be suitable coastal locations for the import/export of these resources. The post-1950s evolution of megacities has relied upon the contained population of the megacity to be the key resource, and these cities have been a centre for the inward influx of natural resources sourced from rural areas and transported to the cities to fuel industry and construction. This change can be seen as a product of improvement of transport networks and greater efficiencies in the mass transport of bulk materials during the late twentieth century (Haff 2013; see also Williams *et al.* 2013). This creation of laterally continuous but temporally distinct deposits may be compared with, say, those created naturally during the progradation (outgrowth) of a delta system.

Secondly, even within the older parts of existing cities, the continuous replacement of the urban fabric, both above and below ground, means that these artificial deposits comprise complex mixtures of pre-Anthropocene and Anthropocene rocks and minerals (and, locally, indeed fossils). The presence of novel materials and minerals in both direct and, to a lesser extent, indirect anthropogenic deposits (Zalasiewicz *et al.* 2013*b*; Ford *et al.* 2014) provides an approach to dating these deposits to decadal level, a resolution far beyond that applicable for previous epochs. This is a rather coarser-grained equivalent of the situation noted above with soils, and again underscores the awkwardness of chronostratigraphy in dealing with short timescales, and complex sedimentary processes and geometries. It is only towns and cities abandoned pre-1950 that may be said to comprise wholly pre-Anthropocene representatives of this deposit type.

Below ground, artificial ground locally deeply extends into underlying strata via the many mineshafts and boreholes sunk to extract resources, with considerable 'halo' effects via such as hydrocarbon extraction (and now, injection of fluids and sand for shale gas extraction). Geologically, this is something of a hybrid, combining features of burrowing, albeit on an enormous scale, with those of

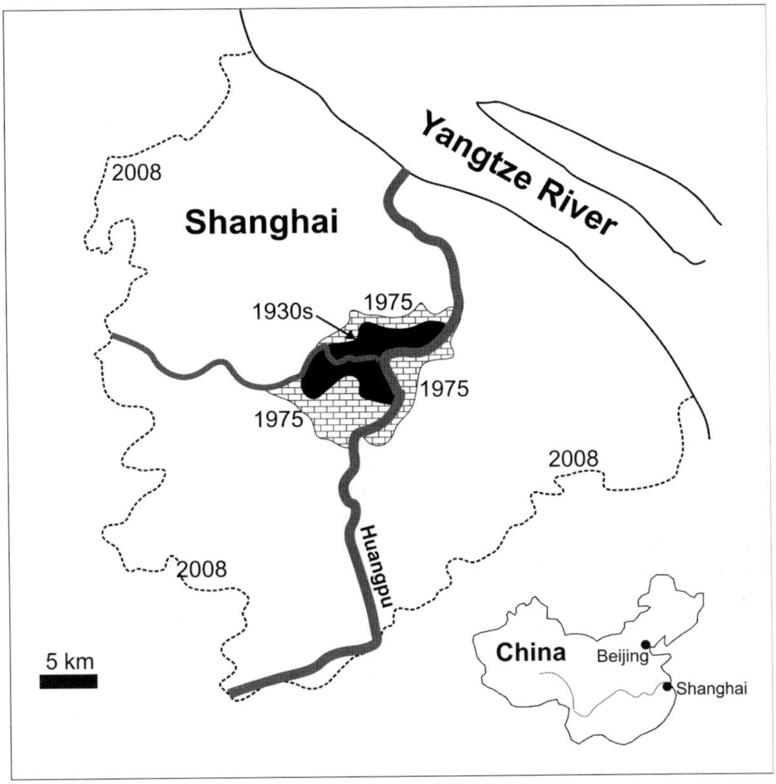

Fig. 2. The rapid mid-twentieth century growth of Shanghai, as an example of the formation of a distinct, extensive sedimentary facies that may be referred to a putative Anthropocene Series. Information from Larmer (2010) and Map of Central Shanghai, printed by the British War Office/US Army Map Service in 1935.

intrusive bodies, showing cross-cutting relationships and even of diagenetic alteration. Neither of the last two phenomena are generally classified within chronostratigraphic units (as they do not show superpositional relationships), as their history may be protracted and only related in general terms to processes acting at the Earth's surface (Ford *et al.* 2014). The subsurface anthropogenic phenomena, by contrast, are very much related to surface activities (and can also impact on the surface, as for instance with aquaculture-related subsidence on the Yellow River delta in China which now reaches 250 mm a^{-1}: Higgins *et al.* 2013). They clearly form a pronounced and temporally constrained event, given the post-war surge in drilling and mining (Ford *et al.* 2014).

Marine settings

Coastal systems

These systems include beaches, tidal flats and deltas. Throughout much of the latter half of the Holocene, these have been commonly progradational, as sediment eroded from the land has accumulated around a coastline more or less fixed as sea level stabilized following its post-glacial rise. Where sediment has built up and built out in this way, then distinct stratal packets that relate to industrialization and land-use change have been recognized and suggested as Anthropocene markers (e.g. Poirier *et al.* 2011). Some are distinctive through their content of heavy metals, organic chemicals and so on (e.g. Allen 1988; Marshall *et al.* 2007; Vane *et al.* 2011; Gałuszka *et al.* 2013), with eutrophication of coastal environments due to influx of excess nitrogen, and these may also be used to help identify an Anthropocene–Holocene boundary. Globally, the overall facies changes are diachronous, but within them some signals (such as distinct chemical markers related to particular industrial processes: Kruge 1999) may provide more or less synchronous marker levels.

Within the last couple of centuries – and, particularly, the last several decades, many coastal systems have seen large-scale change that is clearly relevant to the historical characterization of the Anthropocene, but that complicates the simple progradational picture. For instance, as rivers have been dammed, sediment is temporally stored behind the dams (see above) and does not nourish growing deltas, some of which have as a consequence shrunk back (e.g. Nile, Mississippi: Törnqvist *et al.* 2008). Related phenomena include the draining of coastal wetlands for farmland, resulting in the large-scale loss of such strata as surface peat deposits through desiccation, deflation and oxidation. For instance, some 2000 km^2 of peat up to 4 m thick in the English Fenland alone has disappeared since the eighteenth century, resulting in the exposure of the underlying geology, now itself compacted and oxidized (Smith *et al.* 2010). Attempts at stabilization of coastal erosion rates through construction of coastal defences produces artificial deposits, while impacting upon sediment flux and erosion rates adjacent to protected regions. The Anthropocene boundary here approximates to a regional sedimentary hiatus and disconformity – likely to be buried beneath new coastal sedimentary deposits, a century or two hence, as only a geologically trivial sea-level rise will suffice to trigger marine transgression across such areas.

The expression of the Anthropocene in the environmentally sensitive coastal systems, therefore, represents a diverse patchwork of deposits and lacunae that reflect local interplays of natural and anthropogenic forces.

Shelf/slope marine systems

Human impact on open marine systems has, in general, substantially lagged those on land. The marine fisheries in northern Europe began in earnest in Medieval times, perhaps as a result of technological improvements (e.g. effective drift nets), and their spread across the world has been charted by Roberts (2007). The concomitant, diachronous decline in fish stocks through overfishing changed the structure of marine ecosystems, although impacts on the kind of organisms (e.g. foraminifera, dinoflagellates – much lower in the food chain) used in biostratigraphy have probably been small, even with the dramatic fish declines reported (e.g. Myers & Worm 2003).

More profound physical and chemical impacts on recent marine strata are associated with the industrial age, from about 1800 CE. The greatest physical impact on sediments has been the physical disruption caused by sea-bottom trawling. This is not a modern technique: the fourteenth century saw a petition to regulate the use of the 'wondyrechaun' – essentially a wooden beam trawl used in shallow coastal waters (Roberts 2007) – but open-sea trawling came with steam-powered ships, and has continued to expand markedly in recent decades, moving into slope settings in waters approaching a depth of 1 km.

Sea-bottom trawling now affects some 15×10^6 km^2 each year (Gattuso *et al.* 2009) – representing most of the world's continental shelf area and also including significant areas of deep-water slope (Puig *et al.* 2012) and seamount surface. The process in effect ploughs the seafloor, producing a coarsening-upwards sedimentary signature (Palanques *et al.* 2001; M. Coughlan pers. comm.), with mud swept up into an expanded nepheloid layer and transported more distally, and nutrients redistributed (Dounas *et al.* 2007). Benthic assemblages are altered (Malakoff 2002) and some sensitive ones (e.g. deep-water coral systems) effectively destroyed (Sheppard 2006). Topographical effects may be substantial, with evident smoothing of topographical contours (Puig *et al.* 2012).

More recent extension of 'Worked Ground' into a marine setting can be seen with increased extraction of mineral resources including hydrocarbons and aggregates. It is only since the 1940s that technology and economics has made offshore extraction of hydrocarbons feasible, and it has grown to the point where it currently accounts for about 30% of total global output. Aggregate extraction significantly modifies the marine landscape, causes habitat modification and impacts on benthic communities both within, and downcurrent, of extraction sites, and can significantly change sediment fluxes, potentially starving supplies of sand to coastal areas. Similar concerns are being raised about offshore wind turbine construction, an even more recent and expanding innovation.

Within tropical waters, bleaching of coral reefs in response to rising sea temperatures, in addition to other stressors, such as increased turbidity of marine waters due to runoff, the fishing process of dynamiting reefs and, ultimately, decreasing ocean pH (Tyrrell 2011), may lead to the extinction of whole reef systems, resulting in a drowned reef horizon.

In aggrading sedimentary systems, the resultant facies should have considerable preservation potential. It is of limited diachroneity, given the marked post-1950 expansion of many of the processes involved.

Deep sea

This is usually considered as those areas where water is >200 m deep (i.e. largely below wave base and off the continental shelf edge) and might be simplified into two main systems: the clastic wedges of turbidite fans and contourite drifts that fringe the continental masses; and the slowly accumulating deep-sea oozes that lie beyond. Both systems have been and continue to be affected by physical disturbance (e.g. by trawling, offshore mineral extraction), by input of particulate material ('litter') varying from micron to metre scale in size (and locally indeed larger, in the case of shipwrecks), by chemical contamination with both organic and inorganic substances, by effects associated with atmospheric CO_2 increase and warming (such as variations in pH and dissolved oxygen content), and by biological changes driven by all of the above processes, either directly or indirectly. The extent of these effects – all of which can affect the nature of sediments being deposited – have been qualitatively described but not yet rigorously mapped (Ramirez-Llodra *et al.* 2011). The stratigraphic signal is patchy but locally may be striking. As with the effects of urbanization, local signals go back millennia. Major expansions of activity and, hence, extent of stratigraphic imprint were associated with the Industrial Revolution at around 1800 CE and with the ongoing 'Great Acceleration' that started at about 1950 CE.

The accumulation of litter – material dropped overboard – has reached the level where it rivals the extent of ice-rafted debris (IRD) in scale (Ramirez-Llodra *et al.* 2011), and is now seen in most surveys of the seafloor, where it is easily distinct from the surrounding (mostly very-fine-grained) sediment. We suggest hence terming this material, sedimentologically, as human-rafted debris (HRD) to help characterize a deep-water facies of a putative Anthropocene Series. Given technical progress, it shows the kind of extremely high-resolution 'biostratigraphy' of human artefacts and products also seen on land (Ford *et al.* 2014). Hence, spreads of clinker (combustion products from the coal that powered steam ships) that were universally dumped on the seafloor in the period from about 1800 CE to 1950 CE – now colonized by a specific biota – might be regarded as immediately pre-Anthropocene in our provisional definition, while those with plastics, aluminium and other such more modern materials largely date from after 1950 CE (Ramirez-Llodra *et al.* 2011). In the distal, naturally slow-accumulating parts of the seafloor, such HRD from different centuries will, in effect, fall within and contribute to the same physical layer.

More broadly, within the clastic wedges, the pattern of turbidite/contourite deposition seems not yet to have been substantially affected by human activity; it is not clear that changes in sediment supply caused by large-scale anthropogenic modification of river systems (e.g. Syvitski & Kettner 2011) have yet filtered down to cause substantial change to deep-sea clastic systems, although we regard significant longer-term change as likely (see below). However, local effects include the triggering of turbidity currents (that may also rework HRD into concentrations: Ramirez-Llodra *et al.* 2011) by bottom trawling (Puig *et al.* 2012). In the longer-term, clastic shut-off caused by sea-level rise may be envisaged.

The slowly accumulating deep-ocean oozes beyond will, in addition to such accumulations, be influenced by anthropogenic change, via such signals as a lighter carbon isotopic content in foraminifera shells, from the burning of fossil fuels. Additional chemical signals such as those from anthropogenic organic pollutants or artificial radionuclides are rapidly (e.g. Robison *et al.* 2005) if unevenly (Buesseler *et al.* 2007) transported to the sea via aggregated sinking planktonic debris. However, the very slow accumulation rate over most of this realm means that this material is thoroughly intermixed, by bioturbation, with pre-Anthropocene sediment, precluding recognition at such scale of a distinct Anthropocene Series. Only in regions of significantly more rapid deposition (e.g. Al-Rousan *et al.* 2004) does such a potential Series emerge as a distinct entity with coherent upper (sedimenting) and lower surfaces. However, potential changes to ocean chemistry may result in more extensive anoxia, with eutrophic bottom conditions limiting bioturbation, and changes to the elevation of the Calcite Compensation Depth in response to reduced oceanic pH (Tyrrell 2011), producing a carbonate dissolution layer. In addition, the types of deep-sea mineral extraction planned, if put into practice (of manganese nodules, for instance), will cause widespread and distinct physical and biological modification.

Duration of the Anthropocene: the long-term perspective

The complexities of diachronous event and process boundaries, and scale-dependence effects, visible today, will largely or wholly disappear in any consideration of far future perspective.

We do not consider the Anthropocene as a short transitional phase to some kind of post-Anthropocene interval, even were there to be a catastrophic decrease in the global human population in the near future. Rather, we consider that the future course of geological evolution, with both natural and human feedbacks, will inevitably be shaped by the anthropogenic perturbations that have taken place to date. Thus, the Anthropocene has only just begun and will play out over geological rather than human timescales. The Toarcian and Paleocene–Eocene Thermal Maximum (PETM) (Zachos *et al.* 2005; Cohen *et al.* 2007) events may be regarded as comparable, with an initial perturbation of the carbon cycle, amplification by natural feedbacks including massive carbon release from ground to air, modulated by astronomical pacing (Kemp *et al.* 2005), and slow recovery over the order of 0.1–0.2 Myr. Although each of these events in detail represents a succession of distinct phases, each may also be (and are, in practice) regarded as a whole.

In detail, the Anthropocene departs from the Toarcian and PETM models in a number of ways. It is an incipient 'hyperthermal' in an icehouse rather than greenhouse world, and so the ultimate sea-level rise (barely begun) should give a stronger transgressive signal (Rahmstorf 2007) than that in an essentially ice-free world. Indeed, if the glacial–interglacial cycle is significantly perturbed (Tyrrell 2011) with ice loss that exceeds Quaternary norms, then the geologically rapid transgression that followed the collapse of the end-Ordovician glaciation (Brenchley *et al.* 1994) might be considered as a closer analogue (Zalasiewicz & Williams 2013).

The Anthropocene also has a biotic pattern where perturbations (habitat clearance, predation, transglobal rather than local species invasions) are not simply forced by climate and ocean chemistry; as with previous biotic revolutions, these will be geologically long-lasting quantitatively (i.e. regarding diversity measures) and effectively permanent qualitatively (with new lineages arising from survivors and invaders) (see also Barnosky 2013). This pattern is also unique in modification by unpredictable but likely important feedbacks, both planned and unplanned, within the perturbatory human system (Kellie-Smith & Cox 2011).

One might compare the scale of effects with those recently proposed (the 8.2 and 4.2 kyr events) to subdivide the Holocene Epoch into Ages (Walker *et al.* 2012; see also Gibbard & Walker 2013). As regards global climate, current effects (a $<1\ ^{\circ}C$ global temperature rise since the beginning of the twentieth century) might not be regarded as yet comparing with the 8.2 and 4.2 kyr events in magnitude. However, near-future temperature rises are projected to considerably exceed these (IPCC 2001, 2007), given the unprecedented and ongoing rise in greenhouse gas levels. Other signals (lithostratigraphic, biostratigraphic, chemostratigraphic), though, are already pronounced and, as an ensemble, have no parallel in Earth's stratigraphic history. Debate over the current formal significance of the Anthropocene will need to assess the importance of all the relevant signals, and this is not a trivial task.

Nevertheless, the unprecedented rate of change in its early stages (within a small part of a single interglacial phase) means that the lower boundary to deposits of Anthropocene facies will appear synchronous globally. One may develop the 'super-interglacial' concept of Broecker (1987) by envisaging a variety of stratigraphic signals that vary from 'event beds' (e.g. the urban lithostratigraphic signal), to longer-lasting perturbations of chemical cycles and related effects on global temperature and sea level, to the effectively permanent changes to the course of the Earth's biotic evolution.

Discussion

How might the Anthropocene be characterized? Clearly, it is not simply by the appearance of anthropogenic signals in the stratigraphic record, as these are diachronous, locally dating back to earlier parts of the Holocene and, indeed, into pre-Holocene deposits. Such early records have been used in favour of an 'early Anthropocene' hypothesis that encompasses much of the Holocene (e.g. Ruddiman 2003, 2013) and also in criticism of the attempt to define an Anthropocene unit in stratigraphy at all (e.g. Gale & Hoare 2012; Gibbard & Walker 2013).

The key question seems to be whether the present-day Earth system now has been changed (by whatever agent) sufficiently in scale and permanence to justify a new geological time interval. If that is the case, one also may accept that the change from a putative pre-Anthropocene to an Anthropocene state has taken place non-instantaneously and diachronously. Most changes to the Earth system in our planet's history have been neither instantaneous nor globally synchronous (e.g. Williams *et al.* 2013), and most established geological time boundaries have been compromises – generally vigorously debated – of one sort or another.

The task then becomes one of finding the most effective – or, if one prefers, the least worst – criteria for defining a boundary. Then, one has to decide whether a boundary so defined can function effectively to define both a unit of time (an Anthropocene Epoch) and a body of strata (an Anthropocene Series). This is the question we examine here. We note that the further test for a formal Anthropocene – its use to both geological and arguably wider (Nature 2011; Vidas 2011; Zalasiewicz 2013) communities – falls outside the scope of this paper, as does the question – see above and Zalasiewicz *et al.* (2008, 2011, 2012) and Wolfe *et al.* (2013) – over whether a boundary, if agreed, is best defined by GSSP ('golden spike') or GSSA (numerical date).

It is clear that the material record of a putative Anthropocene Series, even considered with an approximately 1950 CE boundary, is locally distinctive and substantial – a feature reflecting the globally enhanced rates of erosion and sedimentation caused by humans (Hooke 2000; Wilkinson 2005;

Price *et al.* 2011; Syvitski & Kettner 2011). It is in many places also effectively distinguishable from pre-Anthropocene strata, on a decadal or even annual scale of resolution.

Elsewhere, however, the distinction of Anthropocene from pre-Anthropocene strata is less obvious. This may be because there are no significant markers or facies changes (as in desert dune strata, for instance). Or, it might reflect widespread irresolvable mixing of Anthropocene and pre-Anthropocene strata, through non-human bioturbation and other mixing processes (as in the deep ocean). Or there may have been protracted, complex human reworking of the ground (as in long-inhabited cities). Such phenomena prevent the clear, unambiguous and consistent delineation of a laterally continuous 'Anthropocene Series'. We may discuss them in turn as regards comparison with older chronostratigraphic units.

The local inability to unambiguously assign particular units of strata to chronostratigraphic units is a problem as old as is geology. One may take the case of the 'Permo-Triassic', long used as a descriptive bucket label given the difficulty of locating a boundary between Permian and Triassic deposits in 'red bed' deposits that lack fossils, even if it is as sharp and catastrophically founded as that between the Permian and Triassic systems (and between the Palaeozoic and Mesozoic erathems). Even in less stratigraphically opaque strata, chronostratigraphic boundaries, away from the reference 'golden spike' section, can rarely be located within an error bar of less than a few hundred thousand years (Zalasiewicz *et al.* 2013*a*). Most stratigraphic research is based upon the most informative and correlatable sections, but between these there are many stratal units within which major chronostratigraphic boundaries are located only approximately.

Similar uncertainty will certainly apply to an 'Anthropocene Series', with boundaries (now being placed at a decadal/annual scale) being effectively locatable in some places and more uncertainly placed in others. Hence, at least qualitatively, the Anthropocene shares the correlation problems attached to chronostratigraphic units generally, and it is not yet clear whether it possesses these kinds of uncertainties in greater measure than do the established units of the Geological Time Scale.

The problem of the disruption of superposition is rather different. This arises in part out of the exceedingly short timescale of the Anthropocene (to date) and in part out of complex, intermingled sedimentary geometries commonly created by human activity, where clear principles of superposition cannot be applied. This creates situations that archaeologists, for instance, are more familiar with, in discriminating numerous successive historical events within geometrically complex deposits (Edgeworth 2013) and on palimpsest surfaces (where the evidence from different phases of human history is preserved upon essentially two-dimensional surfaces).

The practice of basing chronostratigraphic subdivision upon the principle of superposition reflects the tendency on Earth for thick successions of strata to have built up, virtually since the origin of the planet. For most of the geological record it is an effective means to build and operate the geological timescale, and in older rocks, where stratigraphic uncertainties are measured in millions of years, the superpositional blurring through bioturbation and allied processes may be regarded as negligible. In such circumstances, chronostratigraphy and geochronology have operated in parallel, in their long-established 'dual hierarchy'.

However, at brief geological timescales and/or when extremely fine temporal resolution is sought, disruption of superpositional relationships may become a practical, rather than theoretical problem. This is already the case in the discrimination of high-resolution climate histories from deep seafloor deposits, where those strata with the highest sedimentation rates (and therefore least prone to bioturbational mixing) are actively sought. This phenomenon is, hence, most acutely expressed in the Anthropocene, with its extremely short timescale exacerbated by its peculiarly human-made complex stratal geometries. It might be regarded as a problem as much inherent of chronostratigraphic practice as it is of the Anthropocene.

Nevertheless, despite the complicating effects of these various processes, we propose that a reasonably consistent Holocene–Anthropocene boundary placed at around 1950 CE might be effectively traceable over large areas in both marine and non-marine settings. Attempts to consistently trace and delineate such a unit would reveal the extent to which this proposal is true. They would also help in the understanding of the extraordinary episode of history – whether formalized in stratigraphy or not – which the Earth is currently experiencing.

Conclusions

- A material 'Anthropocene Series' might be defined with a historically recent boundary at approximately 1950 CE, characterized by time proxies such as artificial radionuclides, biostratigraphic changes and human-made novel materials (e.g. plastics and uncombined aluminium). It locally forms substantial, distinct and correlatable sediment bodies in both terrestrial and marine realms.
- Locally, too, Anthropocene deposits so defined are difficult to recognize and correlate for want

of appropriate time markers to fix the boundary. These are analogous to stratigraphically indeterminate deposits in the older stratigraphic record.

- Commonly, also, Anthropocene deposits are difficult to separately recognize as distinct units because of intermixing, for instance by human or non-human bioturbation, reflecting the very short duration of the Anthropocene. This may be regarded as a problem inherent in very high-resolution chronostratigraphy as much as one of the Anthropocene.
- Attempts to better delineate and analyse the material expression of the Anthropocene will increase our understanding of the phenomenon as a whole.

C. Waters publishes with the permission of the Executive Director, British Geological Survey, Natural Environment Research Council. We thank W. Steffen and M. Ellis for thorough and helpful reviews.

References

Allen, J. R. L. 1988. Modern-period muddy sediments in the Severn Estuary (southwestern U.K.): a pollutant-based model for dating and correlation. *Sedimentary Geology*, **58**, 1–21.

Al-Rousan, S., Pätzold, J., Al-Moghrabi, S. & Wefer, G. 2004. Invasion of anthropogenic CO_2 recorded in planktonic foraminifera from the northern Gulf of Aquaba. *International Journal of Earth Sciences*, **93**, 1066–1076, http://dx.doi.org/10.1007/s00531-004-0433-4

Anderson, D. M. 2001. Attenuation of millenial-scale events by bioturbation in marine sediments. *Palaeoceanography*, **16**, 352–357.

Appleby, P. G. 2008. Three decades of dating recent sediments by fallout radionuclides: a review. *The Holocene*, **18**, 83–93.

Autin, W. J. & Holbrook, J. M. 2012. Is the Anthropocene an issue of stratigraphy or pop culture? *GSA Today*, **22**, 60–61.

Bacon, A. R., Richter, D. deB, Bierman, P. R. & Rood, D. H. 2012. Coupling meteoric ^{10}Be with pedogenic losses of ^{9}Be to improve soil residence times on an ancient North American interfluve. *Geology*, **40**, 847–850.

Barnosky, A. D. 2008. Megafauna biomass tradeoff as a driver of Quaternary and future extinctions. *Proceedings of the National Academy of Sciences of the United States of America*, **105**, 11,543–11,548.

Barnosky, A. D. 2013. Palaeontological evidence for defining the Anthropocene. *In*: Waters, C. N., Zalasiewicz, J. A., Williams, M., Ellis, M. A. & Snelling, A. M. (eds) *A Stratigraphical Basis for the Anthropocene*. Geological Society, London, Special Publications, **395**, first published online October 24, 2013, http://dx.doi.org/10.1144/SP395.6

Barnosky, A. D., Matzke, N. *et al.* 2011. Has the Earth's sixth mass extinction already arrived? *Nature*, **471**, 51–57.

Barnosky, A. D., Hadly, E. A. *et al.* 2012. Approaching a state-shift in the biosphere. *Nature*, **486**, 52–56.

Bollhöfer, A. & Rosman, K. J. R. 2000. Isotopic source signatures for atmospheric lead: the Southern Hemisphere. *Geochimica Cosmochimica Acta*, **64**, 3251–3262.

Brenchley, P. J., Marshall, J. D. *et al.* 1994. Bathymetric and isotopic evidence for a short-lived Late Ordovician glaciation in a greenhouse period. *Geology*, **22**, 295–298.

Brown, A. G., Toms, P., Carey, C. & Rhodes, E. 2013. Geomorphology of the Anthropocene: time-transgressive discontinuities of human-induced alluviation. *Anthropocene*, **1**, 3–13, http://dx.doi.org/10.1016/j.ancene.2013.06.002

Broecker, W. S. 1987. *How to Build a Habitable Planet*. Eldigio Press, New York.

Buesseler, K. O., Lamborg, C. H. *et al.* 2007. Revisiting carbon flux through the ocean's twilight zone. *Science*, **316**, 567–570.

Carter, R. M. 2007. Stratigraphy into the 21st century. *Stratigraphy*, **4**, 187–194.

Certini, G. & Scalenghe, R. 2011. Anthropogenic soils are the golden spikes for the Anthropocene. *The Holocene*, **21**, 1269–1274.

Cohen, A. S., Coe, A. L. & Kemp, D. B. 2007. The Late Palaeocene-Early Eocene and Toarcian (Early Jurassic) carbon isotope excursions: a comparison of their time scales, associated environmental changes, causes and consequences. *Journal of the Geological Society, London*, **164**, 1093–1108.

Crutzen, P. J. 2002. Geology of Mankind. *Nature*, **415**, 23.

Crutzen, P. J. & Stoermer, E. F. 2000. The 'Anthropocene'. *Global Change Newsletter*, **41**, 17–18.

Dounas, C., Davies, I. *et al.* 2007. Large-scale impacts of bottom trawling on shelf primary productivity. *Continental Shelf Research*, **27**, 2198–2210.

Edgeworth, M. 2013. The relationship between archaeological stratigraphy and artificial ground and its significance in the Anthropocene. *In*: Waters, C. N., Zalasiewicz, J. A., Williams, M., Ellis, M. A. & Snelling, A. M. (eds) *A Stratigraphical Basis for the Anthropocene*. Geological Society, London, Special Publications, **395**, first published online October 25, 2013, http://dx.doi.org/10.1144/SP395.3

Edwards, K. J. & Whittington, G. 2001. Lake sediments, erosion and landscape change during the Holocene in Britain and Ireland. *Catena*, **42**, 143–173.

Ellis, E. C. 2011. Anthropogenic transformation of the terrestrial biosphere. *Philosophical Transactions of the Royal Society*, **A369**, 1010–1035.

Ellis, E. C., Antill, E. C. & Kreft, H. 2012. All is not loss: plant biodiversity in the Anthropocene. *PLoS ONE*, **7**, e30535.

Elsig, J., Schmitt, J. *et al.* 2009. Stable isotope constraints on Holocene carbon cycle changes from an Antarctic ice core. *Nature*, **461**, 507–510.

Ford, J. R., Price, S. J., Cooper, A. H. & Waters, C. N. 2014. An assessment of lithostratigraphy for anthropogenic deposits. *In*: Waters, C. N., Zalasiewicz, J. A., Williams, M., Ellis, M. A. & Snelling, A. M. (eds) *A Stratigraphical Basis for the Anthropocene*.

Geological Society, London, Special Publications, **395**, first published online January 23, 2014, http://dx.doi.org/10.1144/SP395.12

Gale, S. J. & Hoare, P. G. 2012. The stratigraphic status of the Anthropocene. *The Holocene*, **22**, 1491–1494, http://dx.doi.org/10.1177/0959683612449764

Gałuszka, A., Migaszewski, Z. M. & Zalasiewicz, J. 2013. Assessing the Anthropocene with geochemical methods. *In*: Waters, C. N., Zalasiewicz, J. A., Williams, M., Ellis, M. A. & Snelling, A. M. (eds) *A Stratigraphical Basis for the Anthropocene*. Geological Society, London, Special Publications, **395**, first published online October 24, 2013, http://dx.doi.org/10.1144/SP395.5

Gattuso, J. P., Smith, S. V., Hogan, C. M. & Duffy, J. E. 2009. Coastal zone. *In*: Cleveland, C. J. (ed.) *Encyclopedia of Earth*. Environmental Information Coalition, National Council for Science and the Environment, Washington, DC. (First published in the *Encyclopedia of Earth*, see http://www.eoearth.org/article/Coastal_zone)

Gibbard, P. L. & Walker, M. J. C. 2013. The term 'Anthropocene' in the context of formal geological classification. *In*: Waters, C. N., Zalasiewicz, J. A., Williams, M., Ellis, M. A. & Snelling, A. M. (eds) *A Stratigraphical Basis for the Anthropocene*. Geological Society, London, Special Publications, **395**, first published online October 25, 2013, http://dx.doi.org/10.1144/SP395.1

Gosse, J. C. & Phillips, F. M. 2001. Terrestrial in situ cosmogenic nuclides: theory and application. *Quaternary Science Reviews*, **20**, 1475–1560.

Goudie, A. S. 2009. Dust storms: recent developments. *Journal of Environmental Management*, **90**, 89–94.

Gradstein, F., Ogg, G., Schmitz, M. & Ogg, G. (eds) 2012. *A Geological Time Scale*. Elsevier, Amsterdam.

Haff, P. 2013. Technology as a geological phenomenon: implications for human well-being. *In*: Waters, C. N., Zalasiewicz, J. A., Williams, M., Ellis, M. A. & Snelling, A. M. (eds) *A Stratigraphical Basis for the Anthropocene*. Geological Society, London, Special Publications, **395**, first published online October 24, 2013, http://dx.doi.org/10.1144/SP395.4

Hancock, G. J., Leslie, C., Everett, S. E., Tims, S. G., Brunskill, G. J. & Haese, R. 2011. Plutonium as a chronomarker in Australian and New Zealand sediments: a comparison with ^{137}Cs. *Journal of Environmental Radioactivity*, **102**, 919–929.

Hancock, G. J., Tims, S. G., Fifield, L. K. & Webster, I. T. 2014. The release and persistence of radioactive anthropogenic nuclides. *In*: Waters, C. N., Zalasiewicz, J. A., Williams, M., Ellis, M. A. & Snelling, A. M. (eds) *A Stratigraphical Basis for the Anthropocene*. Geological Society, London Special Publications, **395**, first published online February 28, 2014, http://dx.doi.org/10.1144/SP395.15

Hay, W. W. 2013. *Experimenting on a Small Planet: A Scholarly Entertainment*. Springer, Berlin.

Higgins, S., Overeem, I., Tanaka, A. & Syvitski, J. P. M. 2013. Land subsidence at aquaculture facilities in the Yellow River delta, China. *Journal of Geophysical Research*, **40**, 3898–3902.

Holtgrieve, G. W., Schindler, D. E. *et al.* 2011. A coherent signature of anthropogenic nitrogen deposition to remote watersheds of the northern hemisphere. *Science*, **334**, 1546–1548.

Hooke, R. LeB. 2000. On the history of humans as geomorphic agents. *Geology*, **28**, 843–846.

IPCC 2001. *Climate Change 2001: The Scientific Basis. Contribution of Working Group I to the Third Assessment Report of the Intergovernmental Panel on Climate Change* (Houghton, J. T. Y., Ding, D. J. *et al.* eds). Cambridge University Press, Cambridge.

IPCC 2007. *Contribution of Working Groups I, II and III to the Fourth Assessment Report of the Intergovernmental Panel on Climate Change* (Core Writing Team, Pachauri, R. K., Reisinger, A. eds) *Intergovernmental Panel on Climate Change (IPCC)*, Geneva. Available at: http://www.ipcc.ch/SPM2feb07.pdf

Kaplan, J. O., Krumhardt, K., Ellis, E. C., Ruddiman, W. F., Lemmen, C. & Goldewijk, K. K. 2011. Holocene carbon emissions as a result of anthropogenic land cover change. *The Holocene*, **21**, 775–791.

Kellie-Smith, O. & Cox, P. M. 2011. Emergent dynamics of the climate-economy system in the Anthropocene. *Philosophical Transactions of the Royal Society*, **A369**, 868–886.

Kemp, D. B., Coe, A. L., Cohen, A. S. & Schwark, L. 2005. Astronomical pacing of methane release in the Early Jurassic period. *Nature*, **437**, 396–399.

Kramer, N., Wohl, E. E. & Harry, D. L. 2011. Using ground penetrating radar to 'unearth' buried beaver dams. *Geology*, **40**, 43–46.

Kruge, M. A. 1999. Molecular organic geochemistry of New York Bight sediments. Sources of biogenic organic matter and polycyclic aromatic hydrocarbons. *Northeastern Geology and Environmental Sciences*, **21**, 121–128.

Kulkarni, A. V., Bahuguna, I. M., Rathore, B. P., Singh, S. K., Randhawa, S. S., Sood, R. K. & Dhar, S. 2007. Glacial retreat in Himalaya using Indian Remote Sensing satellite data. *Current Science*, **92**, 69–74.

Larmer, B. 2010. Shanghai Dreams. *National Geographic*, March, 124–141.

Lewin, J. 2012. Enlightenment and the GM floodplain. *Earth Surface Processes and Landforms*, **38**, 17–29, http://dx.doi.org/10.1002/esp.3230

Malakoff, D. 2002. Trawling's a drag for marine life, say studies. *Science*, **298**, 2123.

Mann, M. E. 2002. Little Ice Age. *In*: MacCracken, M. C. & Perry, J. S. (eds) *Encyclopedia of Global Environmental Change. Volume 1, The Earth System: Physical, Chemical Dimensions of Global Environmental Change*. John Wiley, Chichester, 504–509.

Marshall, W. A., Gehrels, W. R., Garnett, M. H., Freeman, S. P. H. T., Maden, C. & Xu, S. 2007. The use of 'bomb spike' calibration and high-precision AMS 14C analyses to date salt marsh sediments deposited during the last three centuries. *Quaternary Research*, **68**, 325–337.

Merritts, D., Walter, R. *et al.* 2011. Anthropocene streams and base-level controls from historic dams in the unglaciated mid-Atlantic region, USA. *Philosophical Transactions of the Royal Society*, **A369**, 976–1009.

Moulin, C. & Chiapello, I. 2006. Impact of human-induced desertification on the intensification of Sahel

dust emission and export over the last decades. *Geophysical Research Letters*, **33**, L18808.
MYERS, R. A. & WORM, B. 2003. Rapid worldwide depletion of predatory fish communities. *Nature*, **423**, 280–283.
NATURE (editorial). 2011. The human epoch. *Nature*, **473**, 254.
NAYAR, A. 2009. When the ice melts. *Nature*, **461**, 1042–1046.
NEW, M. G., LIVERMAN, D. M., BETTS, R. A., ANDERSON, K. L. & WEST, C. C. (eds) 2011. Four degrees and beyond: the potential of a global temperature increase of four degrees and its implications. *Philosophical Transactions of the Royal Society*, **A369**, (1934), 6–19.
PALANQUES, A., GUILLÉN, J. & PUIG, P. 2001. Impact of bottom trawling on water turbidity and muddy sediment of an unfished continental shelf. *Limnology and Oceanography*, **46**, 1100–1110.
POIRIER, C., CHAUMILLON, E. & ARNAUD, F. 2011. Siltation of river-influenced coastal environments: respective impact of late Holocene land use and high-frequency climate changes. *Marine Geology*, **290**, 51–62.
PRICE, S. J., FORD, J. R., COOPER, A. H. & NEAL, C. 2011. Humans as major geological and geomorphological agents in the Anthropocene: the significance of artificial ground in Great Britain. *Philosophical Transactions of the Royal Society*, **A369**, 1056–1084.
PUIG, P., CANALS, M. *ET AL.* 2012. Ploughing the deep sea floor. *Nature*, **489**, 286–289.
RAHMSTORF, S. 2007. A semi-empirical approach to projecting future sea-level rise. *Science*, **315**, 368–370.
RAMIREZ-LLODRA, E., TYLER, P. A. *ET AL.* 2011. Man and the last great wilderness: human impact on the deep sea. *Plos One*, **6**, (8), e22588, http://dx.doi.org/10.1371/journal.pone.0022588
REMANE, J., BASSETT, M. G., COWIE, J. W., GOHRBANDT, K. H., LANE, H. R., MICHELSON, O. & NAIWEN, W. 1996. Revised guidelines for the establishment of global chronostratigraphic standards by the International Commission on Stratigraphy (ICS). *Episodes*, **19**, 77–81.
REVKIN, A. 1992. *Global Warming: Understanding the Forecast*. American Museum of Natural History, Environmental Defense Fund, Abbeville Press, New York.
RICHTER, D. 2007. Humanity's transformation of Earth's soil: pedology's new frontier. *Soil Science*, **172**, 957–967.
ROBERTS, C. 2007. *The Unnatural History of the Sea: The Past and Future of Humanity and Fishing*. Island Press, Washington, DC.
ROBISON, B. H., REISENBICHLER, K. R. & SHERLOCK, R. E. 2005. Giant larvacean houses: rapid carbon transport to the deep sea floor. *Science*, **308**, 1609–1611.
ROCKSTRÖM, J., STEFFEN, W. *ET AL.* 2009. A safe operating space for humanity. *Nature*, **461**, 472–475.
RUDDIMAN, W. F. 2003. The Anthropogenic Greenhouse Era began thousands of Years Ago. *Climatic Change*, **61**, 261–293.
RUDDIMAN, W. F. 2013. The Anthropocene. *Annual Review of Earth and Planetary Sciences*, **41**, 45–68, http://dx.doi.org/10.1146/annurev-earth-050212-123944
SHEPPARD, C. 2006. Trawling the sea bed. *Marine Pollution Bulletin*, **52**, 831–835.
SHOULDERS, S. J. & CARTWRIGHT, J. A. 2004. Constraining the depth and timing of large-scale conical sandstone intrusions. *Geology*, **32**, 661–664.
SMITH, D. M., ZALASIEWICZ, J. A., WILLIAMS, M., WILKINSON, I., REDDING, M. & BEGG, C. 2010. Holocene drainage of the English Fenland: roddons and their environmental significance. *Proceedings of the Geologists' Association*, **121**, 256–269.
STANLEY, D. J. 1996. Nile delta: extreme case of sediment entrapment on a delta plain and consequent coastal land loss. *Marine Geology*, **129**, 189–195.
STEFFEN, W., CRUTZEN, P. J. & MCNEILL, J. R. 2007. The Anthropocene: are humans now overwhelming the great forces of Nature? *Ambio*, **36**, 614–621.
STEFFEN, W.Å. P., DEUTSCH, L. *ET AL.* 2011. The Anthropocene: from global change to planetary stewardship. *Ambio*, **40**, 739–761.
SYVITSKI, J. P. M. & KETTNER, A. 2011. Sediment flux and the Anthropocene. *Philosophical Transactions of the Royal Society*, **A369**, 957–975.
TANAKA, K. L. & HARTMANN, W. K. 2012. Chapter 15. The planetary time scale. *In*: GRADSTEIN, F. M., OGG, J., SCHMITZ, M. & OGG, G. (eds) *The Geological Time Scale 2012*. Elsevier, Amsterdam, 275–298.
TÖRNQVIST, T. E., WALLACE, D. J. *ET AL.* 2008. Mississippi Delta subsidence primarily caused by compaction of Holocene strata. *Nature Geoscience*, **1**, 173–176.
TYRRELL, T. 2011. Anthropogenic modification of the oceans. *Philosophical Transactions of the Royal Society*, **A369**, 887–908.
VANE, C. H., CHENERY, S. R., HARRISON, I., KIM, A. W., MOSS-HAYES, V. & JONES, D. G. 2011. Chemical signatures of the Anthropocene in the Clyde estuary, UK: sediment-hosted Pb, 207/206 Pb, total petroleum hydrocarbon and polychlorinated biphenyl pollution records. *Philosophical Transactions of the Royal Society*, **A369**, 1085–1111.
VIDAS, D. 2011. The Anthropocene and the international law of the sea. *Philosophical Transactions of the Royal Society*, **A369**, 909–925.
WALKER, M., JOHNSEN, S. *ET AL.* 2009. Formal definition and dating of the GSSP (Global Stratotype Section and Point) for the base of the Holocene using the Greenland NGRIP ice core, and selected auxiliary records. *Journal of Quaternary Science*, **24**, 3–17.
WALKER, M. J. C., BERKELHAMMER, M., BJÖRK, S., CWYNAR, L. C., FISHER, D. A., LONG, A. J. & LOWE, J. J. 2012. Formal subdivision of the Holocene Series/Epoch: a discussion paper by a Working Group of INTIMATE (Integration of ice-core, marine and terrestrial records) and the Subcommission on Quaternary Stratigraphy (International Commission on Stratigraphy). *Journal of Quaternary Studies*, **27**, 649–659, http://dx.doi.org/10.1002/jqs.2565
WATERS, C. N., ZALASIEWICZ, J. A., WILLIAMS, M., ELLIS, M. A. & SNELLING, A. M. 2014. A stratigraphical basis for the Anthropocene? *In*: WATERS, C. N., ZALASIEWICZ, J. A., WILLIAMS, M., ELLIS, M. A. & SNELLING, A. M. (eds) *A Stratigraphical Basis for the Anthropocene*. Geological Society, London, Special Publications, **395**, http://dx.doi.org/10.1144/SP395.18
WILKINSON, B. H. 2005. Humans as geologic agents: a deep-time perspective. *Geology*, **33**, 161–164.

Wilkinson, I. P., Poirier, C., Head, M. J., Sayer, C. D. & Tibby, J. 2014. Micropalaeontological signatures of the Anthropocene. *In*: Waters, C. N., Zalasiewicz, J. A., Williams, M., Ellis, M. A. & Snelling, A. M. (eds) *A Stratigraphical Basis for the Anthropocene*. Geological Society, London, Special Publications, **395**, first published online January 31, 2014, http://dx.doi.org/10.1144/SP395.14

Williams, M., Zalasiewicz, J. A., Waters, C. N. & Landing, E. 2013. Is the fossil record of complex animal behaviour a stratigraphical analogue for the Anthropocene? *In*: Waters, C. N., Zalasiewicz, J. A., Williams, M., Ellis, M. A. & Snelling, A. M. (eds) *A Stratigraphical Basis for the Anthropocene*. Geological Society, London, Special Publications, **395**, first published online October 25, 2013, http://dx.doi.org/10.1144/SP395.8

Wolfe, A. P., Hobbs, W. O. *et al.* 2013. Stratigraphic expressions of the Holocene–Anthropocene transition revealed in sediments from remote lakes. *Earth-Science Reviews*, **116**, 17–34.

Yan, P., Shi, P., Gao, S., Chen, L., Zhang, X. & Bai, L. 2002. ^{137}Cs dating of lacustrine sediments and human impacts on Dalian Lake, Qinghai Province, China. *Catena*, **47**, 91–99.

Zachos, J. C., Röhl, U. *et al.* 2005. Rapid acidification of the ocean during the Paleocene-Eocene thermal maximum. *Science*, **308**, 1611–1615.

Zalasiewicz, J. 2013. The epoch of humans. *Nature Geoscience*, **6**, 8–9.

Zalasiewicz, J. & Williams, M. 2013. The Anthropocene: a comparison with the Ordovician-Silurian boundary. *Rendiconti Lincei – Scienze Fisiche e Naturali*, http://dx.doi.org/10.1007/s12210-013-0265-x

Zalasiewicz, J. A., Smith, A. *et al.* 2004. Simplifying the stratigraphy of time. *Geology*, **32**, 1–4.

Zalasiewicz, J., Williams, M. *et al.* 2007. The scale-dependence of strata-time relations: implications for stratigraphic classification. *Stratigraphy*, **4**, 139–144.

Zalasiewicz, J., Williams, M. *et al.* 2008. Are we now living in the Anthropocene?. *GSA Today*, **18**, 4–8.

Zalasiewicz, J., Williams, M. *et al.* 2011. Stratigraphy of the Anthropocene. *Philosophical Transactions of the Royal Society*, **A369**, 1036–1055.

Zalasiewicz, J., Crutzen, P. J. & Steffen, W. 2012. Chapter 32. Anthropocene. *In*: Gradstein, F., Ogg, G., Schmitz, M. & Ogg, G. (eds) *A Geological Time Scale 2012*. Elsevier, Amsterdam, 1033–1040.

Zalasiewicz, J., Cita, M. B., Hilgen, F., Pratt, B. R., Strasser, A., Thierry, J. & Weissert, H. 2013*a*. Chronostratigraphy and geochronology: a proposed realignment. *GSA Today*, **23**, 4–8.

Zalasiewicz, J., Kryza, R. & Williams, M. 2013*b*. The mineral signature of the Anthropocene. *In*: Waters, C. N., Zalasiewicz, J. A., Williams, M., Ellis, M. A. & Snelling, A. M. (eds) *A Stratigraphical Basis for the Anthropocene*. Geological Society, London, Special Publications, **395**, first published online December 19, 2013, http://dx.doi.org/10.1144/SP395.2

An assessment of lithostratigraphy for anthropogenic deposits

J. R. FORD*, S. J. PRICE, A. H. COOPER & C. N. WATERS

British Geological Survey, Environmental Science Centre, Nicker Hill, Keyworth, Nottingham NG12 5GG, UK

**Corresponding author (e-mail: jford@bgs.ac.uk)*

Abstract: The deliberate anthropogenic movement of reworked natural and novel manufactured materials represents a novel sedimentary environment associated with mining, waste disposal, construction and urbanization. Anthropogenic deposits display distinctive engineering and environmental properties, and can be of archaeological importance. This paper shows that temporal changes in the scale and lithological character of anthropogenic deposits may be indicative of the Anthropocene. However, the stratigraphy of such deposits is not readily described by existing classification schemes, which do not differentiate separate phases or lithologically distinct deposits beyond a local scale. Lithostratigraphy is a scalable, hierarchical classification used to distinguish successive and lithologically distinct natural deposits. Many natural and anthropogenic deposits exhibit common characteristics; they typically conform to the Law (or Principle) of Superposition and exhibit lithological distinction. The lithostratigraphical classification of surficial anthropogenic deposits may be effective, although defined units may be significantly thinner and far less continuous than those defined for natural deposits. Further challenges include the designation of stratotypes, accommodating the highly diachronous nature of anthropogenic deposits and the common presence of disconformities. International lithostratigraphical guidelines would require significant modification before being effective for the classification of anthropogenic deposits. A practical alternative may be to establish an 'anthrostratigraphical' approach, or 'anthrostratigraphy'.

Human activity has modified the geological structure of the Earth and continues to do so at an accelerating rate. Humans are now the major driving force behind geological change, responsible for man-made unconformities and the modification of sedimentary patterns. The anthropogenic modification of sedimentary patterns can be attributed to two overlapping processes (Price *et al.* 2011; Zalasiewicz *et al.* 2011): (1) the creation of novel sedimentary environments and sediments (artificially modified ground), and which is the focus of this paper; (2) modifications to natural sedimentary environments through processes such as damming, coastal reclamation or straightening of rivers (Syvitski & Kettner 2011).

In many novel sedimentary environments, such as urban areas, the landscape and shallow subsurface is dominated by anthropogenic processes, including erosion (i.e. excavation) and anthropogenic sedimentation (i.e. deposits of 'made ground'). This presents a range of potential environmental and engineering challenges including unpredictable ground conditions, geological hazards and contamination (Rosenbaum *et al.* 2003). Anthropogenically modified ground can also offer a record of landscape evolution and the impacts of humans on the natural environment. As such, a range of approaches exist to characterize and classify artificially modified ground to inform activities including land-use planning, development and archaeological study (Edgeworth 2013).

Establishing the geometrical shape and spatial relationships of artificially modified ground requires appropriate systems of characterization and classification. Existing stratigraphical classification schemes used for the geological mapping and modelling of artificially modified ground in Great Britain are largely based on morphogenetic attributes (British Geological Survey 1995; McMillan & Powell 1999; Price *et al.* 2004; Ford *et al.* 2010*a*; Price *et al.* 2011). Morphogenetic classification offers a practical and effective means of differentiating broadly defined classes of artificially modified ground, including worked ground, made ground and infilled ground. In contrast, a lithostratigraphical classification is used for bedrock and, increasingly, for natural superficial deposits based essentially on lithological characteristics and spatial relationships (Salvador 1994). Unlike lithostratigraphy, a morphogenetic approach does not generally allow different phases of artificial ground or lithologically distinct anthropogenic deposits to be differentiated, nor allow the changing magnitude of anthropogenic transformation of the landscape to be determined. A lithostratigraphical approach to classifying anthropogenic deposits could contribute to an improved understanding of the role of humans as major geological and geomorphological

From: Waters, C. N., Zalasiewicz, J. A., Williams, M., Ellis, M. A. & Snelling, A. M. (eds) 2014. *A Stratigraphical Basis for the Anthropocene*. Geological Society, London, Special Publications, **395**, 55–89.
First published online January 23, 2014, updated April 8, 2014, http://dx.doi.org/10.1144/SP395.12

agents in the Anthropocene (Price *et al.* 2011). The Anthropocene, if defined, will be a chronostratigraphical unit representing a specified time interval. Throughout the geological column, from the late twentieth century onwards, chronostratigraphical and lithostratigraphical schemes have been developed separately. Chronostratigraphical units have typically been recognized as a response to a significant global event, particularly in the context of changing biotic communities. Consequently, the study of the lithostratigraphy of anthropogenic deposits will not be used to define this new age. However, changes in the nature and extent of such deposits, discernible through lithostratigraphy, could provide one line of evidence to consider in deciding if, and how, such a time unit should be defined.

Novel sedimentary environments and anthropogenic deposits share many similarities with ancient depositional systems, including lithological characteristics and successions that conform locally to the Law (or Principle) of Superposition. This suggests that a lithostratigraphical approach may be applicable. However, several characteristics of anthropogenic deposits present a challenge to lithostratigraphical classification, for example:

- Terrestrial anthropogenic deposits vary greatly in both lateral and vertical extent. Although sequences can attain thicknesses of 65 m or more in Great Britain, component units are commonly only a few metres thick.
- By their nature, many anthropogenic units are strictly allostratigraphical; that is, defined and identified on the basis of bounding discontinuities. However, allostratigraphy has not been popularly applied in Great Britain (Rawson *et al.* 2002). The bounding discontinuities can be defined as unconformities (e.g. artificial deposits resting directly upon bedrock), disconformities (e.g. where there is a time gap between parallel layers of artificial deposits associated with either non-deposition or reworking of the deposits) or the present-day land surface.
- Many anthropogenic units, being surficial deposits, have no overlying strata. However, the associated landform may show characteristic features that can aid the definition of a unit.
- Anthropogenic deposits are commonly lithologically heterogeneous, with bulk compositions characterized by considerable quantities of novel, or manufactured, materials.

Although some of these challenges have been previously recognized in the case of natural superficial deposits, they are more marked in the case of anthropogenic deposits (McMillan 2005; McMillan *et al.* 2011).

The aim of this paper is to determine whether anthropogenic deposits can be classified using the same lithostratigraphical procedures employed for natural deposits. The purpose is not to define a lithostratigraphical classification for anthropogenic deposits, but to consider whether such an approach could be used. It discusses where, or in what circumstances, it would work and where it would fail. It also looks at how elements of a pure lithostratigraphical approach can be adapted or incorporated with other classification schemes to offer the functionality necessary to characterize, quantify and investigate the recent geological record, and contribute to the study of the Anthropocene.

This paper is largely based on the study of anthropogenic deposits in Great Britain, including their lithostratigraphical classification to support systematic survey and three-dimensional (3D) geological modelling. However, the value of lithostratigraphical classification in recording and interpreting anthropogenic successions in sections and boreholes and other 'non-geographical' contexts is also considered. Examples from diverse geographical settings highlight the global relevance of establishing effective systems for their classification.

Novel sedimentary environments, artificially modified ground and anthropogenic deposits

It has been recognized that anthropogenic processes that result in the creation of sediments, erosive features and associated landforms can be broadly classified in two main types (Szabó 2010; Price *et al.* 2011; Zalasiewicz *et al.* 2011). The first type comprises those anthropogenic activities that are deliberate, direct and intentional in their modification of the natural landscape. These activities include the excavation, transport and deposition of natural geological materials related to urban development, mineral exploitation, waste dumping and land reclamation to mention but a few. They are termed 'novel sedimentary environments' by Zalasiewicz *et al.* (2011), and are largely unclassified and poorly recorded or mapped. The second relates to the modification of natural sedimentary environments as an indirect consequence of anthropogenic processes such as agriculture, deforestation, damming and the straightening of rivers (Syvitski & Kettner 2011); these factors are not considered here.

Anthropogenic deposits created in novel sedimentary environments may be coeval with natural deposits and those resulting from modified sedimentary environments. The result is a potentially complex record of interaction and feedback between deliberate anthropogenic processes, natural 'background' sedimentation and processes

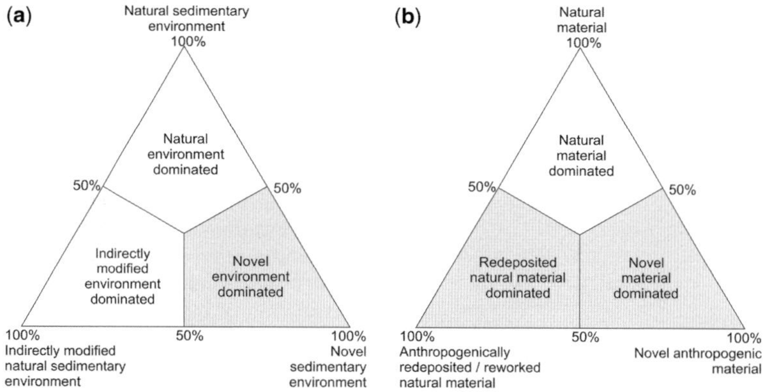

Fig. 1. (**a**) shows the relationship between the natural sedimentary environment, the natural sedimentary environment indirectly modified by human activities and the novel sedimentary environment. Anthropogenic deposits, the focus of this paper, are defined as those dominated by sediment deposited in the novel sedimentary environment (shaded area). (**b**) shows the relationship between novel and natural material in defining the bulk composition of anthropogenic sediments deposited in the novel sedimentary environment. In this diagram, the 'natural material' component reflects input from both natural sedimentary environments and those natural sedimentary environments indirectly modified by human processes. Anthropogenic deposits are defined as those dominated by novel material or redeposited natural material (shaded area).

such as soil formation (Bridgland *et al.* 2006). For the purpose of this paper and the lithostratigraphical classification of anthropogenic deposits, only those deposits that form in dominantly novel sedimentary environments are considered (Fig. 1a).

Novel sedimentary environments are characterized by the anthropogenic alteration of the natural land surface or subsurface through the deliberate creation of sediments (anthropogenic deposits) or voids (excavations) that are together referred to as artificially modified ground (artificial ground). Anthropogenic deposits are taken to include those natural rocks and superficial deposits intentionally moved and redeposited by humans, as well as deposits of novel anthropogenic materials created through manufacture or processing (Fig. 1b) (Norbury 2010). We include in artificial ground natural geological materials that have been displaced, such as spoil from mineral extraction or material moved through landscaping, an interpretation not recognized by Satkūnas *et al.* (2011).

Soils that are created by predominantly deliberate human action, including the addition of novel anthropogenic materials, are included as artificial ground, whereas those that develop through processes such as long-term agricultural cultivation are excluded. Natural materials transformed *in situ* by indirect processes as an unintentional consequence of anthropogenic activity – for example, compaction of soils through construction activity or through contamination by leachates – are not considered here as artificial ground.

Novel sedimentary environments and anthropogenic deposits share many similarities with their natural equivalents. Authors such as Sherlock (1922) propose that 'rocks made by Man' can be classified in a similar way to the classification of natural of rocks to include, igneous, metamorphic and sedimentary types. Examples of igneous anthropogenic rocks include glass, foundry slag and metals. Examples of anthropogenic metamorphic rocks include bricks and ceramics. Examples of anthropogenic sedimentary rocks include concrete. Underwood (2001) refers to those rocks made, modified or moved by humans as 'anthropic rocks' and proposes that they are considered in the rock cycle. Cathcart (2011) recognizes the global significance of anthropogenic erosion and deposition, and proposes that both processes are included in the rock cycle (Fig. 2).

Artificially modified ground within the global rock cycle is significant. It is estimated that the deliberate, annual global flow of anthropogenic sediments is 57 000 Mt (million tonnes), exceeding that of transport of natural sediments to the world's oceans by almost a factor of 3 (Douglas & Lawson 2001). In Great Britain alone it is estimated that over 66 530 Mt of sediment has been moved through the extraction and processing of major mineral resources in around 200 years (Price *et al.* 2011). Locally, Syvitski & Kettner (2011) estimated that approximately 400 Mt of crushed rock and engineering soils were used in one single land reclamation project to construct the Palm Jebal Ali near Dubai in the United Arab Emirates.

It is estimated by Price *et al.* (2011) that around 1.4% of mainland Great Britain is currently covered by significant areas of artificial ground (Fig. 3). By

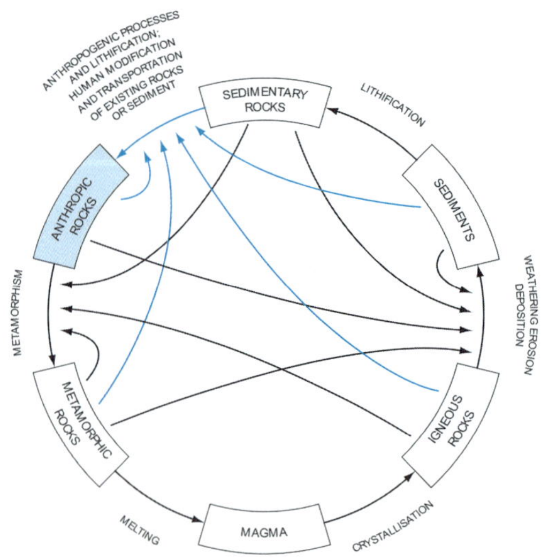

Fig. 2. The Earth's rock cycle including 'anthropic rocks' as suggested by Underwood (2001) and illustrated by Cathcart (2011). Boxes represent materials and arrows represent processes, with anthropogenic processes and deposits shown in blue. Modified after Cathcart (2011).

comparison, mapped alluvium covers only around twice this area when similar map coverage is considered. The extent of artificial ground is typically greatest in urban conurbations where the landscape has been affected by frequent phases of human activity ranging from locally comprehensive cover to an average of 8.2% for the London area (Price *et al.* 2010). About 7% of Great Britain's land cover is designated as urban (UK National Ecosystem Assessment 2011). These areas represent sediment sinks in the novel sedimentary environment, in which humans generate accommodation space for sedimentation by artificially increasing land levels, locally elevating the base level for anthropogenic sediment accumulation and creating environments that potentially favour the preservation of earlier urban strata (Holden *et al.* 2006; Rivas *et al.* 2006; Makedon *et al.* 2009). Rural areas typically show less than 1% artificial ground coverage and represent sediment source areas in the novel sedimentary environment, commonly producing much of the material destined for accumulation in urban and industrial conurbations. These figures, based on British Geological Survey (BGS) data that span several decades, are inevitably underestimates. Prior to the 1990s, artificial ground was not consistently depicted. Since then, the recording and representation of artificial ground on geological maps has generally improved. The current definition of artificial ground as used by the BGS generally refers to landforms and sediments deposited or excavated on or within the shallow ground surface. However, artificial ground is also created at deeper levels in the subsurface. Subsurface artificial ground can be created by processes including underground excavations for mineral extraction, installation of engineered infrastructure and deep burial of wastes. Human intervention and disturbance in the subsurface could be considered a form of 'anthroturbation', similar in nature to other forms of bioturbation.

Although landscaping and earthworks for the construction of buildings are included in this definition of artificially modified ground, as are deposits created from the rubble of former buildings, 'extant' buildings are generally excluded. Arguably, this is an arbitrary distinction as buildings, their foundations and associated earthworks are in physical continuity. This definition recognizes that buildings represent ephemeral sinks and sources of material in the rock cycle, but assumes that their long-term preservation potential is limited. However, in exceptional circumstances, buildings may be preserved relatively intact and form deposits in their own right. Where buildings become preserved in the geological record and structures that are indicative of their function are retained, they may be considered as trace fossils.

The classification of artificial ground including anthropogenic deposits

The importance of understanding the signature of novel sedimentary environments, including the spatial distribution and character of anthropogenic deposits, is recognized by a range of disciplines. For land-use planning, artificial ground represents a potential geological and environmental hazard (Rosenbaum *et al.* 2003). Increasingly, anthropogenic deposits are considered as potential material resources. Reuse and recycling options may include the use of anthropogenic deposits as engineered fill, industrial minerals and feedstock for primary mineral or metal recovery (Waters *et al.* 1996; Lottermoser 2011; Wang & Liu 2012). The value of determining the distribution of historical artificial ground is recognized by archaeological science in the study and management of heritage sites and deposits (Carver 1987; Historic Scotland 2011; English Heritage 2012). The classification and study of artificial ground offers an insight into the distribution and scale of human modification of the landscape and the wider environment.

As a result of this diversity of disciplines and interests, a range of approaches have been used or proposed for the classification of artificial ground. Existing approaches used in geosciences are generally concerned with mapping and modelling the

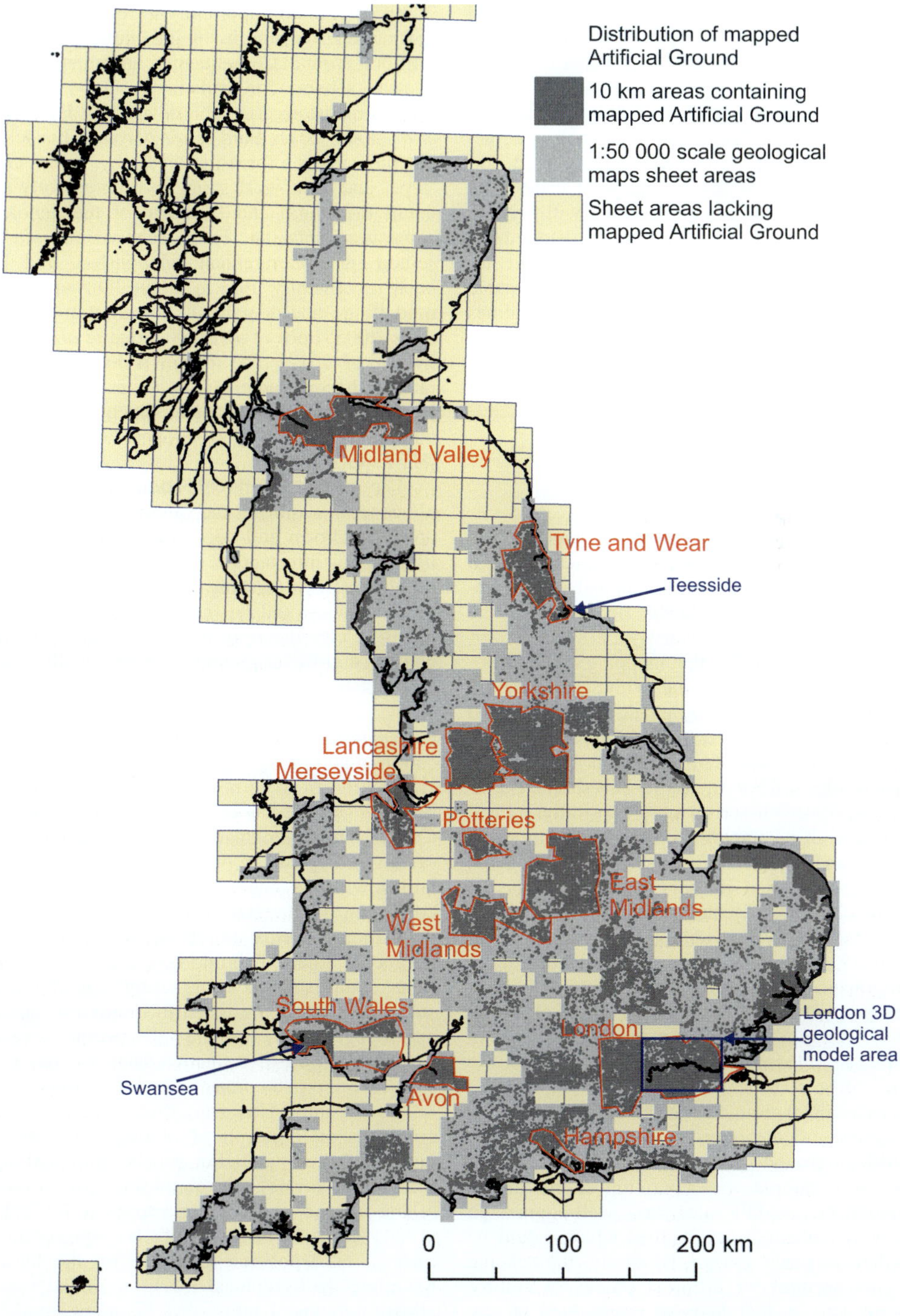

Fig. 3. Maps showing the extent of mapped artificial ground in Great Britain, based on BGS published 1:50 000 scale maps, modified after Price *et al.* (2011). Individual areas of artificial ground have been exaggerated for clarity. Red outlines denote potential domain areas within which artificial deposits may be attributed Group status in any lithostratigraphical classification of anthropogenic deposits. Localities indicated in blue are referred to in the text. DiGMapGB BGS ©NERC. Topographical data ©Crown Copyright Ordnance Survey. All rights reserved.

spatial distribution of artificial ground at a local scale, ranging from 1:10 000 to 1:50 000. Current schemes in use by geosciences are either morphostratigraphical (i.e. morphogenetic) or essentially chronostratigraphical, and no lithostratigraphical classification exists. Topographical surveys depict elements of artificial ground, including road embankments and waste tips, at a range of scales. The requirements of other disciplines vary, with archaeological studies typically operating at a detailed site scale, demanding higher spatial (vertical and lateral) resolution. However, there is considerable overlap between some areas of these disciplines and the geological recording and mapping of artificial ground. The main similarities and differences in the attributes they refer to are discussed below.

Artificial ground classification schemes in use by geoscience

The present British Geological Survey morphostratigraphical approach to artificial ground. Artificial ground has been shown on geological maps published by the BGS for Great Britain since the 1960s (British Geological Survey 1978). However, it was not until the 1990s that mapping of anthropogenic deposits became routine, with the application of a fivefold subdivision into made ground, worked ground, infilled ground, disturbed ground and landscaped ground (McMillan & Powell 1993, 1999; British Geological Survey 1995). This simple classification scheme is based on a morphostratigraphical approach with an emphasis on the landform (morphology) and, to a lesser extent, the anthropogenic process that created it (genesis). Landforms are identified through a combination of field observation and indirect evidence derived from the appraisal of spatial data sources, including aerial photographs, topographical maps including historical maps, contour information and digital elevation models. The combined use of recent spatial data and generations of legacy data allows historical land-use change to be considered and the most appropriate class of artificial ground to be chosen. However, the scheme does not distinguish between different types of artificial ground within each class, and does not allow, for example, the separation of potentially contaminated landfill sites and well-engineered motorway embankments, both of which would be classified as made ground. Similarly, the scheme does not account for different phases of anthropogenic activity that may be represented at any one location.

British Geological Survey enhanced classification scheme. The introduction of digital geological maps and 3D geological models in the 1990s and 2000s offered an opportunity to represent complex spatial relationships involving multiple phases of artificial ground. Consequently, an enhanced classification scheme for artificial ground in two and three dimensions has been devised by BGS (Price *et al.* 2004, 2011; Ford *et al.* 2010*a*). This scheme is structured as a three-tier hierarchy, using 'class', 'type' and 'unit' to describe in progressively more detail the origin and landform of the deposit or excavation (Fig. 4). These subdivisions are presented in a hierarchical relationship, and have been successfully used in urban 2D mapping, 3D modelling and database construction (Waters *et al.* 2005; Ford *et al.* 2008; Price *et al.* 2010; de Beer *et al.* 2012). The scheme is designed to interface with complementary schemes, including the National Land Use Database, which contains information on previously developed land in England (Harrison 2006).

The enhanced classification scheme is morphogenetic, requiring the identification of a diagnostic landform and does not directly incorporate lithological information. Although this scheme is capable of distinguishing different types of artificial ground that would be indivisible using the original BGS classification, it does not allow different phases of artificial ground or lithologically distinct anthropogenic deposits to be differentiated.

Chronostratigraphical classification. A classification of anthropogenic strata that provides the discrimination of materials and boundaries primarily in terms of time has been proposed by Nirei *et al.* (2012). They recognize the following unit types: (a) chronological or 'chrono-layers' of materials laid down in a single event; (b) 'material layers' consisting of materials laid down in one or more depositional events; (c) 'bundles' consisting of a number of adjacent chrono-layers; and (d) 'associations' comprising the whole assemblage of units at a site. The scheme, designed to be used to describe anthropogenic successions at specific localities with detailed site investigation data, provides a hierarchical approach comparable to that used in natural successions. However, this proposed scheme combines the concepts of lithology (the material layer) and time-parallel or geochronological layers (the chrono-layers). The chrono-layer most closely relates to the chronosome proposed by Schultz (1982) for rocks of diverse facies representing an interval of deposition identified on the basis of bounding stratigraphical markers. In the case of allostratigraphical units, that bounding marker is a discontinuity (North American Commission on Stratigraphic Nomenclature 2005). In the context of anthropogenic deposits, it is unlikely such boundary markers could be recognizable between disparate sites. Consequently, this scheme does not lend

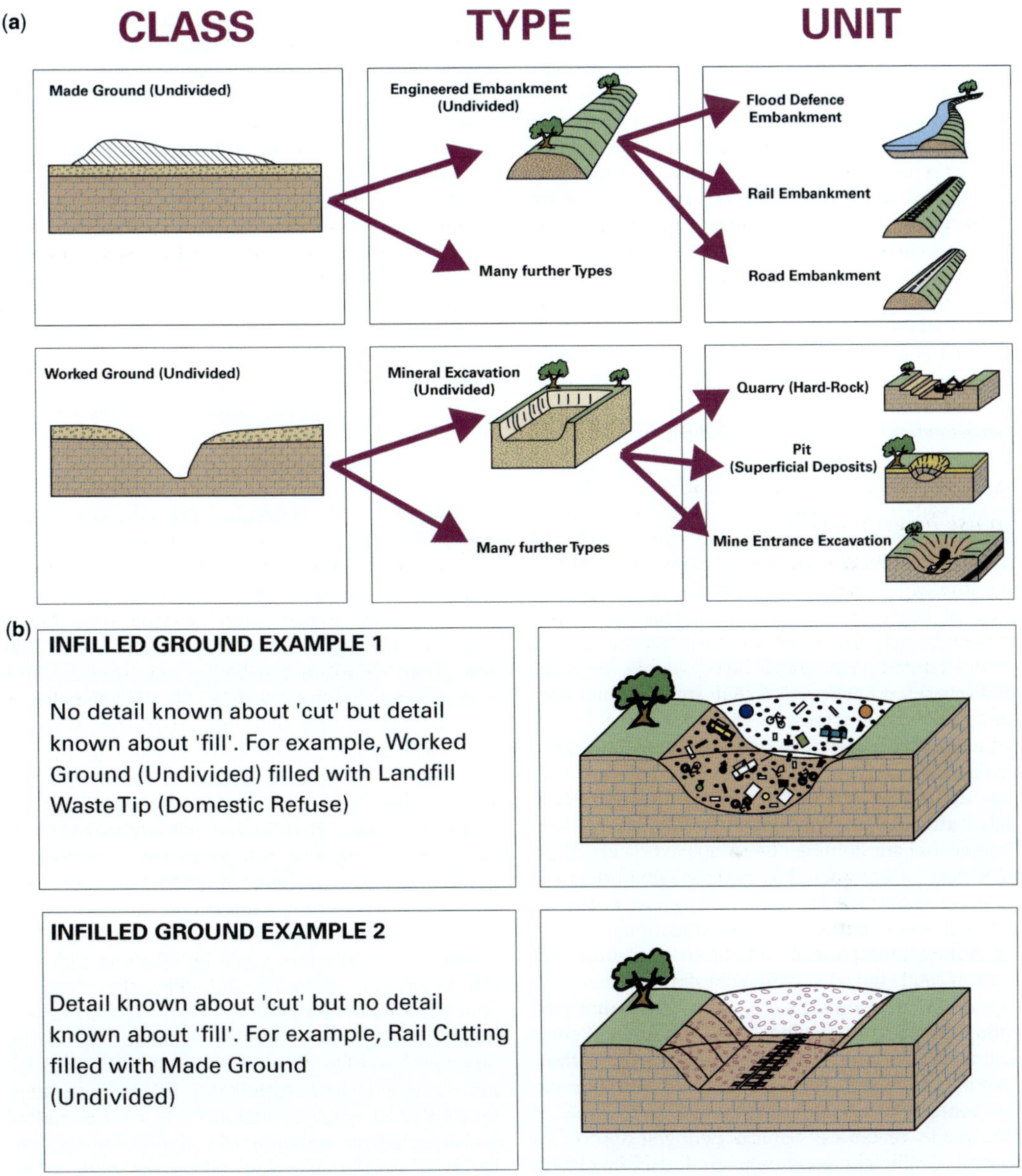

Fig. 4. Schematic diagram showing: (**a**) the structure of the enhanced classification scheme for artificial ground; and (**b**) examples of infilled ground classification derived from the scheme (Ford *et al.* 2010*a*).

itself to mapping and formal stratigraphy, nor the widespread classification of artificial ground.

Artificial ground classification schemes in use by other disciplines

Archaeological recording. Archaeological and geological observation, recording and classification of anthropogenic deposits share many similarities. Archaeologists often refer to the stratigraphical relationship of cultural deposits and structures on the basis of lithological variation, interpreted stratigraphical position and mode of formation. Archaeological description and assessment also considers cross-cutting structures such as foundations, ditches, cellars and basements (Edgeworth

2013). The recognition of assemblages of artefacts, their abundance, and first and last appearance within a sequence of archaeological deposits often provides the basis for the creation of stratification and seriation charts (Carver 1987). Chronology (chronostratigraphy), based on relative or absolute dating, provides the overarching method of correlation. The age of the materials and detail of the artefact content may be referred to as a precise age, an age range, a historical interval (i.e. late Victorian) or a prehistorical interval (i.e. Bronze Age). However, without considerable absolute or relative age information, the application of archaeological classification to the widespread classification of anthropogenic deposits may be problematic.

Classification of urban and industrial soils. The World Reference Base for Soils recognizes two major reference soils groups (RSGs) of anthropogenic soils: Anthrosols and Technosols (IUSS Working Group WRB 2006; Rossiter 2007). Of these, Technosols share some characteristics that are similar to anthropogenic deposits. They comprise at least 20% artefacts in the upper 100 cm of the ground surface (or down to continuous rock or a cemented or indurated layer, whichever is the shallower) or they are sealed by 'technic hard rock' (e.g. pavements and roads). Artefacts comprise novel processed or manufactured anthropogenic materials, and human-transported materials may include redeposited or reworked natural materials extracted from the earth by humans. Where Technosols are dominantly composed of artefacts, they may be considered as anthropogenic deposits in the context of this paper. At a Great Britain national level, hierarchical classification schemes for anthropogenic urban soils have been proposed. Hollis (1992) introduced two classes of urban soils called 'Made-ground soils' and 'Man-modified soils'. Aspects of the genesis and lithological character of Made-ground soils are analogous to anthropogenic deposits. They are formed through the mechanical removal, transport and deposition of man-made materials, natural pedogenic soils or natural geological parent material. Hollis (1992) proposed groups of 'man-made substrates' on the basis of their distinctiveness as anthropogenic artefacts within pedogenic soils, indicating whether they are chemically base-rich or base-poor reflecting their likely environmental impact. Although existing classifications or anthropogenic soils do not lend themselves to stratigraphy, they may provide useful markers in establishing a lithostratigraphy for artificial ground.

Topographical surveys. Topographical surveys with sequential generations of map information provide spatial information about types of artificial ground, ranging from many kinds of excavations and quarries through to engineered and tipped deposits. Maps show the locations and extent of built infrastructure including cuttings and embankments for castles, roads, railways, canals and docks. They show worked ground (quarries) mine and quarry waste (spoil tips), and other land-raising ground areas including waste disposal sites. Used in conjunction with other historical documents they can allow the genesis, and spatial and temporal distribution of some anthropogenic deposits to be delineated. Except for inference from certain types of mineral-processing spoil (i.e. coal or ironstone spoil tips), they provide little lithology information. Hence, topographical surveys offer a useful source of direct and indirect evidence to support the morpho- chrono- and lithostratigraphical classification of artificial ground.

National Land Use Database (NLUD) and Eurostat Land Use/Land Cover Area Frame Statistical Survey (LUCAS) classifications. The National Land Use Database (NLUD) is applicable to England, and characterizes the land into digital map-based polygons dependent on land use (Harrison 2006). Eurostat Land Use/Land Cover Area Frame Statistical Survey (LUCAS) classification is a similar European scheme, but with different subdivisions (Eurostat 2009). Both schemes divide the data into a two-tier hierarchy of Land Use and Land Cover, but have slightly different categories and subdivisions. Both include natural and agricultural land, along with categories for minerals and landfill or waste, transport, industrial and commercial land. Land-use classifications do not indicate the presence of artificial ground as such. Hence, land-use classifications could be of use in identifying areas of anthropogenic deposits, but, in general, do not help with their geological classification. The previous English scheme, called the National Land Use Classification (NLUC), was far more extensive and 'Although it has not been kept up to date, it arguably remains the most complete and detailed presentation of a nationally applicable land use classification' (Harrison 2006, p. 9). Elements of this extensive scheme could be used for subdividing some types of artificial ground, especially those associated with landscaped ground.

English Heritage *Thesaurus of Monument Types.* The *Thesaurus of Monument Types* includes hierarchical groupings and names for many types of building, archaeological structure, modern structure, mineral extraction sites, amongst others (English Heritage 2002). Ford *et al.* (2010*a*) suggested the use of this scheme for recording detail of artificial ground including earthworks and other constructions associated with archaeological sites.

However, it is debateable whether a geological recording scheme needs to subdivide to the same degree as the English Heritage scheme.

The applicability of lithostratigraphy to anthropogenic deposits

Lithostratigraphical procedure is essentially concerned with establishing the classification of rock bodies so that their geometrical shape and spatial relationships can be determined (Rawson *et al.* 2002). Lithostratigraphical units include bodies of sediment (or volcanic material), bedded or unbedded, that are defined and characterized on the basis of their lithological properties and their stratigraphical relationships (Salvador 1994). Globally, lithostratigraphical schemes have been systematically developed to facilitate the study, mapping and 3D geological modelling of bedrock and natural superficial deposits. However, no equivalent scheme exists for the characterization of anthropogenic deposits.

Potential difficulties in applying a lithostratigraphical scheme to anthropogenic deposits have, to some extent, already been faced for onshore natural superficial deposits (McMillan *et al.* 2005, 2011). Like natural superficial deposits, anthropogenic deposits are associated with a wide range of processes, they are commonly discontinuous, variable in thickness and poorly exposed. Furthermore, the regional significance of unconformities and discontinuities seen in sections or boreholes may be poorly understood. It is common that only a single thin unit is present and the deposits do not form a continuous stratigraphical succession. National or regional correlation may not be possible, and the definition of formations may be restricted to districts where correlation is secure. In the latter case, specific provisions have been necessary to account for the challenges associated with natural superficial deposits.

Notwithstanding these difficulties, lithostratigraphy provides an effective means of representing bedrock and natural superficial deposits, allowing different facies and phases of deposition to be represented at any one location and the geological history to be defined. Below, the applicability of lithostratigraphy to anthropogenic deposits is explored.

Lithostratigraphical procedures

If a lithostratigraphical framework is to be established for anthropogenic deposits it needs to conform, as far as possible, to international stratigraphical principles for lithostratigraphical classification (Salvador 1994). Regional application of these guidelines is published by the North American Commission on Stratigraphic Nomenclature (2005), and for Great Britain, most recently, by Rawson *et al.* (2002). The key requirements for a lithostratigraphical framework are:

- *Law of Superposition*: lithostratigraphical units are sedimentary (or volcanic) units that conform to the Law (or Principle) of Superposition, in that an undeformed succession is mainly deposited as layers with each layer younger than the one present beneath.
- *Original lateral continuity*: lithostratigraphical units should have, or originally have had, lateral continuity appropriate to the scale at which they are defined.
- *Lithological distinction*: lithostratigraphical units defined at the mappable scale should have distinguishing lithological characteristics that allow adjacent units to be differentiated. Units need not be lithologically homogeneous, although any variation should be described.
- *Type section*: a lithostratigraphical unit is defined by a type section (stratotype) or by type area. Where possible, the top and base should be defined, but it is recognized that the nature of these boundaries and the bounding deposits may vary laterally.
- *Hierarchical framework*: lithostratigraphical procedure is based on a hierarchical framework in which the 'Formation' is the primary mappable unit. Formations can be amalgamated into 'Groups', and then into 'Supergroups' or subdivided into 'Members', and then into 'Beds'.

Anthropogenic sediments and the Law of Superposition

The vast majority of anthropogenic deposits are formed by human processes that operate on the pre-existing landscape, including the gradual accumulation of material through settlement, industrialization, urbanization and waste disposal. Although evidence exists for submarine deposition (e.g. through the disposal of ship ballast: Boyce *et al.* 2009), anthropogenic sediments are most commonly deposited in terrestrial, subaerial environments.

Anthropogenic sediments may overlie earlier natural or anthropogenic deposits. In common with natural sedimentary deposits, they tend to form by aggradation at their upper bounding surface with successive layers representing younger depositional events (Fig. 5). Although the original lateral continuity, thickness and lithology of these layers may be characteristically different to those of natural sediments (as discussed in the following sections), they conform to the Law of Superposition. This axiom is a basis of archaeological recording and

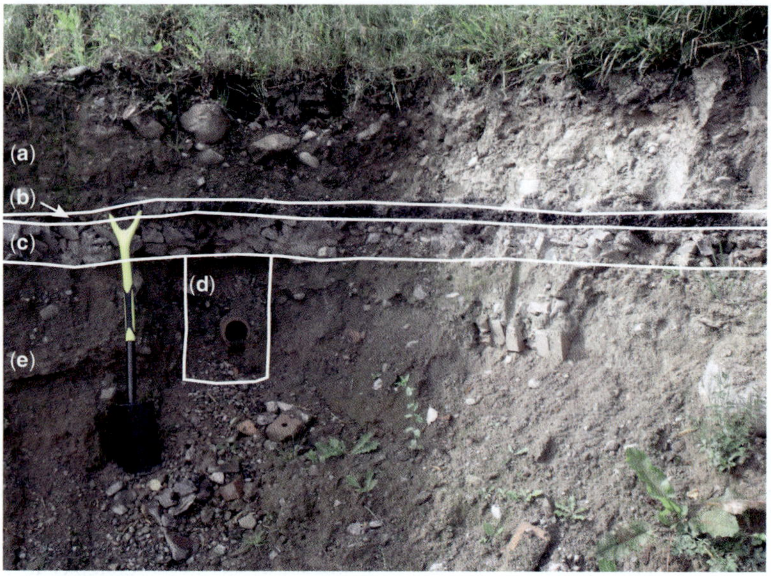

Fig. 5. Section through multiple layers of artificial ground showing superpositional and cross-cutting relationships analogous to those of natural sedimentary deposits. Units (a) and (e) comprise reworked, gravelly and cobbly silt and clay (till), unit (b) is tarmacadam, unit (c) comprises angular cobbles of concrete. Unit (e) is cut by a unit (d) and comprises reworked till used as back-fill for the installation of a clay pipe drain. Inverness area, UK [264135 844495]. Photograph by C. Auton BGS ©NERC 2012. All rights reserved.

interpretation, and is supported by a range of evidence including absolute and relative dating (see Edgeworth 2013).

Although multiple phases of anthropogenic deposition may be emplaced on a common surface, successive phases are often separated by periods of non-deposition, erosion (including excavation and natural down-cutting) and reworking. Although multiple units of anthropogenic deposit may be present at any single location, the preservation of complete or partial stratigraphical sequences is unlikely. The fragmented nature of anthropogenic sedimentary sequences presents a particular challenge to lithostratigraphical classification. Where evidence for relative field relationships is lacking, establishing the spatial relationships between deposits may require the use of chronostratigraphical or documentary evidence, including topographical and land-use information. Where natural and novel sedimentary environments interact, the stratigraphy of any intercalated or bounding natural deposits could be used to constrain the spatial relationships of the corresponding anthropogenic deposits.

The synsedimentary evolution of depositional environments may result in the distinguishing lithology or lithofacies of a single lithostratigraphical unit being deposited at different times across a sedimentary basin. Consequently, natural lithostratigraphical units commonly cut across time planes and boundaries defined by other stratigraphical classifications, including chronostratigraphy and allostratigraphy. Although regional unconformities or major hiatuses are used to separate lithostratigraphical units, local or minor hiatuses, disconformities or unconformities within a sequence are typically ignored (Murphy & Salvador 1999). By their nature, many anthropogenic units are strictly allostratigraphical: individual anthropogenic deposits are typically created in a single phase of deposition and their bounding surfaces, including local hiatuses and disconformities, *are* significant to their definition and identification.

With the exception of reworking and slumping (including the failure of engineered slopes in made ground), few immediate processes are likely to significantly disrupt the original superpositional nature of anthropogenic sediments. However, not all anthropogenic deposits conform to the Law of Superposition. Anthropogenic deposits created in the subsurface through activities such as mining, tunnelling and subsequent back-filling (Figs 5 & 6), or deposits resulting from the transfer of man-made materials into subsurface for storage or disposal, may exhibit complex relationships similar to natural intrusive rocks. By analogy, 'anthropogenic intrusives' that do not conform to the Law of Superposition may be described as lithodemic, and delimited on the basis of their rock characteristics

(Rawson *et al.* 2002). Such deposits may be hosted by earlier natural or anthropogenic deposits.

In modern urban environments, the anthropogenic impact in the subsurface due to the emplacement of material and the construction of foundations and infrastructure may be complex and extensive. This effect on the subsurface may be considered as a form of anthropogenic bioturbation or anthroturbation. Removed from the immediate effects of weathering and constant human reworking, the preservation potential of anthropogenic deposits emplaced in the subsurface may be greater than that of deposits created at the surface. The subsurface environment will be relied upon for a range of resources and services (including physical and thermal resources, space provision and storage), increasing the scale of anthropogenic intervention in the subsurface. This is likely to impart an anthropogenic signature on the subsurface that is characteristic of the proposed new epoch of time, the Anthropocene.

Original lateral continuity of anthropogenic deposits

Lateral continuity is a defining characteristic of lithostratigraphical units and their boundaries (Murphy & Salvador 1999). In the case of bedrock and, to a lesser extent, natural superficial sediments, depositional environments are typically widespread and bounded by physical limits including topographical barriers. The original lateral continuity of natural deposits may be interrupted by subsequent deformation or erosion associated with younger depositional events or topographical incision. However, individual units may commonly be traced over considerable ranges. Conversely, anthropogenic deposits are created in novel sedimentary environments that may be controlled by a combination of human factors in addition to physical constraints. These factors, including social and economic decision making, typically result in anthropogenic deposits that are discrete bodies with complex and largely unpredictable geometries. Deposits of similar character may coalesce to form composite units, or remain isolated but be considered as lateral equivalents. However, few individual anthropogenic deposits have the widespread lateral continuity that typifies and defines most natural lithostratigraphical units. To investigate the applicability of lithostratigraphy to anthropogenic deposits, the extent of natural and anthropogenic deposits can be compared.

Spatial extent of anthropogenic deposits. Based on BGS 1:50 000 scale geological map data of Great Britain, the largest contiguous area of a single bedrock formation (at rockhead) exceeds

Fig. 6. Photograph showing back-filled underground mine workings, exposed by subsequent opencast mining. The remnant pillars of coal are evident as darker areas on the floor of the pit and lowermost part of the quarry face. The intervening stalls represent areas of mainly shale debris partly packed into the voids created by working the coal to help support the workings (often referred to as 'gob' or 'goaf') or the product of roof collapse. An area of such collapse, forming a choked chimney in the right of the photograph, extends about 1 m above the level of the worked coal. Materials emplaced in the subsurface by human processes may be considered as anthropogenic intrusive deposits. The disruption of the subsurface by human activities, such as underground mining and collapse, is described as anthroturbation. Dog and Gun Clay Pit [SE 053 344], Bradford, West Yorkshire. Photograph by C. N. Waters BGS

4600 km^2 (London Clay Formation in the Thames Basin). For natural superficial deposits, the equivalent (at outcrop) is around 2100 km^2 (till of the Lowestoft Formation in eastern England). In contrast, the largest area of artificial deposits (i.e. made or infilled ground) mapped at the same scale is 65 km^2, relating to made ground associated with the industrial legacy of Wolverhampton, central England. Made and infilled ground are typically an amalgamation of different deposits, and may be considered equivalent to group level in natural superficial or bedrock lithostratigraphy. In comparison to the mapping of natural deposits, that of artificial ground in Great Britain is relatively immature. This is reflected in the variable consistency and coverage of artificial ground data (Fig. 3). The map data for Wolverhampton are modern and many of the artificial deposits, including mineral spoil, are characterized by well-defined and readily mappable landforms. In other areas where the land-use history has resulted in extensive artificial deposits with little topographical expression, conventional mapping has often failed to depict these. This is particularly apparent in urban areas, including London, where unmapped artificial deposits are known from borehole records to blanket large areas.

At the other extreme, and perhaps more significant when considering the lithostratigraphical classification of anthropogenic deposits, named formations have been defined for bedrock deposits where the largest contiguous area (at rockhead) approximates to less than 0.007 km^2 (e.g. the Swindale Limestone Formation in NW England), and 0.003 km^2 (at outcrop) for natural superficial deposits (e.g. the Benholm Clay Formation of eastern Scotland). By comparison, the average area of artificial deposits is 0.07 km^2 in extent. For the natural deposits, these are the smallest extremes that have been mapped. They may arguably be seen as either excessive subdivision not representing the 'best practice' application of lithostratigraphical procedure, or they may be records detailing outcrops of unique local deposits.

Thickness and volume of anthropogenic deposits. Lithostratigraphical formations defined for natural deposits generally vary in maximum thickness from a few metres to several hundred metres (Rawson *et al.* 2002). The Lexicon of Named Rock Units (British Geological Survey 2012*a*) contains descriptions, including maximum thickness, of lithostratigraphical units shown on BGS maps. For formation-level lithostratigraphical units recorded in the Lexicon, the maximum thickness for superficial deposits ranges from around 0.7 to over 80 m (averaging 15 m), and around 5 m to over 2 km for bedrock units (averaging 400 m). Greater maximum thicknesses are recorded for bedrock units where tectonic thickening is suspected. The *Lexicon of Named Rock Units* is mainly lithostratigraphical, but it incorporates morphostratigraphical (and lithodemic) schemes, including the BGS artificial ground classifications. However, no information is given for the thickness of anthropogenic deposits.

Borehole records are commonly used in mapping and 3D geological modelling, and offer additional information including the thickness of anthropogenic deposits. The UK National Geoscience Data Centre's collection of onshore boreholes and trial pit records includes over 1.3 million entries (British Geological Survey 2012*b*). The distribution of boreholes and trial pits is generally concentrated in urban areas and centres of extractive industry. About 10% of registered boreholes and trial pits intersect categories of artificial ground on BGS 1:50 000 scale maps that are likely to include anthropogenic deposits. Although many records predate development activities (e.g. construction) that can result in subsequent creation of artificial ground, they provide useful information on the thickness and spatial extent of pre-existing anthropogenic deposits. The greatest intersections of anthropogenic deposits proved in boreholes are typically associated with infilled ground and, specifically, the use of man-made excavations for waste disposal. For example, an intersection in excess of 65 m of back-filled opencast colliery spoil is recorded in the Leeds area of northern England, and is considered to be towards the upper limit of thickness for anthropogenic deposits in Great Britain. Borehole records indicate that these deposits represent a single phase of deposition with a definable lithological character, and may be considered as a single anthropogenic unit. Compaction and diagenesis may reduce the thickness of these deposits, but, if preserved in the rock record, they would present a mappable unit similar in thickness to many natural deposits.

Deeper excavations are recorded in Great Britain, including many that are wholly or partially infilled. However, information on the thickness and lithology of any fill is limited by the age and distribution of borehole records or other evidence. Significant thicknesses of anthropogenic deposits are also likely in the case of back-filled shafts and wells; however, these types of deposit may be more appropriately characterized as anthroturbation or anthropogenic intrusive and subject to lithodemic classification (see the earlier subsection on 'Anthropogenic sediments and the Law of Superposition').

In general, artificial ground is included on BGS 1:10 000 scale maps where the maximum thickness, or depth, is interpreted to be greater than 1 m, based

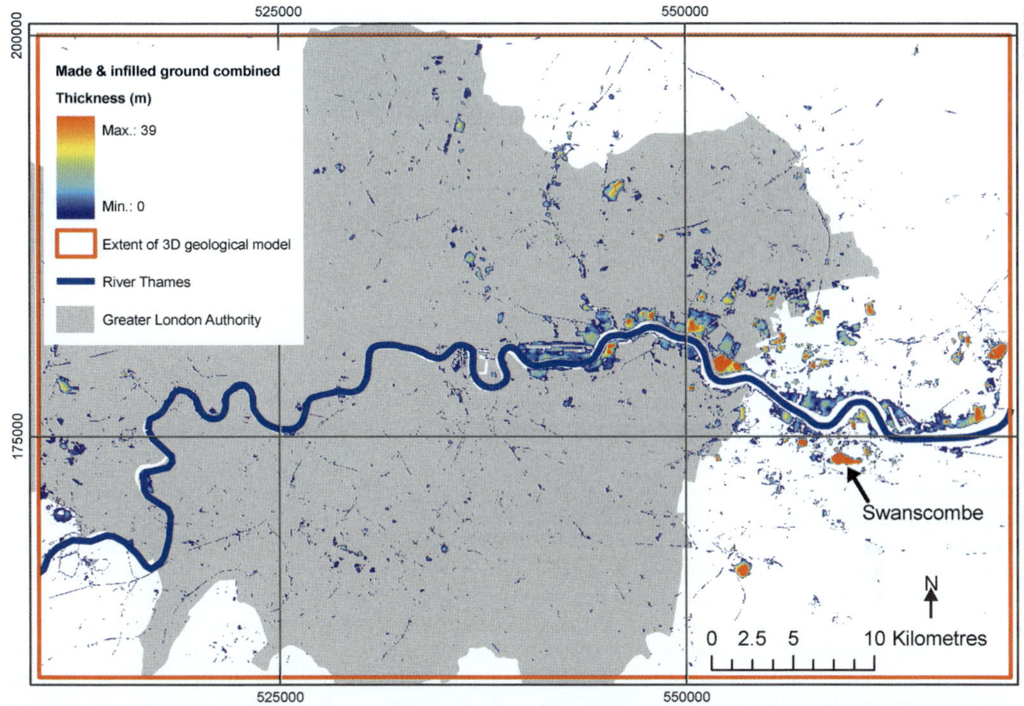

Fig. 7. Map showing the thickness and distribution of artificial ground in the London area based on a 1:50 000 scale 3D geological model (Ford *et al.* 2010*b*). Greater London Authority boundary and River Thames included for reference. Infilled chalk quarries in the Swanscombe area, east of London, show the greatest thickness of anthropogenic deposits.

on landform or borehole evidence. Although geological maps typically give no indication of thickness range, this information can be determined from 3D geological models of the shallow subsurface. For example, the 1:50 000 scale 3D geological model of London (Ford *et al.* 2010*b*) includes major areas of artificial ground (Fig. 7). The greatest concentrations are located in the east of the city and adjacent to the River Thames. An estimate for the mean thickness of undifferentiated anthropogenic deposits based on the model is 1.6 m, reaching a maximum of 39 m in the Swanscombe area of east London where the landscape has been subjected to extensive quarrying of the chalk bedrock and subsequent infilling with material including overburden.

The maximum thickness of anthropogenic deposits is comparable to those derived from the model for natural superficial and bedrock formations including, for example, Pleistocene river terrace deposits (Taplow Gravel Formation, subsequently reclassified as a member in the Maidenhead Formation by McMillan *et al.* 2011) and Palaeogene marine sands (Thanet Sand Formation) (see Table 1). Similarly, the mean thickness of anthropogenic deposits is comparable with the range shown for natural deposits. However, the thickness distribution of anthropogenic deposits differs considerably, being strongly skewed towards lower values (Fig. 8). Almost 90% of the anthropogenic deposits are less than 3 m thick, compared to 6 m for the Taplow Gravel Formation and 20 m for the Thanet Sand Formation. This bias towards relatively thin and discontinuous deposits represents a key challenge in defining a lithostratigraphical classification for anthropogenic deposits.

Three-dimensional geological models allow volumes of material to be assessed. The 1:50 000 scale model indicates that the volume of anthropogenic deposits in London is 168×10^6 m^3. This model excludes deposits of less than 1 m in maximum thickness, including many related to road construction, shallow foundations and landscaping. These deposits may be ubiquitous across much of the Greater London Area (Fig. 7), requiring an average thickness of only 0.1 m across the area to match the figure of 168×10^6 m^3. The spatial characteristics of anthropogenic deposits in London are likely to be representative of other areas with broadly similar economic and social history.

Table 1. *Summary showing thickness information for anthropogenic deposits, and selected examples of natural superficial and bedrock formations in the London area based on 3D geological modelling (Ford* et al. *2010b)*

Unit type	Unit name	Maximum thickness (m)	Mean thickness (m)	Standard deviation	90th percentile	Volume (km^3)
Anthropogenic	Made ground	26	1.4	2.6	–	0.10
	Infilled ground	39	2	3.8	–	0.07
	Made and infilled ground combined	**39**	**1.6**	**3.05**	**3**	**0.17**
Natural superficial	**Taplow Gravel Formation***	**15**	**3.7**	**2.5**	**6**	**0.55**
Bedrock	**Thanet Formation**	**50**	**12.8**	**6.4**	**20**	**22.17**

*Taplow Gravel Formation, subsequently reclassified as a member in the Maidenhead Formation by McMillan *et al.* (2011). The distribution of thicknesses for the units shown in bold are compared in Figure 8.

Temporal changes in volume and extent of anthropogenic deposits. Globally, changes in technology over time have had a dramatic influence on the extent and thickness of anthropogenic deposits. For example, early mining was undertaken largely by hand, but, as steam- and then oil-powered machines were introduced, far greater volumes of anthropogenic sediment could be moved and significantly greater anthropogenic deposits created. In coal-mining areas, increased ratios of coal to overburden are reflected in the size and thickness of associated anthropogenic deposits, including spoil heaps. In Great Britain, ratios of coal to overburden changed from around 1:3 during the early 1940s to 1:18 in 1999 (International Mining Consultants Limited 1999). Similarly, average ratios in US mines changed from around 1:6 to 1:11 between 1946 and 1970 (Boudette *et al.* 1976), and in eastern US mines the ratio changed to between 1:15 and 1:25 by 2002 (US Environmental Protection Agency 2002). Ratios of waste to product for coal, iron ore, most metalliferous and many non-metalliferous minerals have all seen a dramatic rise since World War II. In fact, since 1945, mineral output and related waste production has risen to about 45 km^3 per year, a trend that continues. Similarly, the volume of material excavated and used for major constructions has increased dramatically, driven by widespread land reclamation and infrastructure construction. World cement and aggregate production has followed a similar trajectory to mineral production. Records suggest that around 40 km^3 of limestone, shale and aggregate are currently excavated and moved each year. The total volume of anthropogenic deposits resulting from earthworks will be considerably greater than the volume of construction materials.

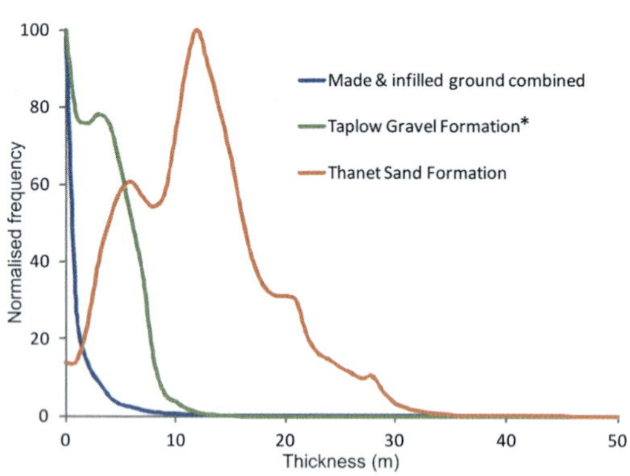

Fig. 8. Chart comparing the thickness distribution of anthropogenic sediments with natural superficial and bedrock units in the London area based on 3D geological modelling (Ford *et al.* 2010*b*). Frequency is normalized to 100 for each unit. *Taplow Gravel Formation, subsequently reclassified as a member in the Maidenhead Formation by McMillan *et al.* (2011).

Along with temporal changes in the volume of other sources of anthropogenic sediment, including landfill waste, changes in the spatial extent of anthropogenic deposits through time may offer a distinguishing characteristic to support the application of lithostratigraphy to anthropogenic deposits. However, on a regional, or even national, scale, significant changes in the volume and extent of anthropogenic deposits may be highly diachronous. The onset of major urban and industrial expansion in Great Britain that resulted in a significant increase in the scale of anthropogenic deposits began around 1750 at the time of the Industrial Revolution. Significant expansions are only now being experienced by some emerging economies of the developing world. Similarly, ancient civilizations across many continents have been responsible for the creation of considerable anthropogenic deposits many millennia before the Industrial Revolution in Great Britain (e.g. Edgeworth 2013).

Lithological distinction of anthropogenic deposits

Establishing a lithological distinction between adjacent units based on observable characteristics is an essential criterion in lithostratigraphical classification. As analogues for natural sediments, anthropogenic deposits exhibit lithological characteristics, and can largely be described in conventional terms including composition, grain size and texture. Anthropogenic deposits may exhibit considerable lateral and vertical lithological variation. Distinguishing characteristics should be apparent at the mappable scale in outcrop, in hand or core samples, or with low-magnification microscopy. Whilst natural exposures of anthropogenic deposits are relatively rare, temporary excavations and trial pits or boreholes for ground investigation can reveal their presence and composition.

Composition of anthropogenic deposits. The bulk composition of anthropogenic deposits may be considered in terms of three components: (1) anthropogenically redeposited or disturbed natural materials that have undergone little or no processing; (2) natural materials deposited from natural or indirectly modified sedimentary systems; and (3) novel materials of an entirely artificial origin that have been processed or manufactured and commonly subjected to heat and/or pressure. A dominant proportion of redeposited or novel material in the bulk composition of a deposit may be definitive of an anthropogenic deposit (Fig. 1b).

Redeposited natural geological materials may be locally derived; for example, in the case of construction site formation where excavated natural material is used nearby as fill. When processing is limited, the composition of the resulting anthropogenic deposit will broadly reflect that of the natural source material. However, the anthropogenic deposit may be distinguished from the parent geological material by the disturbed fabric of the sediment, weathering characteristics and the presence of exotic anthropogenic components (including novel materials) incorporated through mixing. Where multiple phases constructed of locally derived deposit are juxtaposed, the distinction between units on lithological character alone may be problematic.

Redeposited natural geological materials may also be sourced from outside the immediate area in response to economic factors, locally unobtainable quantities or the requirement for material with particular physical characteristics, such as permeability or strength. Imported material is likely to be compositionally distinct from locally occurring natural deposits. For imported natural materials, extensive source areas (or catchments) may result in anthropogenic deposits that are compositionally diverse and readily distinguishable from one another, even where their end use is comparable. However, where restricted or established supply areas exist for particular materials (e.g. major quarries or sand and gravel production areas), multiple phases of deposition may exploit the same parent material and result in sediments that are essentially similar in composition. The transfer through time of sediment supply between source areas as resources are exhausted or new production is introduced may leave a distinct signal in the record of anthropogenic deposits. This is analogous to natural systems where unroofing of a buoyant lithosphere can produce sediment of progressively deeper source being transported into a nearby depositional area.

Novel materials of an entirely artificial origin represent a significant component of many anthropogenic sediments. Volumetrically, the most significant novel materials are glass, plastics, ceramics, concrete, brick and smelting slag (Zalasiewicz *et al.* 2011). Many novel materials are themselves composed of, or derived from, original natural rock, superficial deposits or fluids extracted from the ground. However, their distinguishing feature is that they have been processed, manufactured or refined by human activity in some way to change their original physical and chemical structure. They may have undergone mixing, been chemically formed or, more commonly, altered through being subjected to heat.

Novel materials may occur as component parts of relatively homogenous packages of sediment related to a single, dominant land use (e.g. ceramic waste tips in Great Britain: Wilson *et al.* 1992), or in mixed anthropogenic sediments, or in naturally occurring sediments that have been reworked (Fig. 9). The latter are common where land has been subject to successive phases of use for different purposes. Norbury (2010) recognized that some types of novel materials are strongly related to their geographical and geological location and setting; for example, 'blaes' is a distinctive orange gravel arising from the production of paraffin from combusted oil shale in Scotland, and is commonly used as subgrade for road construction in the area.

Where novel materials are present, the distinction between anthropogenic and natural sediments is generally straightforward. Novel materials may be recognized by characteristics including composition, shape and size of grains, colour and markings. However, novel materials, including plastics, bricks and metal are increasingly present in the natural sedimentary cycle, carried by river systems and incorporated into a range of natural deposits that may subsequently be redeposited as anthropogenic sediments (Fig. 9). Natural materials deposited in modified natural situations include sediments deposited in artificially dammed valleys.

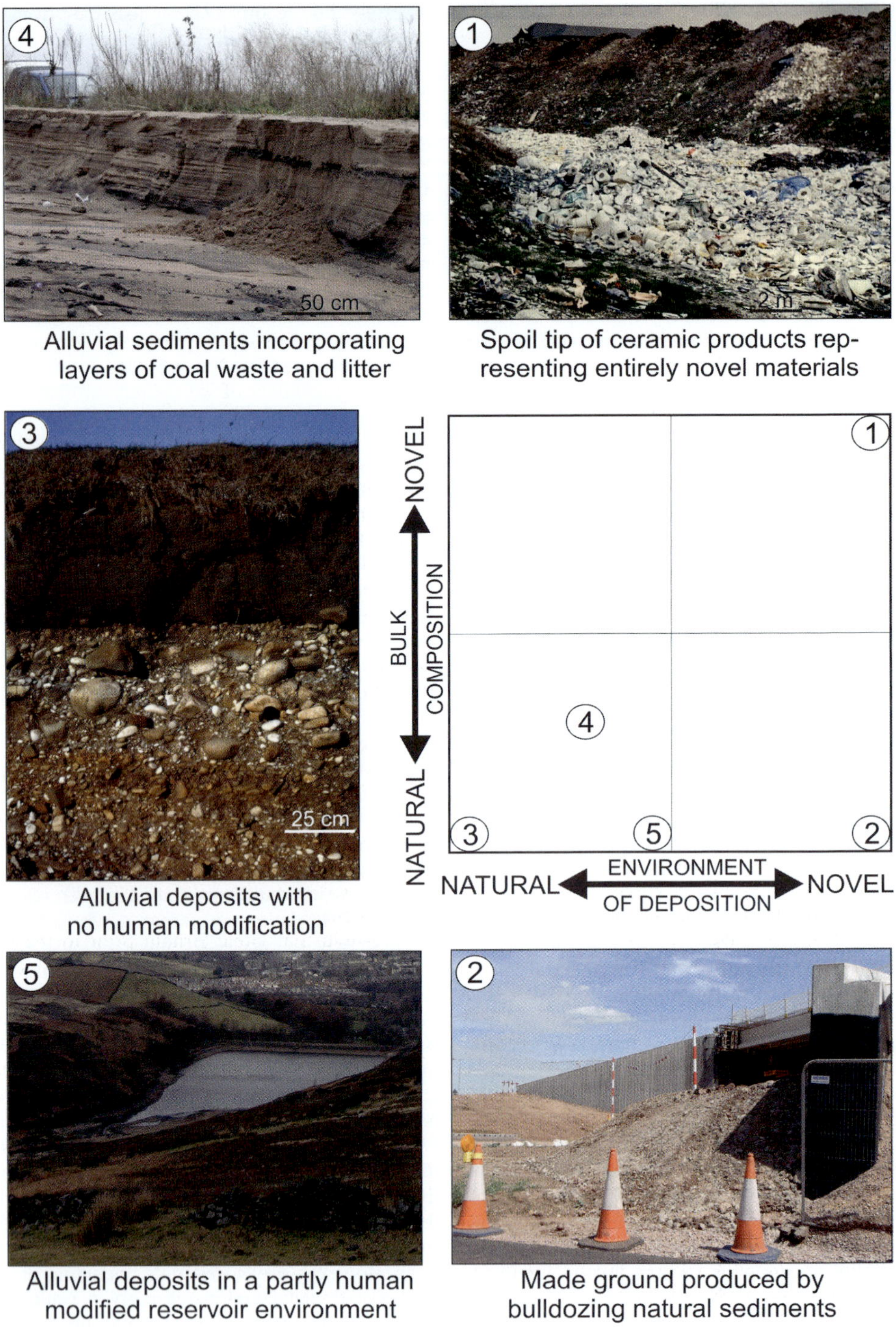

Fig. 9. Relationship between the environment of deposition and bulk composition of sediments including anthropogenic deposits. The range of possible combinations of environment and composition highlights the diverse and extensive nature of deposits that may contribute to the definition of the Anthropocene. The anthropogenic deposits shown in Figures 12 & 13 would plot in similar positions to (1) and (2) respectively. Photography by BGS ©NERC 2012. All rights reserved.

As such, the sediments will be similar to other natural sediments in lithology, texture and grain size, but the geological setting will be modified, and sequences such as those found in deltas and lakes may form in dammed valleys.

Reworking of anthropogenic sediments is widespread in the urban environment and presents a challenge to establishing a lithological distinction between adjacent units. The reuse and redevelopment of land, and the recycling of materials on a site-, local- and regional-scale (the latter now common practice in the construction industry), results in extensive deposits of derived anthropogenic sediment, including novel materials and human-made components. Reworking can result in the homogenization of anthropogenic sediment that may originally have represented multiple phases of deposition. Where natural sediments have been reworked and deposited forming a new structure for an engineering purpose, sediments are likely to be more homogenous than those deposited without engineering control. The extent of reworking in the novel sedimentary environment is arguably much greater than in most natural sedimentary environments. Although documentary evidence (e.g. historical maps) may indicate the reworked nature of anthropogenic sediments, observable characteristics that may allow reworking of natural sediments to be recognized (e.g. disturbed or reorientated fabric) are not generally meaningful for anthropogenic sediments.

Temporal changes in the composition of anthropogenic deposits, including waste. The presence of novel materials in artificial sediments will generally aid the distinction between adjacent units and may represent the defining factor where natural components are not diagnostic; for example, in the case of locally sourced, redeposited natural geological materials. Inevitably, some novel materials become obsolete with time and their relative abundance or assemblages my provide markers for units of a particular age, in a way similar to fossils in the biostratigraphical classification of natural deposits (Fig. 10).

The definition of a lithostratigraphical unit should be based upon the nature of the deposits and can be independent of time constraints. Hence, should a decision be made to define a new Anthropocene Epoch (e.g. at the incoming of plastics around 1950), some units of artificial deposits that are defined lithostratigraphically would be Anthropocene in age, others would span the boundary with the Holocene, many would be entirely Holocene in age. However, the lithological composition of anthropogenic deposits may be influenced by a combination of factors, including technological, societal and economic drivers, and therefore be strongly indicative of their age. Major historical events including conflict and advances in technology have resulted in discernible changes to the composition of anthropogenic deposits. For example, changing practices in methods for the extraction of coal have resulted in not only a rapid increase in the volume of spoil produced, but also the proportions of coal to waste in spoil tips have typically decreased (Waters *et al.* 1996). Understanding temporal changes in composition may support the application of lithostratigraphy to anthropogenic deposits and contribute to the definition of the Anthropocene.

In Great Britain, the transformation from a nomadic hunter–gatherer society to one of settlement, farming and agriculture is associated with the Neolithic (*c.* 4000–2600 BCE). During this time, the scale of direct human transformation of the landscape and the creation of novel sediments was limited by population, availability of raw materials, technology and economic growth. However, localized extraction and processing of metals and subsurface resource exploitation was present on a site scale (Price *et al.* 2011). Settlement and growth of towns following the Roman occupation of Great Britain was associated with an increase in the use of geological materials for construction and engineering, including raw materials for bricks and mortar, crushed rock, dimension stone and roofing slates. It is these materials that form the pre-industrial anthropogenic deposits that are often recognized in the archaeological record, buried beneath the foundations of modern towns and cities. Correspondingly, the pits and quarries used to source these materials may still be evident, partially or wholly back-filled with later anthropogenic deposits.

Pre-industrial Great Britain prior to the eighteenth century was, therefore, largely dominated by exploitation and the use of natural geological materials in construction and engineering as towns began to develop, and people settled in one place. The onset of the Industrial Revolution marked a rapid increase in the rate of exploitation and use of geological materials for construction and engineering, including the placement of these materials in the subsurface as sewerage and underground transport networks were developed. Technological innovation and development, fuelled by coal, resulted in ever-deeper levels of anthroturbation to exploit coal and mineral resources. Subsurface anthropogenic deposits in the form of mineral waste 'goaf' were deposited (Fig. 6), in addition to the creation of voids through shafts, adits and working levels in mines. The deposition of anthropogenic deposits and mineral spoil above ground was common, and often widespread. Manufacturing and processing of iron ore and other metals to provide the raw

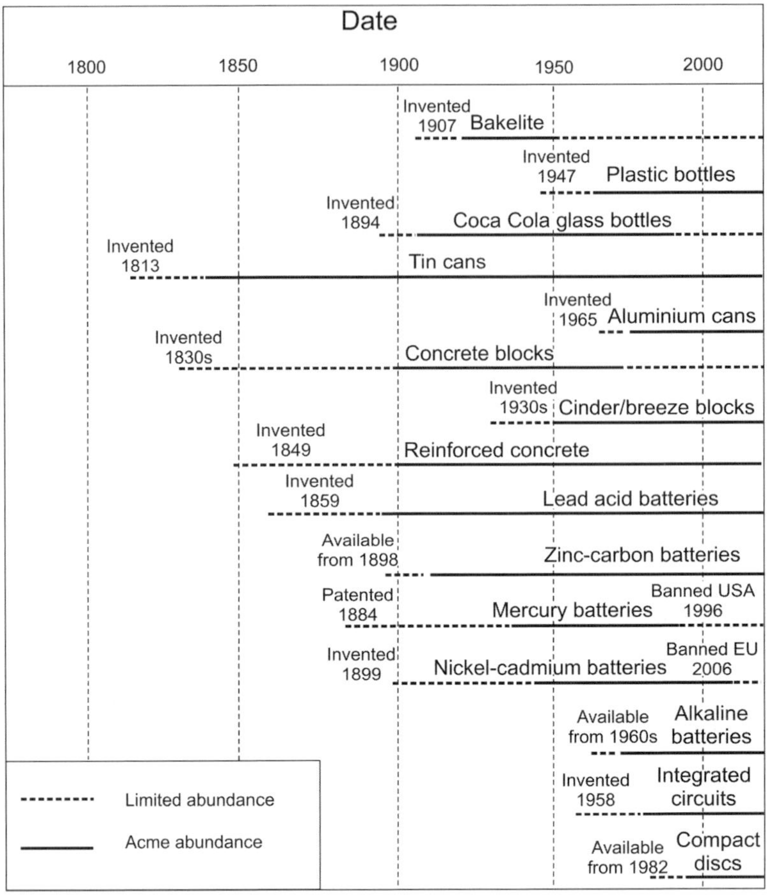

Fig. 10. The first appearance and acme of selected novel materials post-1800 that may, by association, allow the relative age of anthropogenic waste deposits to be determined. The ranges are schematic, based on European/North American landfill deposits, and do not consider global diachroneity.

materials for buildings, machinery and transport infrastructure became widespread. Cities grew as the urban population expanded.

The Industrial Revolution marked a change in the character of anthropogenic sediment deposition in four main ways of potential lithostratigraphical significance. First, the rate of production and use of building and construction materials including bricks and mortar, dimension stone and glass increased. Second, the exploitation of subsurface resources increased and deposition of associated surficial waste tips became widespread. Third, the extraction, processing and use of metals in buildings, transport infrastructure and in manufactured goods increased dramatically, resulting in a significant rise in the proportion and widespread occurrence in the sedimentary environment of novel anthropogenic materials including processed metals and manufactured goods. Fourth, with the increasing exploitation of deeply buried mineral resources, installation of underground transport and drainage infrastructure, anthropogenic materials were commonly deposited below the ground. Arguably, these changes mark the beginning of the acme of anthropogenic deposition.

Subsequent post-industrial development saw rapid population growth increasingly concentrated in towns and cities. In western Europe, this growth and development was disrupted by the outbreak of two world wars. Many cities were partially destroyed as a result. In many cases, building rubble resulting from war damage was widespread and commonly used as fill or, even, ship ballast (Thorpe *et al.* 2011).

The period following World War II in western Europe saw an acceleration in urban growth, technological innovation and increasing use of geological materials. This innovation led to the manufacture and use of a variety of new and novel anthropogenic materials. Oil exploitation and refining produced the feedstock to make many petroleum derivatives such as plastics. The use and implementation of electronic technology expanded, including computers for industrial and personal use. Portable electronic devices including stereos, mobile phones and laptop computers required batteries. In construction, the post-war period saw an acceleration in the use of concrete and the exploitation of the raw material required for its manufacture. The post-war period therefore marks a time during which technological development and rapid urban growth provided the potential drivers and source materials for novel anthropogenic deposits characterized by electronic equipment, extensive concrete manufacture, deep mining and the generation of vast amounts of wastes.

Throughout human history, wastes, including mine spoil, industrial and domestic refuse, comprise the main types of anthropogenic deposit by volume. The lithological composition of waste deposits is largely defined by the source activity. However, other factors including prevailing legislation and the way in which wastes are managed and treated can influence their composition. Consequently, wastes are of particular significance to lithostratigraphical classification as they result in deposits with diverse and distinct lithological characteristics. In the case of municipal solid waste (MSW) deposited as landfill (including historical middens), compositions have changed considerably with time. For example, prior to 1800, volumes of waste in Great Britain were small and dominated by ash, wood, bone, and body and vegetable wastes, with a low proportion of metal due to reuse and recycling (Waste Online 2004). In the 1800s, waste volumes increased and compositions changed to include domestic, industrial, commercial, mine and quarry waste. By the early twentieth century, most houses were fuelled by coal fires, consigning significant volumes of ash to landfill. However, following the implementation of the Clean Air Act in 1956, the use of coal fires in homes, and the proportion of ash and cinder in landfill waste, began to decline (Fig. 11). In the 1960s, with burgeoning production of plastics to replace traditional materials and with increased consumerism and waste production, the volume and composition of wastes changed radically. Subsequent legislation, including the Control of Pollution Act in 1974, and increased reuse and recycling resulted in greater segregation of waste materials, and corresponding changes in waste volume and composition (European Union 1999, 2008). In 2006–2007, the composition of household waste in England included 23% paper and card, 18% food, 16% garden and other organic waste, 10% plastics, 6% glass, 4% metals, 4% wood, 3% textiles, 2% waste electrical and electronic equipment, and 14% 'other' including batteries and hazardous material (Department for Environment Food and Rural Affairs 2009).

In addition to diachroneity related to the volume and extent of anthropogenic deposits, compositional characteristics may also be expected to show contrasting rates of change on a regional and global scale. Whilst the bulk composition of anthropogenic deposits is likely to be most strongly influenced by local natural resources and remain relatively constant through time, aspects such as the specific types and proportion of novel materials may be indicative of an area's particular stage of societal or economic development. However, with increasing globalization in the modern era, the time gap and diachroneity associated with the proliferation of novel materials is, in many cases, considerably reduced.

Preservation potential of anthropogenic deposits. The relative preservation potential of materials comprising anthropogenic sediments, including novel materials and organic content, should be considered when comparing the composition of deposits. Notwithstanding destruction through physical erosion, preservation potential is largely a response of material composition to the physiochemistry of the environment of deposition. Significant environmental parameters that control preservation potential include moisture, temperature, redox potential and pH. For example, bricks may be susceptible to physical degradation in the presence of water and changes in temperature. Processed metals and alloys derived from iron ore are likely to corrode in the presence of oxygen and chloride ions. Plastics, representing synthesized polymers derived from hydrocarbons, may be subject to chemical and physical degradation when exposed to light, oxygen, heat or corrosive fluids. Much like archaeological residues, organic anthropogenic materials may decay after deposition unless they are deposited rapidly in an environment that favours their preservation. There are many cases of exceptional preservation of organic remains in waterlogged cultural deposits beneath cities including York and London in the UK (McCann & Orton 1989; Holden *et al.* 2009). In the case of modern municipal waste, it is estimated that 67% is biodegradable (Department for Environment Food and Rural Affairs 2009). This suggests that if all municipal waste were deposited in landfill, 33% would be available for long-term preservation. Many anthropogenic deposits comprise mixtures of geological raw materials that exhibit differing potentials for preservation. Concrete, for example, is commonly made from siliceous aggregate and cement. Silica

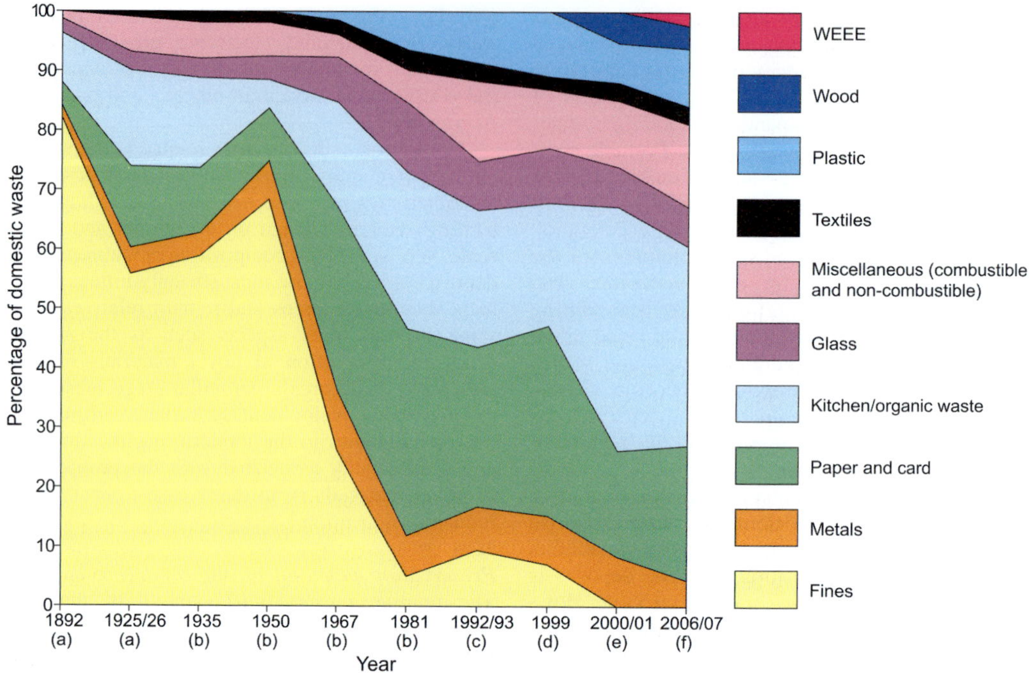

Fig. 11. The change in composition of household Municipal Solid Waste (MSW) in Great Britain. Percentage by weight measurements based on surveys of household collected waste only. The fines category includes dust, ash and undifferentiated screenings. WEEE (Waste Electrical and Electronic Equipment) may be reported in other categories prior to 2002. Based on and incorporating data from: (**a**) Waste Online (2004) and Gandy (1993); (**b**) Bridgewater (1986); (**c**) Burnley (2007), data collected from wheeled bins in England and Wales only; (**d**) Burnley 2001; (**e**) Parfitt (2002); and (**f**) Department for Environment, Food and Rural Affairs (2009), data from England only.

typically provides a robust, preservable material, but the concrete's matrix may be susceptible to chemical and physical degradation, especially in the presence of sulphate- or chloride-rich groundwaters. This change would be a form of diagenesis, equivalent to that seen in bedrock units, with resultant redeposition of calcium carbonate significantly modifying the porosity and permeability of the deposit.

The lithological characterization of anthropogenic deposits. In recognition of the specific engineering and environmental challenges posed by anthropogenic deposits, schemes have been developed for their lithological characterization, including those that make specific provision for novel materials and waste (Hollis 1992; Environment Agency 2005; Norbury 2010).

Norbury (2010) distinguishes three categories of man-made soils for engineering geological purposes which define the lithological composition of anthropogenic deposits:

- Those fine or coarse soils composed of natural materials that have been laid down (redeposited) by man to form new structures like embankments. It may not be easy to recognize such materials without considering other sources of information such as land use, landforms and historical maps.
- Those fine or coarse manufactured or processed materials laid down by man that can be described and tested geotechnically as they are physically or chemically similar to natural soils (e.g. washed materials such as mine tailings from non-coal mines, crushed rock back-fill).
- Those fine or coarse manufactured materials laid down by man that cannot be easily geotechnically described (e.g. domestic refuse in landfill, fly-tipped material and demolition rubble).

The two latter types include deposits that are composed completely or partially of anthropogenic material and may be mixed with reworked natural deposits derived from bedrock or unlithified superficial deposits.

In Great Britain, wastes are classified according to the List of Wastes Regulations 2005 (Environment Agency 2005), which provides codes for all hazardous and non-hazardous wastes based on the process that created them or the material type. It comprises 20 high-level categories of waste with further subdivisions, and includes over 1300 descriptions of common waste types.

The use of these schemes may complement conventional geological description in the lithological characterization of anthropogenic deposits for the purpose of defining lithostratigraphical units. They are applicable where anthropogenic deposits are relatively homogenous as a result of similar land-use processes or the degree of engineering control applied to their deposition. However, anthropogenic deposits are commonly characterized by mixtures of natural and anthropogenic material deposited by more than one type of process. Such processes can include multiple phases of land use during urban development, but it may not be possible to distinguish the individual stages that formed them. For the lithological characterization of artificial deposits that comprise mixtures of anthropogenic and natural materials, existing schemes used in isolation are unlikely to be effective.

Grain size of anthropogenic sediments. The component parts of anthropogenic sediments can be considered as clasts or grains, and their size, shape and sorting can be used to define the lithological character of anthropogenic deposits. Grains may be identified for novel, natural or reworked natural components. The grain size of reworked natural material may reflect the primary deposit; for example, in the case of unlithified gravel. However, where cohesive material including clay or rock is moved intact by mechanical excavation, individual pieces or blocks may be considered analogues to lithic clasts in natural sediments. In this case, the size of the clast, as well as the grain size of the primary material, may be described. Novel materials may be described in a similar fashion, with some materials representing grains of single composition and others representing composites.

The grain size of anthropogenic deposits can be classified by existing schemes that are commonly used to describe natural deposits (e.g. Wentworth 1922; British Standards Institution 1999). Such schemes classify grain size into primary classes of clay, silt, sand, gravel, cobble and boulder. Combinations of these classes can be used to describe sediments of mixed grain size.

For natural clastic sediments, grain size is controlled by factors including environment of deposition, distance travelled from the source area and the nature of the parent material. Sediment transported in many natural fluvial and aeolian environments results in a relatively high degree of sorting. Although manufacturing or processing techniques and the formation of tailings ponds can result in well–sorted anthropogenic sediments, typical transport methods are unlikely to significantly affect grain-size distribution. Consequently, anthropogenic sediments, especially those associated with waste deposits, are typically poorly sorted and exhibit large ranges in grain size, sharing similarities with natural 'boulder clay' or breccia deposits.

The inclusion of novel deposits in artificial ground may require the description of diverse lithologies that are present as clasts within a matrix of natural reworked or novel deposits. For example, demolition spoil may contain large boulders of reinforced concrete, and boulder-sized debris of wood, cobbles of brick and cement in a matrix of sand and gravel. Old landfill sites could include boulders of domestic appliances with cobbles of tins and glass, and plastic bottles plus gravel-sized batteries in a matrix of degraded cardboard and vegetable matter.

The use of lithological descriptors such as clay, sand and gravel has been used extensively within the classification scheme for natural Quaternary deposits (McMillan *et al.* 2011). However, inclusion of a lithological term is discouraged by Salvador (1994) and, because of the inherent heterogeneity of artificial deposits, would not be a suitable adjunct to a lithostratigraphical name.

Sedimentary textures and structures within anthropogenic sediments. Sedimentary structures and textures may be evident within anthropogenic deposits, sharing many common characteristics with natural deposits. They include bedding, lamination, graded-bedding and cross-bedding. The presence of sedimentary structures is controlled by the method of emplacement and the magnitude of subsequent reworking. For example, sedimentary structures such as cross-bedding and lamination can be formed during sedimentation in tailing ponds. Similar structures can also be formed by reworking and hydraulic pumping of sediment from shallow-marine environments onto land reclamation. Large-scale pseudo-cross-bedding can be formed by the tipping and gravitational sorting of materials such as mine waste on spoil heaps (Fig. 12). The way materials are engineered may result in internal sedimentary structures that are diagnostic of particular activities, with types of structure reflecting temporal changes in engineering practices such as old 'end-tipped' railway embankments with little compaction to modern emplaced and compacted road embankments. Similarly landfill sites may show mixed waste deposits that range from old end-tipped materials to modern tipped, spread, compacted and engineered sites formed in lined excavations and capped.

Figure 13 illustrates two examples of sedimentary structures observed during recent terrestrial

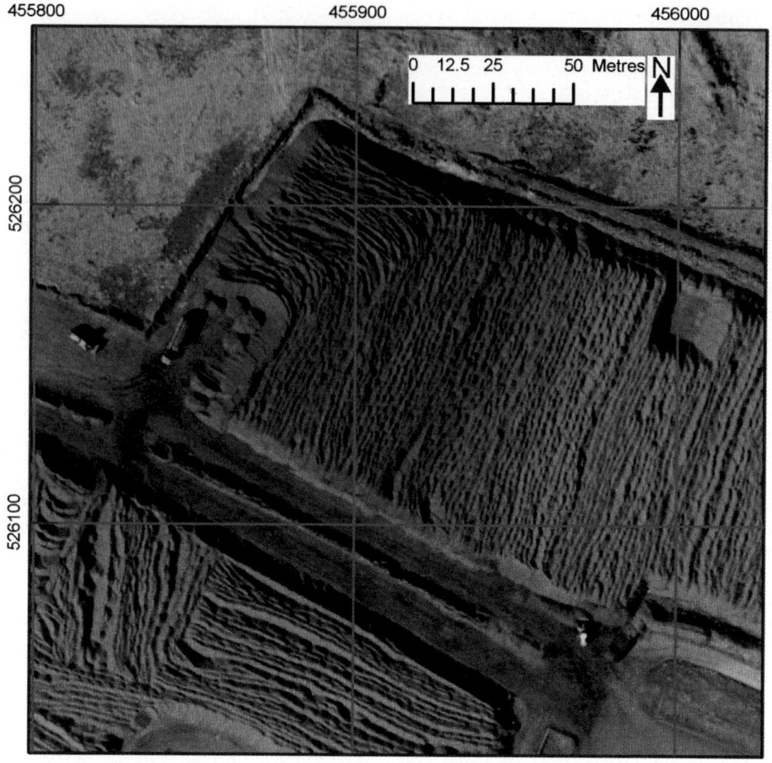

Fig. 12. Aerial photograph showing the geomorphology, process of deposition and sedimentary architecture of a westwards-prograding waste deposit in the NE of England. Waste slag from steel industry is initially deposited by trailer, then pushed up to the front of the deposit by bulldozer to form a series of west-facing ridges comparable in genesis with push-moraines and in overall geometry with natural sedimentary foresets. Aerial photography ©UKP/Getmapping Licence No. UKP2006/01.

landscaping and land reclamation between Dubai and Abu Dhabi in the United Arab Emirates.

Although material structures are not a defining characteristic in lithostratigraphical classification, they can complement lithology and field relationships in providing evidence to help distinguish otherwise similar units and to understand depositional history. As the nature and scale of anthropogenic deposition has changed with time so have the range of structures exhibited by these deposits.

The definition of type sections for anthropogenic deposits

The definition of lithostratigraphical units requires a type section or area to be designated. Type sections provide a formal reference, and should satisfy a range of criteria including being representative of the boundaries of a unit, its lithology or lithological range and, ideally, including a full thickness of the unit in an area where it is at its thickest development. Where these criteria cannot be met by a single type section, composite type sections or type areas may be defined. Type sections should be accessible and have a reasonable likelihood of long-term preservation (Salvador 1994; Rawson *et al.* 2002; North American Commission on Stratigraphic Nomenclature 2005).

As part of the definition of a stratotype, it is necessary to define the base and top of the unit. Where anthropogenic deposits rest directly on natural deposits, the lower bounding surface may be defined as an unconformity. This surface may be the natural land surface or a surface modified by human processes, including excavation. In Japan, the boundary between the artificial and natural deposits is termed the Jinji Unconformity (Nirei *et al.* 2012). In Great Britain, the term rockhead is used to describe the boundary between bedrock and overlying natural and anthropogenic superficial cover (Lawley & Garcia-Bajo 2009). Rockhead in areas of exclusively anthropogenic cover can be used to define the equivalent of the

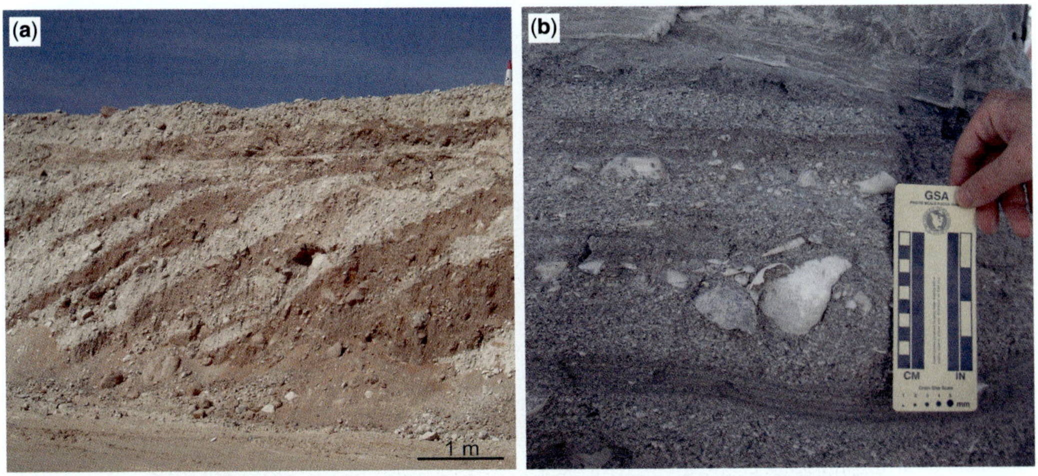

Fig. 13. (**a**) Photograph of anthropogenic deposits showing cross-stratification foresets produced through successive phases of bulldozing aeolian sand and angular fragments of calcareous sandstone during landscaping near Abu Dhabi, United Arab Emirates. Material has been pushed down to the front of the deposit, similar to the accretion of a prograding delta in natural sediments. (**b**) Photograph showing laminated anthropogenic deposits disturbed by the emplacement of an angular sandstone fragment in sediments deposited as slurry during land reclamation near Abu Dhabi, United Arab Emirates. Photographs by BGS ©NERC 2012. All rights reserved.

Jinji Unconformity. Although this unconformity should be readily recognizable, the identification of distinct and regionally correlatable discontinuities *within* artificial successions is more problematic, particularly as hiatuses may be sufficiently short not to be associated with soil development and have little lateral continuity.

In the cases where anthropogenic units are concealed, their upper and lower bounding surfaces may be erosive and characterized by complex cross-cutting relationships. This presents a particular challenge when defining a stratotype as complete sections may not exist, and establishing and correlating reference or composite sections may be complicated by the highly dissected nature of anthropogenic deposits.

In the case of natural deposits, a combination of exposed sections (natural and man-made) and borehole material is used to represent type sections. In some cases, type sections are actively managed to preserve their integrity and permit access (Ellis 2008). In the case of anthropogenic deposits, present-day natural exposures are rare. Exposures that develop naturally (e.g. flood events eroding embankments) or those created by man-made excavations will usually provide only temporary sections. Therefore, defining suitable type sections for anthropogenic deposits based on exposures is likely to be impracticable. Anthropogenic deposits are, however, commonly recovered in borehole samples during ground investigation, especially in areas of regeneration, and such material may offer the best opportunity to observe their lithological characteristics. However, as anthropogenic deposits are often loose or poorly compacted, they are rarely recovered intact and the nature of their bounding surfaces may be unclear. Also, borehole samples are rarely retained beyond the life of the ground investigation and, for the foreseeable future, available borehole material is unlikely to represent all anthropogenic deposits. Arguably, the definition of type sections for anthropogenic deposits may be simpler in the distant future when they are preserved as rocks in the geological record and exposures are available for study.

Applicability of a hierarchical framework and naming of anthropogenic deposits

A hierarchical lithostratigraphical scheme allows the analysis of deposits of comparable characteristics at various scales, from the single trial pit to the entire nation. Without a hierarchy, a plethora of named units would be recognized, with no opportunity to show how these units relate to their neighbours. Hence, lithostratigraphy is more than a classification; it also conveys the geometry of the depositional basins and the processes forming the deposits.

International guidance recommends that each lithostratigraphical unit should be named after an appropriate geographical feature combined with the appropriate unit term (e.g. group, formation)

(Salvador 1994). Inclusion of a lithological term is discouraged by Salvador (1994). However, genetic and lithological descriptors have been used extensively within the classification scheme for natural Quaternary deposits (McMillan *et al.* 2011). Chronostratigraphical or biostratigraphical designations are kept separate as parallel schemes. The use of the same geographical term for units of different status (e.g. formation and member) is not recommended.

Formations. The formation is the primary formal mappable unit of lithostratigraphy (Salvador 1994; Rawson *et al.* 2002; North American Commission on Stratigraphic Nomenclature 2005). A formation is generally defined as the smallest mappable unit that has lithological characteristics which distinguish it from adjacent formations (Rawson *et al.* 2002). However, 'mappability' is a poorly defined criterion, for it depends on the scale of mapping and, in 3D models, scale variations may allow both members and beds to be shown as mappable units. Formations should be regionally significant mappable units, and in Great Britain they are readily represented on 1:50 000 scale maps (McMillan *et al.* 2011).

For natural deposits, formations typically range in thickness from a few metres to several hundred metres, although the definition of formations is scale-independent. In bedrock and superficial deposits, the formation commonly extends over a regional scale, commonly across part or all of a depositional basin or catchment area. For example, a glacial till of Devensian age has different characteristics to a till of Anglian age, and both are classed as formations across the area of former influence of a particular ice sheet. It is important to recognize, in this example, that age is not used as the criteria for recognizing two distinct units, but that the environmental circumstances related to ice sheets of different ages are sufficiently distinctive to allow lithological differences to be recognized between the two formations and their relative stratigraphical position to be determined. Likewise, it can be argued that in Great Britain the broad lithological characteristics of medieval anthropogenic deposits associated with small-scale urban developments are markedly different from Victorian era deposits associated with industrialization and large conurbation development. These are different again from the post-industrialization consumer-driven diversity and bulk of waste products of the present, which may again be different to deposits in a future mineral resource-depleted world. Broad lithological differences with time are described above and are shown in Figure 11.

In Figure 14, an attempt is made to recognize three distinct anthropogenic formations in the Swansea area. The oldest anthropogenic deposits typical of the pre-Industrial Revolution era are of limited lateral extent and are largely associated with building of robust defensive structures, with any habitations of this age not associated with mappable extents of anthropogenic deposits. During the Industrial Revolution, Swansea was a major centre for metal smelting and coal mining, and the extensive anthropogenic deposits are typically related to spoil from extractive or processing industries. After World War II, with the decline of heavy industry in the region, many of the deposits are engineered ground for new houses and roads, and the provision of raised ground to facilitate developments on floodplain areas (Waters *et al.* 2005). These three time periods are responsible for radically different deposits that may be considered to represent single formations. However, where reworking of these deposits has occurred, it may not be possible to recognize to which of these formations a particular deposit belongs. This may be the case in the example provided in Figure 14, where Industrial Revolution era foundry deposits have been reworked and redeposited to form raised fill. If information of the land-use history was not available, it would not be possible to identify within which formation the reworked deposit should be included. The hierarchical nature of lithostratigraphy always allows the unit to be defined as the more inclusive unit, in this case as an undifferentiated part of a group.

It may not be necessary or desirable to force all anthropogenic deposits into a formal lithostratigraphical scheme at formation level. For example, an isolated deposit of uncertain age or origin may not fit easily within such a scheme. Lithogenetic units are locally mappable assemblages of strata, considered without regard to time (Schenck & Muller 1941; Salvador 1994). A lithogenetic unit, mappable or otherwise, is defined by its lithology, morphology and inferred mode of origin (genesis). For Great Britain, BGS classifies lithogenetic units according to the BGS Rock Classification Scheme (RCS) for natural superficial deposits (McMillan & Powell 1999). For anthropogenic deposits, the morphogenetic classification scheme of Ford *et al.* (2010*a*) is a practical mapping and descriptive tool.

Groups and subgroups. Grouping of formations is desirable, although not required, particularly to aid regional mapping (Salvador 1994). Groups and subgroups may or may not be composed entirely of named formations (North American Commission on Stratigraphic Nomenclature 2005), but the establishment of groups without constituent formations should be avoided (Salvador 1994). Groups can be established for formations that reflect the diverse

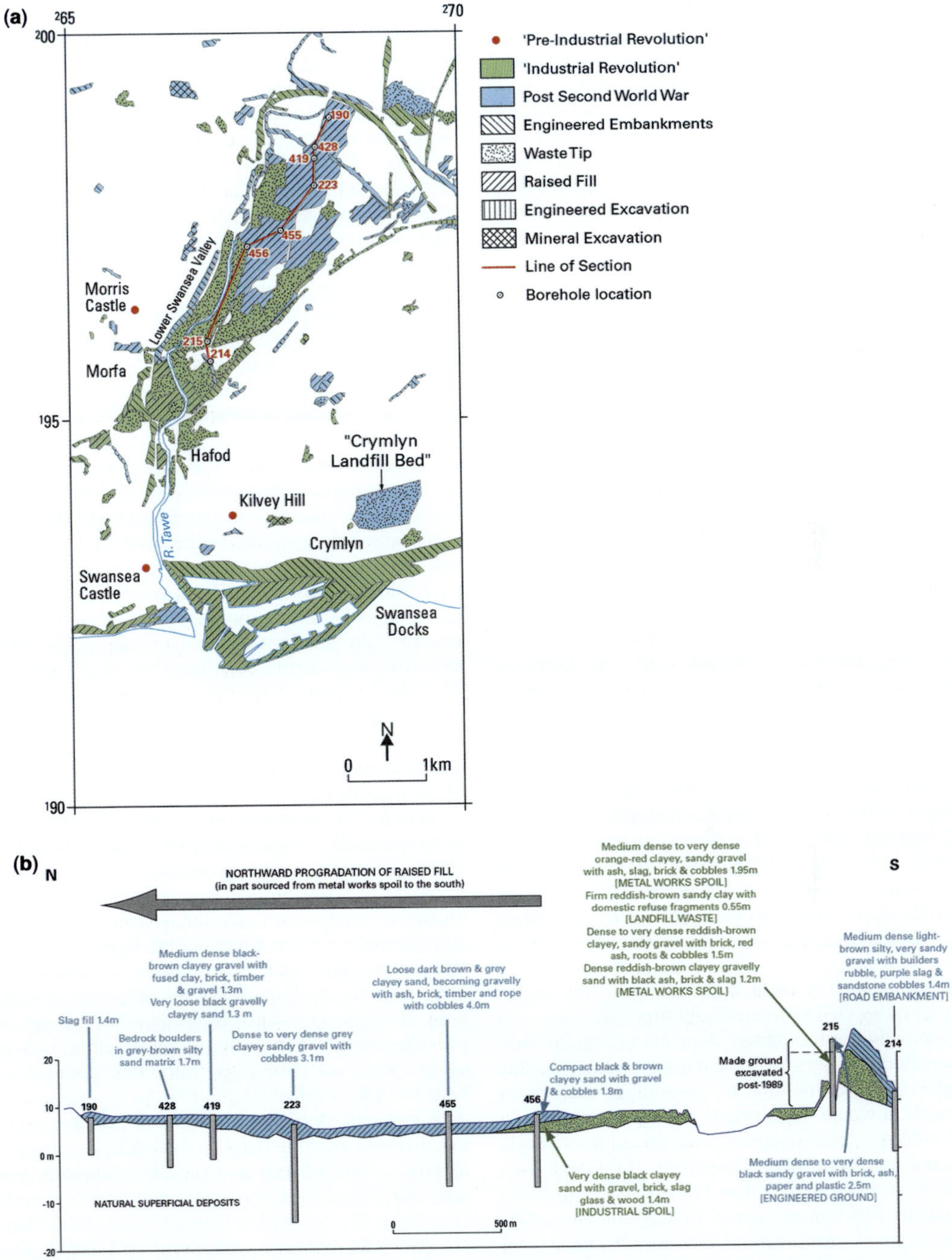

Fig. 14. Example from the Swansea area showing classification of deposits using the existing Ford *et al.* (2010*a*) morphogenetic scheme (ornament) and how they could be reclassified using a lithostratigraphic scheme (colour). The section shows the description of artificial deposits, as recorded in selected boreholes. This demonstrates a problem with lithostratigraphical classification in that the deposits are still mobile, with excavation of metal works spoil deposited originally prior to 1945, to be redeposited in recent years as raised fill in the north of the section. Without an understanding of the history of development of the area it may be difficult to know within which formation the deposits described as 'slag fill' should reside. It also shows that such artificial deposits have an inherent diachroneity in their deposition.

lithologies associated with the principal inferred modes of origin.

Groups permit the mapped formations to be integrated with deposits showing similar lithological or genetic characteristics to allow assessment of successions at broader scales within that basin or catchment area. In the context of natural superficial deposits, it can allow, for example, all glacial deposits to be included in a single group within the area of that ice sheet. This starts to express information on the extent of influence of that ice sheet. In the context of this study, it may be helpful to be able to link all artificial deposits within a given domain of comparable evolution of human activities into a single group. In the example given above of Swansea (Fig. 14), it resides within a broad area of the South Wales Coalfield in which conurbations went through a common history of heavy industries associated with rich mineral resources, followed by decline during the mid-twentieth century and subsequent rejuvenation during recent years. It would be possible to recognize all deposits within the influence of the South Wales Coalfield as part of a single group (Fig. 3), in the same way that domains were erected for superficial deposits to identify catchment groups (McMillan *et al.* 2005). However, in this case the domain is identified by the complex social and economic development of an area, in part controlled by the underlying geology, development of transport networks or simply historical accident. This is a subjective approach, and it is recognized that areas at the margins of a domain may be difficult to define. By recognizing a series of domains that reflect distinct groups, it is envisaged that where anthropogenic deposits are less common in rural areas between the domains, they would be included within formations, but they need not belong to groups.

Supergroups. The most inclusive unit, the Supergroup, as applied to bedrock and superficial deposits is typically used to denote deposits of similar bulk characteristics present at a national scale. In this context, when looking at geological maps and models, it would be helpful to be able to distinguish all anthropogenic deposits, recognized as a single named unit, from natural deposits. For Great Britain, it is suggested that a British Anthropogenic Deposits Supergroup would distinguish the artificial deposits from natural Quaternary superficial deposits of the Great Britain Superficial Deposits Supergroup of McMillan *et al.* (2005) and from bedrock lithostratigraphical designations (Table 2).

Members and beds. More exclusive units, members and beds, may be usefully defined for local to site-specific studies. The lower level units with member and bed status are typically units that may be defined at only a few well-exposed sections or from single boreholes. Such units may not be amenable to systematic and widespread mapping away from their stratotypes.

A member is the formal lithostratigraphical unit next in rank below a formation, and is always part of a formation (Salvador 1994). A member may extend from one formation to another (Salvador 1994). A formation need not be partially or totally subdivided into members. Members are commonly used to define a lithological sequence within a succession of distinctively different lithologies that are representative of the parental formation. For example, a shale member may be recognized within an interval of limestones that dominate a formation. For anthropogenic deposits, a member may be usefully defined to identify deposits of broadly similar genetic origin within a formation. For example, given an identification of a post-World War II succession in the Swansea area as a formation, it would be possible to recognize a member that included areas of engineered land-raising fill above the floodplain, and road and motorway embankments (Fig. 14). These may have similar lithological characteristics that are quite distinct from, say, landfill deposits. In the example provided, such deposits are sufficiently extensive to warrant identification as a member. This would equate with the Raised Fill and Engineered Embankment Type of the BGS enhanced classification scheme (Ford *et al.* 2010*a*), and it is accepted that there would be considerable overlap of the schemes.

A bed is the smallest formal unit in the hierarchy of sedimentary lithostratigraphical units (Salvador 1994). This should not be confused with a sedimentological bed, which identifies sediments deposited during a single event. A lithostratigraphical bed can comprise one or more sedimentological beds. Bed names are commonly applied to distinctive units that may be thin, or known only from a borehole or single exposure. Beds may be selected to provide useful marker units representing single events with one lithology, although, commonly, beds are left as informal and may lack definition of type sections. A bed could represent an extensive fossiliferous marine band in bedrock strata, or a locally developed peat in a superficial deposit. For artificial deposits, a bed could represent a single volume of deposit with a distinct lithological characteristic; for example, the Crymlyn Landfill Bed (Fig. 14). This would equate with the Landfill Waste Tip Unit (Domestic Refuse) of the BGS enhanced classification scheme (Ford *et al.* 2010*a*). Significant units could be delineated based upon morphogenetic and/or lithological characteristics, and named after a prominent location associated with the deposit, but may be left informal. It is not envisaged that all isolated masses of artificial

Table 2. *Examples of lithostratigraphical hierarchies for bedrock and natural superficial deposits (following McMillan et al. 2005) and possible application to anthropogenic deposits (Ford et al. 2008; Waters et al. 2009)*

Lithostratigraphical level	Example sedimentary sequence (extent) [maximum thickness]	Example superficial deposit (extent) [maximum thickness]	Example Artificial ground (extent) [maximum thickness]
Supergroup	Coal Measures Supergroup (Great Britain) [1900 m]	Great Britain Superficial Deposits Supergroup (Great Britain) [200 m]	British Anthropogenic Deposits Supergroup (Great Britain) [65 m]
Group	Pennine Coal Measures Group (central and northern England and North Wales) [1900 m]	Britannia Catchments Group (Great Britain) [100 m]	South Wales Industrial Group (South Wales Coalfield) [12 m]
Formation	Pennine Upper Coal Measures Formation (central and northern England and North Wales) [800 m]	Breighton Sand Formation (Vale of York) [>6 m]	Swansea Industrial Revolution Formation (Lower Swansea Valley) [9 m]
Member	Hemsworth Member (Yorkshire coalfield) [93 m]	Skipwith Sand Member (Vale of York) [2.4 m]	Lower Swansea Valley Foundry Waste Member (Lower Swansea Valley) [9 m]
Bed	Upton Coal (Wakefield) [1.8 m]	Skipwith Peat Bed (Vale of York) [0.2 m]	Hafod Foundry Waste Bed (Crymlyn) [9 m]
Local unit name	Third Cherry Tree Marker (Wakefield) [14.78 m]		Applicable, to local description, but far too many names to formally adopt

deposit should be named as lithostratigraphical beds, in the same way that not all sandstones in a cyclic sedimentary unit are named beds.

A significant problem with the approach expressed above is the inherent separation of deposits with the same name. In bedrock and superficial deposits, subsequent erosion can result in isolation of parts of a named unit, but the assumption is that at the time of deposition they were laterally contiguous. This is clearly not the case in the scheme proposed above, which could result in a series of isolated colliery spoil tips being included within a single member. Either, it is necessary to argue that lithostratigraphy needs to evolve as a classification scheme to consider the unique circumstances of artificial deposits, not currently considered in any scheme, or it be argued that lithostratigraphical definitions are fixed and cannot be modified to incorporate the vagaries of man-made stratigraphy. This is a decision that would need to be made by the International Commission on Stratigraphy.

Indirect evidence

The practical application of lithostratigraphy in geological mapping where exposure is limited by weathering, soil cover, vegetation or buildings involves a range of indirect evidence to support the extrapolation of boundaries between outcrops or lithological boreholes. Lithostratigraphy permits indirect evidence for the identification of units and their boundaries where lithological identity is difficult to determine (Salvador 1994). Compared with natural deposits, anthropogenic deposits typically exhibit fewer exposures because of their function, managed nature and age (i.e. in most cases, natural erosion has not created exposures of artificial deposits). Indirect evidence is, therefore, a key factor when considering a lithostratigraphical approach to the classification and mapping of anthropogenic deposits. Common sources of indirect evidence are described below.

The indirect evidence used most commonly in lithostratigraphical mapping in Great Britain is geomorphology. Many natural deposits are associated with characteristic landforms. These may form as a result of particular physical responses to weathering (e.g. the creation of escarpments or ridges) or they may be constructional, reflecting the original form of the deposit (e.g. a moraine or alluvial terrace). Characteristic slope profiles are sometimes diagnostic of particular bedrock units and useful for geological mapping (Aldiss *et al.* 2012). Constructional landforms are typically used in the characterization and mapping of relatively fresh superficial deposits, and are exhibited by many surficial anthropogenic deposits (McMillan *et al.* 2005).

Landform information for anthropogenic deposits may be obtained by direct or remote observation, including the use of topographical data, aerial photography and digital elevation models (DEMs). In recent years, the increased availability and quality of DEMs, along with the development of visualization systems, has transformed the geomorphological mapping and the study of artificial ground, especially in urban conurbations where high-resolution data are now readily available. DEMs complement topographical data, allowing distinct bodies of anthropogenic deposit to be delineated, and indicative thicknesses and volumes to be assessed (Fig. 15). Where successive DEMs are available, the sequence of deposition and temporal change in volume of artificial deposits may be calculated. Multiple ages of aerial photography or topographical maps can prove equally valuable in assessing geomorphology and human landscape evolution, including evidence of the depositional process that may inform the lithostratigraphical classification of anthropogenic deposits (Fig. 12).

The association between geological deposits and distinct biotas is a form of indirect evidence used to map natural deposits. The physical and chemical properties of deposits and their soils can influence the vegetation that they support, either in terms of species affinity or growth characteristics. Arguably, these responses are more apparent in the case of anthropogenic deposits that typically exhibit properties considerably different to those of natural deposits. The age of anthropogenic deposits may also define vegetation characteristics (Schadek *et al.* 2009). Indirect evidence from vegetation may be of particular value in mapping disused anthropogenic deposits and those with degraded morphologies. Although anthropogenic deposits are known to provide unique habitats for particular species (e.g. metal-loving plants), the practical use of vegetation response in geological mapping relies on non-specialist, directly observable characteristics such as crop type. Indirectly, vegetation response to anthropogenic deposits is also used in the interpretation of aerial photography and remote sensing for the purpose of land-use classification (Slonecker *et al.* 2010).

In many situations, the geophysical properties of anthropogenic sediments are significantly different to those of natural host lithologies. A range of techniques including resistivity imaging, ground penetrating radar (GPR) and gravity surveys are used in archaeological, engineering and environmental studies to delimit and characterize anthropogenic deposits (Fenning & Williams 1997; Guerin *et al.* 2004; Kulessa *et al.* 2006; Boyce *et al.* 2009; Boudreault *et al.* 2010). GPR, in particular,

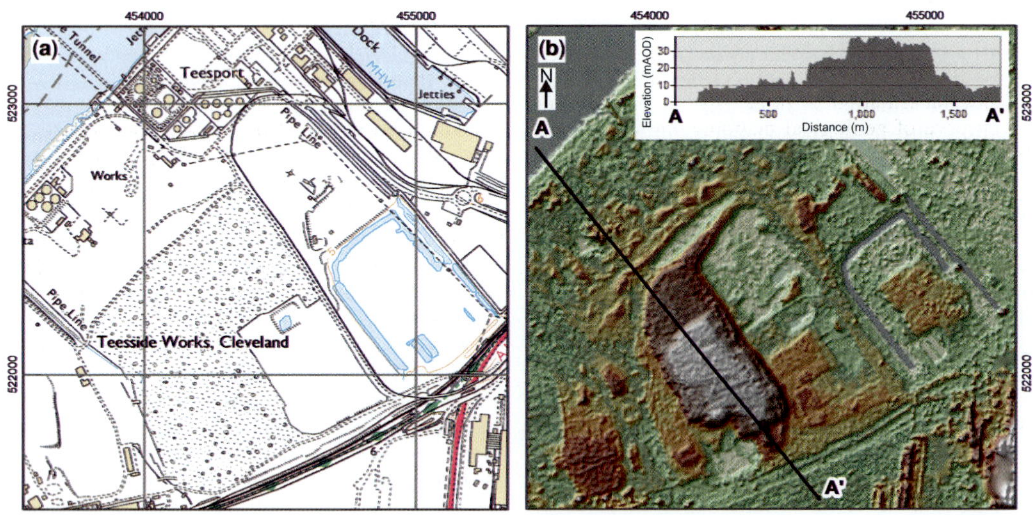

Fig. 15. (**a**) Topographical map and (**b**) DEM data for an area of NE England where the iron and steel industry has resulted in extensive anthropogenic modification of the landscape, including the creation of large waste tips. The DEM data resolve the geomorphology of the anthropogenic deposits in considerable detail, picking out distinct regions within the area shown on the topographical data as a single body of waste (stippled area). The anthropogenic deposits in this area reach around 40 m in elevation (see the inset elevation profile), some 30 m above the surrounding land surface. Topographical data ©Crown Copyright Ordnance Survey. All rights reserved. NEXTMap Britain elevation data from Intermap Technologies.

is used extensively by archaeologists to characterize anthropogenic deposits on a site scale (Bladon *et al.* 2011). Although limited coverage of suitable geophysical data prevents their widespread use in the systematic characterization of anthropogenic deposits, they may be of considerable value in resolving concealed deposits and the effects of anthroturbation including abandoned mine workings (Chambers *et al.* 2007).

Summary

The magnitude of direct anthropogenic transformation of the landscape and the shallow subsurface is globally significant. Human activity associated with urbanization and the exploitation of natural resources is now responsible for a greater flow of materials than natural sedimentary systems. Materials directly and deliberately moved by humans in the novel sedimentary environment are defined as anthropogenic sediments, and their classification is the focus of this paper. Anthropogenic sediments and their deposits are significant in terms of their engineering and environmental properties. They are potential material resources and sources of evidence in archaeological studies and the ongoing debate over the Anthropocene.

Classification schemes created for a diversity of disciplines can be used to varying degrees in the geological study of anthropogenic deposits. Existing classification schemes used by geoscientists in Great Britain for anthropogenic deposits are largely based on morphogenetic attributes. Whilst these approaches offer practical means of depicting the gross distribution of anthropogenic deposits for the purpose of geological mapping and modelling, they do not allow different phases or lithologically distinct deposits to be differentiated. In the case of natural superficial and bedrock sedimentary deposits, a lithostratigraphical approach is used. Lithostratigraphy is based on observable lithological characteristics and the spatial relationship of deposits. This approach distinguishes different phases of deposition, allowing units to be correlated. In the case of anthropogenic deposits, lithostratigraphy could allow the human transformation of the landscape to be charted.

A lithostratigraphical approach to anthropogenic deposits has not previously been attempted. Although natural and anthropogenic sediments share many similarities, they differ in several key respects. Anthropogenic sediments are typically thinner, discontinuous, poorly exposed, lithologically heterogeneous, rich in novel materials and, in many cases, their upper surface is represented by the present-day land surface. Natural superficial deposits can exhibit many similar characteristics to anthropogenic deposits and the challenges these pose to lithostratigraphical classification have been recognized by previous studies. This paper considers whether current lithostratigraphical procedures could be applied to anthropogenic deposits to establish a meaningful and useful classification. A summary of the main findings, with reference to the principles of lithostratigraphy, is presented in Table 3.

Lithostratigraphy is used to classify sediments or volcanic deposits that conform to the Law of Superposition. Many anthropogenic sediments satisfy this requirement; they are typically deposited subaerially, resting on older anthropogenic deposits or, more commonly, sitting unconformably on natural superficial or bedrock deposits. The unconformable boundary between natural and anthropogenic deposits may be relatively easily identified. However, the identification and correlation of significant discontinuities within successions of anthropogenic sediments may be more problematic. Where anthropogenic materials are enclosed by earlier natural or anthropogenic deposits (e.g. in the case of back-filled mine workings), they may be considered as anthropogenic intrusives and classified as lithodemic units. The extensive and increasing interaction between humans and the shallow subsurface may be considered a form of bioturbation or anthroturbation. The preservation potential of anthroturbation in the rock record is considerable and may be a significant factor in recognizing the Anthropocene.

Many anthropogenic units, being surficial deposits, have no overlying strata. Their upper bounding surface is defined as the present-day land surface that may be subject to a combination of natural erosive or depositional processes, or, in the case of many urban areas, frequent anthropogenically driven change. Where the original landforms are preserved, morphology may be used to classify the gross distribution of anthropogenic deposits. However, where landforms are destroyed or indiscernible (e.g. in the case of concealed deposits), a lithostratigraphical approach provides a means by which anthropogenic deposits may be classified and correlated.

Original lateral continuity is a defining characteristic of lithostratigraphical units. However, the lateral continuity and thickness of the majority of anthropogenic deposits is significantly less than that that of natural superficial and bedrock deposits. Social and economic factors dominate physical constraints in controlling the distribution of anthropogenic deposits. As such, individual deposits are usually discrete entities, bounded by limits other than physical constraints. Although geological maps based on existing morphogenetic schemes depict some extensive bodies of made ground, these

Table 3. *Summary of lithostratigraphical criteria applied to anthropogenic deposits*

Lithostratigraphical criteria	Satisfied by Anthropogenic deposits?	Examples of effective situations	Examples of exceptions/challenges
Law of Superposition	Yes – mostly	Most terrestrial deposits, resulting from the sequential emplacement of made ground including heritage deposits	Deposits related to subsurface engineering and storage including back-filled cavities and injected material (i.e. anthropogenic 'intrusives' or 'anthroturbation'); lithodemic classification may be appropriate
Original lateral continuity	Rarely – although 'yes' if broad scales considered	Deposits related to large-scale urban or industrial expansion (e.g. deposits associated with coalfield development); deposits associated with major socioeconomic or environmental drivers including conflict and climate change adaptation (e.g. flood defences)	Most deposits related to site-scale processes; deposits that result from the infilling of surface excavations. NB: in all cases, lateral continuity is a function of the scale of the study and the lithostratigraphical resolution used
Lithological distinction	Yes	Deposits imported from diverse supply areas; deposits incorporating characteristic (ideally time-varying) novel materials; deposits composed of contrasting proportions of natural sediment, novel materials and reworked natural deposits	Multiple deposits composed of locally derived natural material; deposits sourced from lithologically similar supply areas; deposits composed of reworked anthropogenic deposits
Definable type section/area	Rarely	Spatially extensive surficial deposits encountered in subsequent ground investigation boreholes; deposits classified as heritage deposits for which exposures are preserved. NB: type section definition of anthropogenic deposits may be simpler in the geological future	Poor natural exposure of deposits; restricted spatial extent may limit access to type sections/areas; extensive reworking and remodelling of deposits may prevent long-term preservation of type sections/areas
Hierarchical framework and naming	Yes – if broad scales considered	Deposits exhibiting broad lithological and stratigraphical characteristics that are indicative of particular novel environments, and may be grouped accordingly (see Table 2 for examples)	Anthropogenic units require broader ranges of thickness and lithological variability than natural units of equivalent status; anthropogenic deposits do not satisfy stipulation for original lateral continuity

typically represent amalgamations of multiple phases and types of deposit. Where instances exist of extensive anthropogenic sediments resulting from a single phase of deposition, they are typically associated with large-scale mining, industry and infrastructure. Arguably, such deposits also represent amalgamations and may be locally diachronous. However, in the context of a hierarchical lithostratigraphical framework they could be defined as a single unit at an appropriate scale. Anthropogenic deposits associated with a single phase of deposition are considerably thinner than equivalent natural deposits. Establishing lithostratigraphical units for anthropogenic deposits at a mappable scale will, in most cases, require the aggregation of deposits. Key historical events have resulted in significant temporal changes in the volume and extent of anthropogenic material deposited. These changes represent distinguishing characteristics that may support the application of lithostratigraphy to anthropogenic deposits. On a global scale, these changes are highly diachronous.

Lithostratigraphical units defined at the mappable scale should have distinguishing lithological characteristics. The lithology of anthropogenic sediments can be described using conventional descriptive approaches. The bulk composition of anthropogenic deposits is considered in terms of three components: natural materials disturbed by direct human action; natural sediments; and novel materials. Deposits dominated by disturbed and novel material are defined as anthropogenic. The composition and proportion of these components can vary between adjacent deposits, enabling their differentiation. These changes reflect different sediment supply areas, contrasting depositional processes or the relative age of deposits. Where material is locally derived or reworked, lithological distinction is problematic, and a lithostratigraphical approach may be unworkable. However, novel materials can aid the distinction between adjacent units and may provide markers for units of a particular age, akin to the use of fossil assemblages in natural deposits. The original bulk compositions of some types of anthropogenic deposits have varied considerably with time, these changes showing significant regional or global diachroneity. Significant changes often follow key historical events, and waste deposits, in particular, exhibit major temporal changes in their lithological character. As waste deposits are volumetrically the largest type of anthropogenic deposit, this presents a particular opportunity for lithostratigraphical classification in terms of providing distinctive lithologies. However, preservation potential of anthropogenic deposits must be considered when assessing their lithological characteristics as the composition of many deposits, including waste, is subject to significant change. Chemical and physical alteration can be considered analogous to diagenesis in the natural rock cycle.

The definition of a lithostratigraphical unit requires the designation of a type section or area. In the case of anthropogenic deposits, naturally occurring exposures are uncommon, ephemeral, and rarely provide an inclusive depiction of the deposit and its boundaries. Artificial exposures such as trial pits or excavations created during site redevelopment are similarly short lived. The lack of suitable exposures is a significant impediment to the application of conventional lithostratigraphical procedures to anthropogenic deposits. Borehole intersections offer potential type sections for anthropogenic deposits. However, borehole intersections through anthropogenic deposits commonly recover loose samples with disrupted boundaries, few suitable intersections are preserved for future reference and many deposits have little or no borehole coverage.

The hierarchical nature of lithostratigraphy provides a practical means of classifying and rationalizing deposits across a range of scales, with formations defined as the primary mappable unit. In the case of anthropogenic deposits, their thin, discontinuous and diverse nature renders impracticable and undesirable the definition of each individual body or landform as a separate formation. A possible approach to the lithostratigraphical classification of anthropogenic deposits involves the definition of units on the basis of broad lithological and stratigraphical characteristics that are indicative of particular novel environments. These novel environments and deposits may reflect dominant industrial or economic activity, or characteristic associations of activities, and be subject to changes with time. It is the particular lithostratigraphical characteristics of deposits of a particular time, rather than their age as such, that result in a distinctive stratigraphy that may be identified by a range of direct and indirect evidence, and grouped or subdivided accordingly. This approach is similar to that used in the lithostratigraphical classification of natural deposits. However, the range of thicknesses and lithological heterogeneity accommodated by an anthropogenic unit may be considerably greater than that of a natural deposit of a similar lithostratigraphical status, reflecting the inherent variability of anthropogenic deposits. Importantly, this approach allows the inclusion of deposits that were originally spatially isolated within a single unit (e.g. discrete mine waste tips or similar character combined into a single member). Current lithostratigraphical procedures require original lateral continuity and this would have to change if this suggested approach were accepted. Existing classification schemes for anthropogenic deposits may be used in parallel to lithostratigraphy to convey additional information on genesis and landform. In particular, the morphostratigraphical approach of Ford *et al.* (2010*a*) remains a practical mapping and descriptive tool.

Indirect evidence can be used in the identification of lithostratigraphical units and their boundaries. Sources of indirect evidence used routinely in the context of natural deposits are effective in the case of anthropogenic deposits. Aerial photography interpretation and geomorphological mapping, including the appraisal of DEMs, are well suited to delineating anthropogenic deposits with relatively fresh morphologies. Successive generations of imagery or data can be used to chart changing land use and indicate the potential location of concealed deposits. Geophysical surveys, although typically commissioned for purposes other than the study of anthropogenic deposits, can be used to reveal their subsurface geometry. However, the limited coverage of geophysical datasets precludes their widespread use for systematic mapping or modelling.

Conclusion

This study does not set out to define a lithostratigraphy for anthropogenic deposits. Its aim is to assess the applicability of current lithostratigraphical procedures to the classification of anthropogenic deposits. Lithostratigraphy was pioneered in the study of bedrock geology, and has been subsequently adapted for natural superficial deposits. Situations have been identified where anthropogenic deposits are shown to conform to many of the criteria for lithostratigraphy. The scale at which lithostratigraphy may be effectively applied to anthropogenic deposits may be significantly different when compared to natural deposits. However, even if lithostratigraphical units defined for anthropogenic deposits are relatively coarse, this approach affords the end user a level of functionality that is not supported by existing classifications of anthropogenic deposits.

It is clear that much remains to be done to determine the feasibility of any proposal to adopt a lithostratigraphical scheme for anthropogenic deposits. A truly global perspective is essential to ensure that the challenges identified here including scale, diachroneity, and other key differences between natural and anthropogenic deposits are appreciated and resolved. Should a lithostratigraphical approach to the classification of anthropogenic deposits be adopted, the revision of established stratigraphical guidelines would need to be considered. The incorporation of stratigraphical procedures for anthropogenic deposits alongside those for natural deposits in international guidelines would make a significant contribution to the study of anthropogenic deposits and the impact of human processes on the landscape. A consistent lithostratigraphical framework for anthropogenic deposits would, in turn, contribute to the definition and characterization of the Anthropocene.

However, significant changes to existing lithostratigraphical procedures would be required to provide an effective classification of anthropogenic deposits. These changes, and their potential impact across the stratigraphical column, would require careful consideration before being enacted. A practical solution would be to establish a complementary classification scheme designed specifically for anthropogenic deposits. Based on lithostratigraphy and incorporating elements of the current morphogenetic classification, this 'anthrostratigraphical' approach would combine relevant strengths from the existing schemes and could be tailored to accommodate the unique characteristics of anthropogenic deposits.

It is important to remember that chronostratigraphical schemes that define geological time intervals, such as the proposed Anthropocene Epoch, evolve in parallel with lithostratigraphical schemes that aim to characterize the physical attributes of rock masses. The inherently diachronous nature of lithostratigraphy, expressed in the novel sedimentary environment by the growth with time of urban conurbations or the gradual spread of novel processes and lithologies as a product of technical innovations, makes lithostratigraphy a poor criterion for defining the Anthropocene as such.

However, changes in lithostratigraphically discernible characteristics of anthropogenic deposits including rate of accumulation and lithological composition provide some support for the recognition of a significant signature immediately following World War II. Although earlier events and periods in history, including the Industrial Revolution, are associated with national and regionally significant signatures, the period following 1945, or the 'Great Acceleration of the Anthropocene' (Steffen *et al.* 2007), is marked by the first globally expansive change in the lithostratigraphy of anthropogenic deposits.

The authors wish to acknowledge the assistance of colleagues and research partners, including M. Lark and B. Marchant for geostatistical advice, and S. Ward for figure production. Helpful comments from A. Smith, and constructive feedback from S. Davies and J. Zalasiewicz, are also gratefully acknowledged. J. Powell, D. Lowe and H. Kessler are thanked for their contribution to the development of the classification schemes of artificial ground used in 3D geological models and on maps published by the British Geological Survey. This work was funded with support from the British Geological

Survey's Engineering Geology science area. This article is published with the permission of the Executive Director of the British Geological Survey (NERC).

References

Aldiss, D. T., Farrant, A. R. & Hopson, P. M. 2012. Geological mapping of the Late Cretaceous Chalk Group of southern England: a specialised application of landform interpretation. *Proceedings of the Geologists' Association*, **123**, 728–741.

Bladon, P., Moffat, I., Guilfoyle, D., Beale, A. & Milani, J. 2011. Mapping anthropogenic fill with GPR for unmarked grave detection: a case study from a possible location of Mokare's grave, Albany, Western Australia. *Exploration Geophysics*, **42**, 249–257.

Boudette, E. L., Hatch, N. L., Jr. & Harwood, D. S. 1976. *Reconnaissance Geology of the Upper St. John and Allagash River Basins*. United States Geological Survey, Bulletin, **1406**.

Boudreault, J. P., Dube, J. S., Chouteau, M., Winiarski, T. & Hardy, E. 2010. Geophysical characterization of contaminated urban fills. *Engineering Geology*, **116**, 196–206.

Boyce, J. I., Reinhardt, E. G. & Goodman, B. N. 2009. Magnetic detection of ship ballast mounds and anchorages at Caesarea Maritima, Israel. *Journal of Archaeological Science*, **36**, 1516–1526. http://dx.doi.org/10.1016/j.jas.2009.03.007.

Bridgewater, A. V. 1986. Refuse composition projections and recycling technology. *Resources and Conservation*, **12**, 159–174.

Bridgland, D. R., Howard, A. J., White, M. J. & White, T. S. 2006. *The Trent Valley: Archaeology and Landscapes of the Ice Age*. Durham University, Durham, UK.

British Geological Survey 1978. *Sunderland. England and Wales Sheet 21, Solid and Drift Geology. 1:50,000*. Ordnance Survey, Southampton for the Institute of Geological Sciences.

British Geological Survey 1995. *Specification for the Preparation of 1:10 000 Scale Geological Maps*, 2nd edn. Research Report WA/95/64. British Geological Survey, Keyworth, Nottingham.

British Geological Survey 2012*a*. *The BGS Lexicon of Named Rock Units*. http://www.bgs.ac.uk/Lexicon/home.cfm.

British Geological Survey 2012*b*. *Boreholes*. http://www.bgs.ac.uk/products/onshore/SOBI.html.

British Standards Institution 1999. *BS5930: 1999. Code of Practice for Site Investigations*. BSI, London.

Burnley, S. J. 2001. The impact of the European landfill directive on waste management in the United Kingdom. *Resources, Conservation and Recycling*, **32**, 349–358.

Burnley, S. J. 2007. A review of municipal waste composition in the United Kingdom. *Waste Management*, **27**, 1274–1285.

Carver, M. 1987. *Underneath English towns. Interpreting Urban Archaeology*. B. T. Batsford, London.

Cathcart, R. B. 2011. Anthropic Rock: a brief history. *History of Geo and Space Sciences*, **2**, 57–74.

Chambers, J. E., Wilkinson, P. B., Weller, A. L., Meldrum, P. I., Gilvy, R. D. & Caunt, S. 2007. Mineshaft imaging using surface and crosshole 3D electrical resistivity tomography: a case history from the East Pennine Coalfield, UK. *Applied Geophysics*, **62**, 324–337.

de Beer, J., Price, S. J. & Ford, J. R. 2012. 3D modelling of geological and anthropogenic deposits at the World Heritage Site of Bryggen in Bergen, Norway. *Quaternary International*, **251**, 107–116.

Department for Environment Food and Rural Affairs 2009. *Municipal Waste Composition: A Review of Municipal Waste Component Analyses (Project WR0119)*. Defra, London.

Douglas, I. & Lawson, N. 2001. The human dimensions of geomorphological work in Britain. *Journal of Industrial Ecology*, **4**, 9–33.

Edgeworth, M. 2013. The relationship between archaeological stratigraphy and artificial ground and its significance in the Anthropocene. *In*: Waters, C. N., Zalasiewicz, J. A., Williams, M., Ellis, M. A. & Snelling, A. M. (eds) *A Stratigraphical Basis for the Anthropocene*. Geological Society, London, Special Publications, **395**, first published on October 25, 2013, http://dx.doi.org/10.1144/SP395.3.

Ellis, N. 2008. A history of the Geological Conservation Review. *In*: Burek, V. V. & Prosser, C. D. (eds) *The History of Geoconvservation*. Geological Society, London, Special Publications, **300**, 123–135.

English Heritage 2002. *Thesaurus of Monument Types* (updated online). http://thesaurus.english-heritage.org.uk.

English Heritage 2012. *The National Heritage Protection Plan*. http://www.english-heritage.org.uk/professional/protection/national-heritage-protection-plan/.

Environment Agency 2005. *Hazardous Waste. Interpretation of the Definition and Classification of Hazardous Waste*, 2nd edn, Version 2.2. http://www.environment-agency.gov.uk/static/documents/GEHO0603BIRB-e-e.pdf.

European Union 1999. *Council Directive 1999/31/EC of April 1999 on the Landfill of Waste*. http://ec.europa.eu/environment/waste/landfill_index.htm.

European Union 2008. *Directive 2008/98/EC on Waste (Waste Framework Directive)*. http://ec.europa.eu/environment/waste/framework/index.htm.

Eurostat 2009. *LUCAS: Land Use/Land Cover Frame Statistical Survey, Technical Reference Document C-3, Land use and Land Cover Nomenclature*. http://epp.eurostat.ec.europa.eu/portal/page/portal/lucas/documents/Nomenclature_LUCAS2009_C_3.pdf.

Fenning, P. J. & Williams, B. S. 1997. Multicomponent geophysical surveys over completed landfill sites. *In*: McCann, D. M., Eddleston, M., Fenning, P. & Reeves, G. M. (eds) *Modern Geophysics in Engineering Geology*. Geological Society, London, Engineering Geology Special Publications, **12**, 125–138.

Ford, J. R., Cooper, A. H., Price, S. J., Gibson, A. D., Pharaoh, T. C. & Kessler, H. 2008. *Geology of the Selby District: A Brief Explanation of the Geological Map Sheet 71, Selby (England and Wales)*. British Geological Survey, Keyworth, Nottingham.

FORD, J. R., KESSLER, H., COOPER, A. H., PRICE, S. J. & HUMPAGE, A. J. 2010*a*. *An Enhanced Classification of Artificial Ground*. British Geological Survey, Open Report, OR/10/036.

FORD, J. R., MATHERS, S. J., ROYSE, K. E., ALDISS, D. T. & MORGAN, D. J. R. 2010*b*. Geological 3D modelling: scientific discovery and enhanced understanding of the subsurface, with examples from the UK. *Zeitschrift der Deutschen Gesellschaft für Geowissenschaften*, **161/162**, 205–218.

GANDY, M. 1993. *Recycling and Waste. An Exploration of Contemporary Environmental Policy*. Ashgate Publishing, Aldershot.

GUERIN, R., BEGASSAT, P., BENDERITTER, Y., DAVID, J., TABBAGH, A. & THIRY, M. 2004. Geophysical study of the industrial waste land in Mortagne-du-Nord (France) using electrical resistivity. *Near Surface Geophysics*, **2**, 137–143.

HARRISON, A. R. 2006. *National Land Use Database: Land Use and Land Cover Classification*. Office of the Deputy Prime Minister: Queen's Printer and Controller of Her Majesty's Stationary Office, London. http://www.communities.gov.uk/documents/planningandbuilding/pdf/144275.pdf.

HISTORIC SCOTLAND 2011. *Scottish Historic Environment Policy*. Historic Scotland, Edinburgh. www.historic-scotland.gov.uk/index/heritage/policy/shep.htm.

HOLDEN, J., WEST, L. J., HOWARD, A. J., MAXFIELD, E., PANTER, I. & OXLEY, J. 2006. Hydrological controls of in situ preservation of waterlogged archaeological deposits. *Earth-Science Reviews*, **78**, 59–83.

HOLDEN, J., HOWARD, A., WEST, L., MAXFIELD, E., PANTER, I. & OXLEY, J. 2009. A critical review of hydrological data collection for assessing preservation risk for urban waterlogged archaeology: A case study from the City of York, UK. *Journal of Environmental Management*, **90**, 3197–3204, http://dx.doi.org/10.1016/j.jenvman.2009.04.015.

HOLLIS, J. M. 1992. *Proposals for the Classification, Description and Mapping of Soils in Urban Areas*. English Nature/Soil Survey and Land Research Centre, Peterborough.

INTERNATIONAL MINING CONSULTANTS LIMITED 1999. *Prospects for Coal Production in England, Scotland and Wales – DTI Review of Energy Sources for Power Generation*. Her Majesty's Stationery Office, London.

IUSS WORKING GROUP WRB 2006. *World Reference Base for Soil Resources 2006*. World Soil Resources Reports, 103. FAO, Rome.

KULESSA, B., CHIARULLI, B., MCCARTHY, P., HANEY, S. & JONES, K. 2006. Large-scale geophysical reconstruction of man-made ground at former industrial iron-furnace plantations. *Geophysics*, **71**, B55–B61.

LAWLEY, R. & GARCIA-BAJO, M. 2009. *The National Superficial Deposit Thickness Model (Version 5)*. British Geological Survey, Open Report, OR/09/049.

LOTTERMOSER, B. G. 2011. Recycling, reuse and rehabilitation of mine wastes. *Elements*, **7**, 405–410.

MAKEDON, T., CHATZIGOGOS, N. P. & SPANDOS, S. 2009. Engineering geological parameters affecting the response of Thessaloniki's urban fill to a major seismic event. *Engineering Geology*, **104**, 167–180.

MCCANN, B. & ORTON, C. 1989. The fleet valley project. *London Archaeologist*, **6**, 102–107.

MCMILLAN, A. A. 2005. A provisional Quaternary and Neogene lithostratigraphical framework for Great Britain. *Netherlands Journal of Geosciences – Geologie en Mijnbouw*, **84**, 87–107.

MCMILLAN, A. A. & POWELL, J. H. 1993. *BGS Rock Classification Scheme: The Classification of Artificial (Man-made) Ground and Natural Superficial Deposits. Version 2*. British Geological Survey, Technical Report, WG/93/46/R.

MCMILLAN, A. A. & POWELL, J. H. 1999. *BGS Rock Classification Scheme. Volume 4, Classification of Artificial (man-made) Ground and Natural Superficial Deposits: Applications to Geological Maps and Datasets in the UK*. British Geological Survey, Research Report, RR/99/004.

MCMILLAN, A. A., HAMBLIN, R. J. O. & MERRITT, J. W. 2005. *An Overview of the Lithostratigraphical Framework for the Quaternary and Neogene Deposits of Great Britain (Onshore)*. British Geological Survey, Research Report, RR/04/04.

MCMILLAN, A. A., HAMBLIN, R. J. O. & MERRITT, J. W. 2011. *A Lithostratigraphical Framework for Onshore Quaternary and Neogene (Tertiary) Superficial Deposits of Great Britain and the Isle of Man*. British Geological Survey, Research Report, RR/10/03.

MURPHY, M. A. & SALVADOR, A. 1999. International stratigraphic guide- an abridged version (International Subcomission on Stratigraphic Classification of the IUGS; International Commission on Stratigraphy). *Episodes*, **22**, 255–271.

NIREI, H., FURUNO, K., OSAMU, K., MARKER, B. & SATKŪNAS, J. 2012. Classification of man made strata for assessment of geopollution. *Episodes*, **35**, 333–336.

NORBURY, D. 2010. *Soil and Rock Description in Engineering Practice*. Whittles, Caithness.

NORTH AMERICAN COMMISSION ON STRATIGRAPHIC NOMENCLATURE 2005. North American stratigraphic code. *American Association of Petroleum Geologists Bulletin*, **89**, 1547–1591.

PARFITT, J. 2002. *Analysis of Household Waste Composition and Factors Driving Waste Increases*. WRAP Report. Defra, London.

PRICE, S. J., FORD, J. R., KESSLER, H., COOPER, A. H. & HUMPAGE, A. J. 2004. Mapping our impact on the surface of the Earth. *Earthwise*, **20**, 30–31.

PRICE, S. J., BURKE, H. F., TERRINGTON, R. L., REEVES, H. J., BOON, D. & SCHEIB, A. J. 2010. The 3D characterisation of the zone of human interaction and the sustainable use of underground space in urban and peri-urban environments: case studies from the UK. *Zeitschrift der Deutschen Gesellschaft für Geowissenschaften*, **161**, 219–235.

PRICE, S. J., FORD, J. R., COOPER, A. H. & NEAL, C. 2011. Humans as major geological and geomorphological agents in the Anthropocene: the significance of artificial ground in Great Britain. *In*: ZALASIEWICZ, J. A., WILLIAMS, M., HAYWOOD, A. & ELLIS, M. (eds) *The Anthropocene: A New Epoch of Geological Time. Philosophical Transactions of the Royal Society (Series A)*, **369**, 1056–1084.

RAWSON, P. F., ALLEN, P. M. ET AL. 2002. *Stratigraphical Procedure*. Geological Society, London.

RIVAS, V., CENDERO, A. *ET AL.* 2006. Geomorphic consequences of urban development and mining activities; an analysis of study areas in Spain and Argentina. *Geomorphology*, **73**, 185–206.

ROSENBAUM, M. S., MCMILLAN, A. A., POWELL, J. H., COOPER, A. H., CULSHAW, M. G. & NORTHMORE, K. J. 2003. Classification of artificial (man-made) ground. *Engineering Geology*, **69**, 399–409.

ROSSITER, D. G. 2007. Classification of urban and industrial soils in the world reference base for soil resources. *Journal of Soils Sediments*, **7**, 96–100, http://dx.doi.org/10.1065/jss2007.02.208.

SALVADOR, A. 1994. *International Stratigraphic Guide. A Guide to Stratigraphic Classification, Terminology, and Procedure*, 2nd edn. The International Union of Geological Sciences and The Geological Society of America, Denver, CO.

SATKŪNAS, J., GREGORAUSKIENÉ, V., KANOPIENÈ, R., MIKULÈNAS, V., MIKEVIČIUS, V., ŠAČKUS, V. & ŠLAUTERIS, A. 2011. Man-made formations and geopollution: state of knowledge in Lithuania. *Geologija*, **53**, 36–44.

SCHADEK, U., STRAUSS, B., BIEDERMANN, R. & KLEYER, M. 2009. Plant species richness, vegetation structure and soil resources of urban brownfield sites linked to successional age. *Urban Ecosystems*, **12**, 115–126.

SCHENCK, H. G. & MULLER, S. W. 1941. Stratigraphic terminology. *Bulletin of the Geological Society of America*, **52**, 1419–1426.

SCHULTZ, E. H. 1982. The chromosome and supersome-terms proposed for low-rank chronostratigraphic units. *Canadian Petroleum Geology*, **30**, 29–33.

SHERLOCK, R. L. 1922. *Man as a Geological Agent; An Account of his Action on Inanimate Nature*. H F & G Witherby, London.

SLONECKER, T., FISHER, G., AIELLO, D. & HAACK, B. 2010. Visible and infrared remote imaging of hazardous waste: a review. *Remote Sensing*, **2**, 2474–2508, http://dx.doi.org/10.3390/rs2112474.

STEFFEN, W., CRUTZEN, P. J. & MCNEILL, J. 2007. The anthropocene: are humans now overwhelming the great forces of nature? *Ambio*, **36**, 614–621.

SYVITSKI, J. P. M. & KETTNER, A. 2011. Sediment flux in the Anthropocene. *In*: ZALASIEWICZ, J. A., WILLIAMS, M., HAYWOOD, A. & ELLIS, M. (eds) *The Anthropocene: A New Epoch of Geological Time. Philosophical Transactions of the Royal Society (Series A)*, **369**, 957–975.

SZABÓ, J. 2010. Anthropogenic geomorphology: subject and system. *In*: SZABÓ, J., DÁVID, L. & LÓCZY, D. (eds) *Anthropogenic Geomorphology A guide to Man-Made Landforms*. Springer, Dordecht, 3–10.

THORPE, S., BURKE, H. F. & TERRINGTON, R. L. 2011. *The Anthropogenic Land use History and Artificial Ground of the River Fleet*. British Geological Survey, Internal Report, IR/11/042.

UK NATIONAL ECOSYSTEM ASSESSMENT 2011. *The UK National Ecosystem Assessment: Synthesis of the Key Findings*. UNEP-WCMC, Cambridge.

UNDERWOOD, J. R. 2001. Anthropic rocks as a fourth basic class. *Environmental & Engineering Geoscience*, **7**, 104–110.

US ENVIRONMENTAL PROTECTION AGENCY 2002. *Engineering Study*. http://www.epa.gov/region3/mtntop/pdf/appendices/h/REVengineeringcover4-24.pdf

WANG, P. & LIU, D.-Y. 2012. Physical and chemical properties of sintering red mud and bayer red mud and the implications for beneficial utilization. *Materials*, **5**, 1800–1810.

WASTE ONLINE 2004. *History of Waste and Recycling Information Sheet*. http://dl.dropbox.com/u/21130258/resources/informationsheets/historyofwaste.htm.

WATERS, C., WATERS, R. A., BARCLAY, W. J. & DAVIES, J. R. 2009. *A Lithostratigraphical Framework for the Carboniferous Successions of Southern Great Britain (Onshore)*. British Geological Survey, Research Report, RR/09/01.

WATERS, C. N., NORTHMORE, K. *ET AL.* 1996. A Geological Background for Planning and Development in the City of Bradford Metropolitan District. British Geological Survey and Department of the Environment, Technical Report, WA/96/1.

WATERS, C. N., PRICE, S. J., DAVIES, J., TYE, A. M., BROWN, S. E. & SCHOFIELD, D. I. 2005. Urban geology of Swansea–Neath–Port-Talbot. *In*: BASSETT, M. G., DEISLER, V. K. & NICHOL, D. (eds) *Urban Geology in Wales, Volume 2*. National Museum of Wales, Geological Series, **24**, 7–22.

WENTWORTH, C. K. 1922. A scale of grade and class terms for clastic sediments. *Journal of Geology*, **30**, 377–392.

WILSON, A. A., REES, J. G., CROFTS, R. G., HOWARD, A. S., BUCHANAN, J. G. & WAINE, P. J. 1992. *Stoke-onTrent: A Geological Background for Planning and Development*. British Geological Survey, Technical Report, WA/91/01.

ZALASIEWICZ, J., WILLIAMS, M. *ET AL.* 2011. Stratigraphy of the Anthropocene. *In*: ZALASIEWICZ, J. A., WILLIAMS, M., HAYWOOD, A. & ELLIS, M. (eds) *The Anthropocene: A New Epoch of Geological Time. Philosophical Transactions of the Royal Society (Series A)*, **369**, 1036–1055.

The relationship between archaeological stratigraphy and artificial ground and its significance in the Anthropocene

M. EDGEWORTH

University of Leicester, School of Archaeology and Ancient History, University Road, Leicester, LE1 7RH, UK (e-mail: me87@le.ac.uk)

Abstract: This paper investigates the relationship between archaeological stratigraphy (in archaeology) and artificial ground (in geology) and considers their wider application to the investigation and characterization of the Anthropocene. Evidence from two archaeological case studies is used to illuminate key points. The first case study examines stratigraphic sequences from beneath the city of Leicester, UK; the second looks at stratified deposits within the prehistoric settlement mound of Abu Hureyra, Syria. Earthworks, riverworks and cultivation soils are also considered. Archaeological and geological perspectives are combined to develop a unified view of anthropogenic deposits that cover large parts of the surface of the Earth.

Archaeology and geology are related disciplines, sharing common roots in intellectual developments of the eighteenth and nineteenth centuries. The publication of Charles Lyell's *Principles of Geology* (Lyell 1830) was a significant moment in the development of both subjects. Largely as a result of Lyell's book, Nicholas Steno's earlier law of superposition of sedimentary layers and related geological principles were introduced into archaeological thinking. Other major formative influences on archaeology were James Hutton's work on the geological cycle (Hutton 1795) and William Smith's work on the correlation of strata (Smith 1816). Such works opened up the time depth necessary for emergence of archaeology as a modern discipline, and provided key conceptual tools for understanding material evidence encountered in excavation, especially with regard to interpretation of stratigraphy and relative dating of deposits. Detailed analysis of the early historical links between geology and archaeology is not attempted here but excellent accounts are available elsewhere (Daniel 1976; Daniel & Renfrew 1987; Harris 1989).

Despite such common sources and early connections, however, development of stratigraphical analysis in archaeology and geology has to some extent followed divergent paths. This is partly due to increasing academic specialization and division of labour, splitting respective subject matters along disciplinary lines. Although there are areas of overlap and collaboration (such as Holocene studies and geoarchaeology), archaeologists generally work with shallow surface deposits and shallow time depth, leaving the study of underlying rock formations and deep time for geologists. Likewise, geologists have until recently mostly left the study of anthropogenic deposits, and the relatively recent periods associated with their formation, for archaeologists to study.

The emergence of the concept of artificial ground within geology brings the two subjects back onto a potentially convergent path again. Artificial ground started to be shown on geological maps in the 1960s, but it was only in the 1990s that geological mapping of such deposits became standard. At first, investigation of it was mainly concerned with assessing stability and contamination of land for development. More recently – especially in relation to discussion of the Anthropocene as a proposed new geological epoch (Zalasiewicz *et al.* 2008, 2011) – artificial ground has come to be of interest to geologists in its own right (Price *et al.* 2011; Ford *et al.* 2013). There are important differences in how the two disciplines characterize the evidence, but deposits that geologists call artificial ground are broadly similar in principle to those that archaeologists call archaeological stratigraphy. Some consideration of how these different characterizations of the same material evidence might be related or integrated is required.

It is sometimes assumed that most artificial ground was formed in the two hundred years from the onset of the industrial age to the present, and this tends to be supported by examples drawn from once heavily industrialized areas of countries like Great Britain. Thick layers or dumps of ironworks slag, furnace ash and colliery waste, or large quarries containing landfill, might be thought of as typical scenarios. Although taking stock of the accelerated rate of formation of artificial ground in parts of the industrialized world over the last two centuries is important, however, this should not lead to an underestimation of the amount of materials moved or modified by humans in preceding centuries or even millenia, sometimes in the same

From: Waters, C. N., Zalasiewicz, J. A., Williams, M., Ellis, M. A. & Snelling, A. M. (eds) 2014. *A Stratigraphical Basis for the Anthropocene*. Geological Society, London, Special Publications, **395**, 91–108.
First published online October 25, 2013, http://dx.doi.org/10.1144/SP395.3

locations. There is a danger that, if artificial ground is identified with the Anthropocene, and the start of the Anthropocene dated to commencement of the industrial period (Steffen *et al.* 2007), other parts of artificial ground (and the context of industrial deposits within broader sequences of archaeological stratigraphy) would be overlooked.

It is therefore valuable to examine anthropogenic deposits in a range of locations and from different periods. As well as places associated with extensive mining and processing in the industrial period, for example, long established cities such as Leicester, which were important centres of commerce and light industry, might also be considered. Well-known instances of the build-up of anthropogenic deposits in urban contexts from pre-industrial times include the 8 m deep Viking occupation deposits in York (Hall & Hunter-Mann 2003), which formed part of the bid for World Heritage status for the city. However, it cannot be assumed that such processes were confined to historic times. Accumulation of occupation debris is also characteristic of much older sites, such as prehistoric settlement mounds in Syria and other parts of the Middle East.

Artificial ground

The main categories of artificial ground used by geologists in Britain (McMillan & Powell 1999; Rosenbaum *et al.* 2003) are as follows:

- *worked ground*: ground cut away or excavated (quarries, pits, rail and road cuttings, dredged channels, etc.)
- *made ground*: deposited material (engineered fill, flood defences, spoil heaps, coastal reclamation fill; offshore dumping grounds; constructional fill or landraise, etc.)
- *infilled ground*: ground cut away and later infilled (back-filled workings such as pits, quarries, opencast sites, landfill sites, etc.).
- *landscaped ground*: extensively remodelled ground, where it is impossible to delineate areas of worked ground and made ground within it.
- *disturbed ground*: mineral workings where excavations and associated subsidence are complexly associated with each other (collapsed bell pits, shallow mine workings, etc.).

The five main categories of evidence are listed in the BGS Lexicon of Named Rock Units and appear on published 1:10 000, 1:25 000 and 1:50 000 scale maps. Ford *et al.* (2010) provide an enhanced classification scheme, which refines and extends these basic categories, enabling more detailed characterization (Ford *et al.* 2013).

Features shown in Figure 1 are of relatively recent origin, with a general absence of archaeological stratigraphy from earlier periods. This is usual for most geological depictions of artificial ground, perhaps because recognition of it is historically rooted in the identification of geohazards (contamination, pollution, subsidence, instability, etc.). Inevitably, there is a strong emphasis on industrial workings and deposits of recent origin – such as dumps of colliery waste – because these are of most relevance to assessing ground conditions on sites where development or regeneration is to take place. Artificial ground needs to be systematically

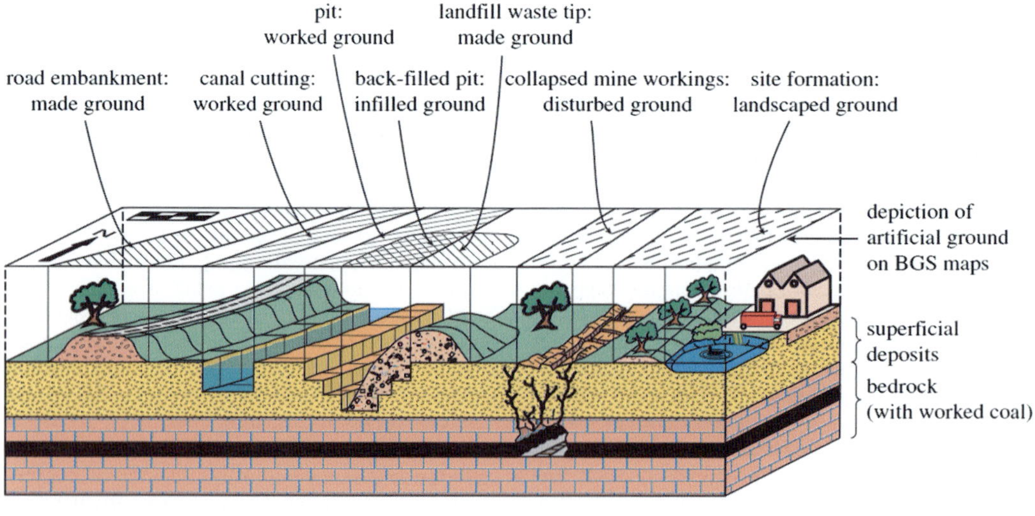

Fig. 1. Examples of the main types of artificial ground (from Ford *et al.* 2010). Reproduced with the permission of the British Geological Survey ©NERC. All rights reserved.

assessed 'to inform the planning process and provide the basis for engineering solutions' (Rosenbaum *et al.* 2003). Much archaeological stratigraphy of earlier periods, however, consists of fairly compacted and stable ground that does not present a geohazard, or at least not of the same order, and accordingly has rarely been the focus of geological investigation. There is therefore a whole raft of potential evidence that may not be shown on geological maps or included in three-dimensional geological models.

Price *et al.* (2011) incorporate points of archaeological knowledge and perspectives into the geological notion of artificial ground, indicating that artificial ground is not simply a product of human activity from the industrial period on, but of much earlier periods too. To demonstrate this, numerous examples are given of landscape-transforming activities dating from the Mesolithic period at the end of the last Ice Age up to the industrial period (Price *et al.* 2011).

The clear implication is that the presence of artificial ground does not in itself provide a stratigraphical marker for the Anthropocene, unless of course the Anthropocene is defined in terms of human modification of the environment in general, its onset dated by the first physical traces of human beings as geomorphological agents in the geological record (in which case all of archaeological stratigraphy would be relevant to the marking and characterization of it). However, if the Anthropocene is defined as having its advent at the start of the industrial period (Zalasiewicz *et al.* 2011 suggest a date of 1800, while Steffen *et al.* 2007 take the industrial era from 1800 to 1945 as Stage 1 of the Anthropocene), then the mere existence of artificial ground by itself is not enough. Stratigraphic markers for the proposed new epoch, thus defined, could however be found *within* the totality of artificial ground.

It follows that, for artificial ground to be reliably used to inform debate on the proposed new epoch, stratigraphical evidence from pre-industrial and even prehistoric periods should be included in its classification. This is the case even if the start of the Anthropocene is dated to 1800 or thereabouts, and industrial age deposits are the main focus of study, for the lower boundaries of these can only be properly defined in relation to what lies below. It is important to understand any component or marker layers *within* the broader stratigraphic sequence of artificial ground as a whole.

Combining notions of artificial ground with archaeological stratigraphy is not easy, however. It effectively amounts to bringing together and merging two very different disciplinary ways of seeing, each with subject-specific scales of analysis. Viewed in relation to geological timelines and deep stratigraphy, artificial ground is a thin and relatively insignificant layer on the surface of more substantial deposits. It is sometimes classed as a single undifferentiated deposit. Viewed in relation to archaeological timelines of the last 10 000 years of human history and prehistory, that same thin layer is perceived as an incredibly detailed and complex record, almost endlessly subdividable into smaller units of analysis.

Archaeological and geological disciplines bring different perspectives to bear on the same evidence, highlighting different parts of it. Deposits originating in the last two hundred years have received little attention in archaeology, often being regarded as 'modern disturbance' or 'overburden'. In geology, concern with suitability of land for contemporary development means that it is precisely these recent deposits that have received most attention, relative to more stable and compacted older anthropogenic deposits. There is no stratigraphic basis for such a disciplinary divide, however. The two incomplete perspectives complement each other, and there is advantage in combining them into a single outlook, in the context of characterizing the stratification of the Anthropocene. To that end, the next section outlines some of the distinctive ways in which archaeologists approach, excavate and understand stratigraphical evidence.

Archaeological stratigraphy

Archaeological conceptions of the typical make-up of urban deposits are illustrated in Figure 2. The depiction can usefully be compared with that in Figure 1.

Despite the use of different terminologies, Figure 2 contains examples of all five categories of artificial ground (worked ground, made ground, infilled ground, disturbed ground and landscaped ground). The material traces of terracing of land, dumping, industrial pollution, cutting of pits, backfilling of voids, and so on, are all illustrated, albeit at a more detailed scale than in the first figure. Prominent here (but noticeably missing from Fig. 1) are layers and features marked as 'intact stratification'. These are typical of the kind of deposits that most field archaeologists would refer to as 'archaeological stratigraphy'.

Detailed methods of recording complex sequences of archaeological stratigraphy have been developed by archaeologists (Roskams 2000, 2001). One of these is the Harris Matrix, an indispensable tool widely used on heavily stratified urban sites in Britain (Harris 1989). In establishing the order in which the various cuts, fills and layers have been formed or laid down, the matrix provides a framework of relative dating that puts all contexts into stratigraphical relation.

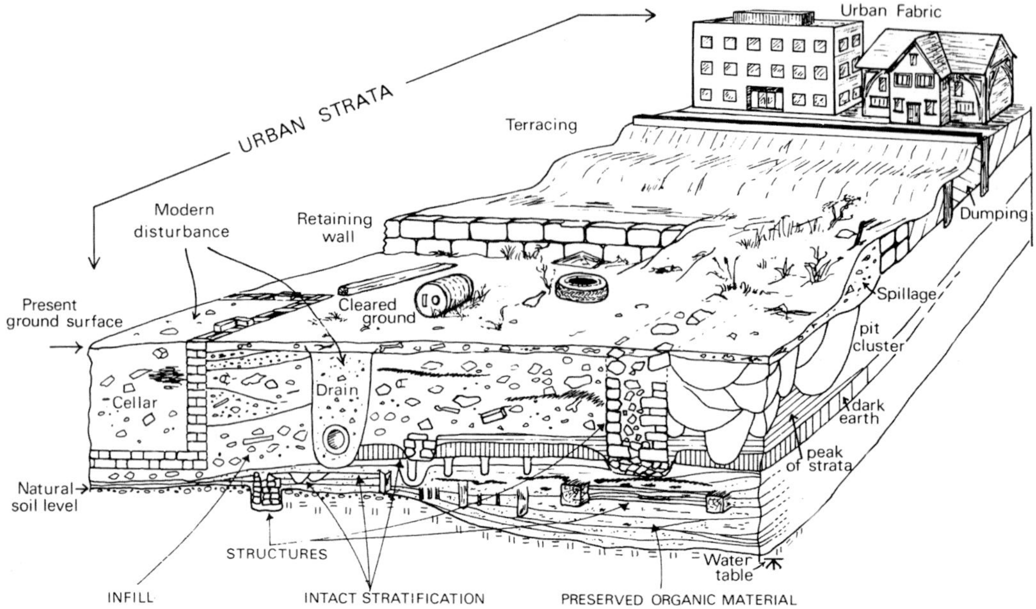

Fig. 2. Different types of archaeological deposits in an urban context (Carver 1987, drawn by Elizabeth Hooper).

Archaeological strata are different from geological strata in important respects. The most obvious of these is that cultural as well as natural forces have been involved in their formation, and must be taken into account in any interpretation of stratigraphy. Patterns of evidence often have as much to do with plans, designs, rationales, purposes, decisions and actions of past human actors as with natural forces of erosion, sedimentation, and so on. Another difference is that the archaeological equivalent of geological 'type-fossils' used to date and correlate layers are usually human artefacts rather than fossils of biological organisms. These have been subject to cultural dynamics rather than processes of natural selection. According to Harris, this means it is not always appropriate to treat archaeological strata in the same way as geological strata: principles of stratigraphy need to be modified to suit archaeological conditions (Harris 1989).

In British archaeology, one of several ways in which past human intervention is recognized in the stratigraphical record is in the form of the 'cut'. Perception of cuts is central to how sites are excavated and recorded (Museum of London 1994). A cut is the physical outline of a pit, posthole, grave, ditch, gully or other feature – a worked surface rather than a deposit. It is similar in some ways to certain kinds of geological unconformity (although there are crucial differences). In terms of the relative dating of archaeological stratigraphy, it is the interface between the fills of a feature and (earlier) layers that the feature cuts through. For the Harris matrix to work, the cut must be given a context number and treated as a stratigraphical unit in its own right (Fig. 3).

What is crucial about the concept of the cut is that it acknowledges the operation of past human agency as a key force, alongside geological and other processes, in the formation of archaeological stratigraphy. The existence of a cut implies a past event – the inferred intentional act of cutting the feature at the time it was made and used. Any application of the concept of the cut onto archaeological evidence, or interpretation of stratigraphy in terms of it, embeds considerations of intentional agency in the perceptual organization of material evidence at a fundamental level.

Of particular interest is the fact that the concept of the cut works at any scale. It is not limited to pits or ditches encountered on archaeological excavations, but is equally applicable to mega-features the width of cities and as deep as skyscrapers are high. The cutting of mineral extraction quarries with giant drag-line excavators and their infilling with landfill or other types of waste, for example, leaves patterns of stratigraphy in the ground not dissimilar to those of Iron Age, Roman or Mediaeval rubbish pits, except on an entirely different order of size (Edgeworth 2013). For more on the archaeological study of landfill deposits, and the study of artefacts and materials they contain, see Rathje & Murphy (1992).

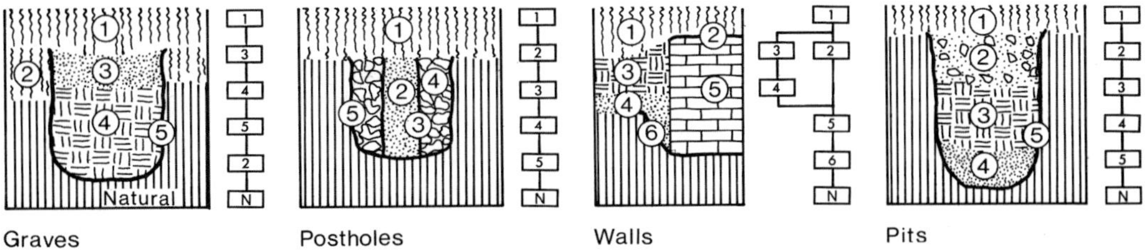

Fig. 3. Examples of archaeological features and fills recorded with the Harris Matrix (Harris 1989). Each feature shown has a 'cut' that contains all the fills or structures within it. Although an interface rather than a deposit, the cut is treated as a stratigraphical unit in its own right. Adjacent matrices indicate stratigraphical relationships, with later units above earlier ones.

The relationship between artificial ground and archaeological stratigraphy is illustrated using two case studies. The first considers the historic core of the city of Leicester in the UK. The second considers the prehistoric settlement mound of Abu Hureyrah in Syria.

Case study 1: Leicester

Like many British cities, Leicester has a long history of development. It probably started as a small Iron Age settlement next to a ford on the River Soar, later becoming the site of a Roman fort in the first century AD, then the town of *Ratae Corieltauvorum*. It had a forum, public baths, basilica, and was enclosed by substantial walls, banks and ditches (Buckley & Lucas 1987). In late Saxon times Leicester was known as one of the Five Boroughs of the Danelaw. It flourished in the mediaeval period with prominent buildings being constructed, including a castle, a friary and numerous churches (Courtney 1998). In post-mediaeval times, up to the mid-twentieth century, Leicester became one of the most prosperous commercial centres in Europe. It featured strongly in the industrial revolution, but not in the same way as the large manufacturing cities of the north like Sheffield or Manchester. Its principal industries were light engineering and the manufacture of hosiery and footwear. There were no rich seams of iron ore or coal, hence no extensive mining, so little dumping of excavated waste materials on the scale of northern slag heaps. Yet, as this case study shows, there is nevertheless a considerable build-up of artificial ground/archaeological stratigraphy in the historic core of the city.

The case study draws from an archaeological deposit model mapped by John Lucas over thirty years ago (Lucas 1981). In compiling it, Lucas drew not only from the results of archaeological excavations but also from geological records, test boreholes by civic engineers, building plans and especially records of public works carried out in the late nineteenth and early twentieth centuries. Foremost of these were data collected during the laying of water mains in the 1890s. Lucas synthesized all this information to produce contour maps of the surface of natural ground, the present ground surface, and the build-up of 'occupational debris' (Fig. 4b).

Figure 4c shows a map of the Roman and mediaeval town, while Figure 4d shows a reconstructed section through deposits along an east–west line *c.* 340 m long, from St Martin's Cathedral in the east down to the canal in the west. It depicts archaeological stratigraphy in the historic core to be on average *c.* 3–4 m deep, up to 5 m deep in places.

The boundary between 'occupational debris' and natural geological deposits is visible. It is highlighted as Boundary A to signify the base of archaeological stratigraphy, or the first surviving physical trace of human modification of the environment at this particular location. In Leicester it dates for the most part from the first century AD, or about 2000 years BP. In some places, underlying the Roman stratigraphy, there are Iron Age pits and postholes, or thin layers of Iron Age occupation deposit, and occasionally earlier prehistoric evidence might be found. In other parts of the city the boundary is dated much later, due to disturbance and truncation of earlier deposits by subsequent activity such as quarrying or levelling of ground – or construction of underpasses for the inner ring road in the 1960s. Typically, however, Boundary A is encountered in undisturbed areas as a sharp transition from natural ground to archaeological stratigraphy, consisting in its lower levels of compacted Roman occupational debris.

Material from the post-mediaeval and industrial periods actually amounts to less than half the total amount of archaeological deposits, except in the central area, where there is up to 4 m of modern layers overlying little more than 1 m of Roman material. The reason for this is instructive. The modern deposits have filled a hollow that was once the

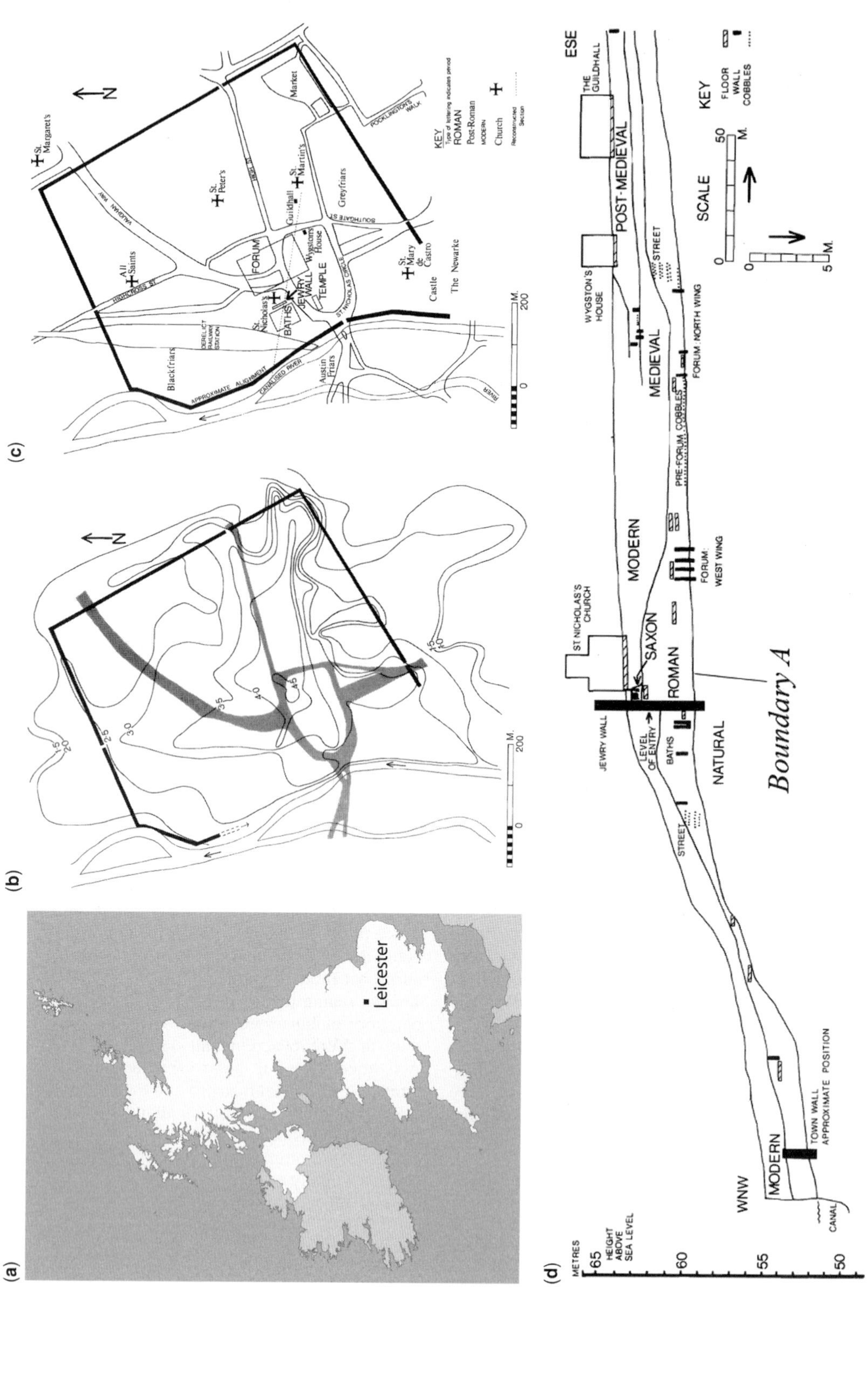
(a)
Leicester
(b)
N
0
200
M.
15
20
25
30
35
40
45
(c)
N
St. Margaret's
All Saints
St. Peter's
St. Martin's
Guildhall
Greyfriars
Market
FORUM
JEWRY WALL
BATHS
TEMPLE
St. Nicholas's
Wygston's House
St. Mary de Castro
Castle
The Newarke
Blackfriars
Austin Friars
DERELICT RAILWAY STATION
APPROXIMATE ALIGNMENT
CANALISED RIVER
RIVER
HIGHCROSS ST
VAUGHAN WAY
HIGH ST
SOUTHGATE ST
ST NICHOLAS CIRCLE
POCKLINGTON'S WALK
KEY
Type of lettering indicates period
ROMAN
Post-Roman
MODERN
Church
Reconstructed Section
0
200
M.
(d)
METRES
HEIGHT ABOVE SEA LEVEL
65
60
55
50
WNW
ESE
CANAL
MODERN
TOWN WALL APPROXIMATE POSITION
STREET
JEWRY WALL
LEVEL OF ENTRY
BATHS
NATURAL
ROMAN
SAXON
ST NICHOLAS'S CHURCH
MODERN
FORUM: WEST WING
PRE-FORUM COBBLES
FORUM: NORTH WING
MEDIEVAL
STREET
WYGSTON'S HOUSE
POST-MEDIEVAL
THE GUILDHALL
SCALE
0
50
M.
0
5
M.
KEY
FLOOR
WALL
COBBLES
Boundary A

main forum area of the Roman city (Fig. 4c, d). Lucas argues that it was the very longevity of use of the forum and its buildings that prevented much build-up of occupational debris (accumulation of which tends to be faster where buildings are short-lived and frequently demolished and replaced). This meant that it accumulated more quickly in areas surrounding the forum, which, by the end of the Roman period, was at a lower level than the rest of the town. It was the existence of this hollow in subsequent periods that created the space for later infill (Lucas 1981). At least in the case of Leicester, then, the accumulation of artificial ground in recent times can be properly understood only by reference to human activities back in the Roman period. Artificial ground of post-mediaeval and industrial date has no special status, but is simply part of a larger stratigraphic sequence, the accumulation of which has been taking place here in different ways and at varying rates of intensity for about 2000 years.

The boundary right at the top of the stratigraphical sequence is noteworthy. It may not even be considered a boundary by some, because there is nothing above it except the standing urban fabric of the present city. Unlike other soil boundaries discussed, this one is provisional – always about to be superseded as further layers of artificial ground are added, or replaced entirely as existing layers are recycled, truncated, cut through, or otherwise transformed. In so far as it is a boundary at all, it is open rather than closed, with the stratum of artificial ground still very much in the process of formation, and the top layer especially subject to intensive reworking as new buildings are constructed over the foundations of the old.

Case study 1: Leicester (detail)

Lucas's 1981 deposit model is extremely broad, dividing occupational debris into basic periods such as Roman, mediaeval and post-mediaeval. In this sense it approximates towards the geological view of artificial ground, which makes relatively few internal differentiations. However, this is only one of many scales on which deposits are characterized in archaeological procedures. The deposit model is based upon, and provides a framework for, much more detailed recording of archaeological stratigraphy.

Figure 5a shows a composite section from an excavation in Causeway Lane, Leicester, located just outside the area of the Roman forum (Conner & Buckley 1999). The length of the section is *c.* 80 m. At this intermediate scale, it shows the profiles of wells, ditches, roads and other features of various periods that make up much of the occupational debris. It is precisely in such complex cases of intercutting features that the concept of the 'cut' (allied with methods of recording that recognize interfaces as well as layers) becomes indispensable for making sense of archaeological stratigraphy, enabling all stratigraphic units to be dated relative to each other.

Figure 5b depicts a small part of this evidence at a still greater level of detail. During excavation out on site, this section drawing would have been drawn at a scale of 1:10. At whatever scale the evidence is viewed, however, the most important boundary is still that between the natural undisturbed deposits and archaeological deposits – Boundary A.

Above Boundary A, note the fractal structure of archaeological evidence, in the sense that what can be represented as a single layer at one scale of analysis (Lucas's deposit model in Fig. 4, or geological mapping of artificial ground) can be seen to consist of numerous cuts, fills and layers when deposits are examined in more detail. There are about 40 recognizable units of stratigraphy in the Roman levels of this 5-m-long section alone. Each unit is potentially subdividable into smaller subunits, depending on the scale of recording.

The soil boundary between Roman and mediaeval deposits can be examined more closely. There is a contrast between the complex stratified Roman stratigraphy below the boundary and the relatively undifferentiated and homogeneous layer above, consisting of dark brown loam up to 3.5 m deep, labelled here as 'garden soil'. The excavators date the latter deposit to gardening and manuring activity from the twelfth century onwards (Conner & Buckley 1999). This activity – as well as leading to accumulation of a large depth of uniform loam – has probably also removed stratigraphy from earlier intervening periods. This is typical of the historic core of Leicester: little evidence survives of Saxon and Viking activity, and it is clear that extensive truncation and reworking of deposits from that period has taken place.

The apparent homogeneity of the brown loam layer is potentially misleading. The layer probably

Fig. 4. Archaeological deposit model of Leicester, UK, by John Lucas: (**a**) location; (**b**) contour map of build-up of 'occupational debris'; (**c**) map of the Roman and mediaeval town, with line of section shown; (**d**) reconstructed vertical section through 'occupational debris' along an east–west line from St Martin's cathedral down to the canal (Lucas 1981). The labelling of Boundary A has been added.

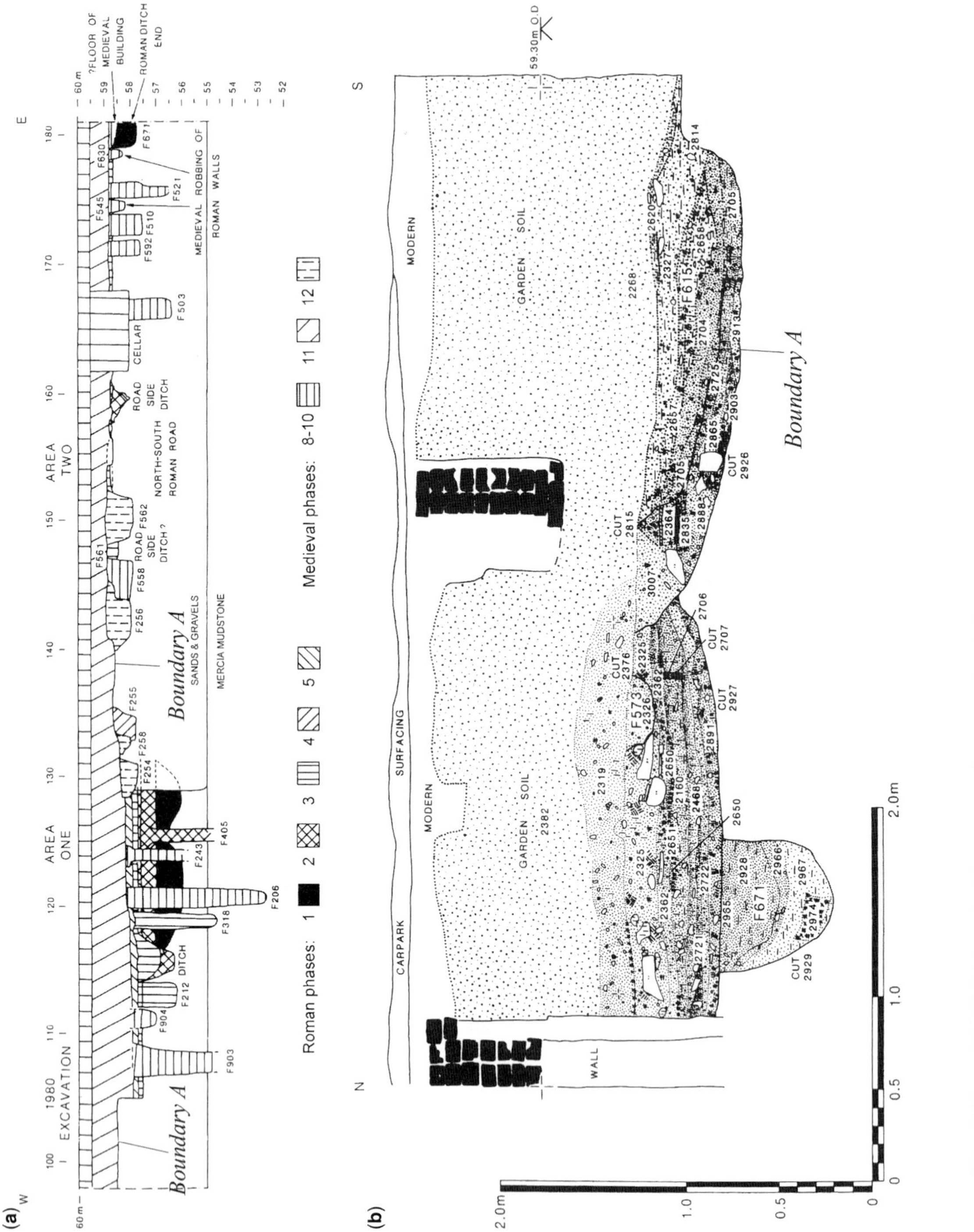
(a)
W
E
1980 EXCAVATION
AREA ONE
AREA TWO
Boundary A
SANDS & GRAVELS
MERCIA MUDSTONE
NORTH-SOUTH ROMAN ROAD
ROAD SIDE DITCH
ROAD SIDE DITCH ?
CELLAR
MEDIEVAL ROBBING OF ROMAN WALLS
?FLOOR OF MEDIEVAL BUILDING
ROMAN DITCH END
DITCH
Roman phases: 1 2 3 4 5
Medieval phases: 8-10 11 12
(b)
N
S
59.30m O.D
CARPARK SURFACING
MODERN
GARDEN SOIL
WALL
CUT 2815
CUT 2926
CUT 2707
CUT 2927
CUT 2929
CUT 2376
F615
F573
F671
2.0m
1.0
0.5
0

accumulated as the result of numerous events, episodes and phases of activity over the course of many centuries, but worm and plant action and processes of cultivation have rendered what might otherwise be an extremely complex deposit – perhaps every bit as complicated as the Roman deposits below – into a single undifferentiated layer. The example illustrates the extent to which, even after deposition, archaeological stratigraphy is subject to a range of cultural and natural transformations (Schiffer 1987).

Modern and industrial layers in Leicester are relatively shallow compared to the occupational debris of earlier periods. The upper boundary shown in Figure 5b is a transient one. Upper layers are particularly subject to reworking through subsequent development. In this case, the thin layer of tarmac surfacing has recently been removed, along with large parts of underlying stratigraphy, by groundworks undertaken in advance of construction of a new office block.

Case study 2: Abu Hureyra

The Neolithic tell of Abu Hureyra in Syria is the site of an agricultural village that grew up on the banks of the River Euphrates. Excavations took place at Abu Hureyra in the 1970s (Moore *et al.* 2000). The settlement mound comprises accumulated debris from about 4000 years of occupation, deriving mostly from the building, subsequent erosion and rebuilding of mud huts, dating right back to the Epi-palaeolithic (Mesolithic) *c.* 11 000 years BP. The site was abandoned in the late Neolithic *c.* 7000 years BP (Moore *et al.* 2000). The example illustrates that extensive and complex deposits of (what might be called) artificial ground do not necessarily date from since the start of the Industrial Revolution, or even from the mediaeval or classic periods of history as at Leicester, but can be much earlier in date.

The tell today has a trapezoidal shape measuring *c.* 400 m × 300 m or *c.* 12 ha, and up to 12 m high, as shown on plan in Figure 6b. It is estimated to weigh over a million tonnes. That represents the material of the settlement mound only. The estimate does not take account of extensively modified anthropogenic deposits related to the former channel and floodplain of the Euphrates, or the former cultivation soils adjacent to and contiguous with the settlement. Much of the material that comprises the settlement mound has been imported from the surrounding area, implying corresponding areas of quarries or truncated ground.

A vertical section through part of the settlement mound shows the build-up of deposits from the Neolithic phases of settlement in some detail (Fig. 6c). There is a high degree of internal structure and detailed patterning, mostly composed in this instance of sequences of occupation levels, with successive buildings often built directly on top of previous buildings. Deposits from later periods of settlement have been lost through erosion in the post-abandonment phase. Features of rectangular profile at the top represent grave-pits in a modern cemetery which has been sited on top of the hill, cutting into more ancient deposits. There is thus a gap of *c.* 7000 years in the upper part of the surviving material record.

Most of the soil boundaries depicted in the vertical section relate to events or phases within the period of occupation. Phases of deposition could have lasted anything from a few moments to hundreds of years. The most important boundary is to be found at the base of the sequence, and its equivalent has already been encountered in the first case study, in the deposit model of Leicester. It is marked as Boundary A to signify once more the soil horizon between 'natural' subsoils below and anthropogenic deposits above, marking the first surviving trace of human impact on the landscape at this location. In this trench it can be dated to about 9500 years BP, although in other trenches on the same site it dates back to 11 000 years BP. Because the settlement mound grew laterally as well as vertically through time, the date of Boundary A varies in different locations at the base of the settlement mound.

The equivalent soil boundary marking the first human modification of natural ground in Leicester, shown in Figures 4 and 5, can be dated (where not truncated by later activity) to *c.* 2000 years BP. Such a vast difference shows that Boundary A is diachronous, varying in date from one location to another, even on the same site.

Stratigraphical sequences from the Abu Hureyra tell span the period of transition from hunter-gathering to farming, and associated shifts in settlement patterns, building types, material culture, domestication of plants and animals, and so on. That means there are extremely important soil horizons and divisions within the sequences of layers, marking significant cultural developments that are wider than the site itself, in the sense that they can be correlated with layers containing similar

Fig. 5. Excavations in Causeway Lane, Leicester, UK, by the University of Leicester Archaeological Services: (**a**) composite section across site; (**b**) more detailed recording of archaeological stratigraphy (Conner & Buckley 1999). The labelling of Boundary A has been added.

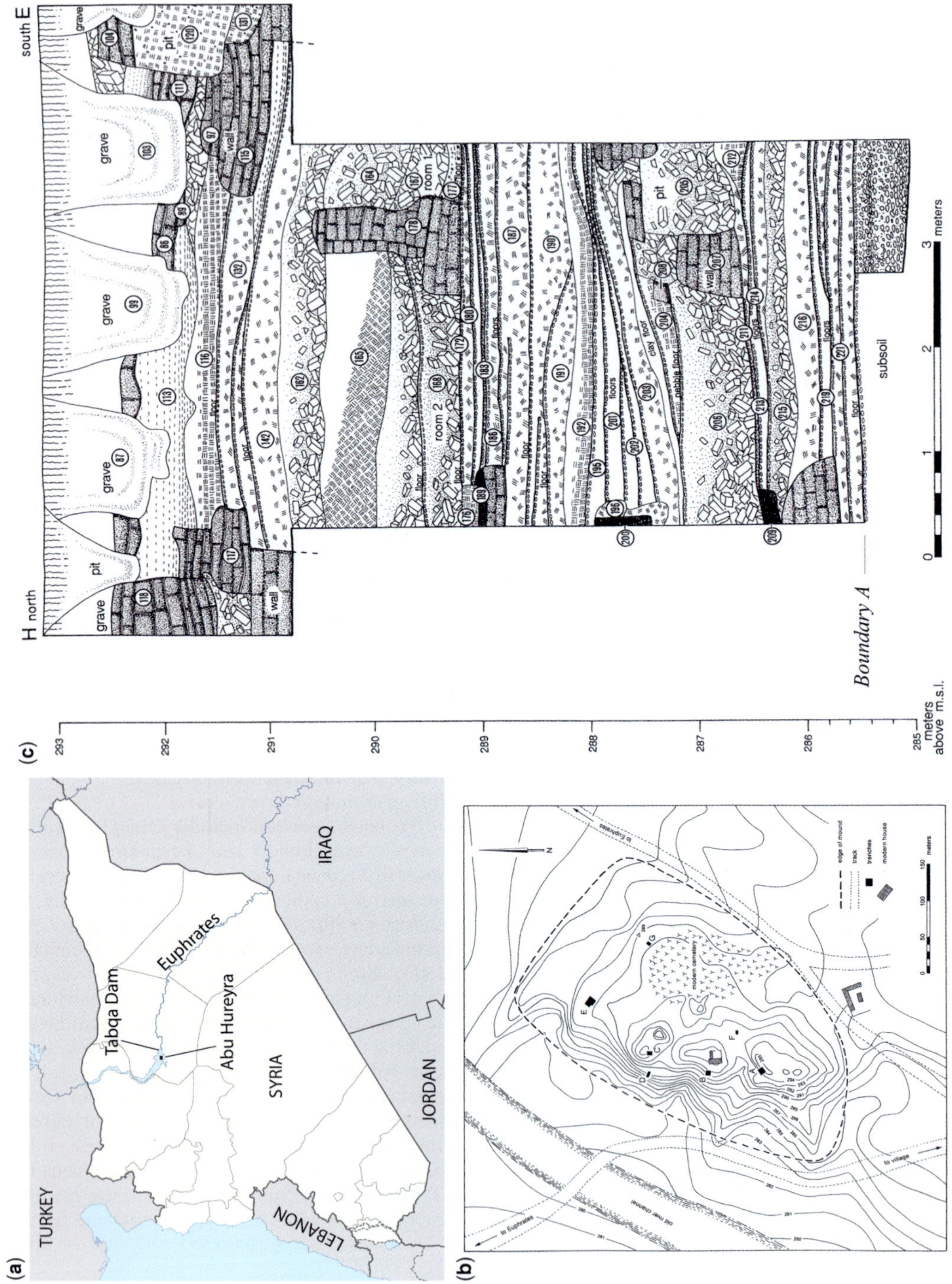
(a)
TURKEY
Tabqa Dam
Euphrates
Abu Hureyra
SYRIA
IRAQ
JORDAN
LEBANON
(b)
(c)
H north
south E
Boundary A
meters above m.s.l.
subsoil
meters

artefacts and configurations of evidence on other sites. Most of the layers in the 8-m-deep sequence of deposits contain no pottery at all, for example. Only at the very top of the prehistoric sequence (in those layers cut by the modern graves) are potsherds present. Soil boundaries dividing non-pottery layers from those containing pottery are taken to divide the Early Neolithic or Pre-Pottery Neolithic from the Pottery Neolithic, a transition that can be dated on this and other sites in the Fertile Crescent region to the eighth millennium BP. However, it is important to realize that this boundary is also diachronous. The transition from pre-pottery to pottery ways of life took place in Britain (where it is taken as a defining feature of the Neolithic period) about a thousand years later in the middle of the seventh millennium BP.

Abu Hureyra is actually an unexceptional site. For mapping of hundreds of settlement mounds like Abu Hureyra in northeastern Syria through thermal imaging of anthropogenic soils from space, and an astonishing map of the distribution of tell-like sites over an area of 23 000 km, see Menze & Ur (2012).

Just as in Leicester, the boundary at the top of the sequence of stratigraphy should be regarded as provisional – open rather than closed – because the deposit is in the process of formation, and accumulation or modification of it in one form or another is likely to continue. In this case it is known that erosion has removed significant amounts of stratigraphy from the top of the mound during the thousands of years since abandonment. More recently, however, there has been a fundamental shift in land-use. An event has taken place that has radical implications for the future geology of the region. Above the line that marks the top of the vertical section there are now several metres of silt deposits. Above that, there is considerable depth of water. That is because this part of the Euphrates valley was flooded when the huge Tabqa dam was constructed in the 1970s. The site now lies under the massive artificial reservoir called Lake Assad, Syria's largest lake, covering a surface area of 610 square kilometres. Many archaeological sites upstream of Tabqa dam, including the Abu Hureyra tell itself, were deeply submerged (for implications of the Anthropocene for heritage sites, see Solli *et al.* 2011).

The boundary between the top of the archaeological deposits shown in Figure 6c and the base of the silts at the bottom of the reservoir could be taken as a marker for the start of the Anthropocene (if the construction of huge dams for hydropower generation is taken as a defining feature of the new epoch). However, that boundary must not be mistaken with Boundary A, which marks the earliest surviving trace in this location of human activity in shaping the face of the Earth.

Riverworks: entanglements of cultural and natural forces

Zalasiewicz *et al.* (2011) point out that anthropogenic modification of sedimentary patterns in rivers by interventions such as damming or straightening, and consequent patterns of sedimentation in estuaries and deltas, are to some extent intergradational with build-ups of artificial ground in cities. This applies not just to recent interventions in hydrological processes; many examples can be drawn from the ancient material record too.

The term 'artificial ground' needs to be examined critically here because it implies that cultural agency is the primary force at work in its production. Artificial ground, by definition, is a product of human action, but most anthropogenic deposits are actually formed as the outcome of a mixture of natural and cultural forces, with human action interleaved with forces of erosion and sedimentation, as illustrated by the following examples.

When a dam is placed across a river to create a reservoir of water behind it, one of the unintended side effects is likely to be retention and accumulation of river silts against the dam. Yet this is a process that can be intentionally utilized and controlled. One of the most widespread techniques for creating cultivatable fields in arid regions of the Near East and North Africa – a method that can be dated back to Neolithic times (Wilkinson 2003) – was to build a low check dam across the valley of a seasonal watercourse with the specific intention of trapping the fertile sediment carried and deposited by floodwater. By building a series of dams along the length of the wadi, whole valley floors could be rapidly terraformed into a stepped series of fields, with dams surviving as retaining walls. Similar techniques could be used on tributary streams to terrace the valley sides (Barker 2002; Edgeworth 2011).

The amount of material moved by such methods – enlisting the forces of gravity, erosion and the flow of water and silts into human cultural projects – adds up to landscape transformation on a vast scale. The total amount of material intentionally moved by such means over many centuries or millennia across the globe has never been

Fig. 6. Excavations at Abu Hureyra, Syria: (**a**) location; (**b**) contour map of settlement mound; (**c**) vertical section through build-up of anthropogenic deposits (Moore *et al.* 2000). The labelling of Boundary A has been added.

quantified, but may well be comparable to estimates for sediment flux caused by the construction of large dams since the 1940s, or the amount of material moved by machines in contemporary terracing work (Syvitski & Kettner 2011). As Richard Periman puts it, 'The Anthropocene begins to emerge when we consider human-environmental activity at a local level, compounded by thousands of years, affecting vast areas of interlocking landscapes' (Periman 2006).

Human engagement with fluvial processes leading to large-scale landscape transformation is often cited as being characteristic of the industrial Anthropocene, although, in fact, such interactions go back far into prehistory. The Yellow River in China provides a useful example. The river derives its colour from the large amount of silt it carries. Left to itself the river would deposit much of its load of silt on the floodplain during periods of flood, some in the form of natural levees, while carrying the rest to the sea. However, in order to prevent flooding of farmland and settlements on the floodplain there is a long tradition of people artificially building up the levees. This has the unintended but unavoidable effect of more sediment being deposited on the riverbed, thereby raising it up and increasing the danger of flooding. To counteract this, the levees have to be heightened yet again, raising the riverbed still further. The result is that the river channel gets elevated high above the floodplain on either side.

According to Syvitski & Kettner (2011), the super-elevated channel of the Yellow River is an outcome of human intervention in river processes since the last major shift of the river and the formation of a new delta in the mid-nineteenth century. However, the authors do not account for similar human–river interactions that have been taking place on previous courses of the Yellow River for millennia, the very shifts of river channels and formations of new deltas being partly the result of earlier human interventions. In this respect, Syvitski & Kettner's geological description of human–river interactions in the industrial-age Anthropocene can be compared with a contrasting archaeological description focusing on earlier interactions (Edgeworth 2011, see also Jiongxin Xu 2003). It is inferred that the origin of super-elevated courses of the Yellow River goes back into prehistoric times.

When flooding of super-elevated rivers does occur, it can be catastrophic. Even as the water is subsiding, it leaves behind a thick deposit of mud that buries fields, farms, villages and cities. The modern city of Kaifeng in China (9 km south of the present elevated course of the Yellow River) provides a spectacular instance of the kind of archaeological stratigraphy that can result from intermittent flooding on this scale. Beneath the level of modern streets and buildings there is a depth of 8 m of river silt, underneath which is the remains of the mediaeval city of Dongjing. Beneath Dongjing there is a 2 m layer of river silt, which in turn overlies heavily compacted material remains of the city of Daliang, dating from the fifth to third centuries BC. Beneath the city of Daliang are further deposits of river sediment (Johnson 1995).

How the mediaeval city of Dongjing was buried under the yellow silts is relevant here. In the context of a peasant revolt of 1642, rebels deliberately breached the levees in order to flood the city, which was under siege at the time. The escaped river deposited metres of silt over vast tracts of floodplain, taking 300 000 lives, burying numerous settlements in the process, changing its course to another channel, shifting the delta and relocating deposition of huge amounts of sediment further along the coast. This partly humanly wrought event was arguably just as much an event of the Anthropocene as the recent building of the Tabqa dam and the creation of the giant reservoir of Lake Assad. It was neither the first nor the last time people were actively involved in changes of course of the Yellow River, but simply part of a long tradition of human–river interaction.

Earthworks: entanglements of materials and ideas

Roderick Salisbury states that earth worked by humans is not a culturally inert material. It gets transformed symbolically as well as physically, with the specific character of the soil playing an active role in the creation of values and meanings (Salisbury 2012). Ways in which the social, ideological, religious and political designs of people get folded into the earth – and, conversely, ways in which earth itself gets folded into cultural systems – are especially apparent in the case of earthworks.

Earthworks range from the highly pragmatic and functional, such as the levees on the Yellow River or terraces of rice paddy fields, to monuments highly charged with symbolic meanings. The 40 m high Neolithic mound of Silbury Hill in Wiltshire (Leary & Field 2010), dated to 2400 BC, consists of *c.* 300 000 cubic metres or an estimated weight of just under 0.5 million tonnes of gravel, chalk and soil, excavated from adjacent quarries. Unlike the settlement mound of Abu Hureyra, composed of the accumulation of occupation deposits over thousands of years, Silbury Hill was designed and built as an intentional transformation of the landscape, probably for ceremonial and religious purposes. It has an intentional structure, quite different from geological formations. It is estimated that it took 500 people about 15 years to construct it (Atkinson 1974).

All earthworks combine practical function with cultural meanings; they are simultaneously material and symbolic entities to varying degrees. This double aspect is particularly evident where earthworks have been constructed to mark boundaries of political territories or sacred areas. Such boundaries exist both 'out there' on the ground and 'in the mind'. They represent not only the material outcomes of physical actions but also the imprints of intentions and cultural symbolisms (Salisbury 2012).

For example, in late Saxon times the small town of Wallingford in Oxfordshire was surrounded on three sides by a massive ditch and rampart, with flowing water re-routed from nearby streams, changing the hydrology of the surrounding area. Enclosing a town with flowing water was not simply for defensive or other utilitarian purposes; it was also a highly symbolic act, creating and expressing, amongst other things, a sense of cultural identity for town inhabitants, and perhaps a sense of exclusion for others. The boundaries around part of the town and the ready supply of flowing water they provided were subsequently utilized in the construction of Wallingford Castle, with its extensive system of moats and ramparts (Creighton *et al.* 2009; Wilkinson *et al.* 2010). The example illustrates, amongst other things, that there is often no clear distinction to be drawn between earthworks and riverworks (with the one often implicating the other). Intentions and cultural meanings are folded into both.

Reworking of earthworks and other anthropogenic deposits can lead to some spectacular inversions of stratigraphy. In the case of Wallingford Castle, for example, each phase of refortification involved movement of large amounts of material, as successive circuits of moats were dug out and other areas backfilled with the resulting spoil. During one such phase, a whole range of thirteenth-century kitchen buildings was buried beneath spoil from a new moat, some of which derived from earlier earthworks. Cob walls of these structures survive to head height under the raised ground level. The castle was later demolished, its earthworks partially flattened, leaving a considerable though highly variable depth of archaeological stratigraphy, much of which has been intensively reworked. Here there is worked ground in the form of excavated features, infilled ground in the form of backfilled features, made ground in the form of demolition debris or imported material used to build up platforms, landscaped ground in the form of heavily truncated earthworks, and disturbed ground in the form of robber trenches dug by subsequent generations in search of building material, all interleaved into detailed and complex stratigraphical sequences (Creighton *et al.* 2009; Wilkinson *et al.* 2010).

Cultivation soils: the extent of 'cultured ground'

It would be a mistake to equate thickness of deposit with significance. In archaeological stratigraphy, the thinnest lens of material in a stratigraphic sequence can sometimes be more informative and relate to longer periods of human activity than the largest and thickest layers. Partly for that reason, some of the less spectacular types of archaeological deposit such as cultivation soils need to be considered here.

Plaggen soils in Orkney, Shetland, Ireland, the Netherlands, Belgium, Scandinavia and other parts of northern Europe are rich soils created (in poor sandy soil regions) through a type of farming called plaggen cultivation. Mostly associated with the mediaeval period, plaggen cultivation has also taken place in prehistoric and post-mediaeval times. Turfs of peat or heather sods were dug from outfield areas to be used as bedding for cattle or sheep. When saturated with slurry, these were spread as manure over infield areas. Anthropogenic soils are full of traces of animals domesticated by people. Plaggen soils are high in phosphates, a chemical signature quite different from non-anthropogenic soils in the vicinity. They can be over a metre deep and often contain abraded pottery sherds and large amounts of black carbonized particles (Holliday 2004; Pears 2009).

Another deposit sometimes taken to be a cultivation soil, common in towns with a long history, is so-called 'dark earth'. Deriving its dark colour from high charcoal content, this type of deposit is often found sealing urban Roman stratigraphy in English towns. Dark earth was first identified in London in 1912 and has been the subject of much discussion amongst archaeologists ever since (Carver 1987). Similar soils have been found elsewhere in northern Europe. In Roman towns in England it frequently overlies deposits dated to the second to fifth centuries AD, and is typically overlaid by deposits dating from the ninth to tenth centuries. It is characterized by a lack of stratigraphic horizons within it, leading some to interpret it as a heavily worked cultivation soil, perhaps representing depopulation of urban centres and a shift to a more rural way of life. Other archaeologists, however, argue that dense occupation of urban areas continued, and that dark earth represents a change in constructional technique and materials – stone being replaced by wood, mud, wattle and daub, with rush matting for flooring and straw for bedding – the resulting accumulation of organic deposits being subject to thorough mixing through subsequent bioturbation and cultivation (Macphail *et al.* 2003).

Often shallower than plaggen soils and dark earth are more extensive ploughsoils that cover large parts of the world. These have been formed through ploughing activity from Neolithic times to the present day. Whereas plaggens and dark earths accumulate primarily through importation and deposition of materials, ploughsoils have been formed mainly through the truncation of underlying subsoil, although there has been some addition of material too. The lower boundary of the ploughzone in England is typically defined by the upper surface of the truncated natural layers (yet another manifestation of Boundary A). The ploughsoil above that line often contains much artefactual material scraped up from truncated archaeological features (fragments of tile, brick, pottery, worked flint, and so on) along with reworked earth from disturbed natural subsoil and earthworm activity, sometimes mixed with additional alluvial and colluvial material. Other artefacts find their way into the ploughsoil through different means, perhaps getting mixed up with manure in farmyards, taken out to the fields and spread on the soil with animal fertilizer (plaggen is an extreme example of a heavily manured ploughsoil). Ploughsoil has been described as 'the greatest depository of archaeological material' (Slowikowski 1995). Some sites are detected from artefact assemblages found in the ploughsoil, and completely ploughed-out sites may have no other existence apart from as ploughzone artefact scatters.

The lack of depth of most ploughsoil is more than made up for by its great horizontal extent. In England there are currently about 5 million hectares of land in arable cultivation, over a third of the total land surface. Even land currently under pasture, wooded or built over has often been under the plough in the past, surviving as a buried ploughsoil layer beneath later accumulations. At its fringes it is intergraded and interleaved with deeper accumulations of artificial ground such as urban deposits at Leicester. Occupational debris of Abu Hureyra too would likewise have been interleaved with surrounding cultivation soils, although these were not a focus of investigation into the settlement mound.

Just as buried ploughsoils and ard-marks are often found beneath Neolithic long barrows (Fowler 1983), so former ploughsoils of more recent origin may underlie later deposits of industrial waste. Some buried ploughsoil should probably be marked onto the schematic in Figure 1, for example, covered over by the embankments and other dumps of material shown, at least where the old land surface has not been cut away by deep workings.

The extent, nature and content of ploughsoils reflect the changes that have taken place in the technology of ploughing and use of fertilizers, as well as the processes of industrialization of the countryside that have occurred as the result of the introduction of steam engines and motor tractors. Thus, nineteenth-century ploughsoils in England are sometimes characterized by inclusions of the coal used by steam ploughs, with networks of field drains laid in trenches cut into the natural subsoil where natural drainage is poor. This is quite different from mediaeval ploughsoil, which has its own distinctive characteristics and inclusions and specific patterns of drainage, often surviving in the earthwork form of ridge-and-furrow.

Ploughsoil can be conceived as a form of earthwork (in which earth has been thoroughly 'worked') and perhaps a form of riverwork too (insofar as irrigation and drainage systems of ploughed fields become integrated parts of river basins and floodplains). Comparison can be drawn with extensive terracing of hillsides associated with rice paddies in parts of Asia, inextricably bound up with control of water. Cultivation soils in general are as relevant to debate on the Anthropocene as accumulations of urban artificial ground or industrial dumps of waste material. Certini & Scalenghe (2011) argue that they can provide useful markers for the base of the Anthropocene, although see Gale & Hoare (2012) for a counter view.

Taking ploughsoil and other cultivation soils into account as integral component parts, the totality of archaeological stratigraphy is less fragmented and discontinuous than might otherwise be assumed. Absent in heavily eroded areas or in places remote from human settlement, there is blanket coverage over intensively cultivated zones. For a world map of anthrosols, see Chesworth (2008).

A unified view: the archaeosphere

This paper has focused on the potential of archaeological approaches for an increased understanding of artificial ground. It also draws from geological perspectives on artificial ground as a single entity, contrasting with archaeological views of its internal complexity. What emerges is the vision of a unified deposit comprising all aspects of archaeological stratigraphy and artificial ground combined, dated by the artificial structures and artefacts within it. This is similar to the 'human event stratum' described by Zalasiewicz (2008), who imagines how it might appear to a hypothetical geologist 100 million years hence. Buried in places, eroded away in others, outcrops of the heavily compacted and distorted stratum are envisaged, containing fossilized structures and artefacts of human civilization. For Zalasiewicz the layer is dominated by urban artificial ground that has accumulated since the industrial revolution. Here, however, it is argued that the concept of the human 'event' (the period during which humans are held to have impacted on the surface of the Earth) should be extended in space and time, and that the idea of the human event stratum should accordingly be broadened out to include Neolithic tells, plaggen soils, sediment built up behind early dams, Roman occupation debris, mediaeval castle earthworks, and so on, together with later industrial age deposits. These are arguably all part of a single, yet almost infinitely subdividable, anthropogenic deposit, comprising both archaeological stratigraphy and artificial ground. This will be referred to here as the archaeosphere.

Useful though it is to envisage the archaeosphere from an imagined vantage point far in the future, from our perspective in the here and now it is a stratum that is still in the process of formation. Although a lower boundary can be found on most sites (the diachronous Boundary A), the upper boundary is entirely provisional, likely to be transformed by later activity or buried under further accumulations, with no one able to say when the human event that produces and recycles it will be closed.

The as-yet-incomplete stratum can be anything from a few millimetres to tens of metres thick, and in places is absent entirely, although it is useful to think of it as a deposit of near-global extent. Parts of it extend deep underground through mines, fracking wells and nuclear waste repositories. It contains underground car parks and railway tunnels. It is not limited to land, but is also found in deltas and estuaries in the form of sediments from humanly modified rivers, and under the seabed as networks of oil-well boreholes and pipelines. It would perhaps be a mistake to suppose that the stratum is limited to Earth, for it is actually interplanetary in extent. Networks of artificial satellites and associated space debris orbiting the Earth invite new conceptions of non-terrestrial orbital stratigraphy (Gorman 2009). Some of the hundreds of satellites currently in geostationary orbit around the Earth, at heights of 36 000 km above the Equator, boosted into higher orbits when decommissioned, will remain in space for tens of thousands of years (National Research Council 1995). Early build-up of more conventional forms of artificial ground – containing artefacts, human footprints, tracks of vehicles, and so on – can now be recognized on the surface of the Moon (and in theory could be correlated with deposits on Earth containing artefacts from the same period). There has also been an exchange of redeposited geological material between the Earth and its satellite, with Moon rocks being brought back to Earth and heavily engineered Earth metals and plastics sent to lunar landing sites in return (Gorman 2005; Zalasiewicz 2008).

With regard to its manifestations on terrestrial surfaces of the Earth, the archaeosphere can be conceptualized as a kind of giant carpet covering large areas, on which the furniture of the human world (its buildings, bridges, monuments, pylons, oil-rigs, telegraph poles, roads, railway viaducts, cities, shanty towns, parks, airports) stands and is supported, and into which it will eventually crumble. Deep-layered in places, threadbare and patchy in others, this carpet of near-global extent provides the surface on which people carry out their lives. Like a carpet, it is so well-used it is taken almost totally for granted (but see Thomas 1958).

The characteristics of the archaeosphere, as formulated here, are as follows:

- It is still in the process of formation, and therefore inherently incomplete. It has a defined (though diachronous) base, but not as yet a permanent or closed upper boundary.
- It was first laid down in small isolated patches, but as human population increased these patches have steadily and exponentially grown in extent, frequency and thickness, joining together to cover larger areas, though often eroded (or interleaved with deposits laid down) by flood or other geomorphological forces.
- It can currently only be studied from the 'inside', that is, by investigators who are themselves agents in its formation. Paradoxically, even the investigation itself will leave its own material traces folded into the deposit (e.g. as cuts and fills of archaeological trenches and geological boreholes).
- Cultural as well as natural forces are involved in its deposition, accumulation, destruction, modification, transformation, truncation, redeposition and recycling.
- It can only be fully understood by reference to human intentions and actions as well as more natural processes. Although some of it is an unintentional byproduct of human activity, much of it has an intentional and symbolic structure. It cannot be entirely explained or described using geological principles of stratigraphy alone.
- Layers and features within the stratum can be accurately dated and correlated with each other by the 'type-fossils' they contain, but these take the form mainly of human artefacts, structures and features, not so much fossils of biological organisms (although there are numerous

concentrated and symbolically ordered deposits of human skeletal material).

- It is not restricted to the Earth, but now has an interplanetary extent, extending into orbital space and onto the surface of the Moon and nearby planets.
- Future formations of the deposit are partly contingent upon present human decisions and actions, including those of the scientific and academic community.

Can the archaeosphere be useful in characterizing or marking the Anthropocene? In answering this question, the way in which the Anthropocene is defined is crucial. If the Anthropocene is defined simply and broadly in terms of the impact of human activity on the surface of the Earth, then the existence of the archaeosphere could be taken as a direct indicator. The statement that 'artificial ground can be interpreted as sedimentary deposits or excavations representing the human geological record during the Anthropocene' (Price *et al.* 2011) could in that case be taken as unambiguously true. Varying considerably in date from one place to another even on the same site, the lower boundary of the archaeosphere, identified as Boundary A, marks the first surviving traces of human activity in the material record at any given location (bearing in mind that earlier manifestations of Boundary A could have been truncated by later ones). According to Edward Harris, this horizon also marks a revolution in stratigraphical processes, with cultural forces – entangled with natural agencies – coming into effect for the first time (Harris 1989).

Most definitions of the Anthropocene refer to more specific thresholds, tipping points or developmental phases that can be taken as starting points for the new geological epoch. The palaeoclimatologist William Ruddiman, for example, argues that the Anthropocene epoch starts with the origins of agriculture. Forest clearance by humans created an increase of carbon dioxide in the atmosphere from 8000 years ago, while cultivation and irrigation techniques led to a huge rise in methane levels *c.* 5000 years ago, with great consequences for climatic patterns (Ruddiman 2003, 2005). Others such as Crutzen & Stoermer (2000), Steffen *et al.* (2007), Crutzen (2009) and Zalasiewicz *et al.* (2011) favour a date for the start of the Anthropocene in the late eighteenth century – the time of invention of the steam engine and onset of a period of exponential growth in world population and expansion of cities, with corresponding increases in the exploitation of mineral, agricultural, water and other resources. In all these cases, it is not the existence of the archaeosphere in and of itself that provides useful stratigraphical indicators to support such claims, but rather particular horizons *within* stratigraphical sequences of the archaeosphere.

It was noted at Abu Hureyra that specific soil horizons mark the introduction or adoption of hitherto unknown forms of technology, such as pottery. Just as the appearance of new-type fossils are taken to indicate the start of geological periods, so the presence of new-type artefacts (potsherds in this instance) are sometimes taken to indicate the start of archaeological periods like the Pottery Neolithic. Different kinds of pottery are classified into type-series, so that layers containing certain types can be correlated with other layers with the same pottery. The same might be done with more recent industrial-age artefacts, from steam-powered machinery to electric turbines, from radio sets to mobile phones, from bottle-tops to Coca Cola cans, or from the first plastics to biro pens and iPod cases. Typological series can be made for all of these. The presence of any artefact or artefact assemblage could be taken to characterize periods or epochs, should it be decided it was relevant and useful to do so.

Deposits associated with particular periods or phases of human activity contain not just artefacts but, depending on conditions of preservation, a plethora of other kinds of evidence. These may include human remains, faunal remains, chemical contamination, cultivated plant remains, pollen and spores, insects, radioactive material, snails, molluscs, isotopes, biomolecules and so on, containing much potential information about the wider environment at the time the deposit was formed. Many of these materials are susceptible to accurate dating techniques such as carbon-14 and thermoluminescence dating.

Recent and contemporary deposits of the archaeosphere are particularly artefact-rich. As well as containing many individual objects of material culture, the stratum can be the matrix for extensive networks of pipes, wires and fibre-optic cables. Flows of oil, electricity, water, gas and sewage run through it. It is inhabited by living bacteria, insects, earthworms, rodents, reptiles and other animals. Subterranean structures within it have their own specific ecologies. Some rivers, formerly on the surface, now flow encased in concrete culverts within the archaeosphere. Other tube-like cavities accommodate the flow of people. Inextricably enmeshed with the biosphere and hydrosphere, the archaeosphere radically influences, enables and constrains the development of myriad forms of life.

It has not been the purpose of this paper to promote any particular definition or proposed date for the start of the Anthropocene, nor to tackle the problem of fitting the proposed epoch within established chronological frameworks of geology, but merely to point out the richness of the archaeosphere as a resource in providing a stratigraphical index for discussion. This applies especially to deposits dated to the last two hundred years, the period arguably of most relevance for current debate on the Anthropocene.

Proximity in time implies a need for a change of scale and focus, looking at the minutiae of evidence as well as the larger picture. Although the search for global signals of the Anthropocene is important, this might be matched by a search for signals on other scales too. Just as networks of human artefacts, and patterns of deposition and accumulation of the archaeosphere, now extend beyond the limits of the Earth and into interplanetary space, they extend also into realms of the very small. Artificial features are taking shape today on nanomaterials little more than a few atoms thick. Developments at scalar extremes, seemingly far apart, are actually inextricably interconnected. Recent advances in nanotechnology, for example, could potentially have enormous effects on systems at a global scale and all intermediate scales between. The Anthropocene, it is suggested here, is more than just a global phenomenon, to be studied on global scales. As a complex intermeshing of natural and cultural forces, it is fractally structured, manifesting on multiple scales at the same time (Edgeworth 2013).

With this in mind, the full potential of investigation of the archaeosphere actually goes far beyond the mere provision of stratigraphical markers for the proposed new geological epoch. It is in reaching a more fine-grained and multi-scalar understanding of the Anthropocene (however it is defined and whatever signals are chosen to mark its beginning) that the detailed study of the archaeosphere may have most value, and there is great potential for archaeologists and geologists to work together to that end.

The author would like to thank S. Price, A. Howard and C. Waters for detailed and useful comments on all aspects of this paper, and to E. Harris for key insights on methodological issues. The British Geological Survey, R. Buckley, A. Moore, M. Carver and Oxford University Press generously gave permission for reproduction of drawings. An early version of the paper was presented at the Yorkshire Geological Society meeting on the Anthropocene at the University of Sheffield, February 2012.

References

ATKINSON, R. J. C. 1974. Neolithic science and technology. *Philosophical Transactions of the Royal Society of London*, **276**, 123–131.

BARKER, G. 2002. A tale of two deserts: contrasting desertification histories in Rome's desert frontiers. *World Archaeology*, **33**, 488–507.

BUCKLEY, R. J. & LUCAS, J. N. 1987. *Leicester Town Defences*. Leicestershire Museums Art Galleries and Records Services, Leicester.

CARVER, M. 1987. *Underneath English Towns: Interpreting Urban Archaeology*. Batsford, London.

CERTINI, G. & SCALENGHE, R. 2011. Anthropogenic soils are the golden spikes for the Anthropocene. *The Holocene*, **21**, 1269–1274.

CHESWORTH, W. (ed.) 2008. *Encyclopaedia of Soil Science*. Springer, New York.

CONNER, A. & BUCKLEY, R. 1999. *Roman and Medieval Occupation in Causeway Lane, Leicester*. Leicester Archaeology Monographs No. 5. University of Leicester, Leicester.

COURTNEY, P. 1998. Saxon and Medieval Leicester: the making of an urban landscape. *Transactions of the Leicestershire Archaeological and Historical Society*, **72**, 110–145.

CREIGHTON, O., CHRISTIE, N., HAMEROW, H. & EDGEWORTH, M. 2009. Wallingford. *British Archaeology*, **35**, 27–43.

CRUTZEN, P. J. 2009. Can we survive the 'Anthropocene' period? Project Syndicate website. http://www.project-syndicate.org/commentary/crutzen1/English.

CRUTZEN, P. J. & STOERMER, E. F. 2000. The Anthropocene. *Global Change Newsletter*, **41**, 17–18. http://www.mpch-mainz.mpg.de/~air/anthropocene/Text.html.

DANIEL, G. 1976. *A Hundred and Fifty Years of Archaeology*. Harvard University Press, Cambridge.

DANIEL, G. & RENFREW, C. 1987. *The Idea of Prehistory*. Columbia University Press, New York.

EDGEWORTH, M. 2011. *Fluid Pasts: Archaeology of Flow*. Bloomsbury Academic, London.

EDGEWORTH, M. 2013. Scale. *In*: GRAVES-BROWN, P., HARRISON, R. & PICCINI, A. (eds) *The Oxford Handbook of the Archaeology of the Contemporary World*. Oxford University Press, Oxford, Chapter 27.

FORD, J., KESSLER, H., COOPER, A. H., PRICE, S. J. & HUMPAGE, A. J. 2010. *An Enhanced Classification for Artificial Ground*. British Geological Survey Open Report **OR/10/036**.

FORD, J. R., PRICE, S. J., COOPER, A. H. & WATERS, C. N. 2013. An assessment of lithostratigraphy for anthropogenic deposits. *In*: WATERS, C. N., ZALASIEWICZ, J. A., WILLIAMS, M., ELLIS, M. A. & SNELLING, A. M. (eds) *A Stratigraphical Basis for the Anthropocene*. Geological Society, London, Special Publications, **395**, http://dx.doi.org/10.1144/SP393.12.

FOWLER, P. J. 1983. *The Farming of Prehistoric Britain*. Cambridge University Press, Cambridge.

GALE, S. J. & HOARE, P. G. 2012. The stratigraphic status of the Anthropocene. *The Holocene*, **22**, 1491–1494.

GORMAN, A. C. 2005. The cultural landscape of interplanetary space. *Journal of Social Archaeology*, **5**, 85–107.

GORMAN, A. C. 2009. Heritage of Earth orbit? Orbital debris – its mitigation and cultural heritage. *In*: DARRIN, A. G. & O'LEARY, B. L. (eds) *Handbook of Space Engineering, Archaeology and Heritage*. CRC Press, Boca Raton, 381–397.

Hall, R. A. & Hunter-Mann, K. 2003. *Medieval Urbanism in Coppergate: Refining a Townscape*. Council for British Archaeology, York.

Harris, E. C. 1989. *Principles of Archaeological Stratigraphy*. Academic Press, London.

Holliday, V. T. 2004. *Soils in Archaeological Research*. Oxford University Press, Oxford.

Hutton, J. 1795. *Theory of the Earth with Proofs and Illustrations*. William Creech, Edinburgh.

Jiongxin, X. 2003. Naturally and anthropogenically accelerated sedimentation in the Lower Yellow River, China, over the past 13.000 years. *Geografiska Annaler Series A*, **80**, 67–78.

Johnson, L. C. 1995. China's Pompeii: twelfth century Dongjing. *Historian*, **58**, 49–68.

Leary, J. & Field, D. 2010. *The Story of Silbury Hill*. English Heritage, Swindon.

Lucas, J. N. 1981. The debris of history: an archaeological survey of Leicester. *Transactions of the Leicestershire Archaeological and Historical Society*, **56**, 1–9.

Lyell, C. 1830. *Principles of Geology, Being an Attempt to Explain the Former Changes of the Earth's Surface, by Reference to Causes now in Operation*. John Murray, London, **1**.

Macphail, R. I., Galinie, H. & Verhaeghe, F. 2003. A future for dark Earth? *Antiquity*, **77**, 349–358.

McMillan, A. & Powell, J. H. 1999. *BGS Rock Classification Scheme Volume 4: Classification of Artificial (Man-made) Ground and Natural Superficial Deposits – Applications to Geological Maps and Datasets in the UK*. British Geological Survey Research Report **RR/99/04**.

Menze, B. H. & Ur, J. A. 2012. Mapping patterns of long-term settlement in Northern Mesopotamia at a large scale. *Proceedings of the National Academy of Sciences of the USA*, **109**, 5146–5147.

Moore, A. M. T., Hillman, G. C. & Legge, A. J. 2000. *Village on the Euphrates: From Foraging to Farming at Abu Hureyra*. Oxford University Press, Oxford.

MUSEUM OF LONDON 1994. *Archaeological Site Manual*. 3rd edn. MOLAS, London.

NATIONAL RESEARCH COUNCIL 1995. *Orbital Debris: A Technical Assessment*. National Academy Press, Washington.

Pears, B. 2009. *The nature, distribution and significance of amended and anthropogenic soils on old arable farms and the elemental analysis of black carbonised particles*. Unpublished doctoral thesis submitted to the School of Biological and Environmental Sciences, University of Stirling.

Periman, R. D. 2006. Visualizing the Anthropocene: human land use history and environmental management. *In*: Aguirre-Bravo, C., Pellicane, P. J., Burns, D. P. & Draggan, S. (eds) *Monitoring Science and Technology Symposium: Unifying Knowledge for Sustainability in the Western Hemisphere*. US Department of Agriculture, Fort Collins, 558–564.

Price, S. J., Ford, J. R., Cooper, A. H. & Neal, C. 2011. Humans as major geological and geomorphological agents in the Anthropocene: the significance of artificial ground in Great Britain. *Philosophical Transactions of the Royal Society*, **369**, 1056–1084.

Rathje, W. & Murphy, C. 1992 *Rubbish! The Archaeology of Garbage*. Harper Collins, New York.

Rosenbaum, M. S., McMillan, A. A., Powell, J. H., Cooper, A. H., Culshaw, M. G. & Northmore, K. J. 2003. Classification of artificial (man-made) ground. *Engineering Geology*, **69**, 399–409.

Roskams, S. (ed.) 2000. *Interpreting Stratigraphy. Papers presented to the Interpreting Stratigraphy Conference*, 1993–1997. Oxford, BAR International Series **910**. Archaeopress, Oxford.

Roskams, S. 2001. *Excavation*. Cambridge University Press, Cambridge.

Ruddiman, W. F. 2003. The anthropogenic greenhouse era began thousands of years ago. *Climatic Change*, **61**, 261–293.

Ruddiman, W. 2005. *Plows, Plagues and Petroleum: How Humans Took Control of Climate*. Princeton University Press, New Jersey.

Salisbury, R. 2012. Engaging with soil, past and present. *Journal of Material Culture*, **17**, 23–41.

Schiffer, M. B. 1987. *Formation Processes of the Archaeological Record*. University of New Mexico Press, Albuquerque.

Slowikowski, A. 1995. 'The greatest depository of archaeological material': the role of pottery in ploughzone archaeology'. *In*: Shepherd, E. (ed.) *Interpreting Stratigraphy*. Witley Press, Hunstanton, **5**, 15–20.

Smith, W. 1816. *Strata Identified by Organized Fossils*. W. Arding, London.

Solli, B., Burstrom, M. *et al.* 2011. Some reflections on heritage and archaeology in the Anthropocene. *Norwegian Archaeological Review*, **44**, 40–88.

Steffen, W., Crutzen, P. J. & McNeill, J. 2007. The Anthropocene: are humans now overwhelming the great forces of nature? *Ambio*, **36**, 614–621.

Syvitski, J. P. M. & Kettner, A. 2011. Sediment flux and the Anthropocene. *Philosophical Transactions of the Royal Society*, **369**, 957–975.

Thomas, W. L. (ed.) 1958. *Man's Role in Changing the Face of the Earth, Vols 1–2*. University of Chicago Press, Chicago.

Wilkinson, I. P., Tasker, A., Gouldwell, A., Williams, M., Edgeworth, M., Zalasiewicz, J. & Christie, N. 2010. Micropalaeontology reveals the source of building materials for a defensive earthwork (English Civil War?) at Wallingford Castle, Oxfordshire. *Journal of Micropalaeontology*, **29**, 1–6.

Wilkinson, T. J. 2003. *Archaeological Landscapes of the Near East*. University of Arizona Press, Tucson.

Zalasiewicz, J. 2008. *The Earth After Us. What Legacy will Humans Leave in the Rocks?* Oxford University Press, Oxford.

Zalasiewicz, J., Smith, A. *et al.* 2008. Are we now living in the Anthropocene? *GSA Today*, **18**, 4–8.

Zalasiewicz, J., Williams, M. *et al.* 2011. Stratigraphy of the Anthropocene. *Philosophical Transactions of the Royal Society*, **369**, 1036–1055.

The mineral signature of the Anthropocene in its deep-time context

JAN ZALASIEWICZ[1]*, RYSZARD KRYZA[2] & MARK WILLIAMS[1]

[1]*Department of Geology, University of Leicester, University Road, Leicester LE1 7RH, UK*

[2]*University of Wrocław, Institute of Geological Sciences, ul. Cybulskiego 30, 50-205 Wrocław, Poland*

**Corresponding author (e-mail: jaz1@le.ac.uk)*

Abstract: The Earth has shown a systematic increase in mineral species through its history, with three 'eras' comprising ten 'stages' identified by Robert Hazen and his colleagues (Hazen *et al.* 2008), the eras being associated with planetary accretion, crust and mantle reworking and the influence of life, successively. We suggest that a further level in this form of evolution has now taken place of at least 'stage' level, where humans have engineered a large and extensive suite of novel, albeit not formally recognized minerals, some of which will leave a geologically significant signal in strata forming today. These include the great majority of metals (that are not found natively), tungsten carbide, boron nitride, novel garnets and many others. A further stratigraphic signal is of minerals that are rare in pre-industrial geology, but are now common at the surface, including mullite (in fired bricks and ceramics), ettringite, hillebrandite and portlandite (in cement and concrete) and 'mineraloids' (novel in detail) such as anthropogenic glass. These have become much more common at the Earth's surface since the mid-twentieth century. However, the scale and extent of this new phase of mineral evolution, which represents part of the widespread changes associated with the proposed Anthropocene Epoch, remains uncharted. The International Mineralogical Association (IMA) list of officially accepted minerals explicitly excludes synthetic minerals, and no general inventory of these exists. We propose that the growing geological and societal significance of this phenomenon is now great enough for human-made minerals to be formally listed and catalogued by the IMA, perhaps in conjunction with materials science societies. Such an inventory would enable this phenomenon to be placed more effectively within the context of the 4.6 billion year history of the Earth, and would help characterize the strata of the Anthropocene.

Deep-time context

Hazen *et al.* (2008; see also Hazen & Ferry 2010) clearly demonstrated the growth in mineral species in the context of the origin and evolution of the Earth. They described a first era that began with the mineral dust that condensed in the aftermath of supernova explosions and drifted in clouds of interstellar dust. This comprises little more than a dozen minerals – diamond with a few carbides, nitrides, oxides and silicates. As our own solar nebula formed, the intense luminosity of the early Sun melted and recondensed the circling dust into primitive chondritic meteorites. Some 50 additional minerals now appeared, among them spinel and members of the pyroxene, feldspar and olivine family (Fig. 1).

The collision-driven growth of larger meteorites and planetesimals fuelled further mineral formation in a second, relatively rapid (from 4.56 to 4.55 Ga) stage of this process. The addition of water from cometary debris enabled low-temperature hydration and aqueous precipitation reactions, and calcite, gypsum, halite, chlorite, serpentine and the first clay minerals appeared. Such minerals can be found in the carbonaceous chondrites that fall to Earth (McCoy 2010), and the total number of minerals now reached *c.* 150. Larger planetesimals were heated (probably by the decay of the short-lived ^{26}Al) to the point at which they began to melt and differentiate into metal-rich cores and silicate exteriors, and fractional crystallization began to produce familiar rock-forming minerals such as quartz and potassium feldspar. By then, the *c.* 250 minerals that can be found in meteorites of all sorts were present. These form the building blocks of an Earth-sized rocky planet.

The Earth's second mineral era began as our planet formed, and differentiated into core, mantle and crust. Classical igneous processes probably dominated in the first half-billion years or so of the Hadean, of which almost no direct record remains. Fractional crystallization and various water-related reactions initially took the number of minerals to some 500 – roughly the same as is probably present on Mars. The Moon, being largely anhydrous, only has approximately 350 (Hazen & Ferry 2010).

A further stage in this era came with crust-related processes, including the kind of differentiation seen in granite bodies, where melting and remelting

From: Waters, C. N., Zalasiewicz, J. A., Williams, M., Ellis, M. A. & Snelling, A. M. (eds) 2014. *A Stratigraphical Basis for the Anthropocene*. Geological Society, London, Special Publications, **395**, 109–117.
First published online December 19, 2013, http://dx.doi.org/10.1144/SP395.2

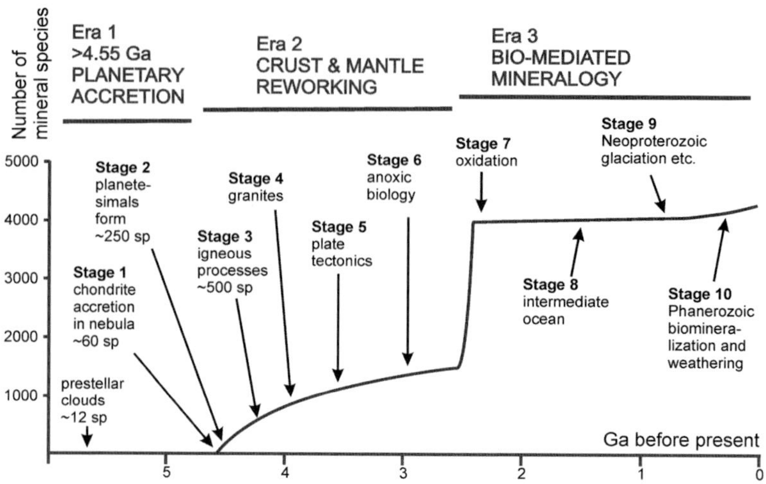

Fig. 1. The growth in mineral species before and through the formation of the Earth; drawn after data in Hazen *et al.* (2008).

drove the composition from basaltic to granitic and beyond, concentrating rare and incompatible elements into a diverse range of new minerals. Pegmatites, an end member of this process, have alone yielded more than 500 minerals.

Some time before (probably well before) 3 Ga, some form of plate tectonics began, and subduction-related processes led to yet more diversification. This stage is associated with the formation of hydrothermal ore bodies, while metamorphism (both regional and contact) took place. On such a dynamic Earth, some 1500 minerals, in total, can be generated. However, the Earth was still biologically dead.

Life appeared on Earth some time before 3 Ga – and probably nearer 4 Ga – and this defined the succeeding stage of the second mineral era. As profound a step as this was, the initial impact on mineralogy was not great. The earliest microbial life-forms soon became abundant, certainly, and generated or mediated abundant and distinctive strata, notably banded iron formations and stromatolitic reefs. However, the number of new mineral species remained small (although this might have included bio-mediated diamonds; Nisbet *et al.* 1994).

Then, at *c.* 2.4 Ga, a biogeochemical innovation appeared that drove a new burst of mineral diversification. The third mineral era began. Photosynthesis led to the development of an oxygenated surface environment, in what is now termed the 'Great Oxygenation Event'. This phenomenon restricted some mineral types. For instance, detrital pyrite disappeared from terrestrial surface deposits (Rasmussen & Buick 1999). Far more importantly, it created a wide range of oxidized products of primary minerals, more than doubling the number of mineral species on Earth to something in excess of 4000 (Hazen *et al.* 2008; Sverjensky & Lee 2010, for instance, describe the diversification of uranium minerals in an oxidized environment).

Further stages of this era have been defined associated with the mid-Proterozoic ocean of 'intermediate oxidation', with the late Proterozoic 'Snowball Earth' events and further oxidation, and with the rise of the biomineralizing metazoa in the Phanerozoic. These further modified the mineral inventory of Earth, although more in terms of pattern, relative abundance and distribution than in any substantial further increase in new minerals. The sophisticated and intricate patterns in biomineralized structures such as bivalve shells, diatom tests and echinoderm plates (Dove 2010), for instance, and the volume and complexity of reef limestones, have made the strata of the last half-billion years of Earth history clearly distinctive.

The next major development in the Earth's mineral storehouse was to await the emergence of organized societies in one recently evolved primate species, *Homo sapiens*. This, current, phase of global mineralogical innovation is one part of the diverse suite of phenomena that may now be said to characterize the 'Anthropocene'.

The Anthropocene

The concept of the Anthropocene – that humans have changed the geological processes operating on the Earth's surface sufficiently to justify a new geological time interval – was first suggested over a century ago by authors such as George Perkins Marsh (1864), Antonio Stoppani (1873 – who proposed the 'Anthropozoic Era') and Vladimir Vernadsky (1945). However, these early suggestions were largely discounted by the geological community (e.g. Berry 1925), which regarded human geological influence as trivial compared to the natural forces of volcanism, mountain building and so on.

The concept of the Anthropocene was restated more recently by Paul Crutzen (Crutzen & Stoermer 2000; Crutzen 2002), at a time when the scientific community was becoming aware that various types of anthropogenic change were geologically significant. The term quickly began to be used in scientific literature (e.g. Steffen *et al.* 2004; Syvitski *et al.* 2005), and was analysed as a potential new, formal addition to the Geological Time Scale by the Stratigraphy Commission of the Geological Society of London (Zalasiewicz *et al.* 2008; see also Zalasiewicz *et al.* 2011), which concluded that a case for formalization could be made. Evidence cited include changes that may be considered as lithostratigraphic (human changes to erosion and sedimentation, both direct and indirect; see Ford *et al.* 2014), biostratigraphic (the current biodiversity changes involving extinctions, species invasions and biomass changes; see Barnosky 2013) and chemostratigraphic (notably modification of the Earth's carbon chemistry, but including various other changes; see Galuszka *et al.* 2013).

An Anthropocene Working Group of the Subcommission on Quaternary Stratigraphy, a constituent body of the International Commission on Stratigraphy, was subsequently set up to examine the question. A body of evidence has already been published (e.g. papers in Williams *et al.* 2011; Zalasiewicz *et al.* 2010; Ellis *et al.* 2010, 2012). There has been considerable interest in this concept in the wider community, both scientific (e.g. Nature 2011; Steffen *et al.* 2011) and popular (e.g. Kolbert 2010). We focus here on a previously neglected aspect of this phenomenon: the increase in the Earth's range of mineral species as a direct product of human activity.

Novel human-made minerals

Humans have long modified the compositions of terrestrial minerals that they have exploited. One of the earliest examples is the production of native metals, generally something of a rarity on a pre-human Earth. In elemental form, only gold, silver, copper, the platinum group elements and iron naturally occur in significant amounts on our planet. Humans have since learnt to separate metals that either seem never to have existed naturally in their native form at any time in Earth history, such as magnesium, calcium, sodium, potassium, vanadium and molybdenum, or those that occur in native form exceedingly rarely, and in tiny amounts, such as aluminium, titanium and zinc (Fig. 2; see the following). Many metal alloys are also produced, and for the most part these – such as bronze, brass and pewter – are also geological novelties, as are the many types of steel (augmented by chromium, molybdenum and other metals).

Other 'synthetic minerals', for smaller-scale purposes, include novel forms of garnet for use in lasers and as artificial gems (Hazen *et al.* 2008), other synthetic crystalline materials used in lasers, including potassium titanyl phosphate (Bierlein & Vanherzeele 1989) and lithium niobate (Kuzminov 1999; Volk & Wohlecke 2008), cubic boron nitride (better known under its brand name Borazon), which is harder than diamond (Wentorf 1993) and now a routinely used abrasive, boron carbide, used in tank armour, bullet-proof vests and nuclear shielding, tungsten carbide, used as the balls in ball-point pens (of which hundreds of billions have been manufactured since 1950 CE) and in armour-piercing shells, various artificial zeolites (van Bekkum *et al.* 2001), graphene, and so on. There seem to be no inventories of new minerals – not least because the current regulations of the International Mineralogical Association (IMA), that validates such names, does not recognize minerals of anthropogenic origin (see the following). However, the combined novel mineral products of the world's laboratories probably now number in the thousands (at least). Most

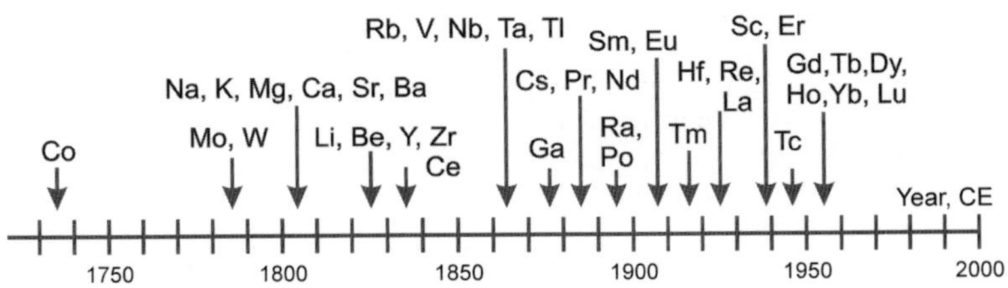

Fig. 2. Timing of the first appearance (by human agency) of pure forms of non-natively formed metals on Earth. Information collated from Wikipedia.

will be rarities (as are, however, most of the recognized natural minerals), while others are more widespread.

One might consider such substances as man-made glass and plastics as new minerals *sensu lato*. Although not strictly stoichiometric in composition, glass (>50 million tons produced annually) and plastics (see Barnosky 2013; Ford *et al.* 2013) have clearly become specific and widely visible parts of the Earth's surface (see below), as have the minerals associated with, say, ceramics, concrete and brick production, and in mining and smelting wastes and slags, which include new mineral forms, and where the petrographic novelty is clear. More widely, there are many solid, liquid and gaseous industrial contaminants that, on release (intentional or accidental) to the external environment, react with other materials, both natural and artificial, to add further diversity to the anthropogenic mineral suite.

It seems clear, therefore, that a significant new epoch in the Earth's mineral evolution has begun, which may emerge as the most striking new development since the Great Oxidation Event of the Precambrian (Fig. 3), and might be considered as a new stage vis-à-vis the classification of Hazen *et al.* (2008). It may, however, be short-lived, unlike the case with natural minerals (and exceedingly short-lived in the case of some of the artificially produced heavy elements made in nuclear reactors). Many man-made minerals will become extinct (i.e. cease to form), either because they come to be obsolete for humans or, ultimately, with the extinction of humans. Nevertheless, there will undoubtedly be a distinctive mineralogical character to the Anthropocene strata of the future, allowing, naturally, for subsequent diagenetic changes (here, archaeological studies of the taphonomy of ancient human artefacts might offer clues as to the nature of the longevity and possible future transformations of anthropogenic minerals). Furthermore, in characterizing this new mineral development, changed abundances may be as important as novel mineralogical composition.

Mineral assemblage changes and characterization of the Anthropocene

Many of the new minerals that humans have made (as with most natural minerals) are rare – borazon, graphene, pure molybdenum and the new garnets, for instance – and so are rarely likely to be of practical use in defining Anthropocene strata. Others (potassium, sodium) are too unstable to survive for any length of time. Overall, the total assemblage of minerals is probably diversifying on something like the scale of the Great Oxygenation Event (or greater). Human action, here, is accentuating the chemical disequilibrium on Earth – a disequilibrium that became a signature of this planet as soon as life evolved on it. Most of the new minerals, however, will not be of practical value to the stratigrapher.

However, the relative proportions of different minerals at the Earth's surface have changed as a result of human activity, and in some cases markedly enough to be used in characterization of strata. Human production of metals now takes place on a large scale, for instance. Annual world production of iron now exceeds a billion tons, that of aluminium exceeds 35 million tons, and of titanium 100 000 tons. Totalling historical world production estimates (USGS 2010), the total amount

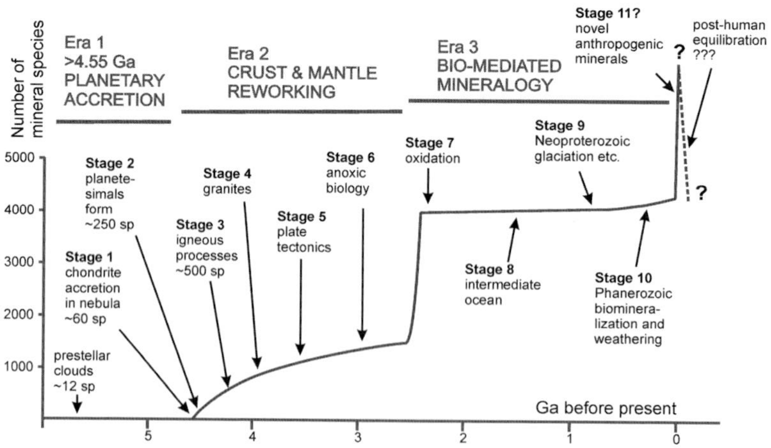

Fig. 3. The growth of mineral species on Earth, drawn after data in Hazen *et al.* (2008), with addition of suggested new Anthropocene mineral stage.

of aluminium produced globally has been about half a billion tons (Fig. 4), assuming a recycling rate of 50%. This is sufficient to place the entire land area of the US (and part of neighbouring Canada) under a layer of standard kitchen aluminium foil of 0.015 mm thickness (Fig. 5). Some 98% of this production has taken place from the 1950s onwards (Fig. 4), emphasizing the scale of the mid-century 'Great Acceleration' of Steffen *et al.* (2007).

The summed global production of iron and steel over the past century, from the same data source (USGS 2011), and again assuming a recycling rate of 50%, has been *c.* 15 billion tons – that is, some 30 times more in terms of mass than aluminium production. Considering that native aluminium is almost unknown geologically, and native iron is rare, this represents a substantial change in the Earth's surface mineral assemblage.

There are other minerals that are rare in pre-industrial geology, but now common at the surface, including mullite. This is found in nature in some contact metamorphic or pyrometamorphic settings, but is now one of the main constituents of fired bricks (e.g. Dunham 1992*a*; global annual production is now >1000 billion bricks annually, Wang 2010) and ceramics, essentially formed by very high-temperature metamorphism of clay minerals. There are also ettringite, hillebrandite and portlandite, in nature obscure hydrated calcium and aluminium silicates that may be found for example in some skarns, but are now key minerals in cement and concrete (with global production of concrete now being some 2 billion tons annually, Crow 2008). Novel 'mineraloids' such as anthropogenic glass are also produced in large amounts.

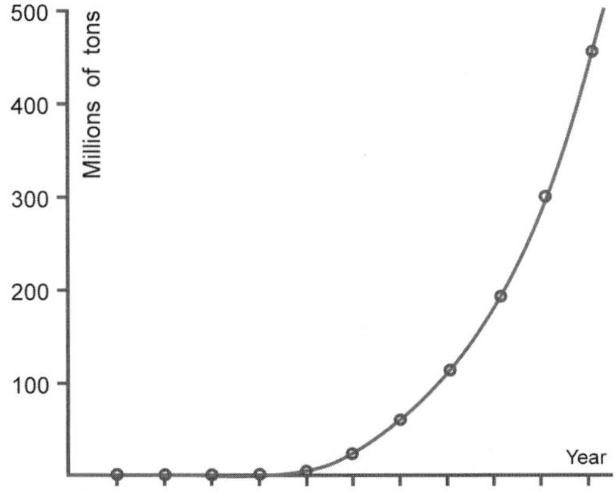

Fig. 4. Cumulative growth of anthropogenic aluminium in the surface environment (adapted from data in USGS 2010, assuming a recycling rate of 50%).

These materials, which characterize our built landscapes, are thus distinctive not only in terms of lithofacies, but also in terms of their mineralogy.

The novel anthropogenic minerals commonly also take new and distinctive forms, or mineral habits – those that reflect our manufacturing processes and the artefacts that result from them. This pattern is in itself a significant step in the elaboration of mineral varieties in Earth history, where primordial magmatic textures began to be supplemented with metamorphic and hydrothermal ones as plate tectonics began, and later by biogenic textures, particularly as a metazoan skeleton-building biota radiated following the 'Cambrian explosion' of life. The new mineral habits created in human laboratories and factories represent an elaboration of mineral patterning that is quite as distinctive as that associated with the emergence of multicellular life. Furthermore, the rate of evolution of these mineral habits exceeds anything produced by pre-human biology (e.g. a wave of aluminium cans and plastic bottles in the 1960s to 1970s largely supplanting the dominant glass bottles of earlier decades: Ford *et al.* 2013).

Even more widely, mineral changes have arisen out of the anthropogenic selective erosion that we call mining. We have now collectively doubled the amount of metals being cycled at the Earth's surface (Rauch & Pacyna 2009; see also Sen & Peucker-Ehrenbrink 2012), with some metals (for instance copper) substantially exceeding that. Some non-metallic elements have increased in a similar manner, such as the phosphorus and reactive nitrogen essential for food production (Canfield *et al.* 2010; Holtgrieve *et al.* 2011; Elser 2011), and these have substantially impacted upon ecosystems, including the creation of seasonal anoxic 'dead zones' in shallow marine waters (Diaz & Rosenberg 2008). Not all anthropogenic changes are additive. Ongoing acidification of the world's oceans is likely to diminish the amount of carbonate precipitated, especially of the more soluble aragonite polymorph (Zeebe *et al.* 2008; Tyrrell 2011).

In terms of such wider impact, it is not only the increased amounts of minerals that are of significance, but their form. For instance, natural windblown mineral dusts (important for fertilization of oceanic plankton-based ecosystems) have only 1–2% of their iron (mostly oxides) in a biologically available form, while the comparable figure for industrial fly ash (iron mostly as sulphates) is *c.* 70% (Schroth *et al.* 2009). These phases, and the changes in mineral assemblages described above, do not often comprise novel minerals *per se*, but their burgeoning may be viewed as akin to the changes in mineral relative abundance and texture associated with Hazen and colleagues' (2008) mineral stages of the Proterozoic–Phanerozoic interval.

Fig. 5. North America with a cover of the global anthropogenic production to date of aluminium expressed as standard-thickness kitchen foil. Image created by Y. Thibault-Picazo.

How might deposits assigned to a putative Anthropocene Series (Zalasiewicz *et al.* 2013) be characterized by their mineral content? It is clear that human artefacts including bricks, some metals, glass and ceramics, and their component minerals, date back to the early archaeological record, from thousands of years ago (e.g. Dunham *et al.* 1992*b*), and thus may be found sporadically in undoubted Holocene strata. This is parallel to the situation where lithofacies (Ford *et al.* 2013; Zalasiewicz *et al.* this volume, in review) and biofacies (Barnosky *et al.* 2013) signals appear locally from at least the early Holocene.

However, lithofacies and biofacies signals show clear subsequent increases from *c.* 1800 CE and again from *c.* 1950 CE. The mineral record, globally, seems to show similar thresholds, both qualitatively and quantitatively. Qualitatively, the history of new mineral formation (as revealed by the history of the development of pure, novel metals, the most accessible source of data) shows that most of these were isolated in the nineteenth and twentieth centuries (Fig. 2). Quantitatively, the production of these metals shows clear and major increases in the mid-twentieth century (e.g. Fig. 4), consistent with the 'Great Acceleration' of Steffen *et al.* (2007), as does the production of concrete, bricks and ceramics (Ford *et al.* 2013) with their distinctive mineral assemblages. The mineral evidence hence adds support to the suggestion that a *c.* 1950 CE Holocene–Anthropocene boundary is, for current and near-future purposes, more clearly definable and correlatable than one at *c.* 1800 CE (Wolfe *et al.* 2013; Barnosky *et al.* 2013; Ford *et al.* 2013).

The likely geological longevity of some modern minerals is known, from both archaeological and geological records. They range from likely short (unmodified iron) to long (mullite). Those of other materials (e.g. corrosion-resistant steel alloys, and aluminium, with its protective surface oxide coating) remain untested geologically. The pattern of modern mineral distribution is distinctive, with large amounts segregated into urban areas and concentrated in rubbish dumps, some more widely dispersed as litter, and yet more deeply buried within foundations, tunnels and boreholes (where preservation potential is high, even if in the far future the relation to the surface deposits may appear enigmatic). Certainly, though, for the present and next few generations of Earth scientists, enough will remain to be geologically useful in characterization.

Human-made minerals, and formal aspects of mineral classification

From the narrative above, there is strong evidence that the Earth's surface mineralogy is changing in terms of both mineral types and diversity, and relative abundance of the main types. The latter is relatively easy to assess, at least approximately (see above). However, as regards new human-made minerals *per se*, an overall view of the Earth's current total mineral diversity is much more difficult to obtain – partly because of current procedures in mineral definition.

Human-made minerals have been specifically excluded from formal recognition by the IMA (Nickel 1995*a*, *b*; Nickel & Grice 1998; http://pubsites.uws.edu.au/ima-cnmnc/). Thus,

> A mineral substance is a naturally occurring solid that has been formed by geological processes, either on earth or in extraterrestrial bodies (Nickel 1995*a*). A mineral species is a mineral substance with well defined chemical composition and crystallographic properties, and which merits a unique name.

Furthermore,

> Anthropogenic substances, that is those made by Man, are not regarded as minerals.

The same restriction applies for biogenic minerals – and usually also amorphous substances (sometimes termed mineraloids) – because of the difficulty of specifying a chemical composition. Following these definitions and guidelines, nearly 5000 mineral species have been formally accepted (http://pubsites.uws.edu.au/ima-cnmnc/). Most are exceedingly rare.

However, the distinction between 'naturally occurring' and biogenic/anthropogenic is not always clear-cut. It is recognized (Nickel & Grice 1998) that human activities such as mining and quarrying can create new minerals, in creating new aqueous/oxidative microenvironments, and the products of this may be included, although not the products of combustion or of reactions with such materials as blasting powder. Furthermore, there is a growing awareness of the role of microbial biology in 'chemical' diagenetic reactions. Also, certain amorphous substances (e.g. georgeite, calcio-uranoite) have been accepted as mineral species by the Commission on New Minerals, Nomenclature and Classification (CNMMN).

The definition of 'natural' as used by the IMA is therefore not straightforward in practice. In philosophy and application, it certainly excludes a wealth of substances that (other than by definition and origin) are minerals in every physical and chemical sense. Thus it does not capture the scale of mineral diversification wrought by biological, and particularly by human action. (Humans, of course, in being linked with the rest of the biological realm – in turn linked with the physicochemical Earth – may be thought of as 'naturally occurring', too.)

Proposed amendment to mineral classification

The scale of this new episode in mineral evolution remains difficult to quantify in terms of the numbers of new minerals that have been created, their chemical and physical structure, and their stratigraphic relationship to the pre-industrial complement of minerals on Earth. Furthermore, many descriptions of novel anthropogenic minerals recently developed, or in current development, may be commercial-in-confidence, or scattered within patent applications. No inventory of new, man-made minerals seems to exist.

This rapidly growing phenomenon, however, has obvious significance for both human society and for the natural environment. We suggest, therefore, that new anthropogenic minerals should be listed and classified in their own right, perhaps as a special category within the IMA's formal listing, and perhaps in conjunction with relevant scientific bodies working on synthetic materials. One need not follow all of the procedures applied to the recognition of new minerals (e.g. each newly created mineral need not be given its own specific name).

One useful development of any combined classification system for natural and synthetic minerals would be a means to combine the traditional definitions used in classical mineralogy with new challenges coming from the rapidly developing materials sciences.

We need somehow to construct bridges connecting natural minerals (including those formed by geological/chemical and biological processes) and human-made substances. The new systematics would apparently need to comprise such major categories as 'natural minerals', 'biominerals', 'anthropominerals' (metals and alloys, ceramics and glasses, cement, concrete, bricks and slags, polymers and plastics, composite materials, semiconductors, synthetic 'minerals', nanomaterials and so on), as well as transitional materials in between. This would be a significant challenge to mineralogists (including technical mineralogists) and materials scientists.

However, the cross-disciplinary links needed here would undoubtedly have the potential to generate new science and new insights. Also, the setting up of some kind of systematic overview of humanity's considerable addition to Earth's mineral inventory will provide yet another prism through which to view the unprecedented changes now taking place on this planet.

Conclusion

The Earth has been host to an increasing diversity of minerals though geological time, mainly via a small number of step-like increases associated with significant global environmental changes. The latest of these increases has been associated with the human transformation of the Earth, and gives deposits of our present epoch a mineralogically, and hence stratigraphically, novel and distinctive character. We suggest that some kind of systematic listing and analysis of these synthetic additions to the mineral kingdom is now appropriate.

We thank M. Kryza for help with drafting Figures 1–4, Y. Thibault-Picazo for preparing Figure 5, and the Mineralogical Society of Poland for stimulating this line of investigation. We are grateful for the constructive comments of the two reviewers.

References

Barnosky, A. D. 2013. Palaeontological evidence for defining the Anthropocene. *In*: Waters, C. N., Zalasiewicz, J. A., Williams, M., Ellis, M. A. & Snelling, A. M. (eds) *A Stratigraphical Basis for the Anthropocene*. Geological Society, London, Special Publications, **395**, http://dx.doi.org/10.1144/SP395.6

Berry, E. W. 1925. The term Psychozoic. *Science*, **44**, 16.

Bierlein, J. D. & Vanherzeele, H. 1989. Potassium titanyl phosphate: properties and new applications. *Journal of the Optical Society of America B*, **6**, 622–633.

Canfield, D. E., Glazer, A. N. & Falkowski, P. G. 2010. The evolution and future of Earth's nitrogen cycle. *Science*, **330**, 192–196.

Crow, J. M. 2008. The concrete conundrum. *Chemistry World*, **March**, 62–66.

Crutzen, P. J. 2002. Geology of mankind. *Nature*, **415**, 23, 17–18.

Crutzen, P. J. & Stoermer, E. F. 2000. The 'Anthropocene'. *Global Change Newsletter*, **41**.

Diaz, R. J. & Rosenberg, R. 2008. Spreading dead zones and consequences for marine ecosystems. *Science*, **321**, 926–929.

Dove, P. M. 2010. The rise of skeletal biominerals. *Elements*, **6**, 37–42.

Dunham, A. C. 1992*a*. Developments in industrial mineralogy: I. The mineralogy of brick-making. *Proceedings of the Yorkshire Geological Society*, **49**, 95–104.

Dunham, A. C. 1992*b*. Developments in industrial mineralogy: II. Archaeological mineralogy. *Proceedings of the Yorkshire Geological Society*, **49**, 105–115.

Ellis, E. C., Goldewijk, K. K., Siebert, S., Lightman, D. & Ramankutty, N. 2010. Anthropogenic transformation of the biomes 1700–2000. *Global Ecology and Biogeography*, **19**, 589–606.

Ellis, E. C., Antill, E. C. & Kreft, H. 2012. All is not loss: plant biodiversity in the Anthropocene. *PLoS ONE, 7*, e30535, http://dx.doi.org/10.1371/journal.pone.0030535

Elser, J. W. 2011. A world awash with nitrogen. *Science*, **334**, 1504–1505.

Ford, J. R., Price, S. J., Cooper, A. H. & Waters, C. N. 2014. An assessment of lithostratigraphy of Anthropocene deposits. *In*: Waters, C. N., Zalasiewicz, J. A., Williams, M., Ellis, M. A. & Snelling, A. M. (eds) *A Stratigraphical Basis for the Anthropocene*. Geological Society, London, Special Publications, **395**, http://dx.doi.org/10.1144/SP395.12

Gałuszka, A., Migaszewski, Z. & Zalasiewicz, J. 2013. Assessing the Anthropocene with geochemical methods. *In*: Waters, C. N., Zalasiewicz, J. A., Williams, M., Ellis, M. A. & Snelling, A. M. (eds) *A Stratigraphical Basis for the Anthropocene*. Geological Society, London, Special Publications, **395**, http://dx.doi.org/10.1144/SP395.5

Hazen, R. M. & Ferry, J. M. 2010. Mineral evolution: mineralogy in the fourth dimension. *Elements*, **6**, 9–12.

Hazen, R. M., Papineau, D. *et al.* 2008. Mineral evolution. *American Mineralogist*, **93**, 1639–1720.

Holtgrieve, G. W., Schindler, D. E. *et al.* 2011. A coherent signature of anthropogenic nitrogen deposition to remote watersheds of the Northern hemisphere. *Science*, **334**, 1545–1548.

International Mineralogical Association. *Commission on New Minerals, Nomenclature and Classification*, http://pubsites.uws.edu.au/ima-cnmnc/

Kolbert, E. 2010. Age of man. *National Geographic*, **March**, 62–83.

Kuzminov, Y. 1999. *Lithium Niobate Crystals: Physicochemical Aspects of Technology*. Cambridge International Science Publishing, Cambridge.

Marsh, G. M. 1864. *Man and nature; or, physical geography as modified by human action*. Charles Scribners, New York. [Reprinted by the Belknap Press of Harvard University Press in 1965.]

McCoy, T. J. 2010. Mineralogical evolution of meteorites. *Elements*, **6**, 19–23.

Nature (Editorial) 2011. The human epoch. *Nature*, **473**, 254.

Nickel, E. H. 1995*a*. Definition of a mineral. *Canadian Mineralogist*, **33**, 689–690.

Nickel, E. H. 1995*b*. Mineral names applied to synthetic substances. *Canadian Mineralogist*, **33**, 1335.

Nickel, E. H. & Grice, J. D. 1998. The IMA commission on new minerals and mineral names: procedures and guidelines on mineral nomenclature. *Canadian Mineralogist*, **36**, 913–926.

Nisbet, E. G., Mattey, D. P. & Lowry, D. 1994. Can diamonds be dead bacteria? *Nature*, **367**, 694.

RASMUSSEN, B. & BUICK, R. 1999. Redox state of the Archaean atmosphere: evidence from detrital heavy minerals in ca 3250–2750 Ma sandstones from the Pilbara Craton, Australia. *Geology*, **27**, 115–18.

RAUCH, J. N. & PACYNA, J. M. 2009. Earth's global Ag, Al, Cr, Cu, Fe, Ni, Pb and Zn cycles. *Global Biogeochemical Cycles*, **23**, GB2001. http://dx.doi.org/10.1029/2008GB003376

SCHROTH, A. W., CRUSIUS, J., SHOLKOVITZ, E. R. & BOSTICK, B. C. 2009. Iron solubility driven by speciation in dust sources to the ocean. *Nature Geoscience*, **2**, 337–240.

SEN, I. S. & PEUCKER-EHRENBRINK, B. 2012. Anthropogenic disturbance of element cycling at the Earth's surface. *Environmental Science and Technology*, **46**, 8601–8609.

STEFFEN, W., SANDERSON, A. *ET AL.* 2004. *Global Change and the Earth System: A Planet under Pressure*. Springer-Verlag, Berlin.

STEFFEN, W., CRUTZEN, P. J. & MCNEILL, J. R. 2007. The Anthropocene: are humans now overwhelming the great forces of nature? *Ambio*, **36**, 614–621.

STEFFEN, W., PERSSON, Å. *ET AL.* 2011. The Anthropocene: from global change to planetary stewardship. *Ambio*, **40**, 739–761.

STOPPANI, A. 1873. *Corsa di Geologia*. Milan.

SVERJENSKY, D. A. & LEE, N. 2010. The great oxidation event and mineral diversification. *Elements*, **6**, 31–36.

SYVITSKI, J. P. M., VÖRÖSMARTY, C. J., KETTNER, A. J. & GREEN, P. 2005. Impact of humans on the flux of terrestrial sediment to the global coastal ocean. *Science*, **308**, 376–380.

TYRRELL, T. 2011. Anthropogenic modification of the oceans. *Philosophical Transactions of the Royal Society*, **A369**, 887–908.

US GEOLOGICAL SURVEY 2010. Aluminium statistics. *In*: KELLY, T. D., MATOS, G. R. (eds) *Historical Statistics for Mineral and Material Commodities in the United States*. US Geological Survey Data Series, **140**. http://pubs.usgs.gov/ds/2005/140/ [accessed 16 December 2012].

US GEOLOGICAL SURVEY 2011. Iron and steel statistics. *In*: KELLY, T. D., MATOS, G. R. (eds) *Historical Statistics for Mineral and Material Commodities in the United States*. US Geological Survey Data Series, **140**. http://pubs.usgs.gov/ds/2005/140/ [accessed 19 December 2012].

VAN BEKKUM, H., FLANIGAN, E. M., JACOBS, P. A. & JANSEN, J. C. (eds) 2001. *Introduction to Zeolite Science and Practice*. Elsevier, Amsterdam.

VERNADSKY, V. I. 1945 (reprinted 2005). Some words about the Noösphere. *21st Century*, **Spring**, 16–21.

VOLK, T. & WOHLECKE, M. 2008. *Lithium Niobate: Defects, Photorefraction and Ferroelectric Switching*. Springer, Berlin.

WANG, X. 2010. *Environmental Pollution from Rural Brick-Making Operations and Their Health Effects upon Workers*. Northwest University for Nationalities, West of China Institute for Environmental Health, Report, March 15 2010–September 15 2010.

WENTORF, R. H. 1993. Discovering a material that's harder than diamond. *R&D Innovator*, **2**(19).

WILLIAMS, M., ZALASIEWICZ, J., HAYWOOD, A. & ELLIS, M. (eds) 2011. The Anthropocene: a new epoch of geological time? *Philosophical Transactions of the Royal Society*, **369A**, 833–1112.

WOLFE, A. P., HOBBS, W. O. *ET AL.* 2013. Stratigraphic expressions of the Holocene–Anthropocene transition revealed in sediments from remote lakes. *Earth-Science Reviews*, **116**, 17–34.

ZALASIEWICZ, J., WILLIAMS, M. *ET AL.* 2008. Are we now living in the Anthropocene? *GSA Today*, **18**, 4–8.

ZALASIEWICZ, J., WILLIAMS, M., STEFFEN, W. & CRUTZEN, P. 2010. The new world of the Anthropocene. *Environmental Science and Technology*, **44**, 2228–2231.

ZALASIEWICZ, J., WILLIAMS, M. *ET AL.* 2011. Stratigraphy of the Anthropocene. *Philosophical Transactions of the Royal Society*, **A369**, 1036–1055.

ZALASIEWICZ, J., WILLIAMS, M. & WATERS, C. N. In review. Can an Anthropocene Series be defined and recognised? *In*: WATERS, C. N., ZALASIEWICZ, J. A., WILLIAMS, M., ELLIS, M. A. & SNELLING, A. M. (eds) *A Stratigraphical Basis for the Anthropocene*. Geological Society, London, Special Publications, **395**.

ZEEBE, R. E., ZACHOS, J. C., CALDEIRA, K. & TYRRELL, T. 2008. Carbon emissions and acidification. *Science*, **321**, 51–52.

Geomagnetic and mineral magnetic characterization of the Anthropocene

IAN SNOWBALL[1], MARK W. HOUNSLOW[2]* & ANDREAS NILSSON[3]

[1]*Department of Earth Sciences, Geophysics, Uppsala University, Villavägen 16, SE-753 26 Uppsala, Sweden*

[2]*Centre for Environmental Magnetism and Palaeomagnetism, Lancaster Environment Centre, Farrer Avenue, Lancaster University, Lancaster LA1 4YQ, UK*

[3]*Geomagnetism Laboratory, School of Environmental Sciences, University of Liverpool, Oliver Lodge Laboratories, Oxford Street, Liverpool L69 72E, UK*

**Corresponding author (e-mail: m.hounslow@lancaster.ac.uk)*

Abstract: Geomagnetic and mineral magnetic data provide geological indices that are both independent of human impact (i.e. geomagnetic) and respond to human-induced environmental impact (i.e. mineral magnetic). We provide the first discussion of such magnetic events for help in defining the Anthropocene. Within the Holocene, a potential geomagnetic marker for the Anthropocene is the low dipole latitude at *c.* 2700 cal a BP, which is associated with distinct palaeosecular variation features in northerly mid- to high-latitude sites. Mineral magnetic records from lake and marine sediments identify major deforestation and soil delivery events from catchment systems in many parts of the world during the last 4000 years. In Europe, clusters of these events occur around both 2600 cal a BP and AD 1100, the former coinciding with a low in geomagnetic field dipole latitude and peak intensity. Mineral magnetic records in peats and lake sediments can reflect particulate pollution from fossil fuel burning. The expansion of major coal burning began *c.* AD 1800 in western Europe and eastern North America, but around AD 1900 this expanded due to more widely distributed coal use, and this event is the most clear mineral magnetic marker for the base of the Anthropocene.

The magnetic properties of most igneous rock, sediments, soils and archaeological materials are dominated by less than a few percent of Fe-oxides, often magnetite. The presence of magnetic remanence-bearing minerals gives rise to the ability of most rocks, sediments and strongly heated human-made materials to record the Earth's magnetic field direction (and sometimes intensity) at or close to the time of their formation. The variation in the direction and intensity of the Earth's magnetic field has changed with time (and location) on timescales of thousands of years, and has given rise to palaeomagnetic (and archaeomagnetic) dating methods, connected to the recording of the geomagnetic field. The palaeomagnetic method, which uses infrequent changes in geomagnetic field polarity, can have age resolutions as detailed as thousands of years (Langereis *et al.* 2010; Macrì *et al.* 2010) but only for materials older than the Holocene. Using the higher-frequency secular variation of the geomagnetic field the age resolution can be decades or potentially years for materials formed in the last few thousand years (Linford 2004). When adequately expressed in natural materials and human-made artefacts, secular variations of the geomagnetic field can provide stratigraphic markers that are independent of human impact on the environment. Such human impacts expressed in sediments are, to a large extent, inevitably diachronous at timescales of decades to centuries, so we need both human-independent and human-dependent tools to define the Anthropocene.

We first discuss the possibilities of using the palaeomagnetic method in helping to define useful regional or global flags for defining the Anthropocene. In addition to recording the geomagnetic field, the abundance, mineralogy and granulometry of Fe-oxides have provided widely used magnetic proxies for identifying environmental change (Maher & Thompson 1999; Evans & Heller 2003). This is because Fe-oxides can be sensitive indicators of changes in soil disturbance, fluvial or aeolian transference and delivery mechanisms of sediments (Dearing 2000), and in the site-specific burning of fuels (Petrovsky & Ellwood 1999; Maher 2009). We also explore in detail how these mineral magnetic indices of environmental change can help define the Anthropocene, as preserved in geological media. Clearly the age of such changes can only be dated by other independent means, but such magnetic measurements do have the ability to understand the linkages related to human impact upon the Earth system.

From: Waters, C. N., Zalasiewicz, J. A., Williams, M., Ellis, M. A. & Snelling, A. M. (eds) 2014. *A Stratigraphical Basis for the Anthropocene*. Geological Society, London, Special Publications, **395**, 119–141.
First published online December 19, 2013, http://dx.doi.org/10.1144/SP395.13

Palaeomagnetic isochrones in the Anthropocene

Datasets available for the geomagnetic field

Shipboard observations of the inclination and declination of the geomagnetic field started in the sixteenth century AD, and magnetic field observatories (which provide measurements of the full geomagnetic field vector) have been available since Carl Friedrich Gauss made the first measurements of field intensity in AD 1832 (Barraclough 1982). These valuable datasets are now complemented by data obtained from satellites and magnetic surveys (Finlay *et al.* 2010). Palaeomagnetists study the natural remanent magnetizations (NRMs), recorded by suitable natural materials and human-made artefacts (e.g. fired ceramics), which provide information about the direction and intensity of the ancient geomagnetic field. In addition, the production of cosmogenic radionuclides in the atmosphere is modulated by the strength of the geomagnetic field, so measurements of their concentrations present in ice cores and sediments provide valuable information about past changes in the field intensity (Muscheler *et al.* 2005). The combination of instrumental observations and palaeomagnetic reconstructions helps us to understand how the core's dynamic magnetic field is generated and changes with time and location.

The observations and recordings indicate that the geomagnetic field at the surface of the Earth varies with location and temporal scale due to internal and external processes. The internal geomagnetic field is generated by convection in the liquid outer core, which causes magnetic induction and significant temporal variations at the Earth's surface over timescales longer than a few months (Jonkers 2007). However, due to the recording mechanism in Earth materials (and dating uncertainties), only changes that take place over longer than decadal timescales are likely to be reconstructed using traditional palaeomagnetic methods. External influences include, at the shorter (<1 s) edge of the geomagnetic spectrum, human-made radio waves and lightning strikes. The effect of solar magnetic activity on the geomagnetic field is most apparent at high latitudes and at timescales equal to or shorter than the 11-year solar cycles, first identified by Schwabe (1844). However, the Sun's magnetic field has probably exhibited variations at frequencies longer than millennial timescales, but assessing these changes is problematic, because current geological proxies used to reconstruct solar activity are also affected by geomagnetic activity, so disentangling them is difficult (Snowball & Muscheler 2007).

In contrast to the majority of markers of environmental change that this volume addresses, including the origin and distribution of 'human moderated' magnetic minerals (dealt with later in this paper), there is no firm evidence that humans have significantly influenced the direction and strength of the Earth's geomagnetic field, except at less than approximately 1 km spatial scales where human-made constructions (e.g. buildings, metal objects) or waste products (e.g. iron working waste) can induce magnetic field anomalies at the Earth's surface. The lack of significant human influence on the geomagnetic field is advantageous for the purpose of identifying the start of the Anthropocene.

At the Earth's surface, the present-day field has a geometry that roughly resembles that of a bar magnet placed at its centre – a geocentric axial dipole (GAD). Changes in the polarity of the Earth's GAD field (i.e. when the geomagnetic North Pole switches place with the geomagnetic South Pole) are global events that form the basis of the geomagnetic polarity timescale, which is a powerful geological dating tool (e.g. Ogg 2012). Less dramatic changes in the dipole field such as the tilt (angular offset between the geomagnetic axis and the geographic axis) and the dipole moment could potentially be used to define the start of the Anthropocene. If the geomagnetic field at the surface was adequately represented by a dipole, then palaeomagnetic measurements from one location could be used to predict, and synchronize, geomagnetic field directions and intensities from anywhere on Earth. However, realistically, we do need to consider the non-dipole components of the field, (including sub-millennial time and location variations), which are more difficult to detect in palaeomagnetic data and do not necessarily have the same kind of signature everywhere on the Earth's surface.

Observations of geomagnetic field change over the last 400 years

Finlay *et al.* (2010) described the most recent version (version 11) of the International Geomagnetic Reference Field (IGRF-11), which is a spherical harmonic model of the field that is constrained by magnetic observatory and satellite data for the time period AD 1900–2010 (at five-year intervals and valid until AD 2015, when the next IGRF should be released). In our contribution to the definition of the Anthropocene, we go back further in time than the start of the IGRF model, which is AD 1900. Observations from a wide variety of historical sources, primarily maritime, form the basis of a longer reconstruction of the geomagnetic field – gufm.1 (Jackson *et al.* 2000) – which spans the period between AD 1590 and AD 1990 (Fig. 1). The gufm.1 predictions of total field,

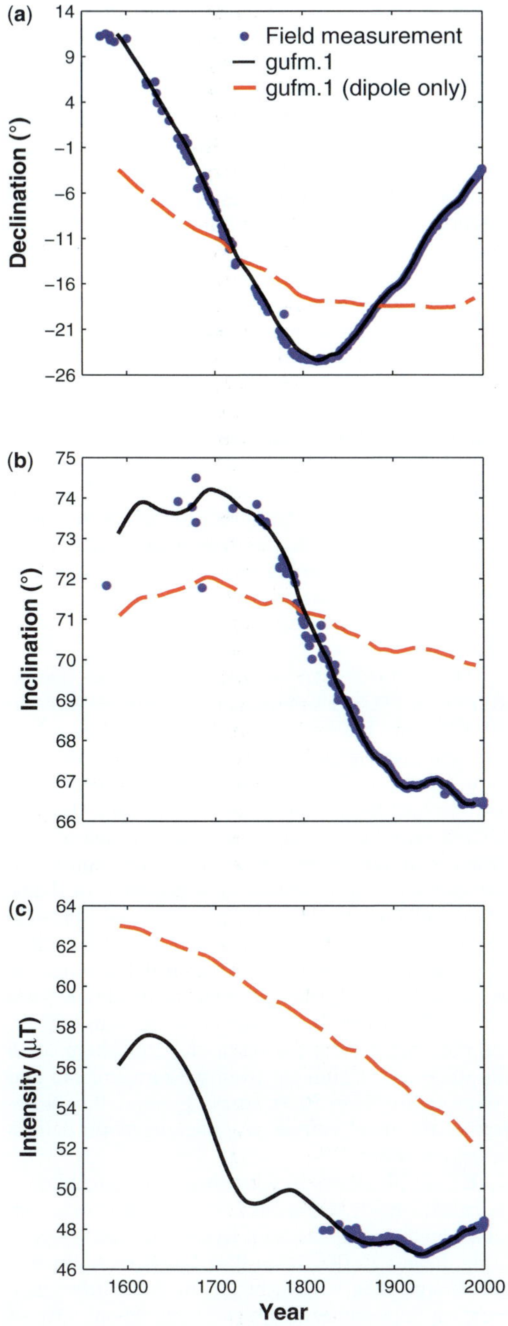

Fig. 1. Historic observations (blue dots) of (**a**) declination, (**b**) inclination and (**c**) geomagnetic field intensity at Greenwich, London, based on data compilations from Malin & Bullard (1981) and Barraclough *et al.* (2000). The geomagnetic field prediction of the gufm.1 model (Jackson *et al.* 2000) is shown as a black line and the dipole field prediction of the gufm.1 model as a dashed red line.

inclination and declination at Greenwich (London) are compared in Figure 1 with historic observations based on data compilations of Malin & Bullard (1981) and Barraclough *et al.* (2000). The observational record (Fig. 1) becomes more detailed and more accurate as time progresses due to the ability to regularly record inclination and intensity. To illustrate the significance of higher degree components, the gufm.1 predictions are also shown based solely on the dipole component of the field (Fig. 1).

To demonstrate the changes in field geometry that have been observed, we show in Figure 2 the gufm.1 predictions of the global radial field intensity on the Earth's surface at three points in time, namely AD 1590, AD 1840 and AD 1990 (Jackson *et al.* 2000). The gufm.1 model predictions illustrate significant deviations from the geocentric axial dipole field model, which would predict the same positive and negative intensities (and inclinations) along lines of equal latitude with zero declination.

A well-known and significant regional difference between the gufm.1 model predictions and the dipole field model field exists in the South Atlantic (Fig. 2), where there is a negative anomaly in field intensity (called the South Atlantic Anomaly). The centre of this anomaly has recently moved westwards at a rate of 0.3° per year (Fürst *et al.* 2009), due to patches of reversed flux at the core–mantle boundary (Gubbins & Bloxham 1985). The growth of this anomaly is considered to be responsible for the approximate 10% reduction in dipole moment that has been witnessed over the past 170 years. This anomaly is one modern example of a non-dipole geomagnetic feature that makes a significant contribution to the overall field geometry, and restricts the ability to use regional palaeofield data for global correlation. Likewise, the surface movement of the north magnetic pole has accelerated to 50 km per year (Newitt *et al.* 2009), while the south magnetic pole is at a relative standstill of a few kilometres per year.

Time-dependent changes in the Earth's magnetic field are commonly and collectively called 'geomagnetic secular variation'. 'Geomagnetic jerks' (a phrase coined by Courtillot *et al.* 1978) are rapid changes in the derivative of the rate of change of the geomagnetic field. These jerks can occur on timescales as short as one or two years and have a variety of origins, mostly invoking torsional oscillations in the outer core (Bloxham *et al.* 2002) with potential couplings to orbital influences (the Chandler wobble) and the length of day (Mandea *et al.* 2010). To detect geomagnetic jerks requires frequent observatory and/or satellite data, which means that records of them are limited to the last half a century, with jerks centred on AD 1969, 1972, 1976, 1982, 1992, 1999 and 2004

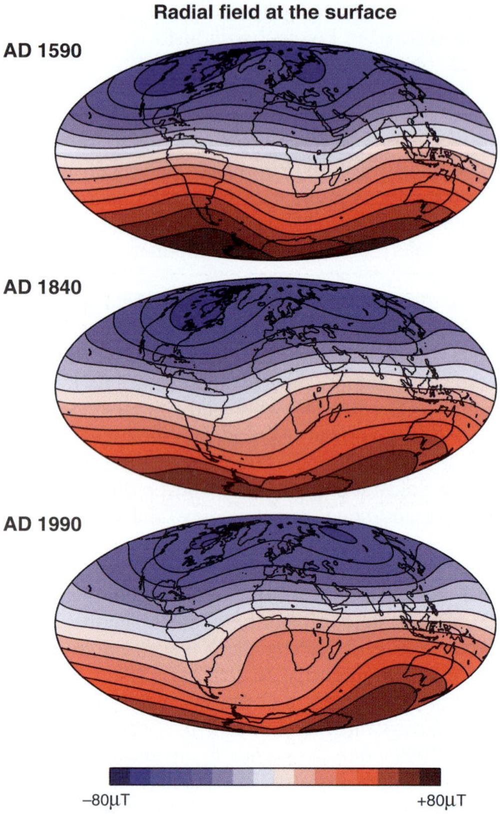

Fig. 2. Contour plots at AD 1590, AD 1840 and AD 1990 of the radial magnetic field at the Earth's surface based on the gufm.1 geomagnetic field model (Jackson *et al.* 2000). Red coloured areas represent areas where the field lines are pointing outwards and blue areas where the field lines point inwards.

(Mandea *et al.* 2010). Owing to their recent discovery, the global nature of geomagnetic jerks and their definition are still being debated (Bloxham *et al.* 2002; Mandea & Olsen 2009). The possibility exists that their predominance in Europe over the past few decades is related to the current field geometry, and that a different geometry could cause them to be more significant in other parts of the world.

Palaeomagnetic field reconstructions and data limitations

On longer timescales, we are required to interpret the less frequent data provided by palaeomagnetic studies, which inevitably have a less than ideal spatial distribution. NRMs are stored in different geological materials (e.g. lavas, marine and lacustrine sediments, and loess) and archaeological archives (e.g. burned ceramics and kilns). Most NRMs are acquired by either (1) cooling of a magnetic material below its Curie (or Néel) temperature – called a thermal remanent magnetization or TRM, or (2) through magnetic mineral grain deposition in water – a depositional or post-depositional remanent magnetization (a DRM or pDRM). Information regarding the intensity and direction of the magnetization are stored until the materials are either remagnetized by natural processes, or demagnetized by laboratory experiments designed to isolate the recorded magnetization. Incorporating palaeomagnetic data into field models requires uncertainties related to the directions and intensities to be understood; these include (1) the precision and reliability of the measurements of the recorded magnetic field, (2) disturbing influences on the magnetic particles following the recording process (e.g. compaction in sediments), and (3) the accuracy and precision of the methods used to date the materials that acquired the TRM or pDRM. Archaeomagnetic data can sometimes be associated with quite precise dates due to associated stratigraphic artefacts or calibration by radiometric dating.

The GEOMAGIA50v2 online database (Donadini *et al.* 2006; Korhonen *et al.* 2008) extends to 50 000 years before present (BP; defined as calendar years before AD 1950), and provides an overview of the available TRMs from igneous rocks and archaeological artefacts (shown for the last 10 000 years in Fig. 3a). These data consist of site-based readings of the direction and/or intensity of the geomagnetic field over a wide range of dates. About 52% of the data for the Holocene provide absolute palaeointensity estimates, but two thirds of these measurements lack directional information because the *in situ* orientation of the specimen is not known (i.e. from fired pottery). More than 95% of the data come from the northern hemisphere, with the majority originating from western Eurasia and covering the last 3500 years. Beyond this timeframe, the main source of palaeomagnetic data is sediment cores.

Figure 3b shows the distribution of palaeomagnetic data from sediments with time and location, based on the compilation of Korte *et al.* (2011), covering the last 10 000 years. The data from sediments are more evenly distributed in both time and location than archaeomagnetic data. About 45% of the data from sediments provides relative palaeointensity (RPI) estimates, which in theory can be calibrated using absolute intensity measurements from archaeological artefacts (or lava samples) from the same region (Donadini *et al.* 2009), or model predictions (Korte & Constable 2006). The data from sediments have the benefit of providing continuous records, which make it possible to build accurate

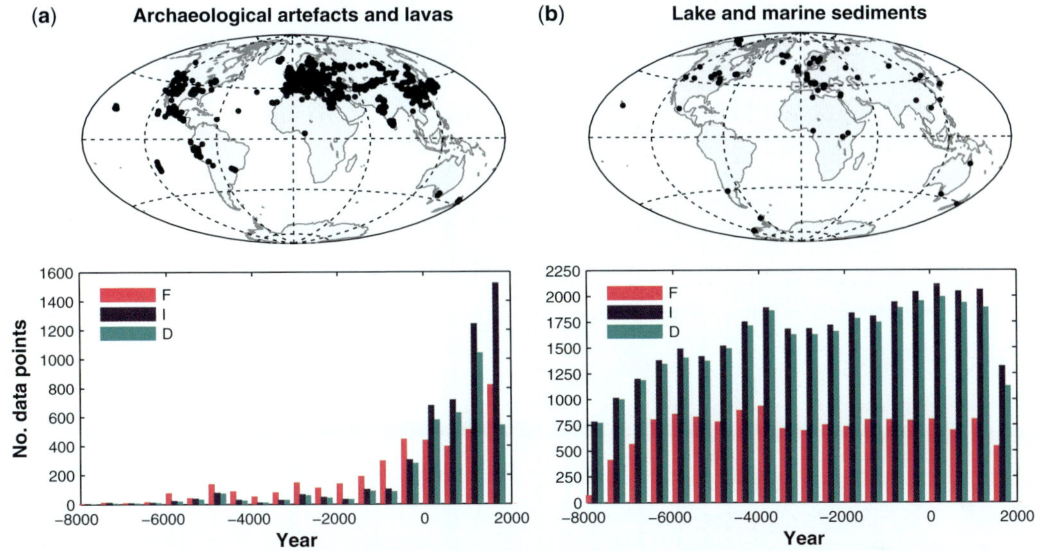

Fig. 3. Location and time distribution (lower panel) of (**a**) archaeomagnetic data and measurements on lavas and (**b**) sedimentary palaeomagnetic data for the last 10 000 years. The data in the lower panel have been grouped in 500-year bins and separated into intensity (F), inclination (I) and declination (D).

and precise age–depth models using other dating techniques. However, many uncertainties surround the mechanisms by which palaeomagnetic signals are 'locked in' to sediments (Snowball *et al.* 2013). One expects the primary geomagnetic signal to be smoothed because the natural remanence is probably acquired over a range of depths. As a consequence, an age difference exists between the deposition of the surface sediments and the acquisition of the NRM, with the size of the age offset inversely related to deposition rate (Roberts & Winklhofer 2004). In addition, weak and unstable magnetizations can lead to substantial noise in the data (e.g. Creer *et al.* 1983). Nevertheless, lacustrine and marine sediments have been used extensively to complement and extend the record of the ancient geomagnetic field, providing information about the direction of the field as well as qualitative changes in intensity.

Europe has the most complete palaeomagnetic datasets, and therefore provides the best palaeofield description (Fig. 3). This description is especially true for the last *c.* 4000 years, for which there is significant overlap in time between archaeomagnetic and sediment-derived data.

As a result of the good location coverage of well-dated archaeomagnetic data in Europe, Gallet *et al.* (2003) discovered the existence of so-called archaeomagnetic jerks, which are sharp turns in the field direction and peaks in the field intensity, which took a couple of centuries to complete. Mandea & Olsen (2009) suggest there should be a clear distinction between 'archaeomagnetic jerks' (associated with changes in the field itself) and geomagnetic jerks because they are associated with changes that occur over very different timescales and are most likely caused by different underlying processes. Prior to a few thousand years ago there are insufficient archaeomagnetic data to detect archaeomagnetic jerks. Snowball & Sandgren (2004) used a sedimentary approach to complement and extend the archaeomagnetic record of these jerks in Europe back to 9000 cal a BP. Thus, for Europe, there is evidence that such jerks took place at 800 BC, AD 200, AD 750–800 and AD 1400 (Gallet *et al.* 2003), with three earlier sedimentary-based jerks at 2700 BC, 2200 BC and 1600 BC (Snowball & Sandgren 2004). These regionally reconstructed jerks are indicated in Figure 4 alongside various alternative predictions of dipole moment and dipole tilt, which we explore in more detail below.

Dipole tilt and dipole moment. Several different approaches have been used to reconstruct global Holocene geomagnetic field variations, with significantly different results, even for the most basic components of the field (Korte & Constable 2005*b*; Knudsen *et al.* 2008; Valet *et al.* 2008; Nilsson *et al.* 2010). As a consequence of directional uncertainties and the restricted location distribution of palaeomagnetic data, each method involves some form of compromise regarding the complexity of the model, the quality of the data used to constrain

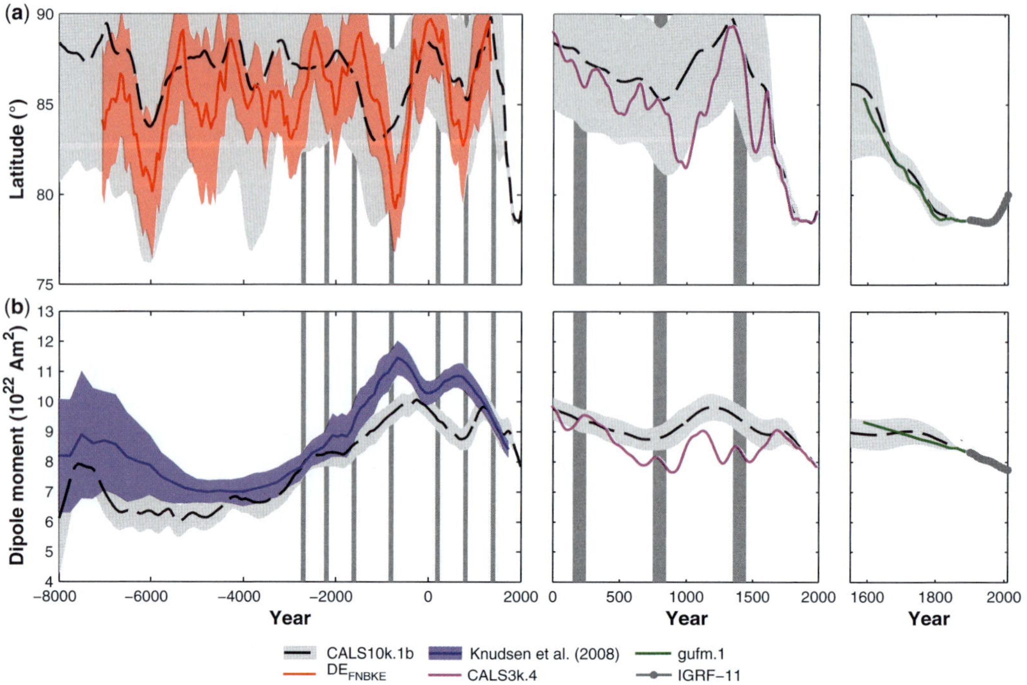

Fig. 4. Dipole tilt, that is, latitude (**a**) and dipole moment (**b**) reconstructions for the last 10 000, 2000 and 400 years respectively. The dipole field estimates are based on CALS10k.1b (Korte *et al.* 2011) (dashed black line and shaded 1σ error bounds), DE$_{FNBKE}$ (Nilsson *et al.* 2011) (solid red line with shaded 1σ error bounds), a dipole moment reconstruction of Knudsen *et al.* (2008) (solid blue line with shaded 2σ error bounds), CALS3k.4 (Korte & Constable 2011) (solid magenta line), gufm.1 (solid green line) (Jackson *et al.* 2000) and IGRF-11 (grey line and dots) (Finlay *et al.* 2010). Vertical grey bars show the occurrence of archaeomagnetic jerks discussed in the text.

the model, and/or the spatial and temporal resolution of the reconstruction. Several models have been produced based on similar approaches using historical data (Hongre *et al.* 1998; Korte & Constable 2003, 2005*a*, 2011; Korte *et al.* 2009, 2011) and these should theoretically provide the most comprehensive description of the field. However, such models require extensive datasets and tend to shift power to higher degree components of the field to accommodate the inconsistencies between different records (Valet *et al.* 2008). Alternative reconstructions with limited complexity (e.g. Genevey *et al.* 2008; Knudsen *et al.* 2008; Valet *et al.* 2008; Nilsson *et al.* 2010) potentially provide better representations of the basic components (e.g. the GAD), because the constraining data are more critically chosen. On the other hand, the less complex models offer limited insight into the details of past changes in the geomagnetic field configuration (Fig. 4). The variation in dipole tilt appears to exhibit a 2700- or 1350-year periodicity, which is resolved in both the DE$_{FNBKE}$, a dipole tilt model (constrained by selected sedimentary records; Nilsson *et al.* 2011), and the newest spherical harmonic model, CALS10k.1b (Korte *et al.* 2011). The reconstructions suggest that dipole latitudes (referring to latitudes of the north geomagnetic pole) less than 80° (similar to the present-day field) are within the confidence interval and may be common features of the field. The various different dipole moment reconstructions all suggest similar long-term developments, with the field strength reaching its Holocene peak around 2000–3000 years ago. The virtual axial dipole moment (VADM) reconstruction (based on a least-squares fit of absolute palaeointensity data from Knudsen *et al.* 2008) shows that the highest dipole moments coincide with the periods of low dipole latitude around 800 BC and AD 800, which the DE$_{FNBKE}$ model predicts). However, the dipole moments based on spherical harmonic models do not show the same relationship, and it could be argued that the high dipole moments in the VADM reconstruction are due to the location bias of the palaeomagnetic data towards Europe (Nilsson *et al.* 2011).

The identified proliferation of the jerks needs to be confirmed by palaeomagnetic data from regions that are still poorly covered, such as Africa and the oceans. It has been argued that they could be associated with variations in dipole tilt (Gallet *et al.* 2005; Nilsson *et al.* 2011). An alternative hypothesis (related to hemispheric asymmetries of the geomagnetic field), suggests that the observed jerks in Europe may coincide with anomalies in the opposite hemisphere (Gallet *et al.* 2009).

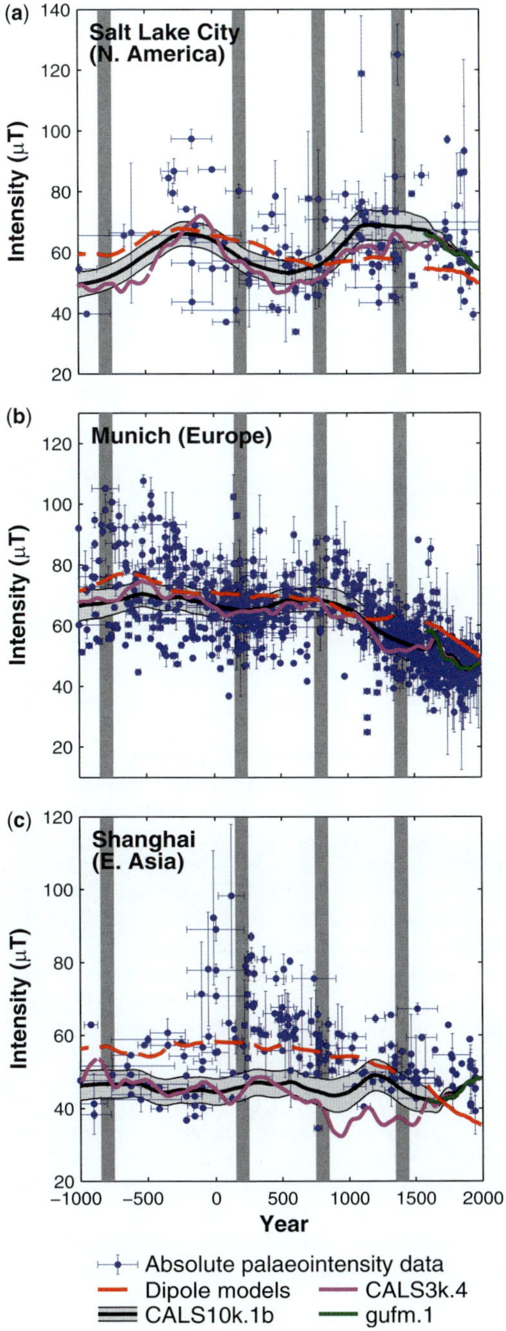

Fig. 5. Geomagnetic field intensity reconstructions for three representative sites in the northern hemisphere: (**a**) Salt Lake City in North America, (**b**) Munich in Europe and (**c**) Shanghai in East Asia. Absolute palaeointensity data (blue dots) were selected from the GEOMAGIA50v2 online database (Donadini *et al.* 2006; Korhonen *et al.* 2008) within a 1500 km radius of each site and relocated assuming a geoaxial dipole. Model predictions from cals10.1b (Korte *et al.* 2011) (solid black line and shaded 1σ error bounds), CALS3k.4 (Korte & Constable 2011) (solid pink line) and gufm.1 (Jackson *et al.* 2000) (solid green line) as well as dipole field model predictions based ^{10}Be data from the GRIP ice core and DE_{FNBKE} (Nilsson *et al.* 2011) combined with gufm.1 (dashed red line). Vertical grey bars show the occurrence of archaeomagnetic jerks discussed in the text.

Absolute palaeointensity data and model estimates for three northern hemisphere regions were chosen (Fig. 5) based on the availability of archaeomagnetic data (Fig. 3). The data were selected from within a 1500 km radius of Salt Lake City, Munich and Shanghai and were relocated assuming an axial dipole model. By adopting the same approach used to produce Figure 1, we show model predictions based on both the entire field description and the dipole field component of the field (Fig. 5). The relatively large directional and intensity uncertainties associated with palaeomagnetic data, and the limited data distribution, means that the resolutions of the models are only fully resolved to the order of spherical harmonic degree 4 in location and 100 years in time (Korte & Constable 2008). The models suggest periods of high geomagnetic field intensity in Europe around 800 BC and AD 800, previously identified as intervals of jerks, and periods of low dipole latitude coinciding with local intensity minima in North America (see also Stoner *et al.* 2013).

To illustrate the directional variations of the Holocene geomagnetic field we have chosen six globally distributed regions/continents: (1) North America, (2) South America, (3) Europe, (4) South Africa, (5) East Asia and (6) Australia (Fig. 6). The five records shown in Figure 6 were selected based on the quality of the data and pre-treated following Nilsson *et al.* (2011). Uncertainties for sedimentary palaeomagnetic data are often difficult to assess and we have chosen not to show them here; instead, all records have been smoothed with similar *c.* 100-year running averages (subject to the original publications). Because no lake sediment data are currently available from South Africa, we show only model predictions for this site. Again, we note a negative correlation between inclination data

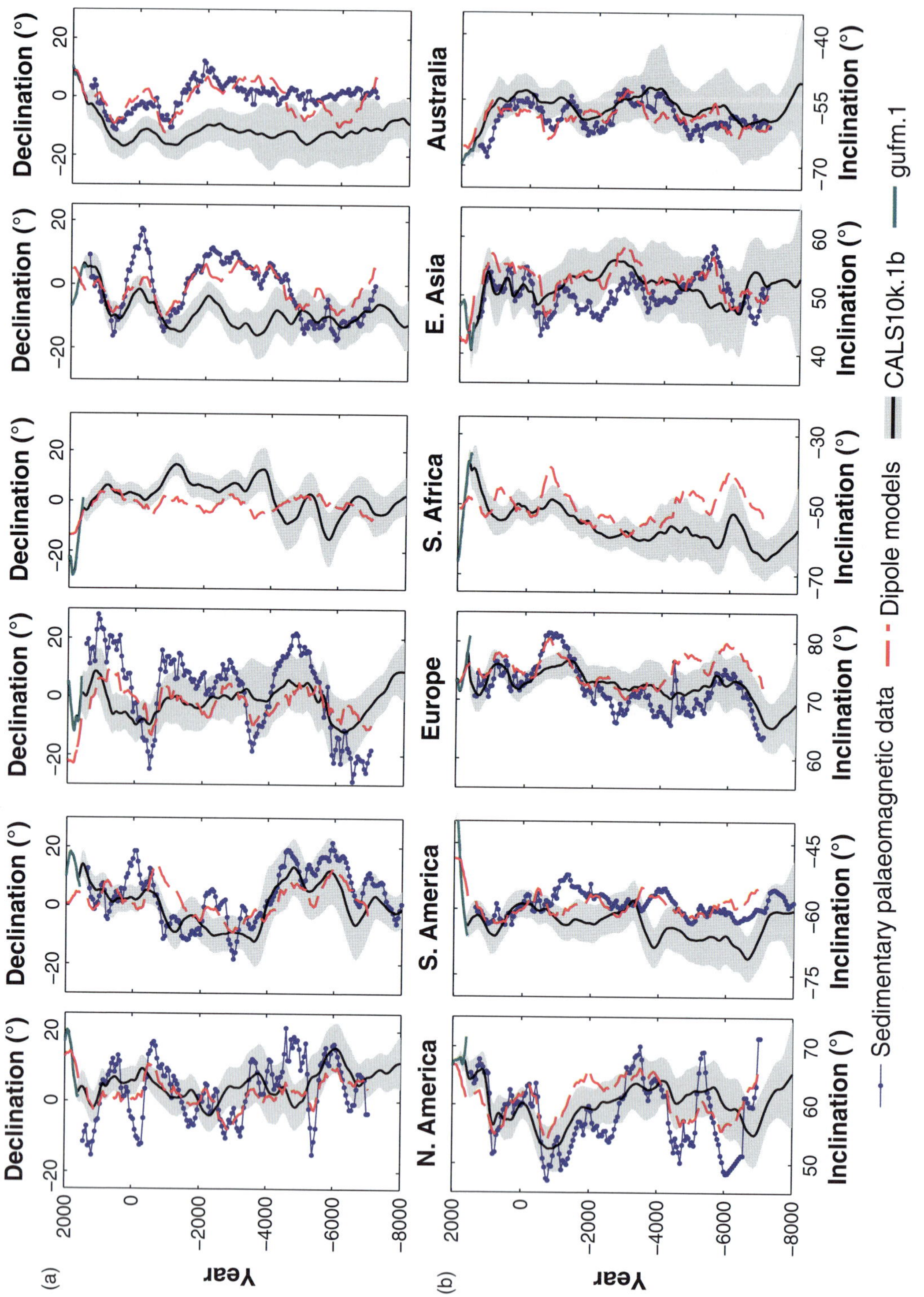
(a)
Year
Declination (°)
(b)
Year
N. America
S. America
Europe
S. Africa
E. Asia
Australia
Inclination (°)
Sedimentary palaeomagnetic data
Dipole models
CALS10k.1b
gufm.1

from Europe and North America between 800 BC and AD 800. Indeed, around these two time periods, much of the data appear to plot closer to the dipole field prediction, suggesting that the field may have had a more dipolar structure. This view is slightly biased by the records that we have chosen to show, and it is not supported by all the available data, as indicated by the deviations of the CALS10k.1b predictions for the same time periods. The relatively similar records from the same longitudes, for example North and South America or East Asia and Australia, are clearly apparent (Fig. 6).

Further back in time than the Holocene, palaeomagnetic data are more scarce and dating uncertainties become larger, but we see the most dramatic changes in the structure of the geomagnetic field, which are manifest as geomagnetic excursions and polarity reversals. Changes in the polarity of the geomagnetic field form the basis of the geomagnetic polarity timescale (e.g. Ogg 2012), but this is not the place for a full description of how and why polarity reversals take place, because we have not experienced a full reversal (as strictly defined by palaeomagnetists) for *c.* 780 ka. We prefer to restrict ourselves to the past 50 ka, because the geomagnetic field underwent at least one geomagnetic excursion during this period. More recent changes in the geomagnetic field need to be considered in this longer context.

Geomagnetic excursions were defined as aborted attempts to reverse polarity (Verosub & Banerjee 1977), but, more recently, there has been a discussion about their origins (Gubbins 1999; Roberts 2008). An excursion was often claimed if the direction of a trustworthy NRM recorded at a specific site could be explained by a virtual geomagnetic pole (VGP) that deviated more than 45° from the geographic pole (Verosub & Banerjee 1977; Merrill & McFadden 1994). However, Vandamme (1994) argued that an excursion could be claimed if the reconstructed VGP was simply outside the ‘normal range’ of geomagnetic secular variation at a site. The majority of excursions coincide with periods of low geomagnetic field intensity (Guyodo & Valet 1999). Laj & Channell (2007) and Roberts (2008) provided succinct reviews on the current state of knowledge about geomagnetic excursions, and the reader is directed to these works for more information.

Our picture of the occurrence of geomagnetic excursions has increased enormously as a result of relatively recent studies of high-resolution sediment sequences (e.g. Laj *et al.* 2000) and ice cores. A problem with initial reporting of geomagnetic excursions was that anomalous palaeomagnetic directions were often based on geological materials in which the NRM had (with hindsight) been influenced by factors other than the geomagnetic field, and so reports of excursions became bountiful. However, to be useful as a global correlation tool, strong evidence is needed of the worldwide extent of an excursion and, in this respect, reconstructions of the production rates of cosmogenic radionuclides (e.g. ^{14}C, ^{10}Be and ^{36}Cl) have proved extremely valuable. Radionuclide records from ice cores can be quantitatively interpreted primarily in terms of the geomagnetic dipole intensity, where higher dipole moment causes the production of radionuclides to decrease (Muscheler *et al.* 2005).

Alternative reconstructions of dipole moments are possible based on (1) the ^{10}Be record from the Summit ice cores converted into absolute values (Muscheler *et al.* 2005), (2) GLOPIS-75, which was an attempt to calibrate RPI estimates from a series of globally distributed marine sediment cores to absolute values (Laj *et al.* 2004), and (3) dipole moments based solely on absolute intensity values obtained from igneous rocks and fired artefacts (Knudsen *et al.* 2008). Figure 7 shows the approximate timings of the Laschamp and Mono-lake excursions, which have been the subject of considerable study. The Laschamp geomagnetic excursion (*c.* 41 ka) was defined first by Bonhommet & Babkine (1967), and has subsequently been found in widely distributed palaeomagnetic datasets, from sediments and igneous rocks, and it is associated with the lowest dipole moment of the past 60 ka. Recently published records indicate that the central part of Laschamp was characterized by fully reversed polarity and a temporary recovery of dipole intensity (Nowaczyk *et al.* 2012). This field geometry and temporary intensity recovery would elevate the status of the Laschamp geomagnetic

Fig. 6. Geomagnetic field reconstructions, declination (**a**) and inclination (**b**), covering the last 10 000 years for six globally distributed sites based on sedimentary palaeomagnetic data (blue dots), model predictions from CALS10k.1b (Korte *et al.* 2011) (solid black line and shaded error bounds) and gufm.1 (Jackson *et al.* 2000) (solid green line) as well as dipole field model predictions based on DE_{FNBKE} (Nilsson *et al.* 2011) combined with gufm.1 (dashed red line). The palaeomagnetic data, selected and smoothed according to Nilsson *et al.* (2011), come from Fish Lake in North America (Verosub *et al.* 1986), Lake Escondido in South America (Gogorza *et al.* 1999, 2002), Lake Nautajärvi in Europe (Ojala & Saarinen 2002), Lake Biwa in East Asia (Ali *et al.* 1999) and Lake Keilambete in Australia (Barton & McElhinny 1981). The coordinates used for the model predictions are the same as the coordinates of the lakes, except for South Africa (latitude, 33°S; longitude, 22°E) from where there are currently no sedimentary palaeomagnetic records.

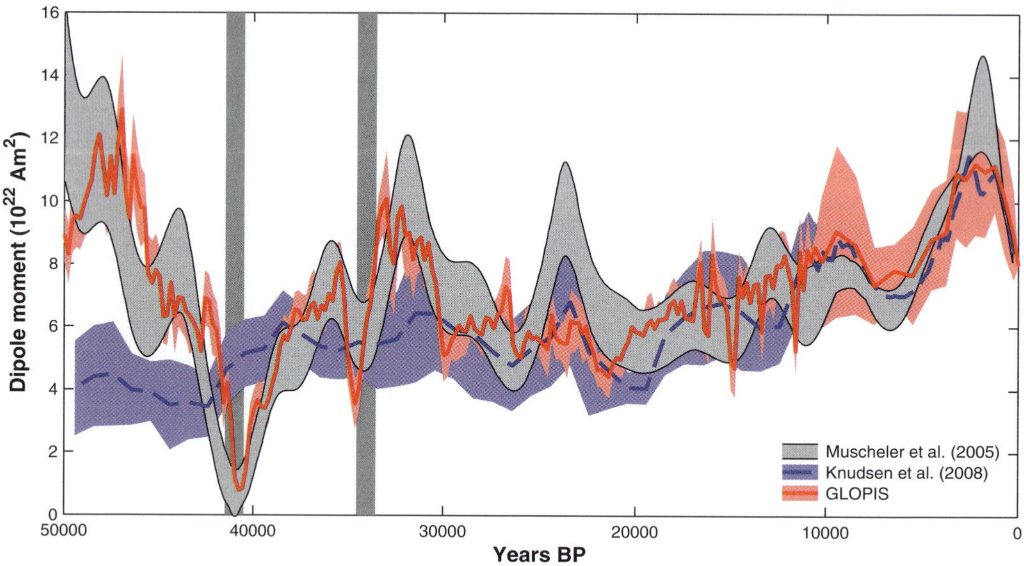

Fig. 7. Dipole moment reconstructions for the last 50 000 years based on ^{10}Be flux data from the GRIP and GISP2 ice cores (Muscheler *et al.* 2005) transferred to the GICC05 timescale (Svensson *et al.* 2008) (shaded grey), a dipole moment reconstruction of Knudsen *et al.* (2008) (dashed blue line and shaded 2σ error bounds) and GLOPIS (Laj *et al.* 2004) (solid red line and shaded 1σ error bounds). Vertical grey bars show the occurrence of the Laschamp and Mono Lake excursions discussed in the text.

excursion to a 'polarity microchron' (Laj & Channell 2007). Results from a recent attempt by Leonhardt *et al.* (2009) to model the geomagnetic field during the Laschamp suggest that the intensity minima observed at different locations during the excursion may have been offset in time by up to 400 years with respect to each other. Their model also predicts a lack of excursional field directions in certain parts of the world, which could explain why some palaeomagnetic data fail to record the Laschamp excursion. These conclusions are in agreement with the results from Brown *et al.* (2007), who induced an excursion/reversal in CALS7k.2 by manipulating the axial dipole moment. The failure of the reconstruction of Knudsen *et al.* (2008) to detect the generally low dipole moment associated with the Laschamp excursion is probably due to the low temporal resolution of the palaeomagnetic data obtained from igneous rocks.

A younger geomagnetic excursion has been reported in a number of records, first defined by anomalous palaeomagnetic directions in a sediment sequence at Lake Mono by Denham & Cox (1971) and Liddicoat & Coe (1979), and was initially assigned a wide range of ages, but is now considered to be 32–31 ka in age (Benson *et al.* 2003; Laj & Channell 2007). A new assessment of the ages of sediment deposited at the type site indicates that the definite 'Mono Lake geomagnetic excursion' may well be the Laschamp excursion (Zimmerman *et al.* 2006; Vazquez & Lidzbarski 2012). Although this 'Mono Lake' debate has not been satisfactorily resolved, several reports of a younger excursion around 30–35 ka have been published (e.g. Lund *et al.* 2006; Nowaczyk *et al.* 2012) and provide reasonable evidence for an excursion during this period. The radionuclide records (^{10}Be and ^{36}Cl) clearly define the Laschamp geomagnetic excursion as a period of low dipole moment, but the younger excursion (Mono Lake) is not clearly identified. Wagner *et al.* (2000) argued that a distinct peak in ^{36}Cl production is evidence of the Mono Lake excursion. When placed on the updated GICC05 timescale (Svensson *et al.* 2008), the low VADM occurred *c.* 34 ka ago.

Guyodo & Valet (1999) argue that values of dipole moment below 4×10^{22} Am^2 are associated with geomagnetic excursions and polarity reversals. Thus, this level represents a threshold at which non-dipolar terms can be expected to appear in palaeomagnetic datasets. Some regional datasets suggest the possibility of younger excursions that are yet to be confirmed by data from more locations. The most conclusive evidence of an excursion more recent than *c.* 31 ka comes from drill cores of lavas on Hawaii, where the Hilina Pali excursion was dated to *c.* 20 ka (Teanby *et al.* 2002). One of the reasons that reliable, sedimentary-based palaeomagnetic records are scarce from many parts of the globe at this time is due to the effects of deglaciation and significant climate change that is manifest in sediment sequences, with the result that (p)DRMs were more influenced by local environmental factors than by the prevailing geomagnetic field.

It is clear from the comparison between the alternative reconstructions of dipole moment (Fig. 7) that the last 5000 years form a period of relatively high dipole moment, with the period between 3000 and 2000 cal a BP possibly the highest moment of the last 45 ka.

A geomagnetic marker to help define the base of the Anthropocene?

Our review of palaeomagnetic data over the last 50 000 years suggests that there will not be a single 'global palaeomagnetic spike' recorded within the last few hundred years that could be used to define the base of the Anthropocene. However, because of the fact that NRMs (particularly TRMs) are extremely stable over time, and also that the changes in the geomagnetic field that they record have not so far been influenced by human activity, palaeomagnetic data can provide valuable isochrones to complement indicators of human impact on the environment. In the near future, geomagnetic field model predictions are likely to be used as tools for archaeomagnetic dating in regions where there are too few groundtruth data to provide a regional secular variation master curve.

In some cases, a short time lag occurs between the formation of the material (which is dated) and the acquisition of the NRM. The severity of this time lag depends inversely on sediment accumulation rates. For igneous rocks the time lag depends upon the cooling rate through the Curie (or Neél) temperatures of the minerals that carry the TRM. Data from heated archaeomagnetic features (i.e. kilns, hearths) generally do not suffer from this problem and the delay may be only a few hours if the material cooled relatively quickly.

The occurrence of the Laschamp, and its potential classification as a polarity microchron, can be used to provide a maximum age for the beginning of the Anthropocene. Although the use of this palaeomagnetic isochron may not be favourable with other advocates, Valet & Valladas (2010) hypothesized that the low dipole moment associated with the Laschamp excursion and the more speculative Mono Lake excursion caused a significant reduction in atmospheric ozone levels as a result of the increased penetration of high-energy solar and galactic particles. One result could have been an increased rate of cancer in subpopulations of fair-skinned Neanderthals (*Homo neanderthalensis*) in Europe, which were competing for resources with modern humans (*Homo sapiens*). Although this hypothesis remains controversial, Valet & Valladas (2010, p. 3892) concluded that 'the synchronism between this low geomagnetic field period and the Neanderthal demise should not be seen as purely accidental'.

There is increasing confidence in the time variation of the geomagnetic field we are able to reconstruct at particular locations. In younger intervals, the next globally recognized palaeomagnetic isochron that may be used in defining the start of the Anthropocene is the relatively low dipole latitude that occurred between 3000 and 2500 cal a BP (Fig. 4), which is associated with distinct palaeomagnetic secular variation (PSV) features across mid-latitude sites (Fig. 6). Although only part of this variation can be attributed to the dipole field component, and some uncertainty remains regarding the chronologies of the individual records, the associated PSV features are extremely well defined in Europe and North America. In their seminal study of PSV in the United Kingdom, Turner & Thompson (1981) labelled the eastwards declination maximum (at 2705 cal a BP) as 'f' and the westerly maximum (at 1950 cal a BP) as 'e'. In terms of a dipole field description (see Fig. 6) we would expect the dipole latitude minima around 3000–2500 cal a BP to coincide roughly with the time of steepest inclination in Europe (at *c.* 2500 cal a BP and labelled ε^1 in the Lake Windermere record). Better resolved studies place the steepest inclination between 2500 and 2600 cal a BP (Gallet *et al.* 2003; Snowball *et al.* 2007; Haltia-Hovi *et al.* 2010). It is natural to expect that regional estimates of the maximum dipole tilt will vary due to the influences of non-dipole field components. However, with the help of field predictions from steadily improving global geomagnetic field models we argue that the distinct PSV variations around this time could be used to provide a regionally applicable chronostratigraphic marker. Clearly 2500 cal a BP marks the boundary between the late Bronze Age and the Pre-Roman Iron Age in southern Sweden, which was characterized by the establishment of permanent fields for cereal (barley) cultivation (Gabrowski 2011).

It is not possible to predict the future geomagnetic field geometry. Palaeomagnetic data do not support extrapolation of the growth of the South Atlantic Anomaly, nor suggest that it is the signal of an impending polarity reversal. In spite of the recently observed decline in dipole moment it is far above the threshold level that palaeomagnetic data suggest is needed for a polarity 'switch' to take place. Given the ups and downs of the last few thousand years, a recovery is just as likely as a continued trend towards lower moment.

Mineral magnetic responses to environmental change

Anthropogenic modification of the Earth surface system has had its most profound and widespread impact on terrestrial systems, particularly those of the soil, riverine and lake systems. From the perspective of human modification of the mineral magnetic signature, terrestrial systems have been the most intensely studied, with relatively few studies examining anthropogenic impacts in the mid or distal continental shelves or the deep marine. As a consequence of this, lake sediments are likely to be the most complete record of anthropogenic-generated mineral magnetic impacts in the long term.

Often, the primary source of Fe-oxides in sediment systems is via the various pathways to which all rocks are subject: weathering, pedogenesis, erosion, transportation and sediment storage at the Earth's surface (Dearing 2000; Dekkers 2003; Evans & Heller 2003). In any sediment archive, at any one place, the derived Fe-oxides will be a location-specific (and rock-type specific) set of Fe-oxides derived through these processes, so that natural sediment systems often have a strong provenance signature imposed on the Fe-oxides generated (Dekkers 2003). In natural sediment systems, changes in climate, vegetation, tectonics and volcanic activity can all impose changes on these pathways, which will impose changes on the resulting magnetic mineral assemblages – hence their ability to provide proxies for palaeoclimatic change (Maher & Thompson 1999).

In addition to these 'rock-bound' pathways, various other processes can generate new kinds of Fe-oxides independent of this rock-bound system. Mineral magnetic studies have identified a number of mechanisms by which additional sets of magnetic minerals can be generated.

Soils

Topsoils, under certain kinds of conditions of drainage and moisture supply, generate new Fe-oxides, either through a reduction process (which generates magnetite) or an oxidation process in moisture-deficient soils (which generates goethite or hematite). Well-drained, seasonally moist soils generate superparamagnetic (<30 nm in size) magnetite, the rate of production of which has been used as a rainfall proxy (Maher 2009). Waterlogged and gleyed soils tend to dissolve Fe-oxides. As a result of these responses, erosion of topsoil clays and silts are transferred into river or lake sediments, where the Fe-oxides can be preferentially concentrated in these fines (Maher & Thompson 1999). Hence, magnetic measurements that can identify these mineral components can be good indicators of topsoil erosion (Dearing 2000; Evans & Heller 2003; Hounslow & Chepstow-Lusty 2004). In dryland soils, because of the oxidation of Fe-oxides by weathering and soil-formation processes, erosion of such soils can produce Fe-oxide components rich in antiferromagnetic minerals (e.g. hematite). Examples include wind-transported dusts derived from sub-Saharan Africa.

Biomineralization of Fe-oxides by bacteria and other organisms

Organisms that produce magnetic minerals (principally magnetite but sometimes greigite) have a generally restricted niche in natural ecosystems (Kopp & Kirschvink 2008), and are mostly dominated by some kinds of bacteria, which generate small particles with restricted grain size distributions (i.e. single-domain) in the size range of *c.* 20–100 nm. These are largely restricted to reduced-oxygen environments in some kinds of lakes and marine sediments. Because of their restricted size and morphology, they can have distinctive magnetic properties (Kopp & Kirschvink 2008).

Natural or anthropogenic biomass burning

Burning of organic matter in wildfires produces new Fe-oxides through the conversion of the trace amounts of Fe in organic matter into Fe oxides (Carrancho *et al.* 2009). In the same way, heating of soils in such burning episodes can produce additional Fe-oxides either through the conversion of weakly or non-magnetic Fe-oxyhydroxides into more strongly magnetic phases, or to the direct breakdown of Fe-bearing silicates (e.g. some clays).

Anthropogenic fossil fuel burning

Burning of fossil fuels generates various kinds of Fe-oxides in the combustion processes, either as a direct byproduct of the fuel itself (e.g. mineral matter in coal) or through the volatilization of the fuel (e.g. vehicles) and subsequent recombination of the Fe in soot and particulates generated by the combustion (Tolonen & Oldfield 1986; Petrovsky & Ellwood 1999; Jordanova *et al.* 2004). The volatilization tends to produce various aluminosilicate glass and Fe-oxides composite particles, often in distinctive fly-ash spherules (Fig. 8), which can be transported into the atmosphere via the hot gases and exhausts from the combustion process. These spherules can be spread over large areas around power plants, metallurgical and smelting sites, or adjacent to roads (i.e. from vehicles) through atmospheric or fluvial dispersal (Petrovsky & Ellwood 1999; Maher 2009). The fly-ash particles commonly

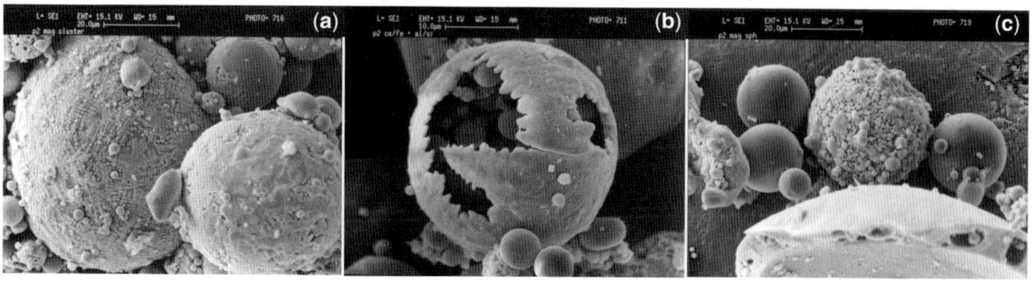

Fig. 8. Morphological variety of magnetic spherules from fuel combustion, shown in scanning electron micrographs. (**a**) Dendritic structures on the surface of *c.* 50 μm diameter spherule on left. (**b**) Hollow spherule with 'baby' spherules internally and externally. (**c**) Spherules with smooth and irregular surfaces with, in lower part of micrograph, part of a broken shell of a larger hollow spherule. All magnetic extracts are from a soil near Gdansk (Poland) consisting of 99% spherules in the extract (original samples courtesy of Jerzy Nawrocki).

contain both magnetite (sometimes Al-substituted) and hematite, but for the most part do not have distinctive granulometric signatures in magnetic measurements because of the often intricate growth of the Fe-oxides in the glassy hosts (Jordanova *et al.* 2004).

Anthropogenic ferrous contamination

Direct ferrous contamination in the modern environment from metallic iron is widespread in the present-day environment, from both modern and archaeological urban and industrial sites and along transport pathways (via rust from vehicles). Although metallic iron is not common in present-day sediment systems, its oxidation products (from rust) such as magnetite can be easily detected. It is not known if such sources have distinctive magnetic measurement signatures (Petrovsky & Ellwood 1999).

Anthropogenic-derived magnetic mineral responses

There are two main mineral magnetic responses to anthropogenically derived environmental forcing mechanisms that have been identified at a large number of locations globally. First, the mineral magnetic record of atmospheric particulate pollution related to fossil fuel burning has left behind a strong record of magnetic Fe-oxides in peat bogs (Oldfield *et al.* 1978; Rothwell *et al.* 2005) and lakes (Dearing 1999). Second, deforestation, soil erosion and sediment delivery into lacustrine and marine systems has left behind strong signatures in the mineral magnetic record (Dearing 1999; Dearing & Jones 2003; Mercuri *et al.* 2012; Jones *et al.* 2013). Identification of both these responses has inevitably relied upon using a multi-method approach, one that often involves integrating mineral magnetic, pollen, historical and often chemical data, combined with detailed chronologies. We explore these two themes in more detail for their relevance in defining the Anthropocene.

Fossil fuel burning. The association between black magnetic spherules and atmospheric pollution generated during the industrial revolution was first clearly demonstrated by Oldfield *et al.* (1978) in peat bogs, which comprise an ideal medium because other contaminating mineral matter that could confound interpretation is largely absent (Richardson 1986). The magnetic response is one that is always displayed as an increase in Fe-oxide abundance expressed by magnetic susceptibility (χ) or the isothermal remanent magnetization (IRM; Table 1). These abundance increases relate to both ferrimagnetic (i.e. magnetite) and antiferromagnetic minerals (i.e. hematite; Table 1) contained in the spherules. Similar interpreted increases in magnetic mineral abundance have been identified in lacustrine (McLean 1991; Power & Worsley 2009), coastal (Plater *et al.* 1998) and marine sediments (Horng *et al.* 2009), where they are often associated with increases in heavy metal concentrations because of the association of these magnetic spherules with industrial pollutants (Locke & Bertine 1986; Dearing 1999; Petrovsky & Ellwood 1999). These particles also add to the contamination load built up in present-day soils (Certini & Scalenghe 2011), but are also clearly concentrated around the source of the pollutants as a long-term load to the soil (Maier & Scholger 2004).

The earliest evidence for the production of these particles in abundance is around AD 1800 in the peat bogs of northern England and New Brunswick in Canada (Table 1; Fig. 9a). Evidence from eastern North America and Europe (Fig. 9c) indicates that the locations of these pollution sources expanded through the 1840s into the 1870s at lake and

Table 1. *Initial rise of magnetic data interpreted as a proxy for fossil fuel burning, particularly evidenced by the occurrence of combustion spherules*

Location	Setting	Year of initial rise	Mineral magnetic change	Ref.
Upper North Grain, Pennines, England	Peat bog	*c.* AD 1800	A+; G+	Rothwell *et al.* (2005)
Various Polish peat bogs	Peat bog	AD 1945–1955	A+(χ)	Strzyszcz & Magiera (2001)
Bloak Moss, Ayrshire, Scotland	Peat bog	*c.* AD 1800	A+ (IRM), H+	Richardson (1986)
Heathwaite Moss, Cumbria, England	Peat bog	*c.* AD 1800	A+(IRM), H−	Richardson (1986)
Lough Fea Bog, Tyrone, N. Ireland	Peat bog	*c.* AD 1850	A+	Oldfield *et al.* (1978)
Bolton Fell Moss, England	Peat bog	*c.* AD 1850	A+	Oldfield *et al.* (1978)
Crummock Water, Cumbria, England	Lake	AD 1905	A+	McLean (1991)
Tees estuary cores, England	Coastal	AD 1925	A+(χ), G−	Plater *et al.* (1998)
Halton, Merseyside, England	Small ponds	*c.* 1900	A+(χ, ARM), G−	Power & Worsley (2009)
Offshore of southwestern Taiwan	Marine	*c.* 1950s	A+(HIRM); H+	Horng *et al.* (2009)
Thane Creek, Mumbai, India	Mudflats	*c.* 1930–1950	A+ (χ, IRM); H−	Blaha *et al.* (2011)
East Lake,Wuhan, China	Lake	1960s	A+ (χ, IRM)	Yang *et al.* (2009)
Regent Street Bog, New Brunswick, Canada	Peat bog	*c.* AD 1800	A+ (IRM)	Tolonen & Oldfield (1986)
Ely Lake Bog, Minnesota, USA	Peat bog	*c.* AD 1890	A+(IRM), H+	Richardson (1986)
Cuyahoga River Basin, Ohio, USA	Impounded dam	AD 1927	A+(χ)	Mann *et al.* (2013)
Del Gato, Argentina	Lake, floodplain	*c.* 1960s	A+, G−	Chaparro *et al.* (2006)
Hamilton Harbour, Lake Ontario, Canada	Lake	AD 1890	A+ (χ)	Pozza *et al.* (2004)
Big Moose Lake, NY, USA	Lake	AD 1900	A+ (IRM); H+	Oldfield (1990)
Lake Lacawac & Lake Giles, PA, USA	Lake	AD 1880–1900 AD	A+(IRM); H+	Kodama *et al.* (1997)

A, magnetic mineral abundance change (single proxy indicator) where + indicates increase and − a decrease in abundance; G, magnetic mineral grain size indicator, where + indicates increase in magnetic grain size and − a decrease; H, antiferromagnetic mineral change (i.e. hematite or goethite), where + indicates a relative increase and − a decrease in antiferromagnetic content. More pollution-based initial rise events are listed in table 7.7 of Dearing (1999).

coastal sites in Europe and North America (Locke & Bertine 1986). The largest number of studied sites containing initial increases in these magnetic pollution particulates date from the 1890s and around AD 1900 – that is, the 'AD 1900 event'. This later interval seems to mark the sedimentary expression of major fossil fuel burning in Europe and eastern North America, probably related to coal burning (Locke & Bertine 1986), and provides a good marker for a 'late base' of the Anthropocene. Later expression of this industrial activity is seen in the sedimentary record in the other continents after the 1920s, or in the 1950s when major industrial expansion took place in eastern Asia (Horng *et al.* 2009; Table 1; Fig. 9c).

Its noticeable that the earliest records of this fuel burning occur in peats; in the same regions, this magnetic pollution signature is not evident in the lake records until *c.* 100 years later (e.g. in Cumbria Crummock Water at AD 1905 and Heathwaite Moss at *c.* AD 1800; Table 1). It is not clear if this time lag represents (1) real regional differences in local fuel burning (or smelting activities), (2) time lags between atmospheric dispersal of particulates and their delivery into lake systems, due to catchment and soil storage of the magnetic particulates, or (3) differences in sediment age model constructions between peats and lake records. This disparity needs to be addressed on a smaller inter-regional scale before this '*c.* AD 1900 event' could usefully be used to define the Anthropocene globally. The authors are not aware of this AD 1900 event being detected so far in marine sediments. This may be a reflection of sediment storage in catchment soils and lakes, which may be transported to the marine environment in the future.

Deforestation and soil delivery. The premise of most mineral magnetic studies of the Holocene is that lake sediments are likely to record details of the interplay between palaeoclimate, ecosystem stability and functioning, and sediment delivery within the catchment of lakes. These linkages are complex and often mostly catchment-specific (Dearing & Jones 2003; Jones *et al.* 2013). Nevertheless, mineral magnetic studies are powerful in discriminating catchment sediment sources (especially topsoils or rock types; Dearing 2000). Deforestation (and vegetation change) events and soil erosion and delivery events are themes that run through most lake sediment mineral magnetic studies (Dearing 1999). Particularly important in many studies are (1) the increase in Fe-oxide abundance (expressed by χ, IRM or anhysteretic remanence) associated with such events (Osleger *et al.* 2008), (2) the increase in the rate of topsoil delivery as expressed by the delivery of finer-grained magnetite (Hounslow & Chepstow-Lusty 2004), and (3) occasionally subsoil contributions locally richer in antiferromagnetic phases (Higgitt *et al.* 1991; Oldfield *et al.* 2003*a*; Fig. 10). If the sediment record is long enough, such events may be multiple within a

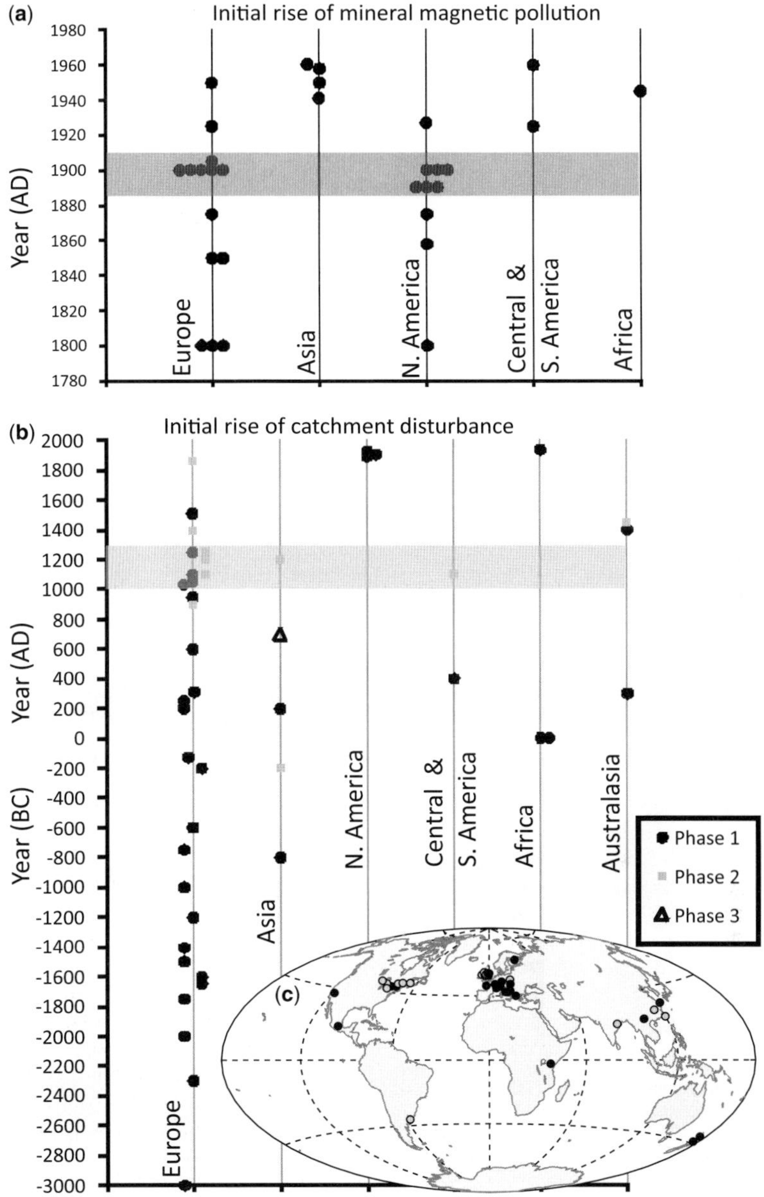

Fig. 9. (**a**) Mineral magnetic pollution events related to initial rise of fossil fuel burning. The data displayed here are provided in Table 1, and include that listed in Dearing (1999). The grey band indicates a concentration of events widely distributed in western Europe and eastern North America. (**b**) Regional timing of mineral magnetic events interpreted as significant shifts in mineral magnetic supply regimes, which are mostly related to deforestation events or major soil erosion events. The data displayed here are provided in Table 2, but also include data from Dearing (1999). Some longer-duration records have multi-phased changes. The grey interval marks a concentration of events, particularly in Europe, in early medieval times associated with agricultural expansion. (**c**) Location of data shown in Tables 1 and 2, with grey-filled circles showing pollution events, and black circles changes in mineral magnetic supply.

lake succession. For example, Gormire Lake (UK; studied by Oldfield *et al.* 2003*a*) had two major deforestation events in the late Iron Age (*c.* 200 BC) and early Medieval phase (*c.* AD 1200), separated by a Dark Ages forest regrowth phase (Fig. 10). In this case, phases with loss of tree cover are

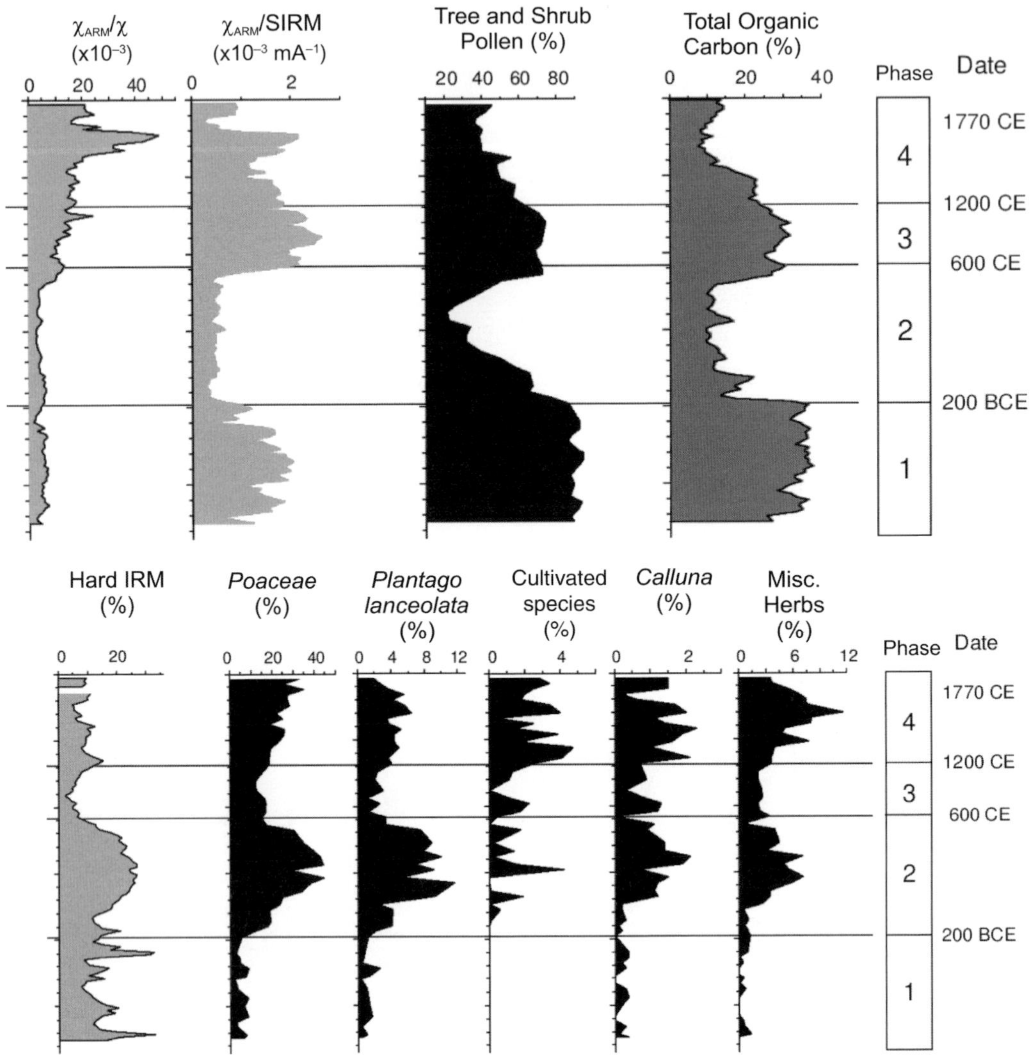

Fig. 10. Mineral magnetic and pollen data (shown as black-filled data) from Gormire Lake (NE England) showing two phases of agricultural expansion, expressed by increases in *Poaceae* (grass) and *Plantago lanceolata* and a matched decline in tree and shrub pollen, starting roughly at 200 BC and AD 1200 (from Oldfield *et al.* 2003*a*). These phases are well discriminated with the hard IRM and χ_{ARM}/SIRM.

associated with an increased input of material with 'hard-IRM' characteristics (i.e. antiferromagnetic material) and an increase in the grain size of the ferrimagnetic fraction (lower χ_{ARM}/Saturation IRM). Like many other similar studies, the integration of pollen, historical and mineral magnetic data is needed to infer such an interpretation.

We have compiled a database of these mineral magnetic events from similar studies in Table 2, which lists the timing of major increases in mineral magnetic supply events, linked to catchment stability changes (often deforestation). This complements the data in the work by Dearing (1999), who tabulated the initial timing of major increases in magnetic susceptibility in lake sediments. For some sites, multiple events are indicated corresponding to several phases of agricultural expansion or catchment change (Table 2; Fig. 9b).

In Europe, at many sites, the earliest catchment instability changes (deforestation) and initial Bronze Age land clearances took place during the interval 2000–1200 BC (Fig. 9), a feature well described from many locations (Van Vliet-Lanoe *et al.* 1992; Edwards & Whittington 2000; Dearing

Table 2. *Anthropogenic impact on delivery of new magnetic mineral signatures into sediment archives*

Location	Setting	Year of initial rise	Mineral magnetic change	Ref.
Morava River floodplain, Czech Republic	Floodplain	*c.* AD 1250	A+(χ)	Kadlec *et al.* (2009)
Petit Lac d'Annecy, France	Lake, floodplain	AD 1000–1100	A+	Dearing & Jones (2003)
Seille Valley, France	Floodplain	50 BC–250 BC	A+ (χ)	Riddiford *et al.* (2012)
Bay of Vilaine, France	Coastal	AD 900–1000	A+ (χ, HIRM)	Sorrel *et al.* (2010)
Butrint, Albania	Coastal marsh	*c.* AD 600	A+ (SIRM), H−; G−	Hounslow & Chepstow-Lusty (2004)
Gormire Lake, Yorkshire, England	Lake	*c.* 200 BC; AD 1200	G−; H+; A− (IRM)	Oldfield *et al.* (2003*a*)
Bassenthwaite Lake, Cumbria, England	Lake	AD 1510; AD 1860*	A+; G+; H−	Hatfield & Maher (2009)
Crummock Water, Cumbria, England	Lake	*c.* AD 1100	A+	McLean (1991)
Lake Holzmaar, West Eifel, Germany	Lake	600 BC; AD 900	A+	Zolitschka (1998)
Unterer Landschitzsee, Austria	Lake	1200 BC, *c.* AD 1400*	A+ (IRM)	Schmidt *et al.* (2002)
Lakes Albano & Nemi, Italy	Lake	2300 BC	A+, H− (%HIRM)	Rolph *et al.* (2004)
Vuonninen, Dvina Karelia, NW Russia	Lake	AD 300	A+ (χ)	Alenius *et al.* (2011)
Adriatic core RF93-30, Italy	Marine	1600 BC, *c.* AD 1100	A− (ARM), H+ (HIRM)	Mercuri *et al.* (2012)
Central Adriatic Sea, Italy	Marine	1650 BC; *c.* AD 1250	A+ (χ); G−; H+	Oldfield *et al.* (2003*b*)
Lake Waynewood, PA, USA	Lake	AD 1895	A+ (SIRM)	Lott *et al.* (1994)
Sharkey Lake, MN, USA	Lake	AD 1905	A+, G−, H−	Geiss *et al.* (2004)
Clear Lake, California, USA	Lake	*c.* AD 1927	A+ (χ, ARM)	Osleger *et al.* (2008)
Laguna Atezca, Molango, Mexico	Lake	*c.* AD 400; *c.* AD 1100	A+ (χ)	Conserva & Byrne (2002)
Lake Erhai, Yunnan, China	Lake	*c.* 800 BC; *c.* 200 BC; *c.* AD 700*	A+ (χ), G−	Dearing *et al.* (2008)
Yangtze Delta, China	Marine, coastal	*c.* AD 300; AD 1200	A+ (χ); G−	Wang *et al.* (2010)
Kondoa District, central Tanzania	Lake	AD 1935–1940	A+; H−	Eriksson & Sandgren (1999)
Lake Pupuke, New Zealand	Lake	AD 1400	A+ (χ)	Striewski *et al.* (2009)
Waipaoa Bay (MD972122), New Zealand	Marine	AD 300, AD 1450	A+, G−	Gomez *et al.* (2012)

A, magnetic mineral abundance change (single proxy indicator) where + indicates increase and − a decrease in abundance; G, magnetic mineral grain size indicator, where + indicates increase in magnetic grain size and − a decrease; H, antiferromagnetic mineral change (i.e. hematite or goethite), where + indicates a relative increase and − a decrease in antiferromagnetic content. *Major impact event. More lake-based initial rise events are listed in table 7.4 of Dearing (1999).

& Jones 2003). These early changes do not seem to be seen so dramatically in other continents, although a few longer records from China suggest that human impacts were important from as early as *c*. 2800 BC (Dearing *et al.* 2008), but did not have a drastic impact on the magnetic mineral signature until *c*. 800 BC, in spite of apparent modification of the river system hydrology *c*. 2300 BC (Fig. 9b).

In Europe, the biggest impacts on the mineral magnetic signature are between *c*. AD 1000 and 1350 (Fig. 9b), with the medieval expansion of agriculture and large monastic-owned estates (Van Vliet-Lanoe *et al.* 1992). Over this same period in the Yunan and Yangtze catchments in China, dramatic increases in the mineral magnetic content and volume of lake and deltaic sediments is coincident with the Nanzhao Kingdom (Dearing *et al.* 2008; Wang *et al.* 2010). The extent of major human impact on mineral magnetic changes during this time interval in other parts of world is not clear, but in a highland site in Mexico the major human impacts are associated with the interval AD 1030–1420 (Conserva & Byrne 2002).

Conclusions

Palaeomagnetic and archaeomagnetic dating methods in the latest Holocene rely upon a precise knowledge of the directional (and intensity) behaviours of the geomagnetic field in the regions of the Earth from which the palaeomagnetic recording sites are found. So far, knowledge of this 'calibration' has been patchy and is strongly concentrated in Europe, eastern Asia and western North America. However, increasingly sophisticated models of the geomagnetic field are being generated, which will allow extrapolation of these field models to parts of the world without a database of observational palaeomagnetic or archaeomagnetic data. This development is likely to make palaeomagnetic dating methods more widely applicable for dating late Holocene and Anthropocene sediments and human-built structures and materials in the future.

The strongest and biggest modification of the geomagnetic field of the past 50 ka is the Laschamp excursion at *c*. 41 ka. This excursion is well recorded in sediments worldwide, but it is practically too old for usefully defining the base of the Anthropocene. Within the last 3000 years one of the strongest global events is the coincident low in the dipole latitude and high dipole moment located at *c*. 2700 cal a BP. This event roughly coincides with agricultural expansion in Europe, and seems to occur at roughly the same interval at which mineral magnetic studies demonstrate some synchronicity of catchment disturbance in many European lake sites. In this interval, the associated secular variation records from northern mid-latitude sites indicate distinctive changes, which in Europe are characterized by the relatively rapid declination swing from the easterly 'f' to westerly 'e' events (Turner & Thompson 1981). In terms of the record of the mineral magnetic relationship to catchment disturbance and erosion of topsoils, the biggest near-synchronous events begin *c*. AD 1100 ($\pm$100 years), when expansion of agriculture in Europe greatly increased the delivery of sediments into lakes and coastal systems. This event also seems to be well marked in eastern Asia, and occurs at a second low in dipole latitude (Fig. 4).

The increased interest in defining the Anthropocene stems from a perspective that climate change primarily caused by burning of fossil fuels has, and will in the future, radically change Earth's environmental systems (Steffen *et al.* 2007). Markers that are left behind in sediments and other archives by burning of fossil fuels are therefore powerful preconditions for defining what we might mean by the Anthropocene. Fossil fuel burning, particularly that of coal, started in the Industrial Revolution and left behind in soils, sediment and peat archives stores of fuel combustion spherules, which can be easily and distinctively identified by physical, chemical and mineral magnetic measurements, and therefore can act as a powerful means to define the base of the Anthropocene. Nations that started industrialization early have the lion's share of blame for our current climate crisis, and therefore it seems natural to measure the start of the Anthropocene from the earliest synchronous interval in these nations at which there is evidence of major fossil fuel burning. Based on the mineral magnetic evidence presented here, this started in the interval AD 1890–1900.

Lacustrine sediments and terrestrial deposits of both natural and archaeological materials have been crucial for understanding the late Holocene evolution of the Earth, and the human impact upon it, certainly more so than studies of marine sediments, which are just beginning to explore sequences that rival the resolution that has been the advantage of lacustrine sequences. It therefore seems natural that a lacustrine succession be used to define what we mean by the start of the Anthropocene.

R. Muscheler secured the ^{10}Be-based VADM reconstruction on the GICC05 timescale. M. Hill provided some additional insight. A. Nilsson was funded by the Natural Environment Research Council, UK (grant no. NE/I013873/1). I. Snowball was supported by The Swedish Research Council (dossier no. 2011-3353). The authors thank J. Zalasiewicz and an anonymous reviewer for their constructive critiques, and M. Ellis and C. Waters for additional comments and editorial handling.

References

ALENIUS, T., SAARNISTO, M., TAAVITSAINEN, J.-P. & LUNKKA, J.-P. 2011. Records of land use and medieval settlement history in Vuonninen, Dvina Karelia, Northwest Russia. *Geoarchaeology*, **26**, 142–163.

ALI, M., ODA, H., HAYASHIDA, A., TAKEMURA, K. & TORII, M. 1999. Holocene palaeomagnetic secular variation at Lake Biwa, central Japan. *Geophysical Journal International*, **136**, 218–228.

BARRACLOUGH, D. R. 1982. Historical observations of the geomagnetic field. *Philosophical Transactions of the Royal Society of London. Series A, Mathematical and Physical Sciences*, **306**, 71–78.

BARRACLOUGH, D. R., CARRIGAN, J. G. & MALIN, S. R. C. 2000. Observed geomagnetic field intensity in London since 1820. *Geophysical Journal International*, **141**, 83–99.

BARTON, C. E. & MCELHINNY, M. W. 1981. A 10 000 yr geomagnetic secular variation record from three Australian maars. *Geophysical Journal International*, **67**, 465–485.

BENSON, L. V., LIDDICOAT, J., SMOOT, J., SARNA-WOJCICKI, A., NEGRINI, R. & LUND, S. P. 2003. Age of the Mono Lake excursion and associated tephra. *Quaternary Science Reviews*, **22**, 135–140.

BLAHA, U., BASAVAIAH, N., DEENADAYALAN, K., BOROLE, D. V. & MOHITE, R. D. 2011. Onset of industrial pollution recorded in Mumbai Mudflat sediments, using integrated magnetic, chemical, ^{210}Pb dating, and microscopic methods. *Environmental Science and Technology*, **45**, 686–692.

BLOXHAM, J., ZATMAN, S. & BUMBERRY, M. 2002. The origin of geomagnetic jerks. *Nature*, **420**, 65–68.

BONHOMMET, N. & BABKINE, J. 1967. Sur la presence d'aimentation inverse dans la Chaine des Puys. *Comptes Rendus Hebdomadaires des Seances de l'Academie des Sciences Series B*, **264**, 92–94.

BROWN, M. C., HOLME, R. & BARGERY, A. 2007. Exploring the influence of the non-dipole field on magnetic records for field reversals and excursions. *Geophysical Journal International*, **168**, 541–550.

CARRANCHO, Á., VILLALAÍN, J. J., ANGELUCCI, D. E., DEKKERS, M. J., VALLVERDÚ, J. & VERGÈS, J. M. 2009. Rock-magnetic analyses as a tool to investigate archaeological fired sediments: a case study of Mirador cave (Sierra de Atapuerca, Spain). *Geophysical Journal International*, **179**, 79–96.

CHAPARRO, M. A. E., GOGORZA, C. S. G., CHAPARRO, M A. E., IRURZUN, M. A. & SINITO, A. M. 2006. Review of magnetism and heavy metal pollution studies of various environments in Argentina. *Earth Planets Space*, **58**, 1411–1422.

CERTINI, G. & SCALENGHE, R. 2011. Anthropogenic soils are the golden spikes for the Anthropocene. *The Holocene*, **21**, 1269–1274.

CONSERVA, M. E. & BYRNE, R. 2002. Late Holocene vegetation change in the Sierra Madre Oriental of Central Mexico. *Quaternary Research*, **58**, 122–129.

COURTILLOT, J., DUCRUIX, J. & LEMOUËL, J. L. 1978. Sur une acceleration récente de la variation séculaire du champ magnétique terrestr. *Comptes Rendus Hebdomadaires des Séances de l'Académie des Sciences Series D*, **287**, 1095–1098.

CREER, K. M., VALENCIO, D. A., SINITO, A. M., TUCHOLKA, P. & VILAS, J. F. A. 1983. Geomagnetic secular variations 0–14 000 yr BP as recorded by lake sediments from Argentina. *Geophysical Journal of the Royal Astronomical Society*, **74**, 199–221.

DEARING, J. A. 1999. Holocene environmental; change from magnetic proxies in lake sediments. *In*: MAHER, B. A. & THOMPSON, R. (eds) *Quaternary Climates, Environments, and Magnetism*. Cambridge University Press, Cambridge, 231–278.

DEARING, J. A. 2000. Natural magnetic tracers in fluvial geomorphology. *In*: FOSTER, I. D. L. (ed.) *Tracers in Geomorphology*. Wiley, Chichester.

DEARING, J. A. & JONES, R. T. 2003. Coupling temporal and spatial dimensions of global sediment flux through lake and marine sediment records. *Global and Planetary Change*, **39**, 147–168.

DEARING, J. A., JONES, R. T. *ET AL*. 2008. Using multiple archives to understand past and present climate–human–environment interactions: the lake Erhai catchment, Yunnan Province, China. *Journal of Paleolimnology*, **40**, 3–31.

DEKKERS, M. J. 2003. Magnetic minerals. *In*: MIDDLETON, G. V. (ed.) *Encyclopaedia of Sediments and Sedimentary Rocks*. Kluwer, Dordrecht, 418–424.

DENHAM, C. R. & COX, A. 1971. Evidence that the Laschamp polarity event did not occur 13,300–34,000 years ago. *Earth and Planetary Science Letters*, **13**, 181–190.

DONADINI, F., KORHONEN, K., RIISAGER, P. & PESONEN, L. J. 2006. Database for Holocene geomagnetic intensity information. *EOS Transactions American Geophysical Union*, **87**, 137–138.

DONADINI, F., KORTE, M. & CONSTABLE, C. G. 2009. Geomagnetic field for 0–3 ka: 1. New data sets for global modeling. *Geochemistry Geophysics Geosystems*, **10**, Q06007, http://dx.doi.org/06010.01029/02008GC002295

EDWARDS, K. J. & WHITTINGTON, G. 2000. Lake sediments, erosion and landscape change during the Holocene in Britain and Ireland. *Catena*, **42**, 143–173.

ERIKSSON, M. G. & SANDGREN, P. 1999. Mineral magnetic analyses of sediment cores recording recent soil erosion history in central Tanzania. *Palaeogeography, Palaeoclimatology, Palaeoecology*, **152**, 365–383.

EVANS, M. E. & HELLER, F. 2003. *Environmental Magnetism: Principles and Applications of Enviromagnetics*. Academic Press, Amsterdam, Boston.

FINLAY, C. C., MAUS, S. *ET AL*. 2010. International geomagnetic reference field: the eleventh generation. *Geophysical Journal International*, **183**, 1216–1230.

FÜRST, F., WILMS, J., ROTHSCHILD, R. E., POTTSCHMIDT, K., SMITH, D. M. & LINGENFELTER, R. 2009. Temporal variations of strength and location of the South Atlantic anomaly as measured by RXTE. *Earth and Planetary Science Letters*, **281**, 125–133.

GABROWSKI, R. 2011. Changes in cereal cultivation during the Iron Age in southern Sweden: a compilation and interpretation of the archaeobotanical material. *Vegetation History and Archaeobotany*, **20**, 479–494.

GALLET, Y., GENEVEY, A. & COURTILLOT, V. 2003. On the possible occurrence of archeomagnetic jerks in the geomagnetic field over the past three millennia. *Earth and Planetary Science Letters*, **214**, 237–242.

Gallet, Y., Genevey, A. & Fluteau, F. 2005. Does Earth's magnetic field secular variation control centennial climate change? *Earth and Planetary Science Letters*, **236**, 339–347.

Gallet, Y., Hulot, G., Chulliat, A. & Genevey, A. 2009. Geomagnetic field hemispheric asymmetry and archeomagnetic jerks. *Earth and Planetary Science Letters*, **284**, 179–186.

Geiss, C. E., Banerjee, S. K., Camill, P. & Umbanhowar, C. E., Jr. 2004. Sediment-magnetic signature of land-use and drought as recorded in lake sediment from south-central Minnesota, USA. *Quaternary Research*, **62**, 117–125.

Genevey, A., Gallet, Y., Constable, C. G., Korte, M. & Hulot, G. 2008. ArcheoInt: an upgraded compilation of geomagnetic field intensity data for the past ten millennia and its application to the recovery of the past dipole moment. *Geochemistry Geophysics Geosystems*, **9**, Q04038, http://dx.doi.org/10.1029/2007GC001881.

Gogorza, C. S. G., Sinito, A. M., Di Tommaso, I., Vilas, J. F., Creer, K. M. & Nuñez, H. 1999. Holocene geomagnetic secular variations recorded by sediments from Escondido Lake (south Argentina). *Earth Planets Space*, **51**, 93–106.

Gogorza, C. S. G., Sinito, A. M., Lirio, J. M., Nunez, H., Chaparro, M. & Vilas, J. F. 2002. Paleosecular variations 0–19,000 years recorded by sediments from Escondido Lake (Argentina). *Physics of the Earth and Planetary Interiors*, **133**, 35–55.

Gomez, B., Carter, L. & Trustrum, N. A. 2012. A 2400 yr record of natural events and anthropogenic impacts in intercorrelated terrestrial and marine sediment cores: Waipaoa sedimentary system, New Zealand. *Geological Society of America Bulletin*, **119**, 1415–1432.

Guyodo, Y. & Valet, J.-P. 1999. Global changes in intensity of the Earth's magnetic field during the past 800 kyr. *Nature*, **399**, 249–252.

Gubbins, D. 1999. The distinction between geomagnetic excursions and reversals. *Geophysical Journal International*, **137**, F1–F4.

Gubbins, D. & Bloxham, J. 1985. Geomagnetic field analysis – III. Magnetic fields on the core–mantle boundary. *Geophysical Journal International*, **80**, 695–713.

Haltia-Hovi, E., Nowaczyk, N. & Saarinen, T. 2010. Holocene palaeomagnetic secular variation recorded in multiple lake sediment cores from eastern Finland. *Geophysical Journal International*, **180**, 609–622.

Hatfield, R. G. & Maher, B. A. 2009. Holocene sediment dynamics in an upland temperate lake catchment: climatic and land-use impacts in the English Lake District. *The Holocene*, **19**, 427–438.

Higgitt, S. R., Oldfield, F. & Appleby, P. G. 1991. The record of landuse change and soil erosion in the late-Holocene sediments of the Petit Lac d'Annecy, eastern France. *The Holocene*, **1**, 14–28.

Hongre, L., Hulot, G. & Khokhlov, A. 1998. An analysis of the geomagnetic field over the past 2000 years. *Physics of the Earth and Planetary Interiors*, **106**, 311–335.

Horng, C.-S., Huh, C.-A., Chen, K.-H., Huang, P. R., Hsiung, K.-H. & Lin, H.-L. 2009. Air pollution history elucidated from anthropogenic spherules and their magnetic signatures in marine sediments offshore of Southwestern Taiwan. *Journal of Marine Systems*, **76**, 468–478.

Hounslow, M. W. & Chepstow-Lusty, A. 2004. A record of soil loss in a coastal setting, using charcoal and mineral magnetism indicators, AD 450 to AD 1200: Butrint, Southern Albania. *The Holocene*, **14**, 321–333.

Jackson, A., Jonkers, A. R. T. & Walker, M. R. 2000. Four centuries of geomagnetic secular variation from historical records. *Philosophical Transactions of the Royal Society of London. Series A, Mathematical and Physical Sciences*, **358**, 957–990.

Jones, R. T., Reinhardt, L. J., Dearing, J. A., Crook, D., Chiverrell, R. C., Welsh, K. E. & Vergès, E. 2013. Detecting climatic signals in an anthropogenically disturbed catchment: the late-Holocene record from the Petit Lac d'Annecy, French Alps. *The Holocene*, **23**, 1329–1339.

Jonkers, A. R. T. 2007. Discrete scale invariance connects geodynamo timescales. *Geophysical Journal International*, **171**, 581–593.

Jordanova, D., Hofmann, V. & Fehr, K. T. 2004. Mineral magnetic characterization of anthropogenic magnetic phases in the Danube river sediments (Bulgarian part). *Earth and Planetary Science Letters*, **221**, 71–89.

Kadlec, J., Grygar, T. *et al.* 2009. Morava River floodplain development during the last millennium, Strážnické Pomoraví, Czech Republic. *The Holocene*, **19**, 499–509.

Knudsen, M. F., Riisager, P., Donadini, F., Snowball, I., Muscheler, R., Korhonen, K. & Pesonen, L. J. 2008. Variations in the geomagnetic dipole moment during the Holocene and the past 50 kyr. *Earth and Planetary Science Letters*, **272**, 319–329.

Kodama, K. P., Lyons, J. C., Siver, P. A. & Lott, A.-M. 1997. A mineral magnetic and scaled-chrysophyte paleolimnological study of two northeastern Pennsylvania lakes: records of fly ash deposition, land-use change, and paleorainfall variation. *Journal of Paleolimnology*, **17**, 173–189.

Kopp, R. E. & Kirschvink, J. L. 2008. The identification and biogeochemical interpretation of fossil magnetotactic bacteria. *Earth-Science Reviews*, **86**, 42–61.

Korhonen, K., Donadini, F., Riisager, P. & Pesonen, L. J. 2008. GEOMAGIA50: an archeointensity database with PHP and MySQL. *Geochemistry Geophysics Geosystems*, **9**, Q04029, http://dx.doi.org/10.1029/2007GC001893

Korte, M. & Constable, C. G. 2003. Continuous global geomagnetic field models for the past 3000 years. *Physics of the Earth and Planetary Interiors*, **140**, 73–89.

Korte, M. & Constable, C. G. 2005*a*. Continuous geomagnetic field models for the past 7 millennia: 2. CALS7 K. *Geochemistry Geophysics Geosystems*, **6**, Q02H16, http://dx.doi.org/10.1029/2004GC000801

Korte, M. & Constable, C. G. 2005*b*. The geomagnetic dipole moment over the last 7000 years – new results from a global model. *Earth and Planetary Science Letters*, **236**, 348–358.

KORTE, M. & CONSTABLE, C. G. 2006. On the use of calibrated relative paleointensity records to improve millennial-scale geomagnetic field models. *Geochemistry Geophysics Geosystems*, **7**, Q09004, http://dx.doi.org/10.1029/2006GC001368

KORTE, M. & CONSTABLE, C. G. 2008. Spatial and temporal resolution of millennial scale geomagnetic field models. *Advances in Space Research*, **41**, 57–69.

KORTE, M. & CONSTABLE, C. G. 2011. Improving geomagnetic field reconstructions for 0–3 ka. *Physics of the Earth and Planetary Interiors*, **188**, 247–259.

KORTE, M., DONADINI, F. & CONSTABLE, C. G. 2009. Geomagnetic field for 0–3 ka: 2. A new series of time-varying global models. *Geochemistry Geophysics Geosystems*, **10**, Q06008, http://dx.doi.org/10.1029/2008GC002297

KORTE, M., CONSTABLE, C., DONADINI, F. & HOLME, R. 2011. Reconstructing the Holocene geomagnetic field. *Earth and Planetary Science Letters*, **312**, 497–505.

LAJ, C. & CHANNELL, J. E. T. 2007. Geomagnetic excursions. *In*: KONO, M. (ed.) *Treatise on Geophysics*. Elsevier, Amsterdam, **5**, 347–416.

LAJ, C., KISSEL, C., MAZAUD, A., CHANNELL, J. E. T. & BEER, J. 2000. North Atlantic paleointensity stack since 75 kal (NAPIS-75) and the duration of the Laschamp event. *Philosophical Transactions of the Royal Society of London*, **358**, 1009–1025.

LAJ, C., KISSEL, C. & BEER, J. 2004. High resolution global paleointensity stack since 75 kyr /GLOPIS-75) calibrated to absolute values. *Geophysical Monograph Series*, **145**, 255–264.

LANGEREIS, C. G., KRIJGSMAN, W., MUTTONI, G. & MENNING, M. 2010. Magnetostratigraphy – concepts, definitions, and applications. *Newsletters on Stratigraphy*, **43**, 207–233.

LEONHARDT, R., FABIAN, K., WINKLHOFER, M., FERK, A., LAJ, C. & KISSEL, C. 2009. Geomagnetic field evolution during the Laschamp excursion. *Earth and Planetary Science Letters*, **278**, 87–95.

LIDDICOAT, J. C. & COE, R. S. 1979. Mono Lake geomagnetic excursion. *Journal of Geophysical Research*, **84**, 261–271.

LINFORD, P. 2004. Archaeomagnetic dating. *Physics Education*, **39**, 145–154.

LOCKE, G. & BERTINE, K. K. 1986. Magnetic sediments as an indicator of coal combustion. *Applied Geochemistry*, **1**, 345–356.

LOTT, A.-M., SIVER, P. A., MARSICANO, L. J., KODAMA, K. R. & MOELLER, R. E. 1994. The paleolimnology of a small waterbody in the Pocono Mountains of Pennsylvania, USA: reconstructing 19th–20th century specific conductivity trends in relation to changing land use. *Journal of Paleolimnology*, **12**, 75–86.

LUND, S., STONER, J. S., CHANNELL, J. E. T. & ACTON, G. 2006. A summary of Brunhes paleomagnetic field variability recorded in Ocean Drilling Program cores. *Physics of the Earth and Planetary Interiors*, **156**, 194–204.

MCLEAN, D. 1991. Magnetic spherules in recent lake sediments. *Hydrobiologica*, **214**, 91–97.

MACRÌ, M., SAGNOTTI, L., DINARÈS-TURELL, J. & ANDREA CABURLOTTO, A. 2010. Relative geomagnetic paleointensity of the Brunhes Chron and the Matuyama–Brunhes precursor as recorded in sediment core from Wilkes Land Basin (Antarctica). *Physics of the Earth and Planetary Interiors*, **179**, 72–86.

MAHER, B. A. 2009. Rain and dust: magnetic records of climate and pollution. *Elements*, **5**, 229–234.

MAHER, B. A. & THOMPSON, R. (eds) 1999. *Quaternary Climates, Environments and Magnetism.* Cambridge University Press, Cambridge.

MAIER, G. & SCHOLGER, R. 2004. Demonstration of connection between pollutant dispersal and atmospheric boundary layers by use of magnetic susceptibility mapping, St. Jacob (Austria). *Physics and Chemistry of the Earth*, **29**, 997–1009.

MALIN, S. R. C. & BULLARD, E. 1981. The direction of the Earth's magnetic field at London, 1570–1975. *Philosophical Transactions of the Royal Society of London. Series A, Mathematical and Physical Sciences*, **299**, 357–423.

MANDEA, M. & OLSEN, N. 2009. Geomagnetic and archeomagnetic jerks: where do we stand? *EOS Transactions of the American Geophysical Union*, **90**, 208–208.

MANDEA, M., HOLME, R., PAIS, A., PINHEIRO, K., JACKSON, A. & VERBANAC, G. 2010. Geomagnetic jerks: rapid core field variations and core dynamics. *Space Science Reviews*, **155**, 147–175.

MANN, K. C., PECK, J. A. & PECK, M. C. 2013. Assessing dam pool sediment for understanding past, present and future watershed dynamics: an example from the Cuyahoga River, Ohio. *Anthropocene*, http://dx.doi.org/10.1016/j.ancene.2013.08.001

MERCURI, A. M., MAZZANTI, M. B. ET AL. 2012. A marine/terrestrial integration for mid–late Holocene vegetation history and the development of the cultural landscape in the Po valley as a result of human impact and climate change. *Vegetation History and Archaeobotany*, **21**, 353–372.

MERRILL, R. T. & MCFADDEN, P. L. 1994. Geomagnetic field stability: reversal events and excursions. *Earth and Planetary Science Letters*, **121**, 57–69.

MUSCHELER, R., BEER, J., KUBIK, P. E. & SYNAL, H.-A. 2005. Geomagnetic field intensity during the last 60,000 years based on ^{10}Be and ^{36}Cl from the Summit ice-cores and ^{14}C. *Quaternary Science Reviews*, **24**, 1849–1860.

NEWITT, L., CHULLIAT, A. & ORGEVAL, J. J. 2009. Location of the north magnetic pole in April 2007. *Earth Planets Space*, **61**, 703–710.

NILSSON, A., SNOWBALL, I., MUSCHELER, R. & UVO, C. B. 2010. Holocene geocentric dipole tilt model constrained by sedimentary paleomagnetic data. *Geochemistry Geophysics Geosystems*, **11**, Q08018.

NILSSON, A., MUSCHELER, R. & SNOWBALL, I. 2011. Millennial scale cyclicity in the geodynamo inferred from a dipole tilt reconstruction. *Earth and Planetary Science Letters*, **311**, 299–305.

NOWACZYK, N. R., ARZA, H. W., FRANKA, U., KINDA, J. & PLESSENA, B. 2012. Dynamics of the Laschamp geomagnetic excursion from Black Sea sediments. *Earth and Planetary Science Letters*, **351–352**, 54–69.

OGG, J. G. 2012. Geomagnetic polarity timescale. Magnetostratigraphy. *In*: GRADSTEIN, F. M., OGG, J. G., SCHMITZ, M. D. & OGG, G. M. (eds) *The Geologic Timescale*. Elsevier, Amsterdam, 85–114.

OJALA, A. E. K. & SAARINEN, T. 2002. Palaeosecular variation of the Earth's magnetic field during the last 10000 years based on the annually laminated sediment of Lake Nautajarvi, central Finland. *The Holocene*, **12**, 391–400.

OLDFIELD, F. 1990. Magnetic measurements of recent sediments from Big Moose Lake, Adirondack Mountains, N.Y., USA. *Journal of Paleolimnology*, **4**, 93–101.

OLDFIELD, F., THOMPSON, R. & BARBER, K. E. 1978. Changing atmospheric fallout of magnetic particles recorded in recent ombrotrophic peat. *Science*, **199**, 679–680.

OLDFIELD, F., WAKES, R. *ET AL.* 2003*a*. The late-Holocene history of Gormire Lake (NE England) and its catchment: a multiproxy reconstruction of past human impact. *The Holocene*, **15**, 677–690.

OLDFIELD, F., ASIOLI, A. *ET AL.* 2003*b*. A high resolution late Holocene palaeo environmental record from the central Adriatic Sea. *Quaternary Science Reviews*, **22**, 319–342.

OSLEGER, D. A., ZIERENBERG, R. A., SUCHANEK, T. H., STONER, J. S., MORGAN, S. & ADAM, D. P. 2008. Clear Lake sediments: anthropogenic changes in physical sedimentology and magnetic response. *Ecological Applications*, **18**, A239–A256.

PETROVSKY, E. & ELLWOOD, B. B. 1999. Magnetic monitoring of air, land and water pollution. *In*: MAHER, B. A. & THOMPSON, R. (eds) *Quaternary Climates, Environments, and Magnetism*. Cambridge University Press, Cambridge, 278–322.

PLATER, A. J., RIDGWAY, J., APPLEBY, P. J., BERRY, A. & WRIGHT, M. R. 1998. Historical contaminant fluxes in the Tees Estuary, UK: geochemical, magnetic and radionuclide evidence. *Marine Pollution Bulletin*, **37**, 343–360.

POWER, A. L. & WORSLEY, A. T. 2009. Using urban man-made ponds to reconstruct a 150-year history of air pollution in northwest England. *Environmental Geochemistry and Health*, **31**, 327–338.

POZZA, M. R., BOYCE, J. I. & MORRIS, W. A. 2004. Lake-based magnetic mapping of contaminated sediment distribution, Hamilton Harbour, Lake Ontario, Canada. *Journal of Applied Geophysics*, **57**, 23–41.

RICHARDSON, N. 1986. The mineral magnetic record in recent ombrotrophic peat synchronised by fine resolution pollen analysis. *Physics of the Earth and Planetary Interiors*, **42**, 48–56.

RIDDIFORD, N. G., BRANCH, N. P., GREEN, C. P., ARMITAGE, S. J. & OLIVIER, L. 2012. Holocene palaeoenvironmental change and the impact of prehistoric salt production in the Seille Valley, eastern France. *The Holocene*, **22**, 831–845.

ROBERTS, A. P. 2008. Geomagnetic excursions: knowns and unknowns. *Geophysical Research Letters*, **35**, L17307, http://dx.doi.org/10.1029/2008GL034719

ROBERTS, A. P. & WINKLHOFER, M. 2004. Why are geomagnetic excursions not always recorded in sediments? Constraints from post-depositional remanent magnetization lock-in modelling. *Earth and Planetary Science Letters*, **227**, 345–359.

ROLPH, T. C., VIGLIOTTI, L. & OLDFIELD, F. 2004. Mineral magnetism and geomagnetic secular variation of marine and lacustrine sediments from central Italy: timing and nature of local and regional Holocene environmental change. *Quaternary Science Reviews*, **23**, 1699–1722.

ROTHWELL, J. J., ROBINSON, S. G., EVANS, M. G., YANG, J. & ALLOTT, T. E. H. 2005. Heavy metal release by peat erosion in the Peak District, southern Pennines, UK. *Hydrological Processes*, **19**, 2973–2989.

SCHMIDT, R., KOINIG, K. A., THOMPSON, R. & KAMENIK, C. 2002. Multi proxy core study of the last 7000 years of climate and alpine land-use impacts on an Austrian mountain lake (Unterer Landschitzsee, Niedere Tauern). *Palaeogeography Palaeoclimatology Palaeoecology*, **187**, 101–120.

SCHWABE, H. 1844. Sonnen-Beobachtungen im Jahre 1843. *Astronomische Nachrichten*, **21**, 233–236.

SNOWBALL, I. & MUSCHELER, R. 2007. Palaeomagnetic intensity data: an Achilles heel of solar activity reconstructions. *The Holocene*, **17**, 851–859.

SNOWBALL, I. & SANDGREN, P. 2004. Geomagnetic field intensity changes in Sweden between 9000 and 450 cal BP: extending the record of 'archaeomagnetic jerks' by means of lake sediments and the pseudo-Thellier technique. *Earth and Planetary Science Letters*, **227**, 361–376.

SNOWBALL, I., ZILLÉN, L., OJALA, A., SAARINEN, T. & SANDGREN, P. 2007. FENNOSTACK and FENNORPIS: varve dated Holocene palaeomagnetic secular variation and relative palaeointensity stacks for Fennoscandia. *Earth and Planetary Science Letters*, **255**, 106–115.

SNOWBALL, I., MELLSTRÖM, A. *ET AL.* 2013. An estimate of post-depositional remanent magnetization lock-in depth in organic rich varved lake sediments. *Global and Planetary Change*, **110**, 264–277. http://dx.doi.org/10.1016/j.gloplacha.2013.10.005

SORREL, P., TESSIER, B. *ET AL.* 2010. Sedimentary archives of the French Atlantic coast (inner Bay of Vilaine, south Brittany): depositional history and late Holocene climatic and environmental signals. *Continental Shelf Research*, **30**, 1250–1266.

STEFFEN, W., CRUTZEN, P. J. & MCNEILL, J. R. 2007. The Anthropocene: are humans now overwhelming the great forces of nature? *Ambio*, **36**, 614–621.

STONER, J. S., CHANNELL, J. E. T., MAZAUD, A., STRANO, S. E. & XUAN, C. 2013. The influence of high-latitude flux lobes on the Holocene paleomagnetic record of IODP Site U1305 and the northern North Atlantic. *Geochemistry Geophysics Geosystems*, **14**, 4623–4646. http://dx.doi.org/10.1002/ggge.20272

STRIEWSKI, B., MAYR, C., FLENLEY, J., NAUMANN, R., TURNER, G. & LUCKE, A. 2009. Multi-proxy evidence of late Holocene human-induced environmental changes at Lake Pupuke, Auckland (New Zealand). *Quaternary International*, **202**, 69–93.

STRZYSZCZ, Z. & MAGIERA, T. 2001. Record of industrial pollution in polish ombrotrophic peat bogs. *Physics and Chemistry of the Earth (A)*, **26**, 859–866.

SVENSSON, A., ANDERSEN, K. K. *ET AL.* 2008. A 60 000 year Greenland stratigraphic ice core chronology. *Climate of the Past*, **4**, 47–57.

TEANBY, N., LAJ, C., GUBBINS, D. & PRINGLE, M. 2002. A detailed palaeointensity and inclination record from drill core SOH1 on Hawaii. *Physics of the Earth and Planetary Interiors*, **131**, 101–140.

TOLONEN, K. & OLDFIELD, F. 1986. The record of mineral magnetic and heavy metal deposition at Regent St. Bog, Fredricton, New Brunswick. *Physics of the Earth and Planetary Interiors*, **42**, 57–66.

TURNER, G. & THOMPSON, R. 1981. Lake sediment record of the geomagnetic secular variation in Britain during Holocene times. *Geophysical Journal of the Royal Astronomical Society*, **65**, 703–725.

VALET, J.-P. & VALLADAS, H. 2010. The Laschamp–Mono Lake geomagnetic events and the extinction of Neanderthal: a causal link or a coincidence? *Quaternary Science Reviews*, **29**, 3887–3893.

VALET, J.-P., HERRERO-BERVERA, E., LE MOUEL, J.-L. & PLENIER, G. 2008. Secular variation of the geomagnetic dipole during the past 2000 years. *Geochemistry Geophysics Geosystems*, **9**, Q01008, http://dx.doi.org/10.1029/2007GC001728

VANDAMME, D. 1994. A new method to determine paleosecular variation. *Physics of the Earth and Planetary Interiors*, **85**, 131–142.

VAN VLIET, B., HELLUIN, M., PELLERIN, J. & VALADAS, B. 1992. Soil erosion in Western Europe: from the Last Interglacial to the Present. *In*: BELL, M. & BOARDMAN, J. (eds) *Past and Present Soil Erosion*. Oxbow Monograph, Oxford, **22**, 101–114.

VAZQUEZ, J. A. & LIDZBARSKI, M. I. 2012. High-resolution tephrochronology of the Wilson Creek Formation (Mono Lake, California) and Laschamp event using $^{238}U-^{230}Th$ SIMS dating of accessory mineral rims. *Earth and Planetary Science Letters*, **357–358**, 54–67.

VEROSUB, K. L. & BANERJEE, S. K. 1977. Geomagnetic excursions and their paleomagnetic record. *Reviews of Geophysics*, **15**, 145–155.

VEROSUB, K. L., MEHRINGER, P. J. & WATERSTRAAT, P. 1986. Holocene secular variation in Western North America: paleomagnetic record from Fish Lake, Harney County, Oregon. *Journal of Geophysical Research*, **91**, 3609–3623.

WAGNER, G., BEER, J., LAJ, C., KISSEL, C., MASARIK, J., MUSCHELER, R. & SYNAL, H.-A. 2000. Chlorine-36 evidence for the Mono Lake event in the Summit GRIP ice core. *Earth and Planetary Science Letters*, **181**, 1–6.

WANG, Z., LI, M. ET AL. 2010. Impacts of human activity on the late-Holocene development of the subaqueous Yangtze delta, China, as shown by magnetic properties and sediment accumulation rates. *The Holocene*, **21**, 393–407.

YANG, T., LIU, Q., ZENG, Q. & CHAN, L. 2009. Environmental magnetic responses of urbanization processes: evidence from lake sediments in East Lake, Wuhan, China. *Geophysical Journal International*, **179**, 873–886.

ZIMMERMAN, S. H., HEMMING, S. R., KENT, D. V. & SEARLE, S. Y. 2006. Revised chronology for late Pleistocene Mono Lake sediments based on paleointensity correlation to the global reference curve. *Earth and Planetary Science Letters*, **252**, 94–106.

ZOLITSCHKA, B. 1998. A 14,000 year sediment yield record from western Germany based on annually laminated lake sediments. *Geomorphology*, **22**, 1–17.

Is the fossil record of complex animal behaviour a stratigraphical analogue for the Anthropocene?

M. WILLIAMS[1]*, J. A. ZALASIEWICZ[1], C. N. WATERS[2] & E. LANDING[3]

[1]*Department of Geology, University of Leicester, Leicester LE1 7RH, UK*

[2]*British Geological Survey, Keyworth, Nottingham, NG12 5GG, UK*

[3]*New York State Museum, 222 Madison Avenue, Albany, NY 12230, USA*

**Corresponding author (e-mail: mri@le.ac.uk)*

Abstract: The base of the Cambrian System is recognized by a characteristic (marine) trace fossil suite assigned to the *Treptichnus pedum* Biozone, which signals increasing complexity of animal behaviour and demarcates the Cambrian from the (older) Ediacaran System (Proterozoic Eonathem). Ichnotaxa of the *T. pedum* Biozone are not the earliest trace fossils, and are preceded in the latest Proterozoic by a progressive increase in the diversity of trace-producing organisms and the communities they comprised, the structural and behavioural complexity of the trace fossils, and even the depth of burrowing in sediments. Parallels can be drawn with the increasing complexity of subsurface structures associated with human cities, which also reflect evolution of an increasingly complex community. Before the nineteenth century, these structures were limited and simple, but beginning with the development of London in the mid-nineteenth century as the world's first mega-city, subsurface structures have become increasingly complex, reflecting the technology-driven behaviour of twentieth- and twenty-first-century humans.

This paper examines the similarity between the evolution of complex animal behaviour whose trace fossil record is used to recognize the base of the Cambrian System 541 $\pm$ 0.63 million years ago (Peng *et al.* 2012), and the complex animal behaviour of the nineteenth to twenty-first centuries that is manifested in the growth of megacities (urban areas with populations greater than 10 million inhabitants). This physical record of human activity could be used in defining, for instance, the base of a formal Anthropocene Epoch/Series.

Definition of the base of the Cambrian

The Cambrian is the earliest period/lowermost system of the Phanerozoic Eon/Eonathem. The base of the system and of its lowermost Terreneuvian Series and very lowest Fortunian Stage (Landing *et al.* 2007) is recognized by the lowest occurrence of a globally widespread trace fossil suite referred to the *Treptichnus pedum* Biozone at its Global Boundary Stratotype Section and Point (GSSP). This GSSP of the Cambrian is defined at a horizon located 2.4 m above the base of the lower Mystery Lake Member of the Chapel Island Formation on the Burin Peninsula of southeastern Newfoundland (Brasier *et al.* 1994; Landing 1994). The subhorizontal, branching burrow systems of *T. pedum* (Fig. 1, also referred to as *Trichophycus pedum* or *Phycodes pedum* in many reports) may represent the feeding and burrowing activity of priapulid worms or of an organism with similar locomotion (Vannier *et al.* 2010).

Treptichnus pedum occurs in rock successions characterized by trace fossils that signal a progressive increase in animal behaviour more complex than that of the preceding (end-Proterozoic) Ediacaran System, and which, for example, include the arthropod scratch marks called *Monomorphichnus* (Jensen 2003). Although *T. pedum* has subsequently been recorded some 4 m below the Ediacaran–Cambrian GSSP (see Gehling *et al.* 2001), the occurrence of this trace fossil nonetheless remains a useful approximate marker for the base of the Cambrian System (MacNaughton 2007). Indeed, recently, Landing *et al.* (2013) redefined the base of the *T. pedum* Assemblage Zone and Cambrian to lie within the lowest range of *T. pedum* and above the highest Ediacaran-type biotas (vendobionts, putative metazoans, and calcareous problematica, such as *Cloudina*). The ichnofauna of the *T. pedum* Biozone form part of the global fossil record of increasing complexity of marine ecosystems and organism behaviour that developed through the late Proterozoic and early Phanerozoic (Fig. 2), with the appearance of the soft-bodied Ediacaran biota some 580 million years ago (e.g. Yuan *et al.* 2011) and the earliest biomineralized fossils 549 million years ago (Wood *et al.* 2002; Zhuravlev *et al.* 2012). The earliest occurrence of *T. pedum* greatly preceded the

From: WATERS, C. N., ZALASIEWICZ, J. A., WILLIAMS, M., ELLIS, M. A. & SNELLING, A. M. (eds) 2014. *A Stratigraphical Basis for the Anthropocene*. Geological Society, London, Special Publications, **395**, 143–148.
First published online October 25, 2013, http://dx.doi.org/10.1144/SP395.8

Fig. 1. Specimen of the ichnofossil *Treptichnus pedum* (Seilacher) showing branches and galleries of this shallow burrow (arrow points to trace fossil). Specimen located precisely at the GSSP selected for the base of the Cambrian System (as well as the Terreneuvian Series and Fortunian Stage; see Landing 1994). The sediment-filled specimen of *T. pedum* is weathered out of very fine-grained sandstone of the lower Chapel Island Formation at Fortune North section, just west of Fortune village, southeastern Newfoundland.

global appearance of diverse mineralized (calcareous and phosphatic) small shelly fossils and, even later, the oldest trilobites (Fig. 2).

The Cambrian was defined with an underlying premise that it was the period during which Bilateria first evolved from simpler metazoan life. However, without preservable hard parts, evidence of such soft-bodied bilaterian organisms is restricted to preservation of their traces (ichnofossils), which are classified using the same binomial Linnaean system (i.e. *Genus species*, as *Homo sapiens*) as body fossils. Hence, definition of the system followed a route of a broad palaeontological concept of increasing biological (ecological and behavioural) complexity at the boundary, where the boundary itself was recognized by the lowest appearance of a single representative ichnofauna.

However, in detail, this fossil record is more nuanced. Bilateria fossils such as *Kimberella* (Fedonkin & Waggoner 1997); bilaterian, possibly arthropod, trace fossils (Martin *et al.* 2000); and trace fossils of bilaterian worms (Dornbos *et al.* 2005) occur coincident with Ediacaran assemblages in strata younger than 560 million years old. Treptichnid trace fossils that record simpler behavioural activity than that in *T. pedum* occur in strata younger than 550 million years (Jensen 2003). What specifically identifies the base of the Cambrian System, and of the Phanerozoic Eonathem is the assemblage of complex behavioural patterns that are coincident with the widespread occurrence of *T. pedum*. The global changes in marine burrowing organism communities record the diversification, and perhaps origination, of many modern coelomate metazoans early in the Cambrian Evolutionary Radiation, as well as a major change in the world ocean, which included the recycling of nutrients back into the water column by burrowing animals (Landing & Westrop 2004; Brasier *et al.* 2011).

It is important to remember that by first recognizing the base of a system through the first

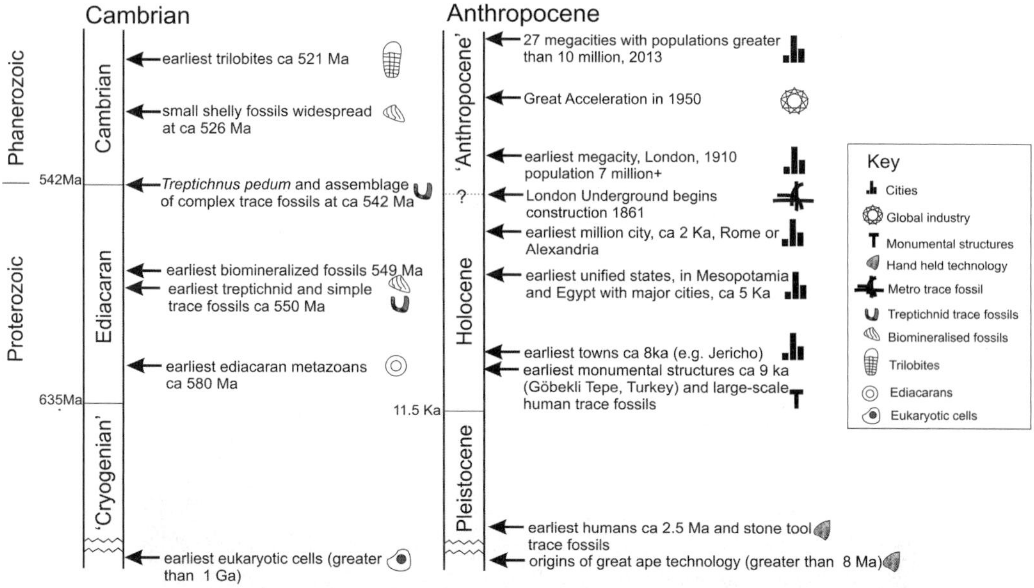

Fig. 2. Biotic events associated with the Proterozoic–Phanerozoic boundary compared with the human events of the past 3 million years. The traditional base of the Cambrian System was early regarded as lying at the lowest appearance of trilobites. The widespread occurrence of small shelly fossils that identify the Siberian 'Tommotian Stage' was also once mooted as the boundary. The boundary approximates the lowest occurrence of the assemblage of trace fossils of the basal Cambrian *Treptichnus pedum* Biozone. The late Proterozoic is also associated with the global appearance of metazoans (the Ediacaran Period), early trace fossils, and very-low-diversity mineralized taxa such as sponges with their spicules, tubular fossils (earliest *Anabarites*), and calcareous problematica limited to the Ediacaran (*Cloudina*, *Namacalanthus*). These evolutionary signatures may be seen as heralding the significant animal behaviour innovations of the earliest Cambrian. The record of events leading to the development of the complex trace fossils of megacities is similarly complex, finding its roots in the technology of great apes, through the stone tools of early humans, to the monumental structures of the past 9000 years, culminating in the megacities. Each step can be seen as a staging post to the main event, in this case taken as the trace fossil systems of the megacities.

appearance of an ichnofauna at a specific locality, there is the possibility that, globally, the onset of the *T. pedum* assemblage is diachronous. In some sections it may be that the first development of this trace fossil assemblage occurs later than at the GSSP. Even more importantly, it is important to understand that even with the best and most thorough study, the recorded range of any fossil invariably underestimates its actual range in every section (Marshall 1990), so even the range of *T. pedum* has proved to appear below the Cambrian GSSP at Fortune Head, eastern Newfoundland (Gehling *et al.* 2001). For this reason, Landing *et al.* (2007) qualified the base of the Cambrian, the Terreneuvian Series and the Fortunian Stage as lying at the base of the *T. pedum* Biozone 'as known at the time of the ratification of the base of the Cambrian at Fortune Head'.

Second, there are limited numbers of sections through the relevant Precambrian–Cambrian boundary, and even fewer that would have had a suitable environment for *T. pedum* to have developed or be preserved. Third, although the accuracy of radiometric dating is rapidly advancing, at such temporally distant periods of time an error of ±0.63 million years for the Precambrian–Cambrian boundary provides a considerable range. The current system by which chronostratigraphy is defined (i.e. the definition of regional and global subdivisions of the rock record) allows for such uncertainties.

Definition of the base of the Anthropocene

If the Proterozoic–Phanerozoic boundary can be recognized by a change in animal behaviour, might not the base of the Anthropocene also be defined by a fundamental change in behaviour and more complex community structure, signalled by the global development of megacities with their extensive signature of trace fossils, including such physical records of human behaviour as roads,

Fig. 3. Construction of the world's first underground railway, the Metropolitan Line in London, 1861. Image freely available from wikicommons at http://en.wikipedia.org/wiki/File:Constructing_the_Metropolitan_Railway.png.

buildings, tunnels, canals and other features? The route to the origins of megacities extends back over 2 million years, to the origins of human technology in the Palaeolithic (Fig. 2).

Megacities have antecedents in the early urban centres of the Fertile Crescent, Nile and Indus valleys, and in the large urban centres of ancient and historical times such as classical Rome in Europe, Alexandria in Africa, Baghdad in the Middle East, Tenochtitlan in the Americas, and Angkor in Asia. These cities have a considerable surface expression, but their classical remains, although including such structures as canals and brick-lined sewers, leave a more limited subsurface record, in effect mimicking the simplicity of the earliest late Proterozoic trace fossil assemblages (Jensen 2003). Compared to earlier cities, the megacities of the twenty-first century exhibit much more complex behaviour.

Megacities have their roots – their First Occurrence Datum, or FAD, as geologists express it in discussing chronostratigraphic units – in nineteenth-century London, which grew to a population of over 3 million in 1850, and over 7 million in 1910 (Emsley *et al.* 2013). Megacities have produced a fossil record of complex burrow systems at and below ground level. These burrow systems, as well as being more extensive and complex than those produced earlier in human history, are orders of magnitude larger and more complex than any non-human burrow systems. Indeed, they represent in effect a new phenomenon in geology, one that is driven by conscious thought and by mechanization. They include the trace fossils of locomotion by people, including metros, canals, both surface and subterranean (i.e. tunnels, walkways, roads and train lines), of power (conduits for electrical cables and gas and steam lines) and of ideas (telecommunications links). They also include faecal traces (sewers and buried rubbish dumps), dwelling structures (underground buildings and foundations), resting traces (underground car parks) and even feeding traces (if quarries and mines, or the basements or foundations of fast food restaurants are considered).

A useful datum for defining the origins of this behavioural and community complexity might be to designate underground metros as a particular style of human trace fossil associated with the mass movement of people (and ideas), and take the world's first underground line – the Metropolitan Line of London (construction began in 1861, see Figs 2 & 3) – as the GSSP (Wolmar 2005). One of the several original metro stations, such as Baker

Street (1863) or Euston Square (1863), might be selected as the locus for a 'golden spike' that would physically define a GSSP for a stage of human development or even the base of the Anthropocene.

The building and maintenance of megacities (and the survival of their human inhabitants) has been responsible, directly or indirectly, for many of the other geological changes associated with the Anthropocene, such as CO_2 emissions, changed patterns of erosion and sedimentation, and biodiversity loss. Megacities have now developed in Europe (e.g. Moscow, London, Paris), Asia (e.g. Shanghai, Beijing, Tokyo, Seoul, Mumbai), Africa (Cairo, Lagos) and the Americas (e.g. Mexico City, New York City, Sao Paulo), with the existence of 27 metropolitan centres exceeding 10 million inhabitants (Brinkhoff 2012). Their spread has been effectively instantaneous from a geological perspective, and the accelerating growth of megacities in the latter part of the twentieth century has covered, in particular, large areas of subsiding coastal plains and river deltas (e.g. Shanghai, see Zalasiewicz *et al.* this volume, in review) where their preservation potential is high. Their physical record is a good proxy for the Great Acceleration of the industrial revolution (Steffen *et al.* 2007, 2011). As one of the most extensive, durable and geologically distinctive aspects of the Anthropocene, megacities have a strong claim to become a significant part of any proposed formal definition.

The evolution of urban areas was diachronous, taking several thousand years to become the preferred habitat of *Homo sapiens*. In 1800, only 3% of the global population lived in cities, but by 2008 more than 50% of the global population was living in cities (UNFPA 2007), covering *c.* 3% of the Earth's land surface (GRUMP 2005). Of this population, *c.* 7% live in the megacities (GRUMP 2005). Cities, and specifically megacities, account for a comparatively small percentage of the global distribution of environments.

However, as seen with the definition of the Cambrian, the indicator trace fossil event is also likely to be diachronous and representative of a particular set of environmental parameters. These do not represent suitable arguments against using the development of large-scale trace fossils in the evolution of megacities as an indicator for the start of the Anthropocene. This definition is comparable to that used to define the base of the Cambrian, in which a significant biological–environmental signature was recognized, then an event was defined – the lowest occurrence of a trace fossil – to specify the base of a chronostratigraphic unit. Evidence from historical records, the introduction of novel minerals (Zalasiewicz *et al.* 2013) or artefacts (Ford *et al.* 2013) make dating of the Anthropocene interval possible to an annual or decadal accuracy, which is five or six orders of magnitude more precise than currently possible for the base of the Cambrian.

We wish to thank the two anonymous referees for their constructive reviews of this paper. C. N. Waters publishes with the permission of the Executive Director, British Geological Survey, NERC.

References

BRASIER, M., COWIE, J. W. & TAYLOR, M. 1994. Decision on the Precambrian–Cambrian boundary stratotype. *Episodes*, **17**, 95–100.

BRASIER, M. D., ANTCLIFFE, J. B. & CALLOW, R. 2011. Evolutionary trends in remarkable preservation across the Ediacaran–Cambrian transition and the impact of metazoan mixing. *In*: ALLISON, P. & BOTTGER, D. J. (eds) *Taphonomy and Bias through Time*. Springer Verlag, Dordrecht, 519–567.

BRINKHOFF, T. 2012. *The Principal Agglomerations of the World*. http://www.citypopulation.de/cities.html

DORNBOS, S. Q., BOTTJER, D. J. & CHEN, J.-Y. 2005. Paleoecology of benthic metazoans in the Early Cambrian Maotianshan Shale biota and the Middle Cambrian Burgess Shale biota: evidence for the Cambrian substrate revolution. *Palaeogeography, Palaeoclimatology, Palaeoecology*, **220**, 47–67.

EMSLEY, C., HITCHCOCK, T. & SHOEMAKER, R. 2013. London history – a population history of London. *Old Bailey Proceedings,* version 7.0. www.oldbaileyonline.org

FEDONKIN, M. A. & WAGGONER, B. M. 1997. The Late Precambrian fossil *Kimberella* is a mollusc-like bilaterian organism. *Nature*, **388**, 868–871.

FORD, J. R., PRICE, S. J., COOPER, A. H. & WATERS, C. N. An assessment of lithostratigraphy for anthropogenic deposits. *In*: WATERS, C. N., ZALASIEWICZ, J. A., WILLIAMS, M., ELLIS, M. A. & SNELLING, A. M. (eds) *A Stratigraphical Basis for the Anthropocene*. Geological Society, London, Special Publications, **395**, http://dx.doi.org/10.1144/SP395.12.

GEHLING, J., JENSEN, S., DROSER, M. L., MYROW, P. M. & NARBONNE, G. M. 2001. Burrowing below the basal Cambrian GSSP, Fortune Head, Newfoundland. *Geological Magazine*, **132**, 213–218.

GRUMP (GLOBAL RURAL URBAN MAPPING PROJECT) 2005. *The Growing Urbanization of the World*. http://www.earthinstitute.columbia.edu/news/2005/story03-07-05.html

JENSEN, S. 2003. The Proterozoic and earliest Phanerozoic trace fossil record: patterns, problems and perspectives. *Interrogative and Comparative Biology*, **43**, 219–228.

LANDING, E. 1994. Precambrian–Cambrian global stratotype ratified and a new perspective of Cambrian time. *Geology*, **22**, 179–182.

LANDING, E. & WESTROP, S. R. 2004. Environmental patterns in the origin and evolution and diversification loci of Early Cambrian skeletalized Metazoa:

evidence from the Avalon microcontinent. *In*: Lipps, J. H. & Wagoner, B. (eds) *Neoproterozoic–Cambrian Biological Revolutions*. Paleontological Society Papers, Boulder, CO, **10**, 93–105.

Landing, E., Peng, S. C., Babcock, L. E. & Moczydłowska-Vidal, M. 2007. Global standard names for the lowermost Cambrian series and stage. *Episodes*, **30**, 283–289.

Landing, E., Geyer, G., Brasier, M. D. & Bowring, S. A. 2013. Cambrian Evolutionary Radiation: context, correlations, and chronostratigraphy – overcoming deficiencies of the first appearance datum (FAD) concept. *Earth-Science Reviews*, **123**, 133–177. http://dx.doi.org/10.1016/j.earscirev.2013.03.008

MacNaughton, R. B. 2007. The application of trace fossils to biostratigraphy. *In*: Miller, W., III (ed.) *Trace Fossils: Concepts, Problem, Prospects*. Elsevier, Amsterdam, 135–148.

Marshall, C. R. 1990. Confidence intervals on stratigraphic ranges. *Paleobiology*, **16**, 1–10.

Martin, M. W., Grazhdankin, D. V., Bowring, S. A., Evans, D. A. D., Fedonkin, M. A. & Kirschvink, J. L. 2000. Age of Neoproterozoic bilaterian body and trace fossils, White Sea, Russia: implications for metazoan evolution. *Science*, **288**, 841–845.

Peng, S., Babcock, L. E. & Cooper, R. A. 2012. The Cambrian period. *In*: Gradstein, F. M., Ogg, J. G., Schmitz, M. & Ogg, G. (eds) *The Geologic Time Scale: 2012 (Volume 2)*. Elsevier BV, Amsterdam, 437–488.

Steffen, W., Crutzen, P. J. & McNeill, J. 2007. The Anthropocene: are humans now overwhelming the great forces of nature? *Ambio*, **36**, 614–621.

Steffen, W. Å. P., Deutsch, L. *et al.* 2011. The Anthropocene: from global change to planetary stewardship. *Ambio*, **40**, 739–761.

UNFPA 2007. *State of World Population 2007: Unleashing the Potential of Urban Growth*. United Nations Population Fund, New York, **108**.

Vannier, J., Calandra, I., Gaillard, C. & Żylińska, A. 2010. Priapulid worms: pioneer horizontal burrowers at the Precambrian–Cambrian boundary. *Geology*, **38**, 711–714. http://dx.doi.org/10.1130/G30829.1

Wolmar, C. 2005. *The Subterranean Railway: How the London Underground was Built and How it Changed the City Forever*. Atlantic Books, London.

Wood, R. A., Grotzinger, J. P. & Dickson, J. A. D. 2002. Proterozoic modular biomineralized metazoan from the Nama Group, Namibia. *Science*, **296**, 2383–2386. http://dx.doi.org/10.1126/science.1071599

Yuan, X., Chen, Z., Xiao, S., Zhou, C. & Hua, H. 2011. An early Ediacaran assemblage of macroscopic and morphologically differentiated eukaryotes. *Nature*, **470**, 390–393.

Zalasiewicz, J. A., Kryza, R. & Williams, M. 2013. The mineral signature of the Anthropocene in its deep-time context. *In*: Waters, C. N., Zalasiewicz, J. A., Williams, M., Ellis, M. A. & Snelling, A. M. (eds) *A Stratigraphical Basis for the Anthropocene*. Geological Society, London, Special Publications, **395**. http://dx.doi.org/10.1144/SP395.2

Zalasiewicz, J. A., Williams, M. & Waters, C. N. In review. Can an Anthropocene series be defined and recognized? *In*: Waters, C. N., Zalasiewicz, J. A., Williams, M., Ellis, M. A. & Snelling, A. M. (eds) *A Stratigraphical Basis for the Anthropocene*. Geological Society, London, Special Publications, **395**.

Zhuravlev, A., Yu., Liñan, E., Gámez Vintaned, J. A., Debrenne, F. & Fedorov, A. B. 2012. New finds of skeletal fossils in the terminal Neoproterozoic of the Siberian Platform and Spain. *Acta Palaeontologica Polonica*, **57**, 205–224.

Palaeontological evidence for defining the Anthropocene

ANTHONY D. BARNOSKY

Department of Integrative Biology and Museum of Paleontology, University of California, Berkeley, CA 94720, USA (e-mail: barnosky@berkeley.edu)

Abstract: Palaeontology formed the basis for defining most of the geological eras, periods, epochs and ages that are commonly recognized. By the same token, the Anthropocene can be defined by diverse palaeontological criteria, in accordance with commonly accepted biostratigraphic practice. The most useful Anthropocene biostratigraphic zones will be assemblage and abundance zones based on mixes of native and non-native species in both the marine and terrestrial realms, although lineage zones based on evolution of crop plants may also have utility. Also useful are human-produced trace fossils, which have resulted in prominent biohorizons that can mark the onset of the Anthropocene, especially the paved road system, widespread through terrestrial regions, and microplastics, ubiquitous in near-shore and deep-water marine sediments. Most of these palaeontological criteria support placing the Holocene–Anthropocene boundary near 1950. Continuation of current extinction rates would produce an extinction biohorizon on the scale of the Big Five mass extinctions within a few centuries, but enhanced conservation measures could prevent making mass extinction an Anthropocene signature. A grand challenge for palaeontologists now is to define Anthropocene biostratigraphic zones rigorously, not only as a necessary precursor to formalizing the epoch, but also to more fully understand how humans have restructured the biosphere.

Geological epochs are geochronological units (Murphy & Salvador 1999; NACSN 2005) – that is, they are '*divisions of time* [italics added] traditionally *distinguished on the basis of the rock record* as expressed by chronostratigraphic units' (NACSN 2005, p. 1583; NACSN as here used is the abbreviation for North American Code of Stratigraphic Nomenclature). Chronostratigraphic units are bodies of rock 'established to serve as the material reference for all constituent rocks formed during the same span of time' (NACSN 2005, p. 1581). Put more simply, epochs, as geochronological units, are abstract slivers of time – you cannot hold them in your hand. Nevertheless, epochs are based on something you can actually touch – rocks – more specifically, the rocks that were deposited during a certain span of time.

These distinctions are important in any attempt to formally recognize the Anthropocene as an epoch in the Geological Time Scale. In order for the Anthropocene to be equivalent to already-defined epochs, it is not enough to simply designate it as a period of time when human impacts were (and are) abundant. In addition, there must be a material basis – something you can touch – that already is and will remain a lasting part of the geological record, and that is distinctive enough to differentiate the chronostratigraphic unit that forms the 'material basis' for the Anthropocene from all others. That 'something you can touch' can include distinctive lithology, chemical signatures or physical indicators of time (such as palaeomagnetic signals), but for all of the epochs that have been formally recognized, the distinctive signature that originally led to their definition came from palaeontological remains – fossils.

In the strict sense of geological nomenclature, fossils are used to define biostratigraphic units, that is, lenses and layers of rock that are characterized by certain types or abundances of fossils. Each biostratratigraphic unit generally also demarcates a certain span of time. Using fossils to divide geological time is possible because the evolution of life has resulted in an irregularly paced turnover of species through Earth's history, meaning that the species that dominate the fossil record of each successive era, period, epoch or age are distinct from those of preceding and later times.

Palaeontology's role in defining recognized epochs

Fossils led Charles Lyell to define the first three geological epochs in 1833 – the Eocene, Miocene and Pliocene – the latter of which he differentiated as Older Pliocene and Newer Pliocene (Lyell 1833). The material bases for Lyell's epochs were mollusc fossils contained in successively higher strata in the geological sequence of Europe, especially Italy and France. Lyell defined the Eocene chronostratrigraphic unit (although he did not use that term as it was not in existence at the time) as those rocks containing an assemblage of mollusc fossils that included about 97% extinct species. The Miocene as Lyell originally defined it encompassed the rocks that had about 83% extinct mollusc species, the Older Pliocene had somewhere

From: Waters, C. N., Zalasiewicz, J. A., Williams, M., Ellis, M. A. & Snelling, A. M. (eds) 2014. *A Stratigraphical Basis for the Anthropocene*. Geological Society, London, Special Publications, **395**, 149–165.
First published online October 24, 2013, http://dx.doi.org/10.1144/SP395.6

between 50 and 67% extinct species, and the Newer Pliocene had only 5% extinct species. Although modern taxonomy now makes those percentages somewhat different, the general principle of fewer extinct species occurring in younger strata still applies.

Similarly, fossils figured prominently in the recognition of subsequently defined geological epochs. In 1839, Lyell (1839) changed the name of his Newer Pliocene to Pleistocene, a term that had previously been applied to geologically young alluvial sediments of the River Po, and in 1846 geologist Edward Forbes (1846) equated the onset of the Pleistocene with the onset of glacial cycles, a connotation that stuck despite the original characterization based on fossils (Walker *et al.* 2008, 2009). Heinrich Beyrich (1854) argued that certain rocks in northern Germany and Belgium contained assemblages of fossils that had fewer modern species than were found in Lyell's Miocene, but more modern species than were characteristic of Lyell's Eocene (Vandenberghe *et al.* 2012). He named the epoch, so represented, the Oligocene, which means 'few recent' in Greek, referring to the lower percentage of modern species relative to the Miocene. The Paleocene ('ancient recent') was carved out of Lyell's Eocene in 1874 by Wilhelm Schimper (1874) on the basis of certain fossil plants that occurred in some older strata of Lyell's Eocene, but not in overlying strata (Vandenberghe *et al.* 2012). The concept for the most recent epoch, the Holocene, also originated with Lyell's fossil-based subdivisions of European strata in 1833. Lyell's 'Recent' was defined as the deposits formed during the time that 'has elapsed since the earth has been tenanted by man' (Lyell 1833). Lyell noted the fossil evidence that would distinguish Recent deposits, notably 'Some *recent* [italics Lyell's] species, therefore, are found *fossil* . . . and others, like the Dodo, may be *extinct*, for it is sufficient that they should once have coexisted with man, to make them referrible [sic] to this era' (Lyell 1833, Ch. V, p. 52). In 1885, the Third International Geological Congress renamed Lyell's 'Recent' as the 'Holocene', a term meaning 'wholly recent' in reference to the percentage of extant species found as fossils (Gibbard & Kolfschoten 2004; Pillans & Gibbard 2012). Paul Gervais (Gervais 1867–1869; Gibbard & Kolfschoten 2004), in his study of fossil vertebrate animals, had coined the word Holocene to categorize deposits that corresponded to post-glacial times. Afterwards, as with the Pleistocene, the climatic association of the Holocene being post-glacial became more prominent than its palaeontological basis.

The definitions for geological epochs have been much refined in the two centuries since they were originally proposed, most recently by precisely stipulating their beginnings with Global Boundary Stratotype Sections and Points (GSSPs), or more informally, 'golden spikes'. Nevertheless, palaeontological criteria still factor into the definition or at least the recognition of most of them. The Paleocene as presently defined begins at the iridium geochemical anomaly that is interpreted to have resulted from a bolide impact about 65.5 million years ago. Important palaeontological markers include a major extinction event (dinosaurs, ammonites, several kinds of foraminifera, etc.), assemblages of echinoids that define the Danian, which is the earliest Stage of the Paleocene, and the assemblages of calcareous nannoplankton, planktonic foraminifera and dinoflagellate cysts that are used to formally subdivide the Danian and other Paleocene Stages into shorter chronostragraphic units through biostratigraphic zonation (Vandenberghe *et al.* 2012). The beginning of the Eocene is also set at a geochemical event, a pronounced negative carbon isotope excursion that indicates a rapid global warming event 55.8 million years ago. Correlative palaeontological criteria, however, include the beginning of the Ypresnian Stage, the base of which 'also defines the base of the Eocene', which is in part recognizable by associations of fossil dinocysts and calcareous nannoplankton (Vandenberghe *et al.* 2012). The beginning of the Eocene likewise is associated with a deep-sea benthic-foraminiferal extinction event, diversifications in planktonic foraminifera, calcareous nannofossils and larger foraminifera, and a mammal dispersal event in North America that defines the base of the Wasatchian North American Land Mammal Age (Vandenberghe *et al.* 2012). The key marker for the beginning of the Oligocene is the extinction of the hantkeninid planktonic foraminifera, and other palaeontological criteria have also figured in its definition, for example, a major transcontinental immigration and extinction event evident in the mammal record, the 'Grande Coupure,' which until recently was regarded as contemporaneous with the beginning of the Oligocene (the definition now places the Grande Coupure slightly above the Eocene–Oligocene boundary) (Vandenberghe *et al.* 2012). The beginning of the Miocene is recognized at the co-occurrence of the calcareous nannofossils *Sphenolithus dephix* and *S. capricornatus*, and with the first appearance of three species of foraminifera (Hilgen *et al.* 2012). The beginning of the Pliocene begins with the first appearance of the nannoplankton species *Coccolithus miopelagicus* and the extinction of *Discoaster kugleri* (Hilgen *et al.* 2012), and the beginning of the Pleistocene with the first appearance of the microfossils *Geophyrocapsa oceanica* and *Globigerinoides tenellus* and extinction of *Discoaster brouweri* (below), *Globigerinoides obliquus extremus* and *Cyclococcolithus macintyrei* (Pillans & Gibbard 2012). The onset of the Holocene is now placed at 11 700 years, formally set at the first signs of climatic warming after the Younger Dryas as reflected in the North Greenland Ice Core Project (NGRIP) core, but also correlating with the last phases of extinction of the Pleistocene megafauna (Bell *et al.* 2004; Walker *et al.* 2008, 2009; Pillans & Gibbard 2012).

Although almost all of the palaeontological changes formally specified to characterize the Cenozoic epochs come from the marine realm, each epoch also coincides with distinct palaeontological changes evident in terrestrial deposits, for example, with the beginning of various land-mammal ages defined for different continents. The beginning of the Paleocene correlates with the beginning of the Puercan North American Land Mammal Age (NALMA), the MP1 Zone of Europe and the Shanghuan Asian Land Mammal Age (ALMA); the Eocene with the Wasatchian NALMA, MP7 and Bumbanian ALMA; the Oligocene with the Orellan NALMA, MP20 and Hsandagolian ALMA; the Miocene with approximately the Arikareean 3 NALMA and MN1; the Pliocene with the Blancan NALMA, MN14 and Yushean ALMA; and the Pleistocene with MN17 (Hilgen *et al.* 2012; Pillans & Gibbard 2012; Vandenberghe *et al.* 2012).

It is probably no coincidence that each of the epoch boundaries correlate with major global climate changes as well as with palaeontological criteria. The Paleocene starts with major climatic disruptions that resulted at least in part from the K–T bolide impact, the Eocene with a major warming event corresponding to the Paleocene–Eocene Thermal Maximum (PETM) carbon-isotope excursion, and both the Oligocene and Miocene with major expansions of the Antarctic Ice Sheet (Hilgen *et al.* 2012; Vandenberghe *et al.* 2012). The Miocene–Pliocene boundary is set at the 'basal Pliocene flooding of the Mediterranean following the (Messinian) salinity crisis' (Hilgen *et al.* 2012), and the Plio–Pleistocene boundary is placed at the first obvious onset of global cooling sufficient to cause northern hemisphere glacial advances (Van Couvering *et al.* 2000; Hilgen *et al.* 2012). Like the formal definition of the Pleistocene (Walker *et al.* 2008, 2009; Pillans & Gibbard 2012), definition of the Holocene also relies on recognition of the geological evidence of global climate change. It seems likely that the major palaeontological differences that help in differentiating each epoch reflect, at least to some extent, the biotic response to these major climatic events, as species contracted or expanded their geographic ranges in concert with shifting habitats, went extinct as habitats disappeared, or kept pace with environmental changes by evolving.

Palaeontological criteria and the Anthropocene

The palaeontological distinctiveness of each existing epoch requires that a newly defined Anthropocene Epoch also be characterized by distinctive organic remains if it is to be equivalent in rank to the Paleocene through Holocene epochs. This is despite the fact that definition of the Paleocene through Holocene epochs followed a different trajectory than has the recognition of the Anthropocene. Past definitions began with recognizing distinctive features of the material rock record, primarily the fossils contained therein and their implications for defining biostratigraphic and chronostratigraphic units. From those stratigraphic entities, geochronological units (epochs) were then recognized. The development of the Anthropocene has gone in exactly the opposite direction. An arbitrary unit of time (a geochronological unit), characterized as the time of intensified human impacts, was first proposed (Crutzen 2002*a*, *b*, *c*; Steffen *et al.* 2007), and now the material 'rock' record – deposits that have accumulated in the past few centuries – is being scoured for distinctive signs that could provide an objective material basis for an epoch (Zalasiewicz *et al.* 2008, 2011*a*, *b*; Merritts *et al.* 2011; Price *et al.* 2011; Steffen *et al.* 2011*a*, *b*; Syvitski & Kettner 2011; Tyrrell 2011; Vane *et al.* 2011; Vince 2011; Syvitski 2012).

Given the importance of fossils in providing that material basis for other epochs, a key question is whether the biotic signals in deposits that accumulated in the past few decades or so can, in fact, be distinguished using palaeontologically significant criteria that parallel those used in characterizing pre-Anthropocene epochs. These criteria demand recognizing and defining biostratigraphic zones and/or prominent biohorizons, such as major extinction episodes or biotically significant events that leave a clear sedimentary signature (a so-called boundary layer). Usually, this requires morphological identification of decay-resistant parts of organisms (for instance, shell or bone for animals, or pollen, seeds and phytoliths for plants) or their trace fossils (tracks, trails, habitation structures such as burrows for animals; leaf or other imprints for plants). In some cases, for fossils less than a few thousand years old, it is also possible to use preserved DNA to obtain highly resolved taxonomic identifications (Willerslev & Cooper 2005).

Anthropocene biostratigraphic zones

Commonly recognized biostratigraphic zones include range zones, lineage zones, assemblage zones, abundance zones and interval zones

Table 1. *Biostratigraphic units useful in characterizing the Anthropocene*

Range zone. The body of strata representing the known stratigraphic and geographic range of occurrence of a particular taxon (a taxon-range zone) or combination of two taxa (a concurrent-range zone) of any rank.

Lineage zone. The body of strata containing specimens representing a specific segment of an evolutionary lineage. It may represent the entire range of a taxon within a lineage or only that part of the range of the taxon below the appearance of a descendant taxon.

Assemblage zone. The body of strata characterized by an assemblage of three or more fossil taxa that, taken together, distinguishes it in biostratigraphic character from adjacent strata.

Abundance zone. The body of strata in which the abundance of a particular taxon or specified group of taxa is significantly greater than is usual in the adjacent parts of the section.

Interval zone. The body of fossiliferous strata between two specified biohorizons. A 'biohorizon' is typically either the highest or lowest stratigraphic occurrence of a given taxon. Interval zones differ from range zones in that they are defined on the first or last occurrences of different taxa, whereas taxon-range zones are defined by the uppermost and lowermost occurrence of a single taxon, concurrent-range zones by the co-occurrence of two taxa, and assemblage zones by the co-occurrence of three or more taxa.

See the North American Stratigraphic Code (NACSN 2005) and the International Stratigraphic Guide (Murphy & Salvador, 1999) for detailed discussions and definitions of these zones.

(Murphy & Salvador 1999; NACSN 2005). Table 1 summarizes the differences between these. Conceptually, assemblage zones and abundance zones formed the basis for Lyell's original recognition of the Eocene, Miocene and Pliocene, and for the subsequent recognition of the Paleocene, Oligocene, Pleistocene and Holocene. Assemblage and/or abundance zones still characterize each epoch, although they are now much refined with respect to those Lyell and his contemporaries used. In recent decades, interval zones and range zones have taken on great importance in both recognizing the boundaries between epochs, and in subdividing epochs. Some, but not all, of these biostratigraphic zones may be useful in identifying Anthropocene strata, as detailed in the following discussion. However, applying the biostratigraphic zone concept to the Anthropocene in many cases would be limited to defining only the lower boundary of the zone, given that many biostratigraphic zones restricted to the Anthropocene would extend into the present. Nevertheless, applying the biostratigraphic zone concept has great utility, in that recognizing a biostratigraphic zone that defines the beginning of the Anthropocene also defines the end of the Holocene and thereby helps to place the Holocene–Anthropocene boundary unequivocably.

Anthropocene range zones

It is not presently possible to use range zones to define or effectively characterize the Anthropocene as it is colloquially conceived – that is, the time that humans have been so abundant and technologically advanced that they are noticeably and irrevocably altering most parts of the planet (Crutzen 2002*a*, *b*, *c*; Steffen *et al.* 2007, 2011a; Price *et al.* 2011). This is because no new species are known to have originated during the past few hundred years, so using any existing species to define a range zone would inevitably place the beginning of the range zone long before the Anthropocene (as the term is now used) is thought to have begun.

For instance, the single most pertinent taxon for defining a range zone in the Anthropocene would be *Homo sapiens*. This range zone by definition would begin with the first appearance of *H. sapiens*, currently dated at around 195 000 years ago (McDougall *et al.* 2005), and has not yet ended, as we are still extant (Fig. 1). This means that the *H. sapiens* range zone would encompass not only the Anthropocene – but also all of the Holocene and part of the Pleistocene. In this context, *H. sapiens* is at best a characteristic taxon of the Anthropocene (but also characteristic of the late Pleistocene and Holocene), rather than a defining taxon.

The co-occurrence of *H. sapiens* and another taxon could in theory be used to define a concurrent-range zone much shorter than the *H. sapiens* range zone (Fig. 2). In practice, candidate species (or subspecies) include those that humans have domesticated, such as *Equus caballus* (horse), *Bos primigenius* (cow), *Capra hurcus* (goat), *Ovis aries* (sheep), *Sus scrofa* (pig), *Canis lupus familiaris* (domestic dog), *Felis catus* (domestic cat) or crop plants such as maize (or corn as it is commonly called in the US) (*Zea mays*). However, all of these domestic species have their first known records extending back to at least eight thousand years ago, which encompasses most of the Holocene and therefore concurrent-range zones based on them would not be restricted to the Anthropocene as it is generally conceived (Mysterud *et al.* 2001;

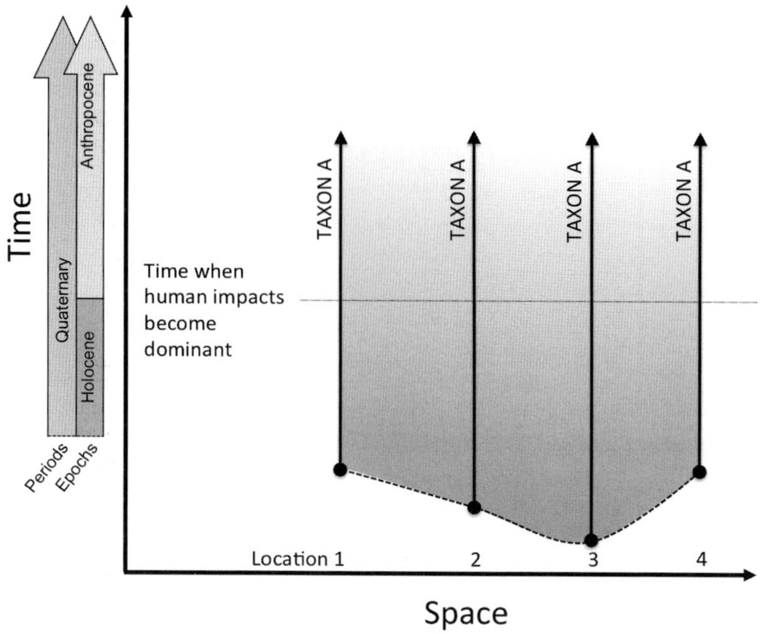

Fig. 1. A range zone based on a single taxon. Usually, these are based on evolutionary first appearances. This type of zone probably has limited utility for defining the Anthropocene, assuming the eventual boundary will be placed sometime within the past 300 years, because no new, widespread, easily fossilizable species are known to have originated within that time. Vertical lines represent ranges of taxa. Dots on the ends of lines indicate first (lowermost) or last (topmost) appearance in a region. Arrows indicate the taxon was already present in a region (downward facing arrow) or continues to persist (upward facing arrow). For most Anthropocene biostratigraphic zones it will only be possible to define a lower boundary (i.e. the Holocene–Anthropocene boundary) because the future is unknown.

Piperno & Flannery 2001; Vila *et al.* 2001; Pedrosa *et al.* 2005; Chen *et al.* 2006; Fernandez *et al.* 2006; Larson *et al.* 2007).

Anthropocene lineage zones

Lineage zones may be more amenable to characterizing the Anthropocene than range zones, but their utility in this regard remains to be proven. Lineage zones do not rely on the origin and/or extinction of species. Rather, they are underpinned by rapid, morphologically significant changes within an evolutionary lineage (typically within species), and can be defined when it is possible to trace the evolutionary sequence that leads from basal members of a species to more derived representatives. Usually the recognition of different parts of the lineage relies on morphological criteria. For example, the varying dimensions of the lower first molar in the muskrat lineage represented by *Ondatra zibethicus* and in species leading up to *O. zibethicus* allow Pleistocene and Holocene strata in which the teeth are fossilized to be subdivided into more resolved chronostratigraphic units (Martin 1993; Mihlbachler 2002).

Defining lineage zones based on crop plants may well be feasible, although this has not yet been done (Fig. 3). A leading candidate is corn, *Zea mays* subspecies *mays*, which is widespread on all continents except Antarctica, and which cannot reproduce in the absence of human cultivation. The earliest evidence of *Zea mays mays* is from archaeological deposits in Mexico that date to nearly 9000 years ago (Piperno & Flannery 2001; Tenaillon & Charcosset 2011). However, since 1847 the most commonly grown strains derive from the so-called Yellow Dent hybrid, so named because of a characteristic dent on each kernel. By 1930, new hybrids began to produce distinctive strains characterized by larger cobs (Wallace & Brown 1988; Troyer 1999, 2004; Duvick 2005). Maize kernels and cobs are commonly preserved in strata for thousands of years, and given their ubiquity now it is likely that fossil remains of today's maize will be preserved in the sedimentary record that is now accumulating. Therefore, should the morphology of kernels or cobs from strains that first appeared after 1800 be shown to be distinguishable from the older varieties, bounding Anthropocene lineage zones by the presence of certain

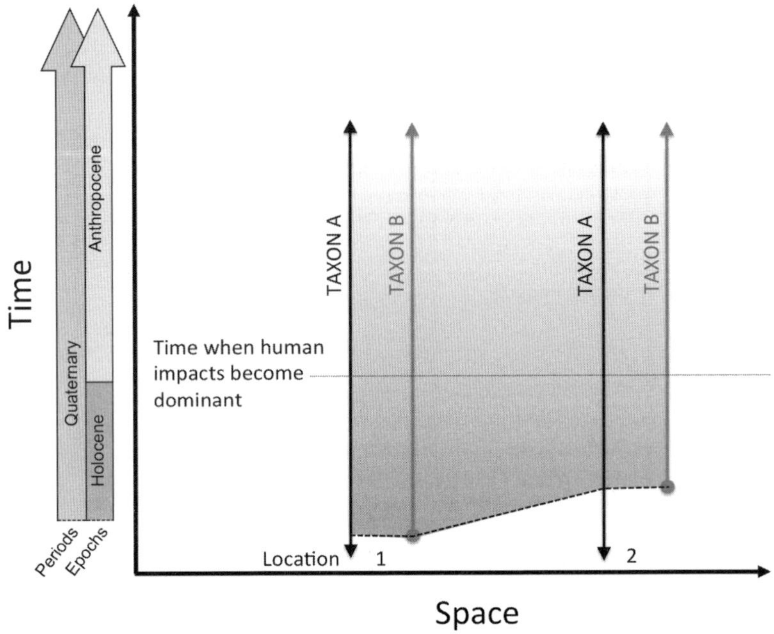

Fig. 2. Concurrent-range zone based on co-occurrence of two taxa. Like range zones, concurrent-range zones have limited utility for defining the Anthropocene because all living taxa have stratigraphic ranges that extend back at least thousands of years. See Figure 1 for further explanation.

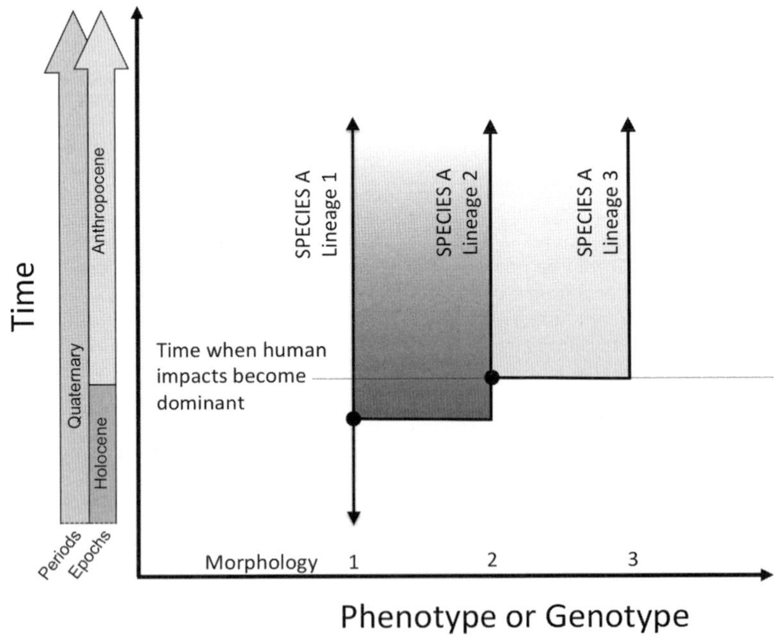

Fig. 3. Lineage zone. Such zones based on the evolution, hybridization and genetic modification of crop plants are potentially useful for characterizing and subdividing the Anthropocene. See Figure 1 for further explanation.

maize strains may well be feasible. Given that molecular biology techniques are already capable of differentiating various varieties of *Zea mays mays*, and that those techniques can be used on fossil material up to several thousands of years old, it will probably be possible for palaeontologists of the future to recognize an unusual genetic mutation that produced supersweet corn (maize) around 1950 (Erwin 1951; Tracy 1997), and the genetically modified varieties of maize that humans began producing and marketing widely in 1998. Similar principles apply to defining lineage zones for other crop plants, such as wheat, rice and soy beans. However, the long-term preservation potential for most crop plants is probably considerably less than for corn.

Anthropocene assemblage zones

Assemblage zones offer a practical way to define the beginning of and characterize the Anthropocene. Defining an assemblage zone depends on recognizing the co-occurrence of three or more fossilizable species that are unique associates with respect to the fossils of previous epochs. Assemblage zones differ in species composition in different biogeographic settings, so are region-specific. However, for the Anthropocene the defining criterion in each region relies on recognizing strata characterized by many human-introduced species (so-called alien species) with respect to underlying strata that are characterized only by taxa native to the region. Such unique associations of introduced aliens plus native species are now widespread as a result of human transport of many species outside their pre-anthropogenic range (Vitousek *et al.* 1997; Pejchar & Mooney 2009).

Most previous epoch definitions originally relied on shallow-water marine assemblages. Parallel criteria for defining the Anthropocene exist in the sea-floor deposits accumulating in proximity to major shipping ports. In these regions, the introduction of alien species has been particularly efficient, because organisms are transported across the oceans when they attach to the hulls of ships or are inadvertently taken in with ballast water, which is then released in the destination port (Bax *et al.* 2003). Such introductions of alien marine species probably started as early as the 1500s, but the first well documented cases date to the 1800s, by which time shipping traffic had increased markedly. For example, in San Francisco Bay (Thompson *et al.* 1997; Cohen & Carlton 1998) off the coast of California, US, the barnacle *Balanas improvisus* was first recorded in 1853 living on ships' hulls, but subsequently it has spread throughout the world. Atlantic oysters (*Crassostrea virginica*), the predatory snail *Urosalpinx cinerea*, the soft-shell clam *Mya arenaria*, the gem clam *Gemma gemma* and the ribbed mussel *Arcuatula demissa* were introduced in 1869. The striped bass (shipped by railroad) *Morone saxatilis* arrived in 1879, Atlantic shipworms (*Teredo navalis*) in 1913, and Japanese oysters (*Crassostrea gigas*), Manila clams (*Venerupis philippinarum*) and Asian mussels (*Musculista senhousia*) in 1932. Beginning in 1940–1950, ship traffic intensified during and following World War II, bringing with it increased rates of introductions of alien species, which were further increased by 1975 as supertanker traffic became common and shortened transoceanic travel times enhanced the survival of exotic organisms in ballast water. Shipping by air also introduced alien species through such avenues as the seaweed used to pack fresh lobsters. As a result, alien species that were introduced to San Francisco Bay between 1940 and 1980 include channelled whelks (*Busycotypus canaliculatus*), oriental prawns (*Palaemon macrodactylus*), red beard sponges (*Microciona prolifera*), European green crabs (*Carcinus maenus*), Chinese mitten crabs (*Eriocheir sinensis*), Asian clams (*Potamocorbula amurensis*), amphipods (*Corophium* sp.) and rough periwinkles (*Littorina saxatilis*). Presently, there are more than 212 exotic species in San Franciso Bay (Thompson *et al.* 1997; Cohen & Carlton 1998; Bax *et al.* 2003). Such chronologies indicate that a geologically unique 'fossil' assemblage began to characterize the floor of San Francisco Bay and environs by the mid-1800s or even earlier, and was dramatically apparent by 1950, providing clear basis for defining Anthropocene assemblage zones there (Fig. 4). The coastal deposits of other port cities, now extending along a large portion of the world's coastal areas, have experienced similar invasions of exotic species, although the species compositions differ among coastlines (Bax *et al.* 2003).

Likewise, opportunities for defining Anthropocene assemblage zones abound in the terrestrial realm, where nearly every ecosystem now incorporates several introduced plant species (Vitousek *et al.* 1997; Ellis 2011; Ellis *et al.* 2012). In the coterminous US, for example, at least 2100 species of plants that were originally found only on other continents are now common constituents of the flora over wide regions – indeed, the aliens make up more than 10% of the flora (Vitousek *et al.* 1997). These introductions began in earnest with arrival of Europeans in the late 1400s and through the 1500s, but accelerated in the 1800s. For instance, in the early 1800s, tamarisk (salt cedar, *Tamarix* spp.) was introduced to the American SE and has since become widespread in riparian areas, where fossilization potential is high for pollen, leaves, seeds and twigs. In the last 50 years, cheat grass (*Bromus tectorum*), originally introduced as cattle feed, has become dominant in

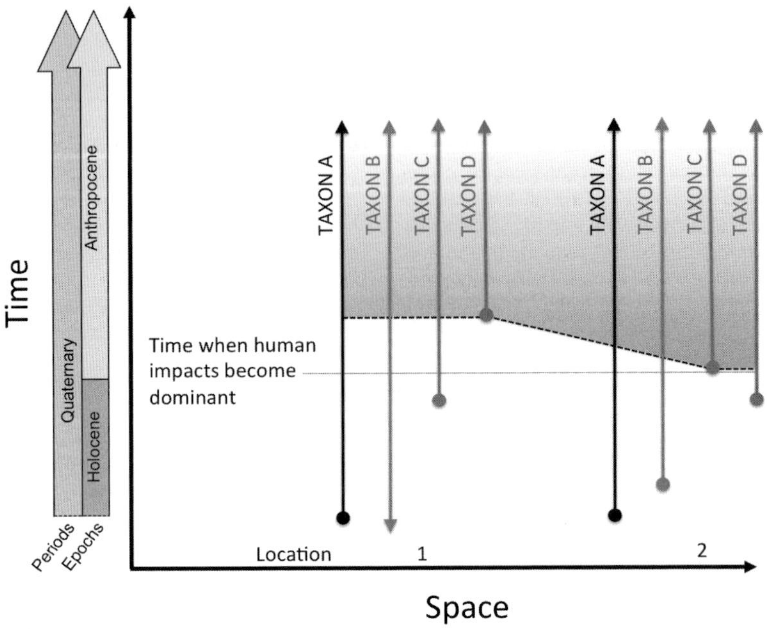

Fig. 4. Assemblage zone. These have great potential for characterizing and defining the Anthropocene, based on clearly specified combinations of alien and native species in various regions. Such zones can already be recognized in many parts of the world. See Figure 1 for further explanation.

many western grasslands and continues to spread. In South America, plants of Eurasian origin such as scotch broom (*Cytisus scoparius*), several species of wild roses (*Rosa* spp.) and thistle (*Onopordum* spp.) were introduced by the mid-1800s and have since become a major constituent of regional floras, even in remote regions. In Chile and Peru alone, there are more than 990 alien plant species, comprising about 13% and 2% of their respective floras (Vitousek *et al.* 1997). In Europe there are more than 6600 recognized introduced plant species (DAISIE 2012; DAISIE is the abbreviation for the database 'Delivering Alien Invasive Species Inventories for Europe'). In Australia, approximately 27 500 introduced plant species now outnumber the native plant species (about 24 000 species) (Thuilier 2012). The net effect of human activities that have moved plant species around the globe is homogenization of the world's flora and alteration of floral composition and abundance patterns.

Such changes are already detectable in the palynological record obtained from lake-sediment cores. The preservation of pollen for thousands and millions of years in such deposits is well documented, so much so that the discipline of palynology has become a vital part of the toolkit that geologists, palaeontologists, ecologists and climatologists use to interpret Earth history, especially for the Pleistocene and Holocene. This means that pollen assemblage zones based on a mix of introduced and native species hold promise in placing the Holocene–Anthropocene boundary in terrestrial settings throughout the world (Coombes *et al.* 2009; Armesto *et al.* 2010).

Successive waves of animal invasions have also characterized the terrestrial realm over the past few hundred years. Since about the 1600s, domestic species such as cattle (*Bos taurus*), pigs (*Sus scrofa*), goats (*Capra hurcus*), sheep (*Ovis aries*), horses (*Equus caballus*), dogs (*Canis lupus familiaris*), cats (*Felis catus*), chickens, Norway rats (*Rattus norvegicus*) and house mice (*Mus musculus*) have become common throughout the world. Since the 1800s, various deer species (such as *Cervus elephus*, *Odocoileus virginianus*, *Odocoileus hemionus* and *Dama dama*) have been introduced for sport hunting into continents where they were previously unknown and subsequently integrated into the regional ecosystem, in some places (like New Zealand) becoming dominant herbivores. All of these vertebrates have great potential to become part of the fossil record.

Anthropocene assemblage zones based on co-occurrence of native and alien species are also readily definable from animal remains in lakes and rivers. For example, zebra mussels (*Dreissena polymorpha*), an eastern European native, were introduced into the US Great Lakes area in 1988, and since then have spread through most of the eastern US waterways and as far west as California (Pejchar & Mooney 2009; USA National Atlas 2012). Similarly, introduced sport fish such as brown trout (*Salmo trutta*) and rainbow trout (*Oncorhynchus mykiss*) have proliferated in many rivers and lakes throughout the Americas, and Atlantic salmon (*Salmo salar*) have been introduced into Pacific waters off the coasts of both North and South America. Fossil remains of such species are particularly likely to be relatively abundant constituents of the future palaeontological record, as they occur in depositional basins where preservation is likely, and because shells, bones, fish scales and otoliths are often identifiable with high taxonomic resolution.

Anthropocene abundance zones

Like assemblage zones, abundance zones are readily definable in sediments that have accumulated over the past two centuries and that are accumulating now. Whereas assemblage zones simply recognize the presence or absence of a taxon, abundance zones rely on recognition of the percentage of specimens that represent a given species in the fossil record. Abundance zones are widely employed in palynological records to identify the boundary between the Pleistocene and Holocene, for example, with Pleistocene zones containing high percentages of pollen from taxa that tend to thrive under cool glacial regimes (for instance, spruce and fir trees in eastern North America), and Holocene zones containing higher percentages of taxa adapted to warmer conditions (various deciduous trees, grasses and shrubs) (Jackson *et al.* 2000; Williams *et al.* 2004). Similar zonations that relied on abundances of native v. introduced genera may well provide a clear record of the onset of the Anthropocene. Such approaches are already proving useful with diatoms preserved in lake sediments, which have been proposed as a marker for the Holocene–Anthropocene transition (Wolfe *et al.* 2012).

Defining abundance zones that specify percentages of native v. introduced taxa also has great utility in the near-shore marine sedimentary record (Fig. 5). In San Francisco Bay, alien species were found to typically comprise 40–100% of the common species, as much as 97% of the total number of individuals, and up to 99% of the biomass (Cohen & Carlton 1998). Another relevant study (Byrnes *et al.* 2007) documents that worldwide, introduced species combined with local extirpations result in skewing the trophic structure, such that the relative percentages of species in highest trophic levels (top predators and carnivores) are reduced relative to percentages of species in lower trophic levels (primary consumers such as macroplanktivores, deposit feeders and detritivores) (Fig. 6). For instance, in the Wadden Sea along the

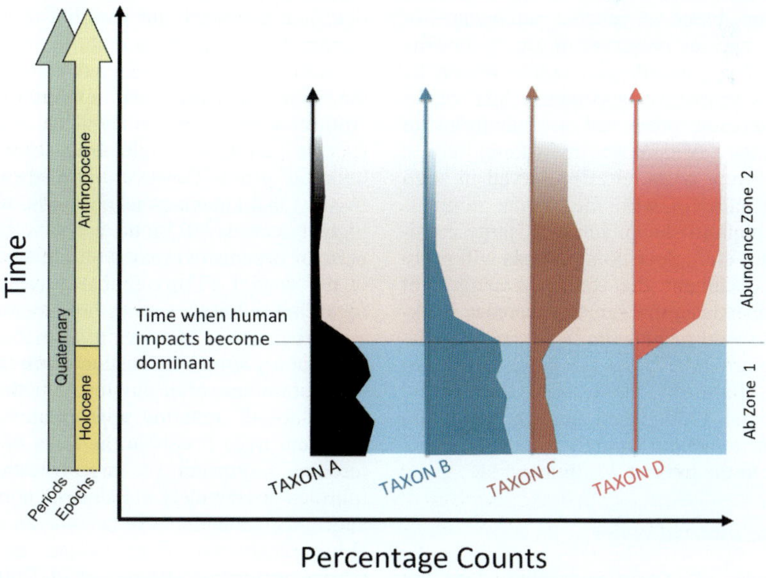

Fig. 5. Abundance zones. These are based on relative percentages of individuals within a taxon and can already be used to characterize sediments formed in the past few centuries and decades. Abundance zones should prove useful in defining and characterizing the Anthropocene. See Figure 1 for further explanation.

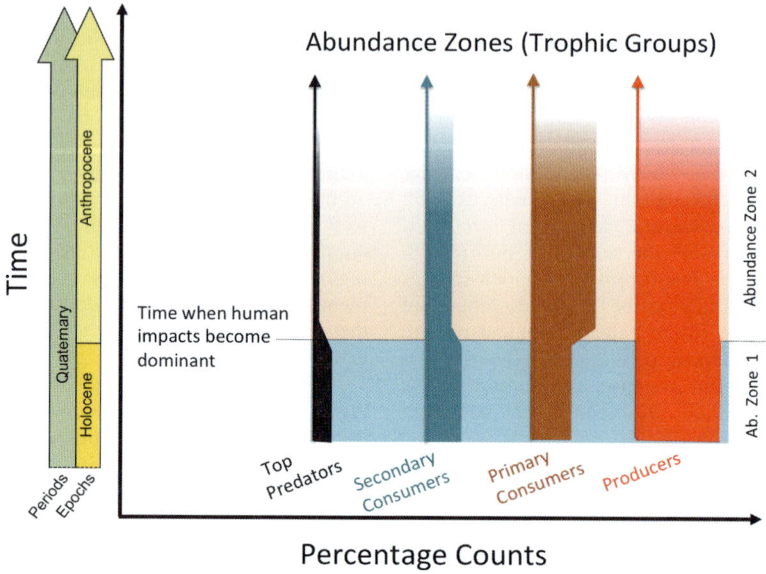

Fig. 6. Abundance zones can be based on the number of species representing major trophic groups that have been documented to have changed recognizably over the last several decades. See Figure 1 for further explanation.

NW coast of Europe, the post-invasion fauna (which could be considered the Anthropocene fauna) is characterized by a 14.0% decrease in top predator/carnivore species, and an 8.6% increase in primary consumer species (Byrnes *et al.* 2007) (Fig. 6).

It is also feasible to define Anthropocene abundance zones based on relative percentages of large-mammal species preserved in the composite fossil record of a given biogeographic region by assessing the number of specimens that represent human remains preserved in cemeteries or elsewhere, remains of domestic megafauna (cows, horses, pigs, sheep, goats, etc.) preserved in sediments, and remains of wild megafauna (animals such as deer, bighorn sheep, antelope, large carnivores, elephants, etc.). Such compilations will probably clearly document that the percentages of human and domestic animal remains increase markedly relative to wild megafauna in two phases, the first beginning around 1750 as the Industrial Revolution began (Barnosky 2008), and the second beginning around 1950, when human population growth began to climb rapidly from about 2 billion people to the present 7 billion.

Anthropocene interval zones

First appearances of certain widespread taxa are typically used to specify interval zones, with each zone defined as the stratigraphic interval between the earliest appearance of an older taxon and the earliest appearance of a younger taxon (Fig. 7). Alternatively, the bounding biohorizon for an interval zone can be an extinction event, for example the last appearance of a taxon. In theory, the interval zone concept is widely applicable to recognizing the Anthropocene–Holocene boundary and to characterizing the Anthropocene, but, in practice, it is probably not possible to use the evolutionary first-appearance biohorizon of a species as an Anthropocene boundary definer. This is due to the same difficulty that applies in recognizing Anthropocene range zones – there are no known species that have originated within the last few hundred years. However, human-produced trace fossils (also known as ichnofossils, which are geological records of biotic activity, but not actual parts of organisms) may well have practical utility in this regard. (There probably is justification for regarding cultural remains produced by humans as a particular kind of trace fossil that would warrant a taxonomy somewhat distinct from other ichnofossils, but for ease of discussion here, the general term 'trace fossil' remains appropriate.) Since 1950, abundant trace fossils in the form of microplastics (eroded fragments of large plastic items, and 'nurdles' used widely in industrial abrasives, exfoliants and cosmetics) have become an easily identifiable constituent of near-shore and deep-water marine sediments (Barnes *et al.* 2009; Fendall & Sewell 2009; Ryan *et al.* 2009; Watters *et al.* 2010; Zarfl & Matthies 2010; Andrady 2011; Cole *et al.* 2011; Zarfl *et al.* 2011). These micro-trace

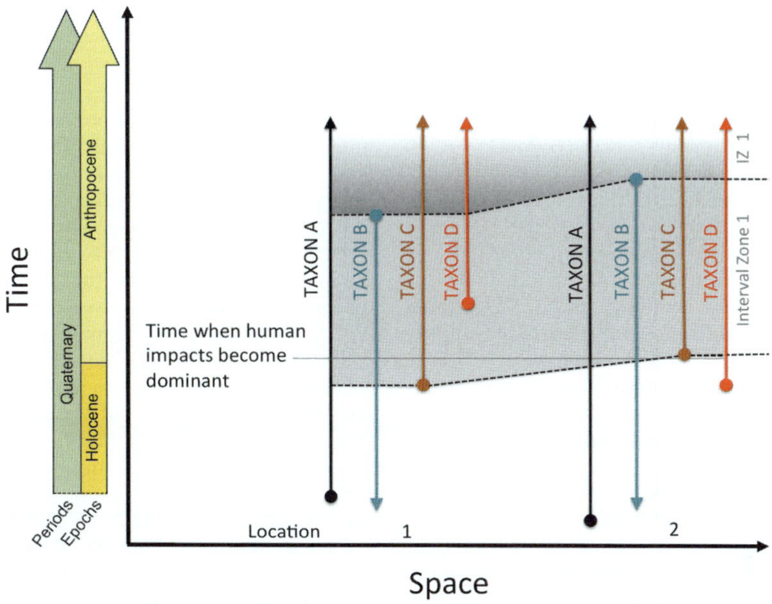

Fig. 7. Interval zones. Useful Anthropocene interval zones can already be defined based on widespread appearances of certain human trace fossils in the sedimentary record since 1950, especially microplastics and the paved road system. In the future it may be possible to define biohorizons useful for bounding interval zones based on extinction horizons, should currently too-high extinction rates not be arrested. So far, extinctions have not been extensive enough to leave a lasting fossil record, although continuation of current trends would ensure a recognizable extinction biohorizon within a few centuries, if not sooner. See Figure 1 for further explanation.

fossils occur in depositional settings that are similar to those that preserve the foraminifera and calcareous nannofossils used to define interval zones corresponding to pre-Anthropocene epochs. Still unknown, however, are whether microplastics or recognizable derivatives will last in the geological record for millions of years, although available information suggests they will persist for at least thousands of years.

Defining an Anthropocene interval-zone boundary based on extinctions may be feasible but is not without complications. More than 900 species have gone extinct since the year 1500 (ESI 2011; ESI abbreviates Endangered Species International), but most of them were not widespread or are species that would have little chance of leaving a fossil record (IUCN 2012; IUCN abbreviates International Union for the Conservation of Nature). Thus, using any single species to define a worldwide Anthropocene biohorizon that was easily recognizable in the fossil record is problematic. Nevertheless, a useful approach may be to use the historically recorded extinctions of different species in different regions to define correlated regional biohorizons that cover much of the world (Fig. 7). Such a composite biohorizon would be most evident within the twentieth century, when approximately 500 extinctions occurred. This represents a clear peak with respect to the number of extinctions that are known in previous centuries: fewer than 50 extinctions per century in the 1500s, 1600s, and 1700s, and about 125 in the 1800s (ESI 2011).

Mass extinction

The K–T mass extinction – the fifth of the so-called Big Five mass extinctions – marks the beginning of the Paleocene Epoch, the Paleogene Period, and the Cenozoic Era (Vandenberghe *et al.* 2012). Recognition that thousands of species are currently threatened with extinction, and that more than 900 have gone extinct in the past five centuries, has led to suggestions that a sixth mass extinction may also be characteristic of the Anthropocene (Leakey & Lewin 1995; Pimm *et al.* 1995; Pimm & Brooks 1997; Pimm & Raven 2000; Wake & Vredenburg 2008; Barnosky *et al.* 2011). Although there is no doubt that extinction rates are elevated, at a minimum 3–12 times above normal background rates, so far less than 1% of the IUCN-evaluated species have actually died out (Barnosky *et al.* 2011), and there is no evidence that species commonly used in biostratigraphy, such as marine calcareous plankton, have disappeared in historic times. Therefore, current extinction levels have not yet approached Big Five mass extinction levels, which are characterized by an estimated 75–96% loss of known species (Barnosky *et al.* 2011). Neither are present levels yet at the magnitude of the Late Quaternary Megafauna Extinction, which took place near the Pleistocene–Holocene transition and resulted in the loss of about 50% of the large-bodied mammals in the world, or 4% of all mammal species (Barnosky 2008). Thus, there is presently no justification for associating a mass extinction horizon with the Anthropocene.

However, avoiding introducing a mass extinction biohorizon in the foreseeable future will require enhancing and accelerating biodiversity conservation measures. If currently elevated extinction rates continue, the sixth mass extinction (75% species loss) would occur within three to five centuries (Barnosky *et al.* 2011). Even sooner than that, it is likely that without enhanced conservation effectiveness, an extinction threshold exceeding the late Quaternary Megafaunal Extinction would occur, given that currently 22% of mammals, 14% of birds, more than 30% of amphibians, and 29% of evaluated reptiles are threatened with extinction according to IUCN criteria.

Boundary layers

The bolide impact that contributed to the K–T mass extinction resulted in a so-called boundary clay (Alvarez *et al.* 1980; Schulte *et al.* 2010). This thin geological marker, which usually contains an iridium spike and other distinctive features, is recognizable in geological sections in many parts of the world, and is a key feature in defining the end of the Mesozoic and beginning of the overlying epoch, period and era (Vandenberghe *et al.* 2012). Even more widespread are the trace fossils of humans, in the form of roads, cities, open pit mines, dams and levees. The network of roads alone now extends through more than two-thirds of Earth's ice-free land surface (Globaïa 2011). The paved portion of this network, comprising a crushed rock and gravel foundation, cement and asphalt, will certainly be preserved in the geological record worldwide. Other trace fossils that will stand the test of time include buildings, the steel bars used to reinforce concrete (rebar), and the myriad metal and plastic items discarded in refuse dumps associated with every human settlement. The net effect of these trace fossils will be to produce a boundary layer that is more widespread than the iridium-rich clay used to recognize the K–T boundary.

Placement of the Holocene–Anthropocene boundary

Most discussions on the placement of a potential Holocene–Anthropocene boundary advocate beginning the Anthropocene near the year 1800 or the year 1950 (Crutzen 2002*a*, *b*, *c*; Crutzen & Steffen 2003; Crutzen & Ramanathan 2007; Steffen *et al.* 2007, 2011*a*, *b*; Zalasiewicz *et al.* 2008, 2011*a*, *b*). The former would reflect the acceleration of the Industrial Revolution, the latter the dramatic increase in human impacts worldwide that occurred with globalization and human population growth following World War II. The approach of selecting a calendar year at which to place the Holocene–Anthropocene boundary differs conceptually from dating the boundaries between most other epochs, and indeed between most divisions of geological time (notable exceptions being some divisions of the Precambrian that have numerical definitions). For pre-Anthropocene epochs of the Cenozoic, the geological and palaeontological breaks were first noted to identify where in the rock record the boundaries occurred, then geological-dating techniques were applied to estimate the age of the boundary. Those age estimates were used to define the limits of the corresponding geochronological units. Further refinement of age estimates and precise definition of boundaries came after seeking and deciding upon the most continuous, best stratigraphic section to use for the Global Boundary Stratotype and Point. Because no geological dating technique can pinpoint an exact number-of-years-before-present, the dates for all epoch boundaries are in fact approximations, each with their own error bar, rather than precise points in time. Similar principles can be applied to define the Holocene–Anthropocene boundary, and in fact are necessary if the Anthropocene is to be equivalent in status to other epochs, even though it is possible to tie the Holocene–Anthropocene boundary to a precise age because the boundary would fall within historic time.

A separate problem is designating a physical representation of the geochronological boundary that is temporally equivalent throughout the world. This point is particularly relevant in thinking about the utility of biostratigraphy in the Anthropocene compared to older boundaries. In cases where biostratigraphic zones form the basis for recognizing a geochronological boundary, the physical manifestation (where the fossils occur in the rocks) of the boundary is in theory always time-transgressive to some extent, because a new taxon always has to have a single place of origin, and then spreads out from there. Even though, by definition, biostratigraphically useful taxa are those that are common and spread rapidly, that spread can take centuries to millennia or longer – instantaneous in geological time, but long with respect to pinning a boundary to a given year. Furthermore, even though the goal of a biostratigraphic zone is in many cases to pinpoint the 'first' or 'last' occurrence of a taxon, in reality this is probably never achieved, because the fossil record is far from complete (Signor & Lipps 1982; Wang & Marshall 1994; Marshall 2010). Most taxa are characterized by very low abundance (i.e. few individuals of that taxon exist, and usually in a restricted geographic area) for a period of time after they make their first evolutionary appearance. They may then increase in abundance and geographic range until they are prolific and widespread (Vrba & DeGusta 2004). Usually prior to extinction, both abundance and geographic range decline over some period of time until eventually the last individual dies. Because individuals are so rare and geographically restricted in the earliest and latest parts of their evolutionary history, the chances of fossilization then are extremely small. Therefore, the 'first' and 'last' occurrences of biostratigraphically useful taxa in practice most likely record the first and last times a taxon occurred with enough abundance in a given area to be incorporated into the fossil record (Fig. 8). It is in this sense that 'first' and 'last' appearances are so useful in characterizing GSSPs.

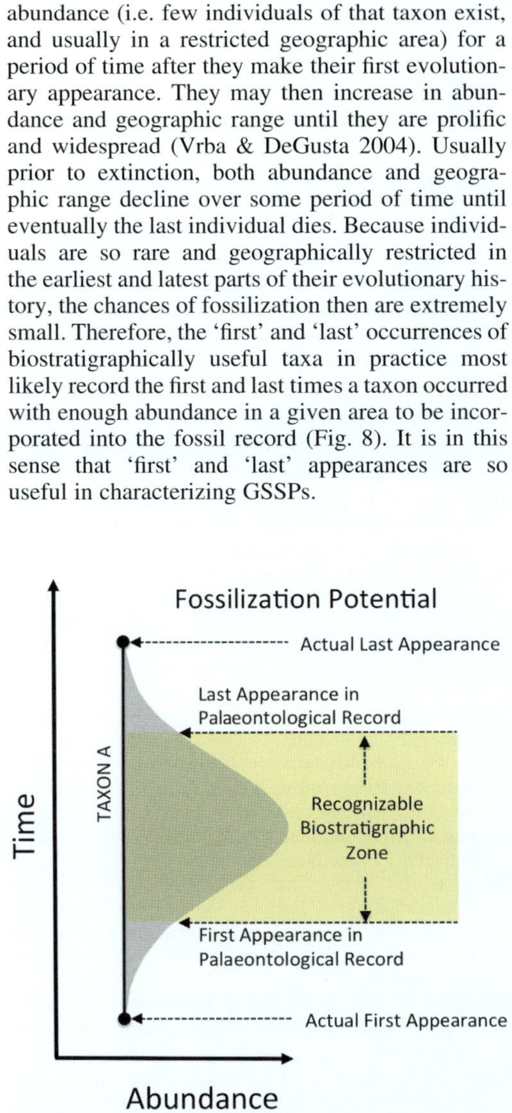

Fig. 8. Biostratigraphic zones typically define the time a taxon or taxa occur in high enough abundance to become a common part of the fossil record, rather than representing the actual first or last occurrences. This consideration is relevant to deciding where to place the boundary for the Anthropocene – the biostratigraphic signal should be obvious and widespread.

Applying these principles and considerations to defining the Holocene–Anthropocene boundary suggests the most robust palaeontologically based boundary would fall closer to 1950 than to 1800. The most useful and easily recognized biostratigraphic zones would be assemblage and abundance zones based on mixes of introduced and native species. The introduction of non-native species began centuries before 1800 in many of the world's ecosystems coincident with early ship-based exploration, but there is no definitive cluster of introductions that would distinguish the 1800s from the few preceding centuries. However, intensified shipping traffic and air travel beginning in the mid-twentieth century correspondingly accelerated the number of introduced species, causing a relatively sudden jump (over a few decades) in both the percentages of introduced v. native species, and in the abundance of individuals of introduced species. These intensified introductions started around 1940, and accelerated through the latter half of the century. Taking into account that the physical manifestation of a biostratigraphic boundary is likely to vary slightly in age from geographic region to geographic region for the reasons given above, and that the 'first' fossil records of taxa typically record the time a taxon's abundance grew to critical levels, it is likely that 1950 would closely approximate the time at which mixes of native and non-native species became widespread in the sedimentary record.

Likewise, the most viable biohorizons that can be used to define Anthropocene interval zones – based on correlation of several regional extinctions, or on widespread distribution microplastic trace fossils – also date to the twentieth century. Finally, the trace-fossil boundary layer that was emplaced with the development of the worldwide paved road system was deposited during the mid-twentieth century, and most of it after automobiles became dramatically more abundant after 1950.

For lineage zones, a case might be made for placing the boundary nearer to 1800, from recognizing a Dent-Corn (Maize) Lineage Zone, which would begin in 1847. However, other significant dates in maize evolution include 1920–1930, when strains that produced larger cobs were developed, near 1950, with mutations that produced supersweet corn, and in 1998, when genetically engineered strains were first marketed and became widespread. Other crop plants would feature more change after the Green Revolution intensified around 1960 than before that time.

Conclusions

All previously defined geological epochs either incorporate palaeontological criteria into their definition, or are characterized by distinctive aspects of the palaeontological record. Therefore, for the Anthropocene to be equivalent in rank, it too should be definable and characterizable by its palaeontological signature. The same principles of biostratigraphic zonation can be applied to the Anthropocene as have been applied for earlier epochs. The most useful Anthropocene zones will probably be assemblage and abundance zones based on mixes of native and non-native species. These may be especially useful in near-shore marine environments, in as much as marine fossils have been so critical in defining and characterizing other epochs. Other useful biohorizons that could clearly demarcate the lower boundary of Anthropocene interval zones can be based on the first occurrence of various kinds of human-produced trace fossils, especially microplastics, which have become widespread in marine sediments worldwide. The boundary layer now in place from the world's paved road system is more extensive than the boundary clay produced by the K–T asteroid strike and will form a lasting marker in the geological record. All of these biostratigraphic criteria are most compatible with placing the beginning of the Anthropocene near 1950. It is also possible to recognize lineage zones based on crop plants, which might provide weak justification for placing the Anthropocene boundary somewhat earlier.

An Anthropocene extinction biohorizon may emerge from the composite record of at least 500 extinctions that are known to have occurred in the twentieth century, but given the kinds of species involved and their geographic distributions and fossil record, it is not clear that there would be a lasting palaeontological signature of their disappearance. Current extinction rates are elevated far above background and would result in an extinction biohorizon equivalent to those that characterize the Big Five mass extinctions within a few centuries, should those rates not be slowed. Therefore, intensive biodiversity conservation efforts will be required to ensure that an obvious mass extinction signal does not eventually characterize the Anthropocene, even though such a biohorizon is not yet evident.

A remaining challenge for palaeontologists is to document the nascent palaeontological evidence that has accumulated in the sedimentary record of the past few hundred years in order to develop a biostratigraphic zonation capable of rigorously characterizing the Anthropocene. This is essential not only to define the Anthropocene in a way that is parallel to other epoch definitions, but, even more importantly, to understand the extent to which humans have restructured the biosphere since they became the dominant species on Earth.

I thank C. Waters and M. Williams for inviting me to contribute this chapter to the volume. Discussions with the following individuals were helpful in formulating ideas: E. A. Hadly, M. Holmes, R. Kirchholtes, E. Lindsey, K. C. Maguire, A. W. Poust, M. Allison Stegner, B. Swartz, J. Swift, S. Tomiya, N. A. Villavicencio, V. Walbot and G. O. U. Wogan. I gratefully acknowledge the Stanford University Department of Environmental Earth System Sciences (EESS), which provided financial support and facilities for this work during my tenure as Cox Visiting Professor in EESS.

References

Alvarez, L. W., Alvarez, W., Asaro, F. & Michel, H. V. 1980. Extraterrestrial cause for the Cretaceous–Tertiary extinction. *Science*, **208**, 1095–1108.

Andrady, A. L. 2011. Microplastics in the marine environment. *Marine Pollution Bulletin*, **62**, 1596–1605.

Armesto, J. J., Manuschevich, D., Mora, A., Smith-Ramireza, C., Rozzia, R., Abarzúaa, A. M. & Marquet, P. A. 2010. From the Holocene to the Anthropocene: a historical framework for land cover change in southwestern South America in the past 15,000 years. *Land Use Policy*, **27**, 148–160.

Barnes, D. K. A., Galgani, F., Thompson, R. C. & Barlaz, M. 2009. Accumulation and fragmentation of plastic debris in global environments. *Philosophical Transactions of the Royal Society of London Series B*, **364**, 1985–1998.

Barnosky, A. D. 2008. Megafauna biomass tradeoff as a driver of Quaternary and future extinctions. *Proceedings of the National Academy of Sciences of the USA*, **105**, 11 543–11 548.

Barnosky, A. D., Matzke, N. *et al.* 2011. Has the Earth's sixth mass extinction already arrived? *Nature*, **471**, 51–57.

Bax, N., Williamson, A., Aguero, M., Gonzalez, E., Geeves, W. & Charcosset, A. 2003. Marine invasive alien species: a threat to global biodiversity. *Marine Policy*, **27**, 313–323.

Bell, C. J., Lundelius, E. L. Jr. *et al.* 2004. The Blancan, Irvingtonian, and Rancholabrean mammal ages. *In*: Woodburne, M. O. (ed.) *Late Cretaceous and Cenozoic Mammals of North America*. Columbia University Press, New York, 232–314.

Beyrich, E. 1854. Über die Stellungder hessischen Tertiärbilldungen. Berichte der Verhandlungen der königlichen. *Preussischen Akademie der Wissenschaften, Akademie der Wissenschaften zu Berlin*, **1854**, 640–666.

Byrnes, J. E., Reynolds, P. L. & Stachowicz, J. J. 2007. Invasions and extinctions reshape coastal marine food webs. *PLoS One*, **3**, e295.

Chen, S.-Y., Duan, Z.-Y., Sha, T., Xiangyu, J., Wu, S.-F. & Zhang, Y.-P. 2006. Origin, genetic diversity, and population structure of Chinese domestic sheep. *Gene*, **376**, 216–223.

Cohen, A. N. & Carlton, J. T. 1998. Accelerating invasion rate in a highly invaded estuary. *Science*, **279**, 555–558.

Cole, M., Lindeque, P., Halsband, C. & Galloway, T. S. 2011. Microplastics as contaminants in the marine environment: a review. *Marine Pollution Bulletin*, **62**, 2588–2597.

Coombes, P. M. V., Chiverrell, R. C. & Barber, K. E. 2009. A high-resolution pollen and geochemical analysis of late Holocene human impact and vegetation history in southern Cumbria, England. *Journal of Quaternary Science*, **24**, 224–236.

Crutzen, P. J. 2002*a*. The 'anthropocene'. *Journal de Physique IV*, **12**, 1–5.

Crutzen, P. J. 2002*b*. The effects of industrial and agricultural practices on atmospheric chemistry and climate during the Anthropocene. *Journal of Environmental Science and Health A*, **37**, 423–424.

Crutzen, P. J. 2002*c*. Geology of mankind. *Nature*, **415**, 23–23.

Crutzen, P. J. & Ramanathan, V. 2007. Atmospheric chemistry and climate in the Anthropocene. *In*: Bindé, J. E. (ed.) *Making peace with the Earth: what future for the human species and the planet?* Berghahn Books, New York, 113–120.

Crutzen, P. J. & Steffen, W. 2003. How long have we been in the Anthropocene era? *Climatic Change*, **61**, 251–257.

DAISIE 2012. Delivering Alien Invasive Species Inventories for Europe. European Commission under the Sixth Framework Programme through the DAISIE project. http://www.europe-aliens.org [accessed 4 October 2013].

Duvick, D. N. 2005. The contribution of breeding to yield advances in maize (*Zea mays* L.). *Advances in Agronomy*, **86**, 83–145.

Ellis, E. C. 2011. Anthropogenic transformation of the terrestrial biosphere. *Philosophical Transactions of the Royal Society A*, **369**, 1010–1035.

Ellis, E. C., Antill, E. C. & Kreft, H. 2012. All is not loss: plant biodiversity in the Anthropocene. *PLoS One*, **7**, e30535.

Erwin, A. T. 1951. Sweet corn: mutant or historic species? *Economic Botany*, **5**, 302–306.

ESI 2011. *Extinct species charts*. Endangered Species International, Total Extinctions. http://www.endangeredspeciesinternational.org/overview5.html [accessed 12 December 2012].

Fendall, L. S. & Sewell, M. A. 2009. Contributing to marine pollution by washing your face: microplastics in facial cleansers. *Marine Pollution Bulletin*, **58**, 1225–1228.

Fernandez, H., Hughes, S., Vigne, J.-D. & Helmer, D. 2006. Divergent mtDNA lineages of goats in an Early Neolithic site, far from the initial domestication areas. *Proceedings of the National Academy of Sciences of the USA*, **103**, 15375–15379.

Forbes, E. 1846. On the connexion between the distribution of the existing fauna and flora of the British Isles, and the geological changes which have affected their area, especially during the epoch of the Northern Drift. *Memoirs of the Geological Survey of Great Britain*, **1**, 336–432.

Gervais, P. 1867–1869. *Zoologie et Paleontology Générales, Nouvelles Recherches sur les Animaux Vertétebrés et Fossiles*, Bertrand, Paris, **2**.

Gibbard, P. & Kolfschoten, T. V. 2004. The Pleistocene and Holocene epochs. *In*: Gradstein, F. M., Ogg, J. G. & Smith, A. G. (eds) *A Geologic Time Scale 2004*. Cambridge University Press, Cambridge, 441–452.

GLOBAÏA 2011. *A cartography of the Anthropocene*. globaïa.org. http://globaia.org/en/anthropocene/ [accessed 12 December 2012].

Hilgen, F. J., Lourens, L. J. *et al.* 2012. The Neogene period. *In*: Gradstein, F., Ogg, J., Schmitz, M. & Ogg, G. (eds) *The Geologic Time Scale 2012*. Elsevier, Oxford, Amsterdam, Waltham, 923–978.

IUCN 2012. *IUCN Red List, extinct species*. International Union for the Conservation of Nature. http://www.iucnredlist.org/search [accessed 12 December 2012].

Jackson, S. T., Webb, R. S., Anderson, K. H., Overpeck, J. T., Iii, T. W., Williams, J. W. & Hansen, B. C. S. 2000. Vegetation and environment in Eastern North America during the Last Glacial Maximum. *Quaternary Science Reviews*, **19**, 489–508.

Larson, G., Albarella, U. *et al.* 2007. Ancient DNA, pig domestication, and the spread of the Neolithic into Europe. *Proceedings of the National Academy of Sciences of the USA*, **104**, 15 276–15 281.

Leakey, R. C. & Lewin, R. 1995. *The Sixth Extinction: Patterns of Life and the Future of Humankind*. Anchor Books, New York.

Lyell, C. 1833. *Principles of Geology: Being an Attempt to Explain the Former Changes of the Earth's Surface, by Reference to Causes Now in Operation*, Vol. III. John Murray, London. Reproduced at http://www.esp.org/books/lyell/principles/facsimile/

Lyell, C. 1839. *Nouveaux Éléments de Géologie*. Pitois-Levrault, Paris.

Marshall, C. R. 2010. Using confidence intervals to quantify the uncertainty in the end-points of stratigraphic ranges. *In*: Alroy, J. & Hunt, G. (eds) *Quantitative Methods in Paleobiology*. The Paleontological Society Papers, Boulder, CO, **16**, 291–316.

Martin, R. A. 1993. Patterns of variation and speciation in Quaternary rodents. *In*: Martin, R. A. & Barnosky, A. D. (eds) *Morphological Change in Quaternary Mammals of North America*. Cambridge University Press, Cambridge, 226–280.

McDougall, I., Brown, F. H. & Fleagle, J. G. 2005. Stratigraphic placement and age of modern humans from Kibish, Ethiopia. *Nature*, **433**, 733–736.

Merritts, D., Walter, R. *et al.* 2011. Anthropocene streams and base-level controls from historic dams in the unglaciated mid-Atlantic region, USA. *Philosophical Transactions of the Royal Society A*, **369**, 976–1009.

Mihlbachler, M. C. 2002. Morphological chronoclines among Late Pleistocene muskrats (*Ondatra zibethicus*: Muridae, Rodentia) from Northern Florida. *Quaternary Research*, **58**, 289–295.

Murphy, M. A. & Salvador, A. E. 1999. International Stratigraphic Guide – an abridged version, International Subcommission on Stratigraphic Classification of IUGS, International Commission on Stratigraphy. *Episodes*, **22**, 255–271.

Mysterud, A., Stenseth, N. C., Yoccoz, N. G., Langvatn, R. & Steinheim, G. 2001. Nonlinear effects of large-scale climatic variability on wild and domestic herbivores. *Nature*, **410**, 1096–1099.

NACSN 2005. North American Stratigraphic Code, North American Commission on Stratigraphic Nomenclature. *AAPG Bulletin*, **89**, 1547–1591.

Pedrosa, S., Uzun, M., Arranz, J.-J., Gutiérrez-Gil, B., Primitivo, F. S. & Bayón, Y. 2005. Evidence of three maternal lineages in near eastern sheep supporting multiple domestication events. *Proceedings of the Royal Society B*, **272**, 2211–2217.

Pejchar, L. & Mooney, H. A. 2009. Invasive species, ecosystem services and human well-being. *Trends in Ecology and Evolution*, **24**, 497–504.

Pillans, B. & Gibbard, P. 2012. The Quaternary period. *In*: Gradstein, F., Ogg, J., Schmitz, M. & Ogg, G. (eds) *The Geologic Time Scale 2012*. Elsevier, Oxford, Amsterdam, Waltham, 979–1010.

Pimm, S. L. & Brooks, T. M. 1997. The sixth extinction: how large, where, and when? *In*: Raven, P. H. & Williams, T. (eds) *Nature and Human Society: The Quest for a Sustainable World*. National Academy Press, Washington, 46–62.

Pimm, S. L. & Raven, P. H. 2000. Extinction by numbers. *Nature*, **403**, 843–845.

Pimm, S. L., Russell, G. J., Gittleman, J. L. & Brooks, T. M. 1995. The future of biodiversity. *Science*, **269**, 347–350.

Piperno, D. R. & Flannery, K. V. 2001. The earliest archaeological maize (*Zea mays* L.) from highland Mexico: new accelerator mass spectrometry dates and their implications. *Proceedings of the National Academy of Sciences of the USA*, **98**, 2101–2103.

Price, S. J., Ford, J. R., Cooper, A. H. & Neal, C. 2011. Humans as major geological and geomorphological agents in the Anthropocene: the significance of artificial ground in Great Britain. *Philosophical Transactions of the Royal Society A*, **369**, 1056–1084.

Ryan, P. G., Moore, C. J., Franeker, J. A. V. & Moloney, C. L. 2009. Monitoring the abundance of plastic debris in the marine environment. *Philosophical Transactions of the Royal Society of London Series B*, **364**, 1999–2012.

Schimper, W. P. 1874. *Traité de paléontologie végétale*, Vol. **3**. J. B. Ballière, Paris.

Schulte, P., Alegret, L. *et al.* 2010. The Chicxulub asteroid impact and mass extinction at the Cretaceous–Paleogene boundary. *Science*, **327**, 1214–1218.

Signor, P. W. I. & Lipps, J. H. 1982. Sampling bias, gradual extinction patterns, and catastrophes in the fossil record. *In*: Silver, L. T. & Shultz, P. H. (eds) *Geological Implications of Impacts of Large Asteroids and Comets on the Earth.* Geological Society of America, Special Papers, Boulder, CO, **190**, 291–296.

Steffen, W., Crutzen, P. J. & McNeill, J. R. 2007. The Anthropocene: are humans now overwhelming the great forces of nature. *AMBIO: A Journal of the Human Environment*, **36**, 614–621.

Steffen, W., Grinevald, J., Crutzen, P. & McNeill, J. 2011*a*. The Anthropocene: conceptual and historical perspectives. *Philosophical Transactions of the Royal Society A*, **369**, 842–867.

Steffen, W., Persson, Å. *et al.* 2011*b*. The Anthropocene: from global change to planetary stewardship. *AMBIO: A Journal of the Human Environment*, **40**, 739–761.

Syvitski, J. P. M. 2012. Anthropocene: an epoch of our making. *Global Change (International Geosphere–Biosphere Programme)*, **78**, 12–15.

Syvitski, J. P. M. & Kettner, A. 2011. Sediment flux and the Anthropocene. *Philosophical Transactions of the Royal Society A*, **369**, 957–975.

Tenaillon, M. I. & Charcosset, A. 2011. A European perspective on maize history. *Comptes Rendus Biologies*, **334**, 221–228.

Thompson, J., Parchaso, F. *et al.* 1997. *The history and effects of exotic species in San Francisco Bay*. US Geological Survey Water Resources Division. http://sfbay.wr.usgs.gov/benthic_eco/exotic_species/index.html [accessed 12 December 2012].

Thuilier, C. 2012. Introduced plants outnumber natives. *Australian Geographic*, 14 August 2012. http://www.australiangeographic.com.au/journal/invasive-plants-outnumber-australian-natives.htm [accessed 12 December 2012].

Tracy, W. F. 1997. History, genetics, and breeding of Supersweet (shrunken2) sweet corn. *Plant Breeding Reviews*, **14**, 189–236.

Troyer, A. F. 1999. Background of U.S. hybrid corn. *Crop Science*, **39**, 601–626.

Troyer, A. F. 2004. Background of U.S. Hybrid Corn II: Breeding, Climate, and Food. *Crop Science*, **44**, 370–380.

Tyrrell, T. 2011. Anthropogenic modification of the oceans. *Philosophical Transactions of the Royal Society A*, **369**, 887–908.

USA NATIONAL ATLAS 2012. *Zebra mussels*. National Atlas of the United States. http://www.nationalatlas.gov/articles/biology/a_zm.html [accessed 12 December 2012].

Van Couvering, J. A., Castradori, D., Cita, M. B., Hilgen, F. J. & Rio, D. 2000. The base of the Zanclean Stage and of the Pliocene Series. *Episodes*, **23**, 179–187.

Vandenberghe, N., Hilgen, F. J. *et al.* 2012. The Paleogene period. *In*: Gradstein, F., Ogg, J.,, Schmitz, M. & Ogg, G. (eds) *The Geologic Time Scale 2012*. Elsevier, Oxford, Amsterdam, Waltham, 855–921.

Vane, C., Chenery, S., Harrison, I., Kim, A., Moss-Hayes, V. & Jones, D. 2011. Chemical signatures of the Anthropocene in the Clyde estuary, UK: sediment-hosted Pb $^{207/206}$Pb, total petroleum hydrocarbon, polyaromatic hydrocarbon and polychlorinated biphenyl pollution records. *Philosophical Transactions of the Royal Society A*, **369**, 1085–1111.

Vila, C., Leonard, J. A. *et al.* 2001. Widespread origins of domestic horse lineages. *Science*, **291**, 474–477.

Vince, G. 2011. An epoch debate. *Science*, **334**, 32–37.

Vitousek, P. M., D'Antonio, C. M., Loope, L. L., Rejmanek, M. & Westbrooks, R. 1997. Introduced species: a significant component of human-caused global change. *New Zealand Journal of Ecology*, **2**, 1–16.

Vrba, E. S. & DeGusta, D. 2004. Do species populations really start small? New perspectives from the Late Neogene fossil record of African mammals. *Philosophical Transactions of the Royal Society of London B*, **359**, 285–293.

Wake, D. B. & Vredenburg, V. T. 2008. Are we in the midst of the sixth mass extinction? A view from the world of amphibians. *Proceedings of the National Academy of Sciences of the USA*, **105**, 11 466–11 473.

Walker, M., Johnsen, S. *et al.* 2008. The Global Stratotype Section and Point (GSSP) for the base of the Holocene Series/Epoch (Quaternary System/ Period) in the NGRIP ice core. *Episodes*, **31**, 264–267.

Walker, M., Johnsen, S. *et al.* 2009. Formal definition and dating of the GSSP (Global Stratotype Section and Point) for the base of the Holocene using the Greenland NGRIP ice core, and selected auxiliary records. *Journal of Quaternary Science*, **24**, 3–17.

Wallace, H. A. & Brown, W. L. 1988. *Corn and Its Early Fathers*. Iowa State University Press, Ames.

Wang, S. C. & Marshall, C. R. 1994. Improved confidence intervals for estimating the position of a mass extinction boundary. *Paleobiology*, **30**, 5–18.

Watters, D. L., Yoklavich, M. M., Love, M. S. & Schroeder, D. M. 2010. Assessing marine debris in deep seafloor habitats off California. *Marine Pollution Bulletin*, **60**, 131–138.

Willerslev, E. & Cooper, A. 2005. Ancient DNA. *Proceedings of the Royal Society B*, **272**, 3–16.

Williams, J. W., Shuman, B. N., Webb, T. III, Bartlein, P. J. & Leduc, P. L. 2004. Late Quaternary vegetation dynamics in North America: scaling from taxa to biomes. *Ecological Monographs*, **74**, 309–334.

Wolfe, A. P., Hobbs, W. O. *et al.* 2012. Stratigraphic expressions of the Holocene–Anthropocene transition revealed in sediments from remote lakes. *Earth-Science Reviews*, **116**, 17–34.

Zalasiewicz, J., Williams, M. *et al.* 2008. Are we now living in the Anthropocene? *GSA Today*, **18**, 4–8.

Zalasiewicz, J., Williams, M. *et al.* 2011*a*. Stratigraphy of the Anthropocene. *Philosophical Transactions of the Royal Society A*, **369**, 1036–1055.

Zalasiewicz, J., Williams, M., Haywood, A. & Ellis, M. 2011*b*. The Anthropocene: a new epoch of geological time? *Philosophical Transactions of the Royal Society A*, **369**, 835–841.

Zarfl, C. & Matthies, M. 2010. Are marine plastic particles transport vectors for organic pollutants to the Arctic? *Marine Pollution Bulletin*, **60**, 1810–1814.

Zarfl, C., Fleet, D. & Fries, E. 2011. Microplastics in oceans. *Marine Pollution Bulletin*, **62**, 1589–1591.

Coral reefs in the Anthropocene: persistence or the end of the line?

O. HOEGH-GULDBERG

Global Change Institute (Building 20), University of Queensland, Saint Lucia, 4072 QLD, Australia (e-mail: oveh@uq.edu.au)

Abstract: Tropical coastal ecosystems such as coral reefs have provided food, income and resources to humans for millennia. The first interactions that people had with coral reef ecosystems left little signature or impact, most probably due to the restricted access, as well as the challenges and ephemeral technologies that people used to exploit these important ecosystems. As human populations expanded along tropical coastal areas, however, the influence of coastal people on coral reefs grew rapidly. Deforestation and coastal agriculture reduced coastal water quality, with many fringing coral reefs disappearing as impacts grew. Increasing numbers of fishers with increasingly advanced technologies exploited tropical coastal fisheries so that many marine species declined dramatically. These activities have removed some functional groups to the point where ecological transformations away from coral-dominated communities increasingly occurred. Shipping and marine pollution, as well as ocean warming and acidification from the burning of fossil fuels, have added further stress on tropical marine ecosystems. The latter represents a major threat with even small amounts of change potentially driving the ecological extinction of coral reefs and other marine ecosystems. Given inaction on the core drivers of these changes, the future does not look bright for coral reef ecosystems as we move into the critical phase of the Anthropocene Epoch.

Charles Darwin was fascinated by coral reefs, many spectacular examples of which he encountered in his long journey around the globe (Darwin 1842). For Darwin, coral reefs presented an evolutionary paradox given their high diversity and productivity within the extremely low-productivity tropical oceans in which they were found. These same reefs were also growing in relatively stable waters that were significantly cooler and more alkaline than those of today. Numerous times in his book on coral reefs (Darwin 1842), Darwin remarked on the complexity and repeating patterns associated with coral reefs, no doubt in agreement that coral reef ecosystems were vast and relatively robust to change. While humans lived around coral reefs during this time, the impact of their activities was limited and involved hunting and gathering along shorelines that were relatively pristine.

Fast forward to today's coral reef ecosystems and one might predict Darwin's despair. Over the past two centuries, human activities have dramatically changed coastlines, polluted coastal waters, and have warmed and acidified the ocean at global scales directly impacting coral reefs. Fishing activities have escalated to a point where many reef-based fisheries have collapsed or are facing collapse. Tropical organisms have moved hundreds of kilometres toward higher latitudes as oceans have warmed and acidified, creating novel, although probably temporary, assemblages of organisms. Coral reefs are in rapid decline, with the abundance of reef-building corals decreasing by at least 50% over the past 40–50 years (Hughes 1994; Gardner *et al.* 2003; Bruno & Selig 2007; De'ath *et al.* 2012). Comparing recent events to those of the past indicates that the current rates of change are several orders of magnitude higher today than they have been for tens of millions of years.

As a consequence of these growing threats, the future of coral reefs and other tropical coastal ecosystems hangs in doubt. In this chapter, the persistence of coral reefs in the context of the Anthropocene Epoch is explored. Three 'phases' of the Anthropocene are identified. An early phase of reef perturbation may be defined from the first impacts of humans on natural systems that became visible about 8000 years ago, soon after the beginning of the Holocene (117 ka BP) and the stabilization of sea levels after the last ice age (this phase approximates to the 'Early Anthropocene' of Ruddiman 2003 and others). This period continues through to the beginning of a 'mid' phase, which began with the Industrial Revolution and continues through to today (Crutzen 2002). As we continue beyond today, we enter a critical 'late' phase, a period dominated by accelerating climate change and by a level of unpredictability not seen for hundreds of millions of years.

From: Waters, C. N., Zalasiewicz, J. A., Williams, M., Ellis, M. A. & Snelling, A. M. (eds) 2014. *A Stratigraphical Basis for the Anthropocene*. Geological Society, London, Special Publications, **395**, 167–183.
First published online March 20, 2014, http://dx.doi.org/10.1144/SP395.17

The distribution and abundance of coral reef ecosystems

Coral reefs are persistent features of shallow-water habitats throughout the tropical regions of the planet. Despite occupying less than 0.1% of the world oceans, they provide home for 25% of all marine species (Spalding & Grenfell 1997). At the heart of these ecosystems are reef-building or hermatypic corals that live in a close symbiotic relationship with photosynthetic dinoflagellates (Genus *Symbiodinium*), which provide their host with organic carbon in return for access to inorganic nutrients such as ammonia and phosphate (Muscatine & Porter 1977). The resulting access to abundant energy enables reef-building corals to precipitate large quantities of calcium carbonate as aragonite and build complex skeletons. Over time, these skeletons build up and are glued together by organisms such as calcareous algae to create the reef framework (Fig. 1). The three-dimensional (3D) complexity of reef frameworks subsequently supports fisheries and tourism by providing habitat as well as coastal barriers, which reduce the energy of waves and then provide protection to coastlines and human infrastructure (Moberg & Folke 1999). The ecosystem services provided by coral reefs have an estimated value of US$375 billion (Costanza *et al.* 1997), a monetary value that is probably an underestimate given the difficulty of estimating the value of many ecological services such as sand production and coastal protection. Around 850 million people live within 100 km of coral reefs, deriving significant benefits from the food and income provided by these productive ecosystems (Burke *et al.* 2011).

Environmental controls on reef development

Coral reefs are found between latitudes 30°N and 30°S, and are associated with shallow (>100 m) areas where waters are generally clear, low in inorganic nutrients and particles, and relatively alkaline (pH > 8.0). The concentration of calcium and carbonate ions relative to the saturation constant of aragonite, which is the principal form of calcium carbonate within coral skeletons (Aragonite Saturation State, SS_{Arg}), is critical to the production and maintenance of carbonate reef system. The SS_{Arg} affects both the calcification and decalcification (erosion and dissolution) of coral reef frameworks. Geographically, carbonate reef systems are strongly associated with aragonite saturation states that exceed 3.3 (Kleypas *et al.* 1999; Hoegh-Guldberg *et al.* 2007). Under these conditions, coral reefs grow and calcify, producing substantial deposits of calcium carbonate that are visible from space, and which result in massive geological deposits of limestone and related materials within the Earth's crust.

Carbonate coral reefs are largely absent from turbid areas where sediment and nutrients have reduced light levels. Communities of corals, however, can exist without the build-up of significant amounts of carbonate materials (Kleypas *et al.* 1999). In these cases, biological and physical erosion exceeds the rate at which dead coral skeletons

Fig. 1. Reef-building corals play a major role in establishing the 3D nature of reef frameworks, which provide habitat to fisheries as well as coastal barriers against wave stress. In this example, Saunders Reef on the Great Barrier Reef (11°30′41″S, 144° 4′27″E), healthy reef-building corals are providing habitat for the five-line snapper, *Lutjanus quinquelineatus* and many other reef species. Photograph: author.

are produced by communities of corals, usually as a consequence of the three key parameters (i.e. temperature, light and SS_{Arg}) being suboptimal. High-latitude populations of corals (25–30°N/S) (Harriott & Banks 1995 2002) and corals growing in the low SS_{Arg} of equatorial upwelling regions (Manzello *et al.* 2008) are two prominent examples. In these cases, the much reduced 3D structure associated with these coral communities dictates reduced biological diversity and fisheries potential.

Coral reefs over time

Modern (Scleractinian) corals appeared in the fossil record around 237 Ma ago (Fig. 2) which was 40 Ma after the Permian mass extinction event (Stanley & Fautin 2001). While modern corals did not form significant carbonate reef structures for the first 14 Ma of their existence, their presence within the fossil record is characterized by periods in which carbonate reef systems were present and other intervals of time, often millions of years in length, where carbonate reef systems disappeared from the fossil record ('reef gaps'). More generally throughout the Phanerozoic, reef gaps typically occurred after the 'big five' major mass extinction events and were often associated with high levels of CO_2 which built up over thousands and, sometimes, millions of years in the Earth's atmosphere (Berner & Kothavala 2001; Berner 2003; Veron 2008). As a result of these periods, reef-building corals, as with many other organisms, underwent a significant reduction in diversity and abundance (Stanley & Fautin 2001). In addition to the five mass extinction events, there have been other periods of relatively rapid environmental change such as the Younger Dryas event (*c.* 11–10 ka BP), which probably had major impacts on tropical coastal ecosystems, although our ability to detect ecological changes during these relatively short periods is limited by the precision of current methods.

Over the past 420 ka, coral reefs have experienced small shifts in temperature, pH and carbonate ion concentrations, with changes generally happening over thousands of years (Hoegh-Guldberg *et al.* 2007). Conditions changed significantly as the world moved from glacial to interglacial phases, with typical coral reef locations experiencing an increase in temperature of 4–6 °C and a potential decrease in pH of 0.1–0.2 units over an interval of 10–20 ka (Fig. 3). Major changes would have occurred to sea levels at the same time, which varied by as much as 180 m, transforming coastlines and ecosystems across the planet. Crucially, however, many organisms living in these changing oceans would have had time (thousands of years) in which to shift their distributions or adapt to the changing environmental conditions.

Early phase of human influences (8 ka BP–AD 1760)

Modern humans first appeared in Africa around 200 ka ago and underwent a number of distinct dispersal events that began approximately 70 ka BP (McEvoy *et al.* 2011) and reached every continent by 20 ka BP, with the exception of Antarctica (Stewart & Stringer 2012). During this time, modern humans interacted with coastlines as part of habitats that they regularly visited during their hunter-gathering activities. As sea levels stabilized

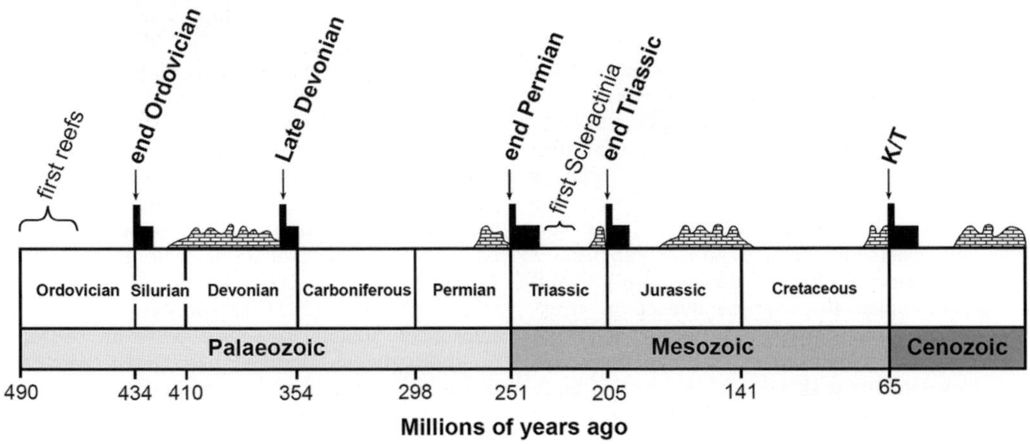

Fig. 2. Geological timeline illustrating the five mass extinction events and their relationship to peak periods of reef-building activity. 'Reef gaps' are indicated by black rectangles that follow mass extinction events. Reprinted with permission of Veron (2008).

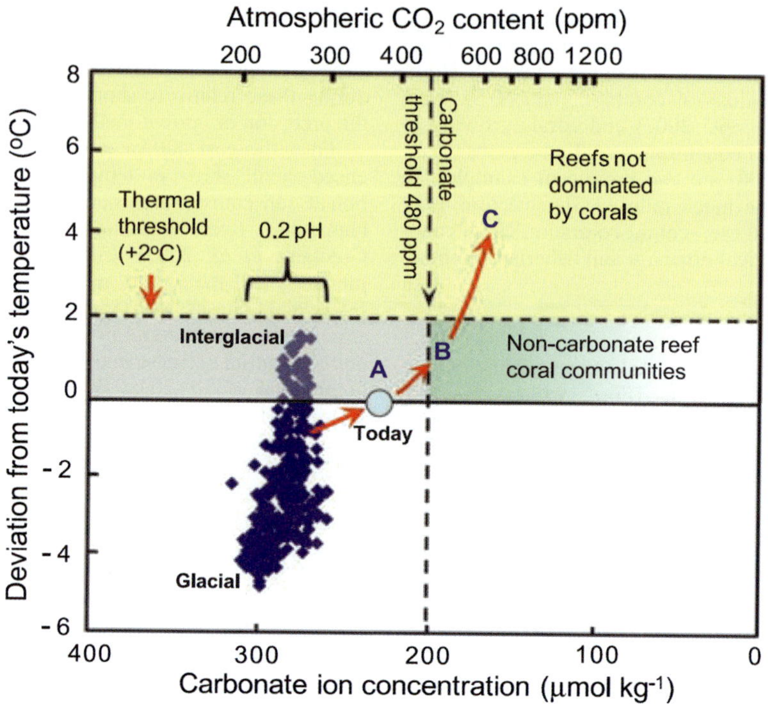

Fig. 3. Temperature, atmospheric CO_2 and carbonate ion concentration reconstructed for the past 420 ka using the Vostok Ice Core dataset (Petit *et al.* 1999). Assumptions include constant salinity (34 ppt), mean sea temperature (25 °C) and total alkalinity (2300 mmol kg^{-1}). Acidity of the global ocean varies by ± 0.1 pH units over the past 420 ka (individual values not shown), and thresholds for major changes to coral communities are indicated for thermal stress (+2 °C) and carbonate-ion concentrations ([carbonate] = 200 µmol kg^{-1}; approximate aragonite saturation *c.* $\Omega_{aragonite} = 3.3$; $[CO_2]_{atm} = 480$ ppm). Further details can be found in Hoegh-Guldberg *et al.* (2007; with permision of Science Magazine) and its accompanying supplementary online material.

after the last glacial period (*c.* 11.7 ka BP), evidence of the domestication of plants and animals began to appear at a number of separate locations worldwide, heralding the advent of the first agricultural groups and societies, and the Neolithic Revolution (Barker 2009; Bocquet-Appel 2011). The relative environmental stability of the Holocene set the scene for the rapid spread of agriculture and technology, and eventually of townships and cities (Diamond 2006). Evidence of the impacts of humans on ecosystems is minimal during this early phase owing to the small number of humans, and their relatively ephemeral artefacts and technologies.

While climate change and the proliferation of modern humans had a large influence on terrestrial organisms and ecosystems (Zalasiewicz *et al.* 2008), impacts within coastal and marine systems have developed at a slower rate during this early phase. Over much of this time, humans locally exploited marine life, albeit at a small scale relative to the enormity of ocean resources. In addition to coastal hunting and gathering, humans steadily improved boating and fishing technologies during the Holocene, leading to a small but steadily growing impact on coastal ecosystems (Pandolfi *et al.* 2003). Evidence of the impact of humans on tropical coastal areas is seen in the changing composition of middens (Álvarez *et al.* 2011), which in some cases has yielded important information on how humans impacted marine resources. The spread of humans in the tropical Pacific, for example, is associated with a shift from early diets that were dominated by molluscs to those consisting of finfish and agricultural products, presumably due to the reduced availability of the former as coastal populations and exploitation levels grew (Dalzell 1998; Jackson *et al.* 2001).

Pandolfi *et al.* (2003) explores evidence for the impact of humans on coral reefs at 14 locations worldwide, starting at 20 ka BP and ending in the modern period. In this analysis, they rate the impact of humans on seven coral reef guilds through time, estimating the frequency of reef sites within 14 locations that were pristine, abundant, depleted, rare or ecologically extinct for seven cultural periods. Not surprisingly, large herbivores (e.g. parrotfish, turtles, whales) and carnivores (e.g. sharks, large pelagic fish) experience the earliest and greatest impacts, beginning around 3.5 ka BP (Pandolfi *et al.* 2003, table 1, SOM) (Fig. 1), and accelerating at the close of this early phase as shipping and fishing technology rapidly developed (Jackson *et al.* 2001). Changes in other guilds such as small herbivores and carnivores, corals, suspension feeders and seagrass showed changes that were smaller in scale and which occurred later in time (Fig. 4). However, the pace of all of these changes were heightened by a rapidly growing human population in an increasingly connected world (Pandolfi *et al.* 2003). As a result, we enter the mid-phase of reef history, with widespread growing impacts on almost every aspect of coastal and oceanic ecosystems, and environments.

Mid-phase: the great disruption (AD 1760–present)

While humans had a significant and growing influence over the preceding 8 ka, nothing can compare with the impact of humans on ecosystems that followed the first (1760–1830) and second (1830–1870) Industrial Revolutions (Wyatt 2009). The two periods involved the rapid transition to new manufacturing processes, which included hundreds if not thousands of simultaneous breakthroughs in almost every aspect of life, including textile manufacturing, metallurgy, agriculture and many other industrial processes. The Second Industrial Revolution (also referred to as the Technological Revolution) built on these gains, adding increased momentum from the adoption of steam-powered boats, ships, railways, telecommunications and other technologies that increasingly globalized the influence of modern humans to a point never seen before. Among these changes was an increasing suite of activities that had mostly negative impacts on coastal processes, sustainable fishing, and the temperature and chemistry of the ocean.

Declining water quality from modified coastal processes

Coral reef ecosystems are extremely sensitive to small changes in temperature, salinity, pH, light, sediment and nutrient levels among a range of other factors (Bryant *et al.* 1998; Burke *et al.* 2011), all of which are influenced by coastal processes and the link between land and sea. For example, removal of coastal forests invariably increases the movement of sediment and nutrients down rivers and into coastal areas, reducing light levels, as well as increasing sedimentation and increasing eutrophication. In tropical areas, these changes have led to the decline of coastal coral reefs across the tropics, such as seen along the Queensland coastline over the last 100 years (Fig. 5). In this case, evidence from isotopic signals in long-lived corals and other environmental archives indicate that the amount of sediment moving down rivers such as the Burdekin River in Queensland increased 5–10-fold following the beginning of European settlement and land-use practices within the associated catchments. These changes preceded the loss of coral on inshore fringing coral reefs along the coast of Queensland (Wachenfeld 1997). The impact of changing coastal land use through deforestation and agriculture has driven the loss of coral and coral reefs across the tropics (Bryant *et al.* 1998; Burke *et al.* 2011), and continues as a major challenge to coral reefs in the Anthropocene. As world population continues to grow, and more and more people move to coastal areas, modification of coastal processes will only increase, although awareness of this threat may lead to the adoption of practices such as planting riparian strips and regulating the water quality flowing off agricultural areas into coastal waters.

Over-harvesting of coral reef resources

As with many marine systems (Jackson *et al.* 2001; Myers & Worm 2003; Worm *et al.* 2007), coral reef ecosystems are increasingly being harvested for food and resources (Pandolfi *et al.* 2003). As the Industrial Revolution took off, limitations in terms of powering fishing fleets, developing new fishing technologies, or processing and transporting catch were reduced dramatically. From the 1950s onward the global impact of fishing has expanded to take in most coastal and ocean resources (Fig. 6) (Watson *et al.* 2011). These changes have dramatically altered fisheries across the planet with typical industrialized fisheries reducing community biomass by 80% within 15 years of beginning exploitation (Myers & Worm 2003). The rise of super-trawlers and factory fleets has continued to intensify the pressure on the fisheries across the world oceans, often violating the territorial waters of countries and making management of fisheries increasingly difficult (Berkes *et al.* 2006; Watson *et al.* 2011).

In addition to the rise of industrialized fishing, artisanal fishing by coastal communities has also increased (Hawkins & Roberts 2004). This is particularly important along the tropical coastal areas, where an estimated 500 million often impoverished people live in coastal areas, deriving food and income from coral reef ecosystems. In addition to deriving food, artisanal fishing can also represent an important source of income, especially with

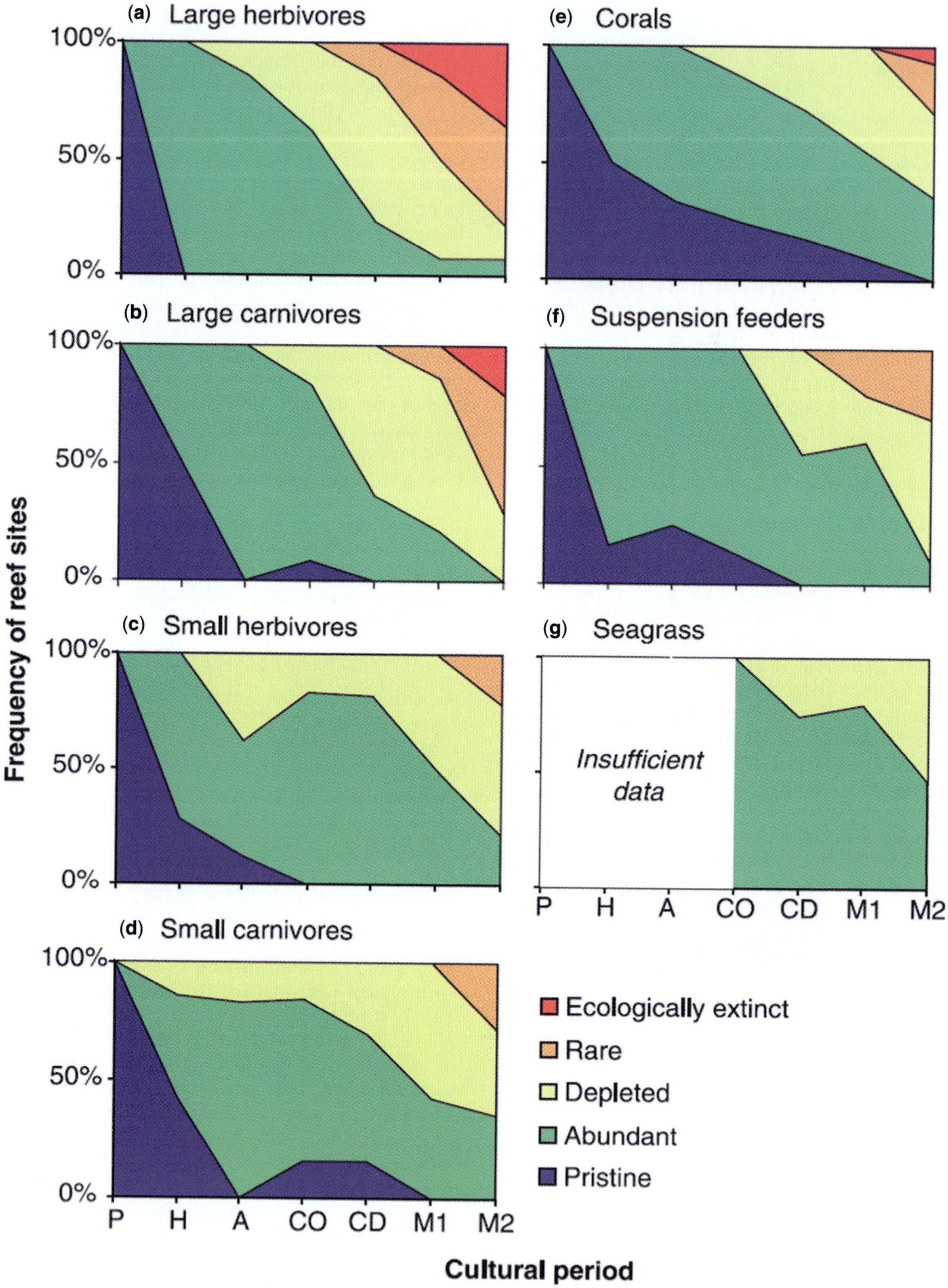

Fig. 4. Changes in the ecological structure of coral reef guilds (expressed as per cent of reefs within a region) over time. Fourteen sites were examined from the tropical Western Atlantic, Red Sea and Northern Australian regions. Cultural periods are as follows: P, Prehuman impacts; H, Hunter-gatherer; A, Agricultural; CO, Colonial Occupation; CD, Colonial Development; M1, early Modern; M2, late Modern–present. Adapted from Pandolfi *et al.* (2003, with permision of Science Magazine).

Fig. 5. Coastal reef fringing Stone Island near Bowen on the Great Barrier Reef, photographed by (**a**) Saville-Kent in 1893 and (**b**) A. Elliot in 1994. This pair of photographs documents the dramatic loss of coral communities that appears typical across most of the Queensland coastline during the last 50–100 years. Photographs compiled by D. Wachenfeld, Great Barrier Reef Marine Park Authority (Wachenfeld 1997, with permision of GBRMPA).

respect to specialized products such as sea cucumbers, shellfish and shark fin (Fig. 7) (Clarke *et al.* 2006). In the latter case, sharks have disappeared from almost every ocean, and many shark species are being exploited at rates that are unsustainable, and probably face extinction during the present century (Worm *et al.* 2013).

As human populations have increased, it is clear that we are increasingly 'fishing down the food web' (Pauly *et al.* 1998), and changing ecological structure and energy flow within the ocean. These successive removal of predators from marine ecosystems appears likely to produce 'surprises', such as the appearance of food webs dominated by jellyfish and other organisms (Pauly *et al.* 2008). These trends paint intriguing pictures of how the next phase of the Anthropocene Epoch might unfold.

A key concept in understanding the changes that are occurring during the mid-Anthropocene is the impact of fishing on ecological functional groups that are critical to maintaining ecosystem states, such as those dominated by reef-building corals (Bellwood *et al.* 2004). Removing parrotfish from coral reefs, for example, reduces the

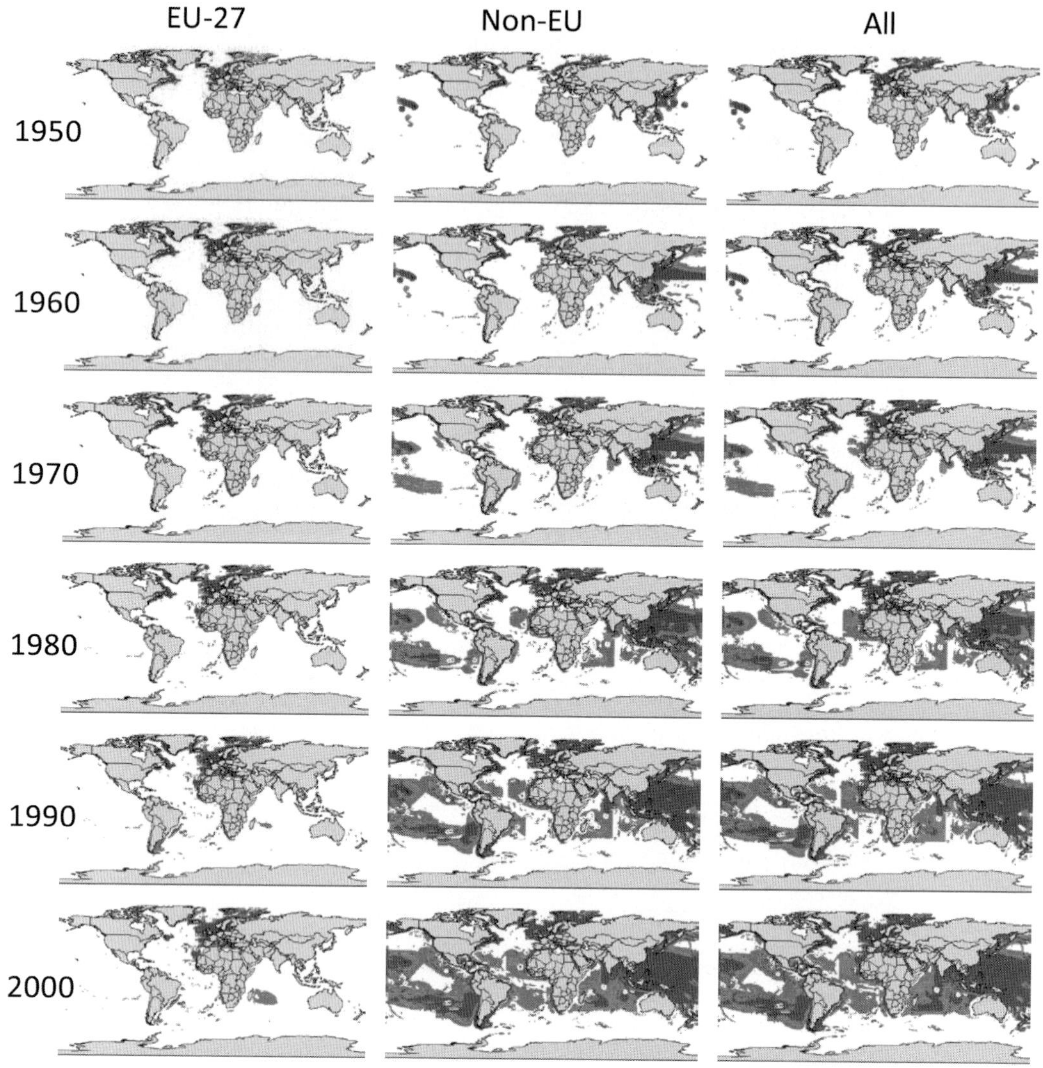

Fig. 6. Proportion of primary productivity taken through landings by fishing boats associated with the European Union (EU-27), non-European union fishing fleets, and all fishing fleets in 1950, 1960, 1970, 1980, 1990 and 2000. Intensity of grey indicates the threshold primary production required (PPR) levels (from 0 to 30%). From Watson *et al.* (2011).

important contribution that these organisms make to the control of seaweed through their herbivorous activities (Bellwood *et al.* 2004). The importance of preserving functional groups on coral reefs is highlighted by the events that occurred in the Caribbean during the 1970s and 1980s (Hughes 1994). As human populations have expanded in the Caribbean, so has fishing pressure. In the case of the Caribbean, the removal of herbivorous fish resulted in the grazing pressure on seaweeds coming down to a single species of sea urchin, *Diadema antillarum*. After hurricane impacts and the appearance of a disease that caused the mass mortality of *D. antillarum*, Caribbean reefs underwent a dramatic shift from reefs that were largely dominated by reef-building corals to those mostly covered with macroalgae (seaweeds). These events highlight the importance of maintaining functional groups on coral reefs as well as the important interactions that can occur between multiple ecological drivers or stresses.

Shipping and marine pollution

Physical and chemical marine pollution places additional stressors on coral reef ecosystems.

Fig. 7. Sharks for sale in Kota Kinabalu, Sabah, Malaysia. Photograph: author.

Land-based sources of pollution include chemicals, sewerage and marine debris, such as plastic bags and other objects that have their origin either from cities and towns or from shipping activity (Derraik 2002). In the latter case, abandoned fishing nets ('ghost nets') can continue to trap fish and other wildlife long after they have been discarded by fishing fleets (Gunn *et al.* 2010). Shipping has expanded rapidly since the Industrial Revolution, with dramatic increases in the size, number and range of shipping traffic. Over 80% of merchandise traded globally is carried by the ocean. Shipping has direct and indirect impacts on the ocean. In addition to being a significant source of chemical and physical waste, international shipping also facilitates the movement of organisms (many impacting destination ecosystems) throughout the global ocean (Keller *et al.* 2011). The International Maritime Organization (IMO) has also estimated that global shipping contributes 2.7% of global CO_2 emissions, in addition to other greenhouse gases, such as nitrous oxide (Agarwal 2009).

Ocean warming and acidification

One of the defining characteristics of the mid-phase of reef change are the extraordinary changes that began to occur to global temperature and climate as the legacy of the Industrial Revolution started to take hold. The rapid expansion of manufacturing and consumerism in combination with the rise of oil, coal and gas as a fuel source began to drive unprecedented changes to the atmospheric concentration of CO_2 and other greenhouse gases. As fossil fuels were burned in ever-increasing amounts, humans began also to modify the terrestrial landscape by removing trees, creating pasture and agricultural enterprises on an unprecedented scale. These changes have resulted in rapid and unprecedented (in 40 Ma at least) changes to the concentration of key greenhouse gases, such as CO_2, methane (CH_4) and nitrous oxide (N_2O) (Fig. 8a). As a result of these changes, average global temperature (including land, atmosphere and ocean) has increased by 0.8 °C above the preindustrial value (Fig. 8b). Most of the extra heat (>90%) that has been trapped as a result of these changes has gone into the ocean. As the ocean has got warmer, it has expanded. Expansion of the ocean together with the increased melting of ice sheets, such as those in Greenland and the Western Antarctic, have combined to increase sea levels. Warmer waters have changed the direction and strength of surface winds, as well as increasing the strength of storm systems in some oceans. Changes in wind stress along with the warming of the upper layers of the ocean have reduced water column mixing in some cases, and increased it in others (IPCC 2007*a*, *b*).

The increased flux of CO_2 into the oceans has decreased its pH by 0.1 units and the concentration of carbonate ions by around 25% (Fig. 8c). These changes in chemistry represent a fundamental departure from the chemical state of the oceans seen for at least the last 1 Ma (Figs 3 & 9) and, if our current rate of emission of CO_2 continues, ocean chemistry will be outside conditions seen over the past 40 Ma at least (Raven *et al.* 2005),

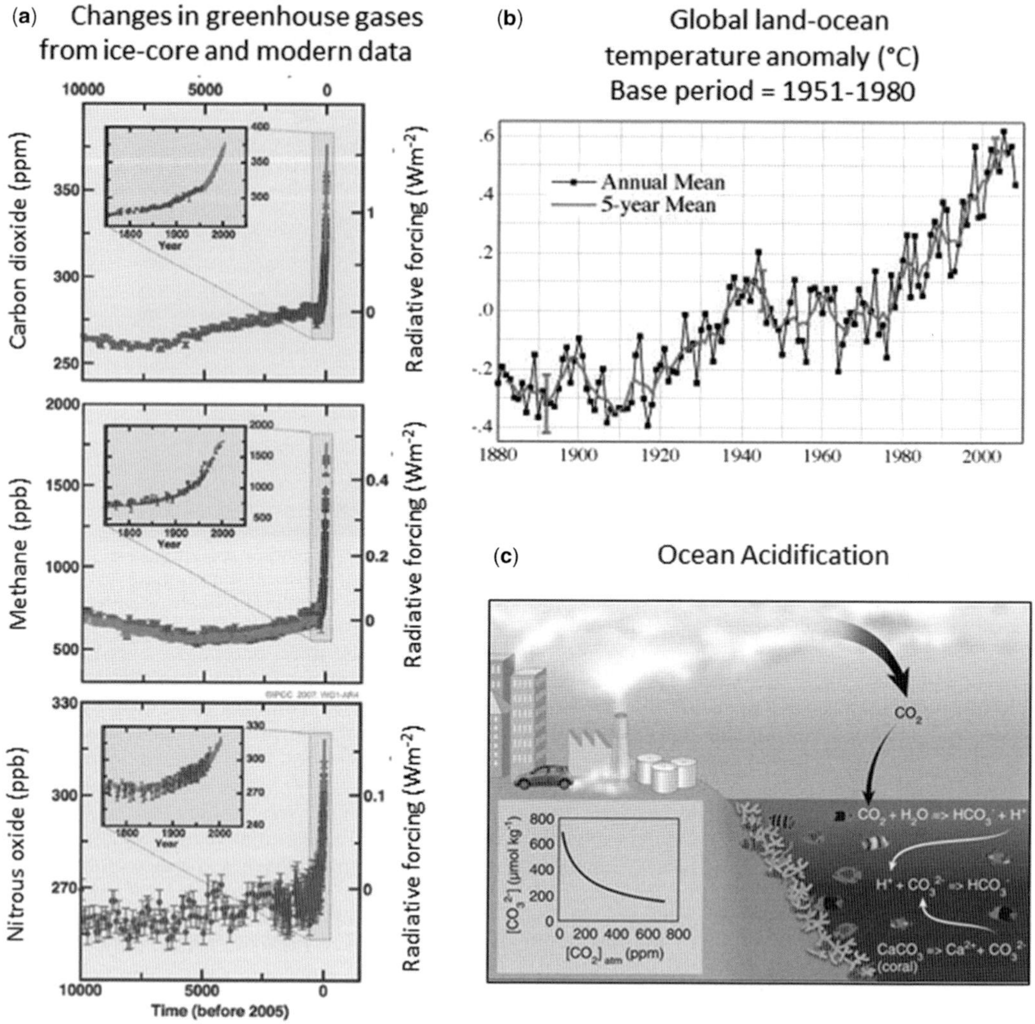

Fig. 8. (**a**) Changes in greenhouse gases from ice-core and modern data (IPCC 2007*a*, *b*). (**b**) Global land–ocean temperature anomaly (°C). Base period = 1951–1980 (GISS/NASA). (**c**) Ocean acidification and its potential impacts on coral reef ecosystems. (Adapted from Hoegh-Guldberg *et al.* 2007, with permision of Science Magazine.)

(Fig. 10). These changes will take many tens of thousands of years to reverse (Doney *et al.* 2009).

The changes to the physical and chemical nature of the oceans have resulted in extensive changes to the distribution and abundance of species and ecosystems within the ocean. As a result of these changes, marine organisms have been moving to higher latitudes consistent with the imposed warming trends. Poloczanska *et al.* (2013) used the 'vote counting' methodology (Parmesan & Yohe 2003; Root *et al.* 2003; Rosenzweig *et al.* 2008) to demonstrate a coherent response by marine organisms and ecosystems to recent climate change. Studies within this analysis were pre-examined to ensure that they were long enough and sufficiently well designed to test the notion that climate change was or was not the driver of the changes observed. Of these studies, 81% were in a direction consistent with climate change, demonstrating a statistically significant response of biological systems within the ocean to warming trends. Poloczanska *et al.* (2013) showed that changes to temperature also potentially altered the phenology or timing of key life-history events, such as plankton blooms, migratory patterns, and spawning in fish and invertebrates over recent decades. The extent of the

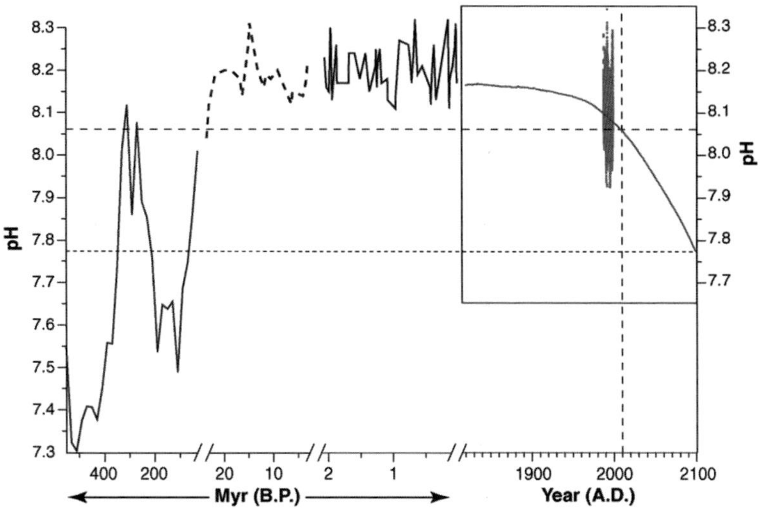

Fig. 9. Trends in pH: past, present and future. Values for today's oceans are compared to palaeo-pH values obtained using boron isotopes and other methodologies. The degree of ocean acidification obtained under business as usual (BAU) scenarios (right-hand side) will decrease below any pH value seen for at least the last 40 Ma. Modified and reprinted with permission from Pelejero *et al.* (2010).

changes in these aspects of marine life-cycles was found to differ according to the taxa or life-history stages examined. Not surprisingly, highly mobile organisms, such as fish, pelagic crustacea and zooplankton assemblages travelled much further than more sessile organisms over the same period of time.

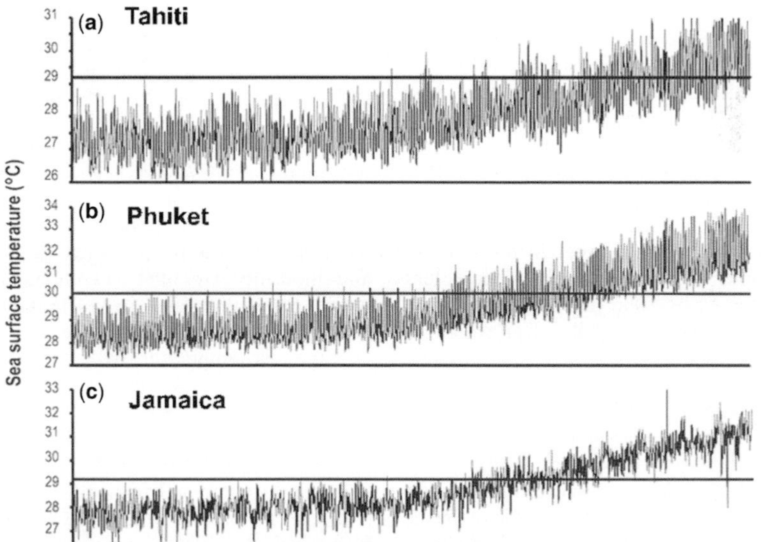

Fig. 10. Sea surface temperatures generated by a global coupled atmosphere–ocean–ice model (ECHAM4/OPYC3, details in Hoegh-Guldberg 1999) from 1860 to 2100. Horizontal lines indicate the thermal thresholds of corals at each sites: Tahiti (17.5°S, 149.5°W), Phuket (7.5°N, 98.5°E), Jamaica (17.5°N, 76.5°W). It is notable that winter temperatures in all three cases become so warm that they exceed the thermal tolerance of reef-building coral communities (with permision of Marine and Freshwater Research).

The impact of changing ocean temperature and chemistry on sessile ecosystems, such as coral reefs, has already been very significant. Reef-building corals and their symbionts are sensitive to changes in environmental conditions, such as sudden changes in temperature, light or the concentration of chemical toxins (e.g. cyanide) (Hoegh-Guldberg & Smith 1989; Jones & Hoegh-Guldberg 1999). When exposed to these conditions, the symbiosis between corals and their brown dinoflagellate symbionts tends to fall apart, creating the appearance of 'bleaching' in the coral hosts. Beginning in 1979, large communities of corals began to undergo mass coral bleaching during episodes of warmer than normal conditions. Over the past 30 years, the frequency and severity of mass coral bleaching and mortality has increased as ocean temperatures have increased. Over the past 15 years, there have been record mass coral bleaching events in almost every part of the world (Hoegh-Guldberg *et al.* 2007), including the Caribbean (Eakin *et al.* 2010), Western Indian Ocean (McClanahan *et al.* 2007) and Great Barrier Reef (Berkelmans *et al.* 2004).

The combined pressures have decreased the resilience of coral reef ecosystems to shocks that coral reefs would normally experience. Recent surveys in many parts of the world's tropics have revealed that coral reefs are in decline. The coral cover of Caribbean reefs has declined from 50–70% in the mid-1970s to 1–20% by the mid-1990s. In an analysis of coral reefs across the Western Pacific, SE Asia and the Great Barrier Reef over the past 40 years, Bruno & Selig (2007) show a 50% drop in coral cover since the early 1980s (De'ath *et al.* 2012). Similar changes in the abundance of corals on the Great Barrier Reef have also shown a 46% decrease since 1983. In the latter case, extrapolating the trend in declining coral abundance would suggest that the average coral reef in the world's largest continuous reef system might have less than 10% coral cover by 2050 (De'ath *et al.* 2012). In addition to changes in coral cover, there is growing evidence of a decrease in calcification rates. De'ath *et al.* (2009) investigated the growth rates of 328 colonies of the massive coral *Porites* from 69 reefs within the Great Barrier Reef and found that calcification rates have declined by 14.2% since 1990. The decrease in calcification rate was found universally across the Great Barrier Reef, and was unprecedented in at least the last 400 years. While it was not possible to identify whether this was a consequence of elevated temperature or of the change in the chemistry of seawater bathing the Great Barrier Reef, it was clearly not due to local factors such as declining water quality and fishing given that the decrease was found in corals both near to and far from the coast across approximately 2000 km of latitudinal range. Smaller-scale studies have also found decreases in calcification of a similar magnitude (Tanzil *et al.* 2009). Given the central role that corals play in providing the habitats of fisheries and a multitude of other organisms, as well as providing ecosystem services such as coastal protection, these changes that are currently occurring make for a troubling future.

Late phase: coral reefs as a fading signal

Understanding how the Anthropocene is likely to unfold in the near future depends on the large number of variables that will influence global change over the coming decades and centuries. Changes in technology, global population and climate change play highly influential roles in determining whether future worlds will be similar to today or as radically different as those depicted in the films *Mad Max* (directed by George Miller, 1979) and *Blade Runner* (directed by Ridley Scott, 1982). Our inability to predict near-term events, such as the Facebook revolution or the rise of Twitter (Porter *et al.* 2011), suggests that there is the need for considerable caution in making statements about how our planet and defining Epoch (the Anthropocene) will continue to develop and unfold. Natural ecosystems, however, allow some insight into the scale of change that is likely to unfold over the coming century. There is an urgent need to understand this latest phase of the Anthropocene, and to gain some understanding of how decisions made today are likely to play out in the long-term for both the viability of natural ecosystems and *Homo sapiens*.

The sensitivity of coral reefs to elevated temperature provides an opportunity to understand how the frequency of impacts such as coral bleaching and mortality are likely to change in a warmer world. Small increases in sea temperature of 1–2 °C for several weeks above the long-term summer average sea temperature will cause coral communities to undergo significant and noticeable mass coral bleaching (Hoegh-Guldberg 1999). This relationship has underpinned the ability of remote sensing technologies, such those used by the National Oceanic and Atmospheric Administration (NOAA) to predict when and where mass coral bleaching will occur, based on the temperature anomaly alone (Strong *et al.* 2004, 2011; Eakin *et al.* 2010). Using projections of sea temperature from general circulation models for a doubling of CO_2 reveals that even small changes in sea temperature will rapidly exceed the known tolerance of corals to bleaching and mass mortality on an annual basis by 2030–2050 (Hoegh-Guldberg

1999). Under these scenarios, there is not enough time for corals to recover between disturbances (they need at least 10–20 years), which has the catastrophic outcome that coral-dominated habitats more or less disappear by the middle of the current century. Several other studies have used increasingly sophisticated models to explore these rather challenging conclusions, and have either confirmed the conclusions of Hoegh-Guldberg (1999) or have come to even more pessimistic conclusions about the fate of coral reefs in a rapidly warming and acidifying world (Donner *et al.* 2005; Frieler *et al.* 2013).

The progressive changes to tropical marine ecosystems, such as coral reefs, provide a stark reminder of the linkage between climate change, natural ecosystems and the well-being of people. As we move from today, where coral-dominated ecosystems still exist in the world and the multiple benefits of coral reefs still flow to people, we are already seeing the progressive replacement of coral-dominated reefs with reefs that have no coral cover (Fig. 11). Reefs such as those found in the Western Pacific, for example, will increasingly resemble Caribbean reefs, where coral-dominated reefs have been replaced in many regions by reefs that have a mixed assemblage of non-coral organisms. As corals become less abundant and their ability to calcify decreases, reef frameworks will become increasingly brittle and prone to damage during storms (which are likely to increase in severity; Emanuel 2005). This in turn will lead to a decrease in biodiversity and habitat for key fisheries species worldwide (Hoegh-Guldberg *et al.* 2007; Bell *et al.* 2011*a*, *b*, 2013). As well, and perhaps less understood, there would be an increasing risk to coastal communities and infrastructure from carbonate reef barriers, which will eventually be unable to protect coastlines from wave and storm impacts. Our understanding, however, of how long a non-calcifying reef framework in a relatively more acidic ocean will survive is an important, yet unanswered, question.

Under the current rate of emission of greenhouse gases, such as CO_2, to the atmosphere (>2 ppm a^{-1}), we will exceed concentrations of greenhouse gas equivalents of 800 ppm (Fig. 12a). At this point, surface temperatures (land and ocean) will be 4–6 °C warmer than those of the preindustrial period (Fig. 12b, c). These changes exceed, and will continue to do so, the ability of the long-lived organisms, such as corals (which have generation times from 5 to over 100 years in length), to respond evolutionarily to the rapid pace of change (Hoegh-Guldberg 2012) with the expectation that coral reefs will be ecologically extinct in the very near future. Given the huge inertia of ocean temperature and chemistry, we will have entered at this point another 'reef gap' that is likely to last for millions of years (Veron 2008). As we continue on this emissions pathway, we will see average global temperatures eventually reaching 6–12 °C above the Preindustrial Period.

Fig. 11. Extant examples of reefs from the Great Barrier Reef used here as analogues for the ecological conditions anticipated for different combinations atmospheric concentrations of CO_2 and temperature. (**a**) Reef slope communities at Heron Island are typical of current conditions; (**b**) Mixed algal and coral communities associated with inshore reefs around St Bees Island near Mackay resemble algal-dominated reefs resulting from the reduced growth rates and competitive ability reef-building corals. (**c**) The inshore reef slope around the Low Isles near Port Douglas showing reefs that no longer have coral or seaweeds associated with them, and are in a condition of net erosion and dissolution. (Adapted from Hoegh-Guldberg *et al.* 2007, with permision of Science Magazine.)

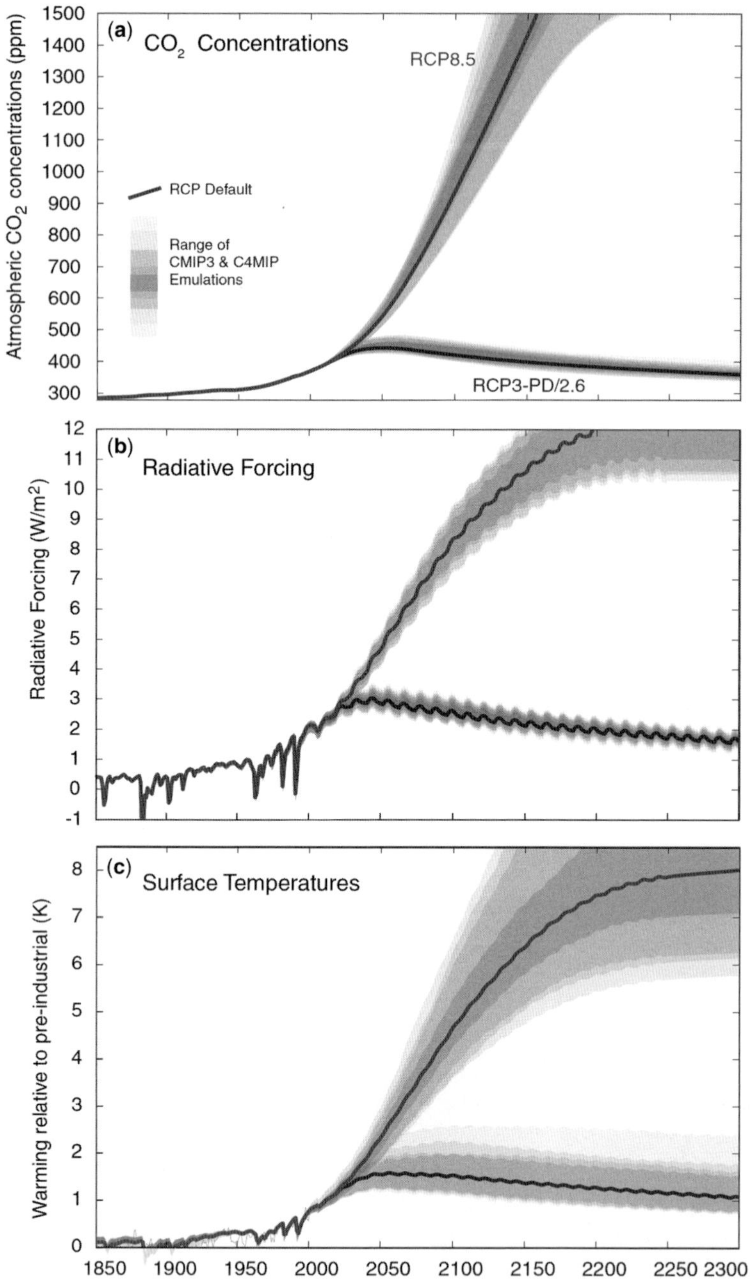

Fig. 12. Atmospheric conditions from 1852 to 2300. Solid lines represent ensemble means for: (**a**) concentrations of CO_2; (**b**) total radiative forcing (natural and anthropogenic); and (**c**) global mean surface temperatures for RCP2.6 (also known as RCP3-PD) and RCP8.5 compared across the full range of CMIP3 GCM and C4MIP carbon-cycle model emulations (shaded areas). Reprinted with permission from Meinshausen *et al.* (2011).

It is unlikely that many organisms and ecosystems that we have on our planet today will be able to survive these conditions. Given the current impacts of a world that is only 0.8 °C warmer, it is very likely that we will experience, at this point, the sixth mass extinction on Earth.

Given unmanageable risks associated with the current Business-As-Usual (BAU) emissions trajectory, it is extremely unlikely that we will (through choice or ability) go down this emissions pathway. Global society, as a consequence, is likely to move towards non-CO_2-emitting technologies and practices at an increasing pace, which will be driven by the ecological and economic impacts that are and will develop over the next few years. Under this scenario (RCP3/2.6; Fig. 12) involving a rapid, non-BAU shift to zero emissions over the next 20 years, average global temperature is still likely to increase by another 1 °C but would begin to trend downwards by the middle–late part of the current century (Fig. 12c). At this point, the flux of CO_2 into the ocean would diminish rapidly, stabilizing and possibly reversing (in combination with mixing CO_2 into deeper layers) ocean acidification. Impacts on marine ecosystems, such as coral reefs, would still be very significant over this century. However, reef-building corals are likely to persist on some reefs, albeit at lower levels of abundance. Yet, as ocean temperature and chemistry begin to stabilize and then reverse, these remaining coral communities are likely to rebound over the subsequent decades and centuries and reform the magnificent coral reefs of the past.

Conclusion

The Anthropocene represents a fascinating yet challenging period on planet Earth, especially as we consider how humans are changing environments and ecosystems as vast as those in the ocean. Escalating coastal populations, accelerating technology and climate change are driving species and ecosystems towards ecological collapse and extinction. Coral reef ecosystems provide a graphic example of how human-driven global change can directly impact our well-being through its pernicious influence on the world around us. Perhaps one of the greatest ironies is that *Homo sapiens* have only been around for 200 ka yet are set to eliminate an ecosystem, upon which they depend, for potentially millions of years. Being optimistic, this story and irony should surely galvanize global society in preventing the Anthropocene ending in such a disastrous way. Perhaps we should be mindful of the words of J. Robert Oppenheimer, father of the atomic bomb, as we stand on the brink of this latest phase of the Anthropocene. In reference to the new-found destructive power of humans through the atomic bomb, he was compelled to cite the Hindu God, Vishnu, from the *Bhagavad Gita*: 'Now I am become Death, the destroyer of worlds' (Rhodes 1977).

The author would like to acknowledge the Queensland Smart State Premier's Fellowship and the ARC Centre for Excellence in Coral Reef Studies for support during the development and writing of this chapter.

References

Agarwal, S. K. 2009. Mitigating Greenhouse Gas (GHG) emissions from International Shipping in Post-Kyoto Climate Policy: legal issues and challenges. *Maritime Affairs: Journal of the National Maritime Foundation of India*, **5**, 73–96.

Álvarez, M., Briz Godino, I., Balbo, A. & Madella, M. 2011. Shell middens as archives of past environments, human dispersal and specialized resource management. *Quaternary International*, **239**, 1–7.

Barker, G. 2009. *The Agricultural Revolution in Prehistory: Why did Foragers Become Farmers?* Oxford University Press, New York.

Bell, J. D., Johnson, J. E. & Hobday, A. J. 2011*a*. Implications of climate change for contributions by fisheries and aquaculture to Pacific Island economies and communities. *In*: Bell, J. D., Johnson, J. E. & Hobday, A. J. (eds) *Vulnerability of Tropical Pacific Fisheries and Aquaculture to Climate Change*. Secretariat of the Pacific Community, Noumea, New Caledonia, 733–801.

Bell, J. D., Johnson, J. E. & Hobday, A. J. (eds). 2011*b*. *Vulnerability of Tropical Pacific Fisheries and Aquaculture to Climate Change*. Secretariat of the Pacific Community, Noumea, New Caledonia.

Bell, J. D., Ganachaud, A. *et al.* 2013. Mixed responses of tropical Pacific fisheries and aquaculture to climate change. *Nature Climate Change*, **3**, 591–599, http://dx.doi.org/10.1038/nclimate1838

Bellwood, D. R., Hughes, T. P., Folke, C. & Nyström, M. 2004. Confronting the coral reef crisis. *Nature*, **429**, 827–833.

Berkelmans, R., De'ath, G., Kininmonth, S. & Skirving, W. J. 2004. A comparison of the 1998 and 2002 coral bleaching events on the Great Barrier Reef: spatial correlation, patterns, and predictions. *Coral Reefs*, **23**(1), 74–83.

Berkes, F., Hughes, T. *et al.* 2006. Globalization, roving bandits, and marine resources. *Science*, **311**, 1557–1558.

Berner, R. A. 2003. The long-term carbon cycle, fossil fuels and atmospheric composition. *Nature*, **426**, 323–326.

Berner, R. A. & Kothavala, Z. 2001. GEOCARB III: Aa revised model of atmospheric CO_2 over Phanerozoic time. *American Journal of Science*, **301**(2), 182–204.

Bocquet-Appel, J.-P. 2011. When the World's Population Took Off: Tthe Springboard of the Neolithic Demographic Transition. *Science*, **333**, 560–561.

Bruno, J. F. & Selig, E. R. 2007. Regional decline of coral cover in the Indo-Pacific: timing, extent, and subregional comparisons. *PLoS One*, **2**(8), e711.

Bryant, D., Burke, L., McManus, J. & Spalding, M. 1998. *Reefs at Risk: A Map-Based Indicator of Threats to the World's Coral Reefs*. World Resources Institute, Washington, DC.

Burke, L., Reytar, K., Spalding, M. & Perry, A. 2011. *Reefs at Risk Revisited.* World Resources Institute, Washington, DC.

Clarke, S. C., Magnussen, J. E., Abercrombie, D. L., McAllister, M. K. & Shivji, M. S. 2006. Identification of shark species composition and proportion in the Hong Kong shark fin market based on molecular genetics and trade records. *Conservation Biology*, **20**(1), 201–211.

Costanza, R., d'Arge, R. *et al.* 1997. The value of the world's ecosystem services and natural capital. *Nature*, **387**, 253–260.

Crutzen, P. J. 2002. Geology of mankind. *Nature*, **415**, 23–23.

Dalzell, P. 1998. The role of archaeological and cultural-historical records in long-range coastal fisheries resources management strategies and policies in the Pacific Islands. *Ocean & Coastal Management*, **40**(2), 237–252.

Darwin, C. R. 1842. *The Structure and Distribution of Coral Reefs.* Smith Elder, London.

De'ath, G., Lough, J. M. & Fabricius, K. E. 2009. Declining coral calcification on the Great Barrier Reef. *Science*, **323**, 116–119.

De'ath, G., Fabricius, K. E., Sweatman, H. & Puotinen, M. 2012. The 27-year decline of coral cover on the Great Barrier Reef and its causes. *Proceedings of the National Academy of Sciences of the United States of America*, **109**(44), 17,995–17,999.

Derraik, J. G. 2002. The pollution of the marine environment by plastic debris: a review. *Marine Pollution Bulletin*, **44**(9), 842–852.

Diamond, J. M. 2006. *Collapse: How Societies Choose to Fail or Succeed.* Penguin, New York.

Doney, S., Fabry, V., Feely, R. & Kleypas, J. 2009. Ocean acidification: the other CO_2 problem. *Annual Review of Marine Science*, **1**, 169–192.

Donner, S. D., Skirving, W. J., Little, C. M., Oppenheimer, M. & Hoegh-Guldberg, O. 2005. Global assessment of coral bleaching and required rates of adaptation under climate change. *Global Change Biology*, **11**(12), 2251–2265.

Eakin, C. M., Morgan, J. A. *et al.* 2010. Caribbean corals in crisis: record thermal stress, bleaching, and mortality in 2005. *PLoS ONE*, **5**(11), e13969.

Emanuel, K. 2005. Increasing destructiveness of tropical cyclones over the past 30 years. *Nature*, **436**, 686–688.

Frieler, K., Meinshausen, M., Golly, A., Mengel, M., Lebek, K., Donner, S. D. & Hoegh-Guldberg, O. 2013. Limiting global warming to 2 °C is unlikely to save most coral reefs. *Nature Climate Change*, **3**, 165–170.

Gardner, T., Côté, I., Gill, J., Grant, A. & Watkinson, A. 2003. Long-term region-wide declines in Caribbean corals. *Science*, **301**, 958.

Gunn, R., Hardesty, B. D. & Butler, J. 2010. Tackling 'ghost nets': local solutions to a global issue in northern Australia. *Ecological Management & Restoration*, **11**(2), 88–98.

Harriott, V. & Banks, S. 1995. Recruitment of scleractinian corals in the Solitary Islands Marine Reserve, a high latitude coral-dominated community in Eastern Australia. *Marine Ecology Progress Series*, **123**(1), 155–161.

Harriott, V. & Banks, S. 2002. Latitudinal variation in coral communities in eastern Australia: a qualitative biophysical model of factors regulating coral reefs. *Coral Reefs*, **21**(1), 83–94.

Hawkins, J. P. & Roberts, C. M. 2004. Effects of artisanal fishing on Caribbean coral reefs. *Conservation Biology*, **18**(1), 215–226.

Hoegh-Guldberg, O. 1999. Coral bleaching, climate change and the future of the world's coral reefs. *Marine and Freshwater Research*, **50**(8), 839–866.

Hoegh-Guldberg, O. 2012. The adaptation of coral reefs to climate change: is the Red Queen being outpaced? *Scientia Marina*, **76**(2), 403–408.

Hoegh-Guldberg, O. & Smith, G. J. 1989. The effect of sudden changes in temperature, light and salinity on the population-density and export of zooxanthellae from the reef corals *Stylophora pistillata* Esper and *Seriatopora hystrix* Dana. *Journal of Experimental Marine Biology and Ecology*, **129**(3), 279–303.

Hoegh-Guldberg, O., Mumby, P. J. *et al.* 2007. Coral reefs under rapid climate change and ocean acidification. *Science*, **318**, 1737–1742.

Hughes, T. P. 1994. Catastrophes, phase shifts, and large-scale degradation of a Caribbean coral reef. *Science*, **265**, 1547.

IPCC. 2007*a*. *Climate Change 2007: The Physical Science Basis. Contribution of Working Group I to the Fourth Assessment Report of the Intergovernmental Panel on Climate Change, 2007* (Solomon, S., Qin, D. *et al.* eds). Cambridge University Press, Cambridge.

IPCC. 2007*b*. *Climate Change 2007: Impacts, Adaptation and Vulnerability. Contribution of Working Group II to the Fourth Assessment Report of the Intergovernmental Panel on Climate Change* (Parry, M. L., Canziani, O. F., Palutikof, J. P., van der Linden, P. J., Hanson, C. E. eds). Cambridge University Press, Cambridge.

Jackson, J. B., Kirby, M. X. *et al.* 2001. Historical overfishing and the recent collapse of coastal ecosystems. *Science*, **293**, 629–37.

Jones, R. J. & Hoegh-Guldberg, O. 1999. Effects of cyanide on coral photosynthesis: implications for identifying the cause of coral bleaching and for assessing the environmental impacts of cyanide fishing. *Marine Ecology Progress Series*, **177**, 83–91.

Keller, R. P., Drake, J. M., Drew, M. B. & Lodge, D. M. 2011. Linking environmental conditions and ship movements to estimate invasive species transport across the global shipping network. *Diversity and Distributions*, **17**(1), 93–102.

Kleypas, J. A., McManus, J. W. & Menez, L. A. B. 1999. Environmental limits to coral Rreef development: where do we draw the line? *American Zoologist*, **39**, 146–159.

Manzello, D. P., Kleypas, J. A., Budd, D. A., Eakin, C. M., Glynn, P. W. & Langdon, C. 2008. Poorly cemented coral reefs of the eastern tropical Pacific: possible insights into reef development in a high-CO_2 world. *Proceedings of the National Academy of Sciences of the United States of America*, **105**(30), 10,450–10,455.

McClanahan, T. R., Ateweberhan, M., Graham, N. A. J., Wilson, S. K., Ruiz Sebastián, C., Guillaume, M. M. M. & Bruggemann, J. H. 2007. Western Indian Ocean coral communities: bleaching responses and susceptibility to extinction. *Marine Ecology Progress Series*, **337**, 1–13.

McEvoy, B. P., Powell, J. E., Goddard, M. E. & Visscher, P. M. 2011. Human population dispersal "Out of Africa" estimated from linkage disequilibrium and allele frequencies of SNPs. *Genome Research*, **21**(6), 821–829.

Meinshausen, M., Smith, S. J. *et al.* 2011. The RCP greenhouse gas concentrations and their extensions from 1765 to 2300. *Climatic Change*, **109**(1–2), 213–241.

Moberg, F. & Folke, C. 1999. Ecological goods and services of coral reef ecosystems. *Ecological Economics*, **29**(2), 215–233.

Muscatine, L. & Porter, J. W. 1977. Reef corals: mutualistic symbioses adapted to nutrient-poor environments. *Bioscience*, **27**, 454–460.

Myers, R. A. & Worm, B. 2003. Rapid worldwide depletion of predatory fish communities. *Nature*, **423**, 280–283.

Pandolfi, J. M., Bradbury, R. H. *et al.* 2003. Global trajectories of the long-term decline of coral reef ecosystems. *Science*, **301**, 955.

Parmesan, C. & Yohe, G. 2003. A globally coherent fingerprint of climate change impacts across natural systems. *Nature*, **421**, 37–42.

Pauly, D., Christensen, V. V., Dalsgaard, J., Froese, R. & Torres, F., Jr. 1998. Fishing down marine food webs. *Science*, **279**, 860–863.

Pauly, D., Graham, W., Libralato, S., Morissette, L. & Deng Palomares, M. L. 2008. Jellyfish in ecosystems, online databases, and ecosystem models. *Hydrobiologia*, **616**(1), 67–85.

Pelejero, C., Calvo, E. & Hoegh-Guldberg, O. 2010. Paleo-perspectives on ocean acidification. *Trends in Ecology & Evolution*, **25**(6), 332–344.

Petit, J. R., Jouzel, J. *et al.* 1999. Climate and atmospheric history of the past 420 000 years from the Vostok ice core, Antarctica. *Nature*, **399**, 429–436.

Poloczanska, E. S., Brown, C. J. *et al.* 2013. Global imprint of climate change on marine life. *Nature, Climate Change*, **3**, 919–925.

Porter, A. L., Cunningham, S. W., Banks, J., Roper, A. T., Mason, W. & Rossini, F. A. 2011. *Forecasting and Management of Technology*. Wiley, Chichester.

Raven, J., Caldeira, K. *et al.* 2005. *Ocean Acidification Due to Increasing Atmospheric Carbon Dioxide*. Royal Society Policy Document, **12/05**. The Royal Society, London.

Rhodes, R. 1977. *"I Am Become Death ...": The Agony of J. Robert Oppenheimer*. American Heritage Publishing Co., Rockville, MD.

Root, T. L., Price, J. T., Hall, K. R., Schneider, S. H., Rosenzweig, C. & Pounds, J. A. 2003. Fingerprints of global warming on wild animals and plants. *Nature*, **421**, 57–60.

Rosenzweig, C., Karoly, D. *et al.* 2008. Attributing physical and biological impacts to anthropogenic climate change. *Nature*, **453**, 353–357.

Ruddiman, W. F. 2003. The anthropogenic greenhouse era began thousands of years ago. *Climatic Change*, **61**(3), 261–293.

Spalding, M. & Grenfell, A. 1997. New estimates of global and regional coral reef areas. *Coral Reefs*, **16**(4), 225–230.

Stanley, G. D. & Fautin, D. G. 2001. The Origins of Modern Corals. *Science*, **291**, 1913–1914.

Stewart, J. R. & Stringer, C. B. 2012. Human evolution out of Africa: Tthe role of refugia and climate change. *Science*, **335**, 1317–1321.

Strong, A. E., Liu, G., Meyer, J., Hendee, J. C. & Sasko, D. 2004. Coral Reef Watch 2002. *Bulletin of Marine Science*, **75**(2), 259–268.

Strong, A. E., Liu, G., Skirving, W. & Eakin, C. M. 2011. NOAA's Coral Reef Watch program from satellite observations. *Annals of GIS*, **17**(2), 83–92.

Tanzil, J. T. I., Brown, B. E., Tudhope, A. W. & Dunne, R. P. 2009. Decline in skeletal growth of the coral *Porites lutea* from the Andaman Sea, South Thailand between 1984 and 2005. *Coral Reefs*, **28**(2), 519–528.

Veron, J. E. N. 2008. Mass extinctions and ocean acidification: biological constraints on geological dilemmas. *Coral Reefs*, **27**(3), 459–472.

Wachenfeld, D. 1997. Long-term trends in the status of coral reef-flat benthos — the use of historical photographs. *In*: Wachenfeld, D., Oliver, J. & Davis, K. (eds) *State of the Great Barrier Reef World Heritage Area Workshop, 1997*. Great Barrier Reef Marine Park Authority, Towsville, Queensland, 134–148.

Watson, R., Zeller, D. & Pauly, D. 2011. *Spatial Expansion of EU and Non-EU Fishing Fleets into the Global Ocean, 1950 to the Present*. Fisheries Centre, University of British Columbia, Vancouver.

Worm, B., Barbier, E. B. *et al.* 2007. Biodiversity loss in the ocean: How bad is it? Response. *Science*, **316**, 1282–1284.

Worm, B., Davis, B. *et al.* 2013. Global catches, exploitation rates, and rebuilding options for sharks. *Marine Policy*, **40**, 194–204.

Wyatt, L. T. 2009. *The Industrial Revolution (Greenwood Guides to Historic Events, 1500–1900)*. Greenwood Press, Westport, CT.

Zalasiewicz, J., Williams, M. *et al.* 2008. Are we now living in the Anthropocene? *GSA Today*, **18**(2), 4–8.

Microbiotic signatures of the Anthropocene in marginal marine and freshwater palaeoenvironments

I. P. WILKINSON[1,2*], C. POIRIER[3], M. J. HEAD[4], C. D. SAYER[5] & J. TIBBY[6]

[1]*British Geological Survey, Nottingham NG12 5GG, UK*

[2]*Department of Geology, University of Leicester, Leicester LE1 7RH, UK*

[3]*Littoral Environnement Sociétés, Université de La Rochelle, CNRS, 2 rue Olympe de Gouges, 17000 La Rochelle, France*

[4]*Department of Earth Sciences, Brock University, St. Catharines, Ontario L2S 3A1, Canada*

[5]*Environmental Change Research Centre, Department of Geography, University College London, London WC1E 6BT, UK*

[6]*Department of Geography, Environment and Population, University of Adelaide, South Australia 5005, Australia*

**Corresponding author (e-mail: ipw@bgs.ac.uk)*

Abstract: The term 'Anthropocene' has been proposed to indicate a geological interval characterized by global anthropogenic environmental change. This paper attempts to recognize a method by which the Anthropocene can be defined micropalaeontologically. In order to do this, microfloras and microfaunas (diatoms, macrophytes, dinoflagellate cysts, foraminifera and ostracods) from nearshore waters through to paralic and freshwater aquatic milieux are considered, and biotic variability with an anthropogenic causation identified. Microbiotic change can be related to anthropogenically induced extinctions, pollution-related mutation, environmentally influenced assemblage variability, geochemistry of carapaces/tests, floral change related to lacustrine acidification, faunal and floral correlation to industrial and agricultural signatures and introduction of exotic species via shipping. The influence of humanity on a local scale can be recognized in assemblages as far back as 5000 years BP. However, widespread anthropogenic change took place in Europe and America, particularly in the nineteenth and twentieth centuries, although in Asia (e.g. Japan) it cannot be observed prior to the twentieth century. Profound and global biotic change began in the mid-twentieth century and, if the Anthropocene is to be defined in this way, then the period 1940–1945 might encompass the biotic base of the interval.

The increasing impact of human activity on the biosphere has been acknowledged for more than 140 years, beginning with Stoppani's (1873) reference to the 'anthropozoic era'. More recently, Crutzen (2002) postulated that, at some point in the recent past, the Holocene had ended and a new epoch had commenced, defined by global environmental change caused by anthropogenic activity (Fig. 1). The concept of the Anthropocene is gathering favour among scientists, although it has neither been formally proposed nor assigned a rank as yet. Zalasiewicz *et al.* (2008, 2010, 2011) pointed out possible methods by which the interval might be recognized and the problems associated with the concept.

It is clear that a fundamental change has taken place in the evolution of freshwater systems and fluxes during the last 50–200 years (Meybeck & Vörösmarty 2005). During the Holocene, change was related to endorheism, fluctuations in glacial cover, and exposure of the continental shelf, together with filters such as wetlands, lakes, floodplains and estuaries. During the last 50–200 years, anthropogenic water management resulted in a reduction of runoff, and anthropogenic filters (e.g. irrigation and dammed reservoirs) largely replaced natural ones. The increased disturbance and management of catchment areas has also resulted in accelarated erosion (Merritts *et al.* 2011) and changes in the sediment flux (Syvitski & Kettner 2011), both of which impinge on the freshwater microfloras and faunas as well as biotas in the coastal zone, around estuaries and in deltas. Some fluxes (sodium, chloride, sulphate, nitrogen, phosphorous, heavy metals) have increased globally, whereas others are stable (calcium, bicarbonate, sediments) or reduced (dissolved silica), and future change is likely to be a result of anthropogenically

From: Waters, C. N., Zalasiewicz, J. A., Williams, M., Ellis, M. A. & Snelling, A. M. (eds) 2014. *A Stratigraphical Basis for the Anthropocene*. Geological Society, London, Special Publications, **395**, 185–219.
First published online January 31, 2014, http://dx.doi.org/10.1144/SP395.14

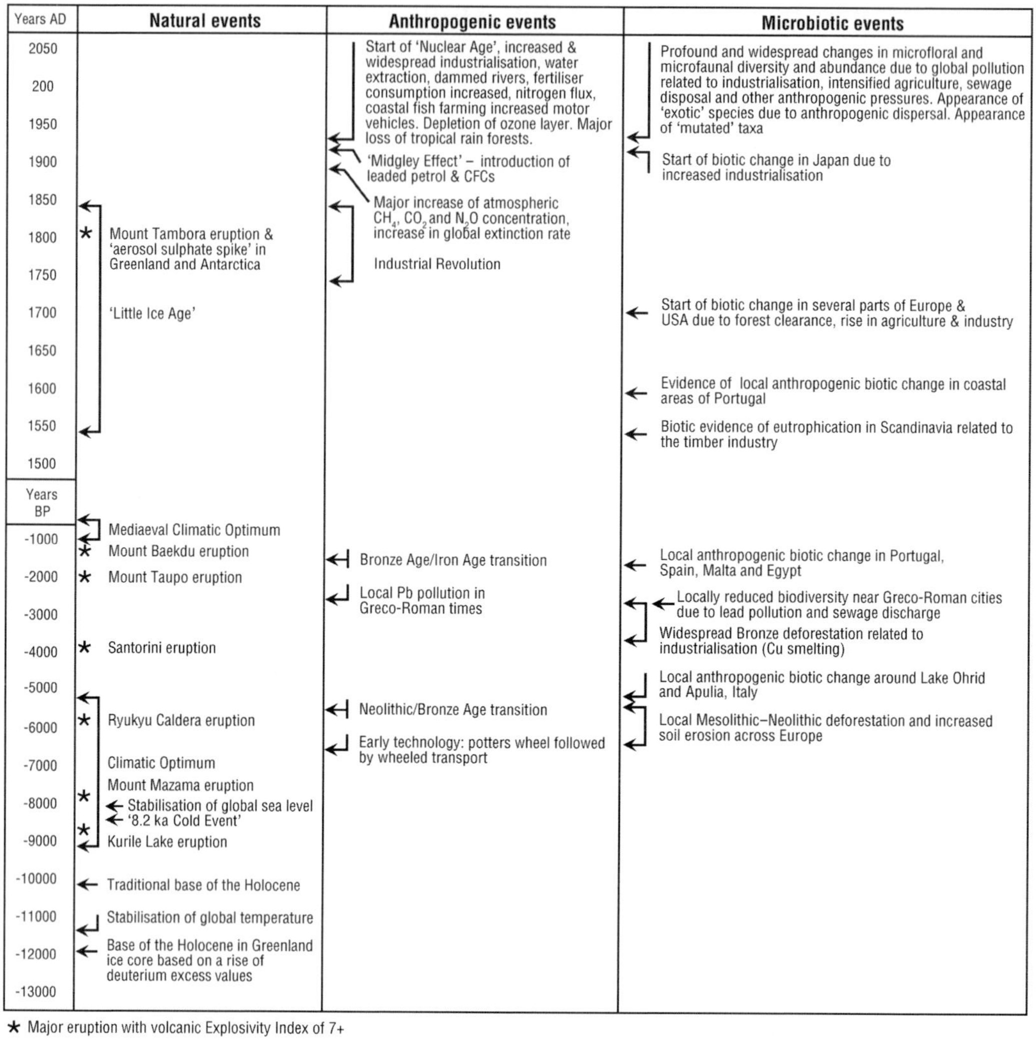

Fig. 1. Natural and anthropogenic events related to microbiotic events through the Holocene and Anthropocene. Years BP are based on radiocarbon dating based on 1950 being 0 years BP.

induced soil erosion, pollution and water abstraction for irrigation and consumption. Natural habitats of lacustrine and fluviatile communities have, therefore, come under extreme pressure.

Anthropogenic activities also impact on the coastal strip. As more water is abstracted or dammed, so the natural flow of water, nutrients and sediments to the coastal zone has been interrupted. The coastal zone is also under pressure because much of the world's food production, industry and habitation is focused on this area, and anthropogenic impacts on coastal ecological systems dynamics have resulted in greater environmental stress over time (e.g. Kremer & Crossland 2002). The interrelationship of the different parts of the hydrosphere means that anthropogenic activities impacting on freshwater bodies go on to influence estuarine and coastal biotas, habitats, communities and trophic structure. In addition, the discharge of pollutants originating from agricultural and industrial processes has resulted in mutation and extirpation.

Traditional biostratigraphical methods employed in defining chronostratigraphical units in the geological column involve the recognition of zones and subzones, based on key indices and defined by inceptions, extinctions, concurrent ranges or acmes of key taxa. However, since the end of the

Pleistocene, traditional biostratigraphical methods are unavailable except at local or regional levels.

Anthropogenic influences have caused extinctions among terrestrial animal and plant species throughout the Holocene, and are now affecting marine and brackish marine milieux more profoundly. The rate of biotic change is increasing, however, and it has been argued that a major extinction event on a par with the K–T Boundary event may result (e.g. Wilson 2002; Barnosky *et al.* 2011).

The present paper examines the nature, timing and causes of anthropogenic biotic change in microfloral and microfaunal communities, from marginal marine through to freshwater palaeoenvironments. We concentrate on diatoms, dinoflagellate cysts, foraminifera and ostracods, in order to determine whether the Anthropocene can be identified and, if so, when it began. Although the environmental databases of the different biotic groups discussed herein are variable in extent and detail, we believe that they have collectively recorded major trends.

Diatoms

Diatoms are photosynthetic algae with a siliceous frustule that live in aquatic and semi-aquatic environments and are the biological group most commonly used in palaeolimnological studies (Smol & Stoermer 2010). They tend to be well preserved in sediment cores and are known to be sensitive indicators of in-lake environmental conditions, especially water chemistry, habitat availability and the dominant location for benthic v. pelagic algal primary production (Battarbee *et al.* 2001; Bennion *et al.* 2010). Consequently, diatoms have been widely used to detect human impacts upon lakes, making them highly suitable for attempting a delineation of the Anthropocene. Recognizing this, a number of studies have explicitly used diatoms to help define the Holocene–Anthropocene transition, notably Holmgren *et al.* (2010) for four lakes in the Svalbard archipelago, Arctic Norway, and Wolfe *et al.* (2013) for several remote Arctic and alpine lakes in the Northern Hemisphere. Furthermore, a number of European studies focusing on the detection of ‘reference conditions’ as required by the European Union Water Framework Directive (EU-WFD; European Union 2000) indirectly provide much useful information regarding the timing of major human-induced changes in lakes.

In our review of diatom evidence for the Anthropocene, we focus on regions of the world where there are sufficient records to generalize on the timing of major human-induced diatom changes, importantly, Northern Europe (several individual countries), North America and Australia. For some of these regions, previous metastudies and syntheses have already summarized diatom responses to anthropogenic forcing. Other parts of the world are also considered, especially with regard to early human impacts. Key anthropogenic drivers commonly cited as having induced changes to diatom assemblages are nutrient-enrichment due to catchment-scale increases in nutrient loading (N and P) and atmospheric deposition of N, acidification, metal pollution, alkalization and climate change.

Early regional evidence for eutrophication: 5000 years BP to 1800 AD

There is considerable evidence for early human impacts on lakes in many parts of the world, extending back at least 5000 years to the beginning of the Subboreal period. This is especially true of long-term palaeolimnological studies undertaken in lowland Europe, where diatom changes in generally alkaline lakes have been linked to catchment-scale disturbances, especially forest removal and early agricultural activities. For example, at Diss Mere, England, Fritz (1989) detected early nutrient-enrichment impacts from around 5000 years BP, with a major Bronze Age acceleration of this process after 2500 years BP. Similar early evidence for cultural eutrophication comes from Wales (>2750 years BP in Llyn Padarn; Elner & Happey-Wood 1980), southern Sweden (>4000 years BP and >2700 years BP in Lakes Sämbosjön (Digerfeldt & Håkansson 1993) and Bussjösjön (Håkansson & Regnéll 1993), respectively) and Denmark (>3700 years BP in Dallund Sø; Bradshaw *et al.* 2005), to name just a few locations of Holocene-scale studies of land–lake relationships.

The signature for change in the aforementioned lakes has typically been increases in mesotrophic–eutrophic planktonic diatoms, especially *Asterionella formosa*, *Fragilaria crotonensis* and small centric species in the genera *Stephanodiscus* and *Cyclostephanos*. Similar issues are also evident for upland, low-alkalinity lakes in northern Europe since *c.* 2500 years BP. These lakes are thought to have been impacted by early eutrophication and/or alkalization. For example, Renberg *et al.* (1993*a*, *b*) inferred a pH increase from *c.* 5.5 to 6.5 for several southern Swedish lakes. This was explained by the release of base cations and nutrients from the forest biomass and upper soils to surface waters linked to early forest grazing and local-scale agricultural activities.

Early human impacts on lakes are not restricted to Europe, however. For example, at Crawford Lake in Canada, Ekdahl *et al.* (2007) revealed a dramatic shift to eutrophic plankton taxa in *c.* 1250 AD due to the early horticultural activities of native

Iroquoian populations. In this study there was some evidence of lake recovery after 1490 AD, until the intensive agricultural activities of European settlers accelerated eutrophication impacts once again post-1870 AD. In Lago de Pátzcuaro, Mexico, Bradbury (2000) observed that early Mesoamerican agriculture resulted in increases in eutrophic planktonic diatom taxa such as *Aulacoseira ambigua* and *Cyclostephanos* spp. from 3.2 ka due to irrigation runoff and, to a lesser extent, the fertilizing of crops. These impacts, which accelerated after 0.8 ka, but subsided with the Spanish occupation (Bradbury 2000) are some of the earliest known for the region (see Ortega *et al.* 2010).

There is also much evidence, as is perhaps expected, for widespread and accelerating eutrophication impacts in the European Medieval period (Fritz 1989; Bradshaw *et al.* 2005). For example, Hübener *et al.* (2009) inferred increasing nutrient inputs to several alkaline German lakes since 1400 AD, and in other parts of Europe substantial eutrophication is known to have occurred some time before the 1800s (e.g. Räsänen *et al.* 2006; Bjerring *et al.* 2008; see below).

In the Southern Hemisphere, the vast majority of diatom-based studies of human impact have been undertaken in southeastern Australia in the catchment of the Murray River. As a result, most published diatom sequences have been extracted from riverine wetlands, although there are also records from crater lakes. Although most Australian sequences extend for more than a millennium, few cover the whole Holocene. For the long-term studies that do exist (e.g. D'Costa *et al.* 1989; Gell *et al.* 1994; Reid *et al.* 2002; Gell *et al.* 2005*a*, *b*; Reid *et al.* 2007; Tibby *et al.* 2006), there is little unambiguous evidence for early impacts from Aboriginal activities, and for the most part environmental change outside the realm of natural variability is thought to be mostly restricted to post-1850 AD. By contrast, in New Zealand, there is clear evidence of landscape modification associated with Maori settlement. For example, in Diamond Lake, a human-driven fire event (inferred from charcoal) after 1320 AD is thought to have promoted the planktonic diatom *Cyclotella stelligera* (McWethy *et al.* 2010).

Widespread diatom changes post-1800 AD

Thousands of sediment records in the world are suitable for assessing diatom responses to anthropogenic impacts after 1800 AD. Since then, increased agricultural activities, industrialization and air pollution on a global scale have all had severe impacts on the biological structure and function of lakes across the globe.

A key diatom metastudy is that by Battarbee *et al.* (2011), who described temporal patterns of environmental change using the 'LakeCores' database, which includes 947 European lakes, especially in the UK, Scandinavia, The Netherlands and France. The aim of this study was the identification of 'reference conditions' prior to major impacts due to acidification and eutrophication. Although not explicitly focused on identifying an Anthropocene signature, this study is ideally designed for its assessment across a large geographical area and covering multiple lake types: deep, shallow, small, large and of low and high alkalinity. To summarize, diatom responses in the cores, suitable diatom-pH and diatom-total P transfer functions were applied to low- and high-alkalinity lakes, respectively. With respect to low-alkalinity lakes, it was concluded that 'conditions prior to 1800, and in most cases 1850' (Battarbee *et al.* 2011, p. 408), could be considered as representing the reference state prior to human-induced acidification. Indeed, Battarbee *et al.* (2011) suggested that the first evidence for eutrophication in most European lakes (95 sites assessed) occurred over the period 1850–1900.

A reference conditions cut-off of 1850 AD has been applied in many parts of Europe (based on multi-lake studies undertaken in the UK (Bennion *et al.* 2004), Ireland (Leira *et al.* 2006), Denmark (Bjerring *et al.* 2008)), with this date thought to predate the major onset of industrialization and intensive agriculture. Nonetheless, it is widely recognized that many European lakes were already subject to significant anthropogenic influence prior to this due to catchment-scale eutrophication impacts in particular (Räsänen *et al.* 2006; Hübener *et al.* 2009; Gottschalk 2011). For example, 17 of 21 studied Danish lakes were already eutrophic before 1850 AD (Bjerring *et al.* 2008).

In terms of anthropogenic-induced acidification of low-alkalinity upland lakes, several studies have shown impacts since around 1850 AD in the UK (Battarbee 1984; Jones *et al.* 1989), with later impacts (post-1900 and especially post-1930s) evident in North America and Scandinavia, including Sweden, Norway and Finland (Battarbee & Charles 1987; Renberg *et al.* 1993*b*). The unprecedented nature of recent changes in pH in the context of natural Holocene-scale fluctuation was clearly shown by Jones *et al.* (1989) for the Round Loch of Glenhead, Scotland. In this lake, changes in diatom-inferred pH were relatively minor until *c.* 1850 (pH $\sim$ 5.4), after which a reduction by almost 1 pH unit was evident over the next 130 years due to atmospheric sulphur deposition in a base-poor catchment.

A key difference between pre- and post-1850 AD in terms of diatom-inferred human impacts on lakes is evidence for the earliest human influences on many remote High Arctic, Northern European and American alpine lakes, due to global warming and atmospheric contamination, especially by N. For example, in a review of 55 palaeo-records from the circumpolar Arctic, Smol *et al.* (2005) observed a widespread reorganization of diatom assemblages after *c.* 1850 AD. In this study, climate warming and a lengthening of the summer growing season were postulated as causal. Indeed, in a confident statement, Smol *et al.* (2005, p. 4402) announced that their data provided 'distinctive biostratigraphic signatures of the Holocene–Anthropogenic transition for the most climatically-sensitive subregions of the Arctic'. Despite this, however, there has been some debate on the causes and timing of diatom changes within different Arctic lakes and regions.

Considerable research has focused on Svalbard, Arctic Norway. For 23 Svalbard lakes, Birks *et al.* (2004) showed marked increases in the rate of biological change (diatoms, chrysophyte cysts, chironomids) over the last 50–200 years, with several factors hypothesized as responsible: atmospheric deposition of nutrients and pollutants (spheroidal carbonaceous particles (SCPs), polycyclic aromatic hydrocarbons (PAHs), PCBs and possibly Pb), recent warming, and an interval of elevated precipitation between 1965 and 1975. In a combined diatom and N stable isotope study of four Svalbard lakes, Holmgren *et al.* (2010) showed evidence for the first anthropogenic impacts in the 1900s (two lakes) and 1920s (two lakes) linked to warming, but the most pronounced compositional changes occurring in the latter half of the twentieth century correlated with a decline in the $\delta^{15}N$ of sediment organic matter indicative of enhanced deposition of atmospheric N. Thus, a strong possibility exists that biological changes in the Arctic have been driven by regionally variable combinations of climate change and N enrichment. It is also noteworthy, however, that not all studies of remote lakes have shown human impacts, this being true of 23 lakes in west Greenland where post-1850 diatom changes were well within the natural range of variability (Perren *et al.* 2009).

In parallel with this Arctic-based work, several studies have focused on mid-latitude Alpine lakes including lakes in the Rocky Mountain National Park, USA (Wolfe *et al.* 2001, 2003). Again, similar to some Arctic lakes, diatom responses indicative of eutrophication have been found to coincide with recent inferred increases in atmospheric N deposition since *c.* 1950. Interestingly, the same diatom species replacements observed in studies of early human impacts of lakes (as discussed above) were evident, notably expansions of the planktonic taxa *A. formosa* and *F. crotonensis*. With a clear focus on defining the Holocene–Anthropocene transition for remote Arctic and Alpine lakes in the USA, Canada, west Greenland and the Norwegian High Arctic, Wolfe *et al.* (2013) reviewed the diatom (in addition to algal pigment and N isotopes) data available for 42 sediment records, covering 25 lakes. The study showed change as early as 1850 AD for one site, but for most lakes the highest rates of change occurred during two discrete intervals: 1950–1970 AD and after 1980 AD (Fig. 2).

In the Southern Hemisphere, again the most extensive palaeolimnological datasets for assessing post-1800 diatom changes in lakes come from southeastern Australia. Several studies of shallow river cut-off lakes (known as billabongs) show considerable change associated with European settlement after *c.* 1850, suggesting severe ecosystem degradation including a widespread loss of aquatic macrophytes (Reid *et al.* 2002, 2007; Gell *et al.* 2005*b*). The mechanisms behind the observed changes remain the subject of debate, and a combination of factors was likely important in different regions, including clearance of catchment vegetation and accelerated soil erosion, nutrient enrichment and salinity increases. Not all studies have been able to reliably radiometrically date these initial anthropogenic disturbances, but Gell *et al.* (2005*b*) and Reid *et al.* (2007) used the presence of exotic pollen markers to show unprecedented changes to the diatom communities of Tareena and Hogan's billabongs in the late 1800s, while Leahy *et al.* (2005) showed a significant change in Bollin Billabong after *c.* 1840.

A setting more remote from major settlements, Lake Euramoo (in the wet tropics of Queensland, northeastern Australia), has been subjected to two diatom-based studies. It is one of the best dated records that span the European boundary in Australia (Haberle *et al.* 2006) and extends to the Last Glacial Maximum (Tibby & Haberle 2007). Diatoms shifted outside the range of natural variability (Morgan *et al.* 1994) after 1880 in response to catchment clearance, as registered in pollen and historical records. As with most other palaeo-studies in Australia, the changes registered in this record suggest a greater importance of local catchment processes than regional–global impacts such as climate change or atmospheric pollution.

Discussion

Our review could not possibly consider the vast number of diatom sequences that have been produced. In addition, coverage of the Southern Hemisphere is poor and patchy. However, the broad spread of sites and the existence of several metastudies and syntheses allow us to make some general conclusions on the temporal position of the Anthropocene. First, there is widespread evidence for early

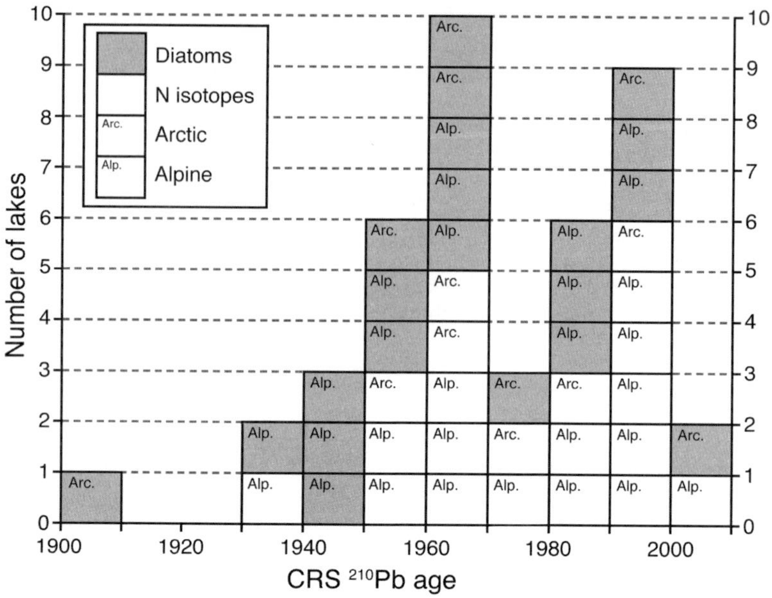

Fig. 2. Intervals of maximum diatom change for 21 nitrogen isotope and 21 diatom records from remote North American and Northern European arctic and alpine lakes (after Wolfe *et al.* 2013).

human impacts in Europe well before the 1800s, and there is every reason to suggest that most natural lowland European lakes were disturbed many millennia ago. Nonetheless, it is clear that the change in diatoms becomes globalized certainly after 1850 AD due to global warming and atmospheric pollution. The timing of the change is, however, very patchy, varying between regions and in different lakes in the same region depending on the nature of human impacts and the individual characteristics of lakes. While many lakes started to change post-1850, there is also much evidence for a great acceleration of the Anthropocene post-1945. Keatley *et al.* (2011), for example, showed that some 33% of 67 diatom-based palaeolimnological studies in Europe and North America suggested an acceleration of nutrient enrichment after 1945, while many Arctic and Alpine lakes also exhibited pronounced change in this era. Wolfe *et al.* (2013, p. 31) also suggested that the Great Acceleration and a Holocene–Anthropocene boundary at *c.* 1950 AD best 'reconciles the primary data' for remote lakes in their review.

Dinoflagellate cysts

Dinoflagellates (Division Dinoflagellata) are mostly single-celled, free-living, biflagellated organisms classified as alveolate protists (Fensome *et al.* 1993; Adl *et al.* 2012). They are among the most important primary producers in the oceans, but at least 50% of species are heterotrophs and mixotrophs. Dinoflagellates are also found in brackish, freshwater and many other aquatic environments, and some are benthic (Taylor & Pollingher 1987). As a group, therefore, dinoflagellates are sensitive indicators over a broad environmental range.

With very few exceptions, the resting cyst (generally regarded as a hypnozygote) is the only geologically preservable part of the life cycle, and only around 13–16% of dinoflagellate species produce a resting cyst (Head 1996). Nonetheless, in neritic environments, especially in higher latitudes where a benthic resting stage offers advantages for survival, dinoflagellate cysts (hereafter 'dinocysts' or 'cysts') have an abundant and relatively diverse record in the sediments. This makes them potentially useful for assessing anthropogenic impact (Dale 1996, 2001, 2009; Dale & Dale 2002), providing certain conditions are met. The sediments containing the cyst record need to be undisturbed and well dated, allowing historical records of human population growth and industrial and other pollution to be linked to changes in the cyst record. The background variability of the cyst record must also be understood, to distinguish between natural variations of the ecosystem and events resulting from human impact.

The durability of dinoflagellate resting cysts has also permitted their inadvertent transport around the world in ships' ballast tanks, resulting in the introduction of invasive species (Casas-Monroy *et al.* 2011 and references therein) that should leave a stratigraphic signature of anthropogenic influence.

Surprisingly, relatively few dinocyst studies have addressed the environmental impacts of humans over the past centuries, and these studies are primarily located in Norway, Sweden, Italy, Massachusetts (USA), the St. Lawrence Estuary, Canada, and Japan in marginal marine areas known to have been strongly affected by anthropogenic activities. A selection of studies is highlighted below, along with a pre-human impact study from the Kattegat showing the importance of recognizing background variability during the late Holocene.

Skagerrak and Kattegat, Norway: record before human impact

Cyst studies have revealed that the bloom-forming dinoflagellate *Gymnodinium nolleri* invaded the Kattegat at *c.* 6000 years BP in response to an increase in water temperature. It attained a minor peak at *c.* 4500 years BP, was present in low abundances during 2000–400 years BP, and reached a peak approximately corresponding to the Medieval Warm Period (*c.* 1000–700 years BP). It finally went into decline after *c.* 500 years BP, with a local disappearance at *c.* 300 years BP probably in response to the Little Ice Age (as *Gymnodinium catenatum* in Dale *et al.* 1993; Dale & Nordberg 1993). Subsequent studies from the Skagerrak and Kattegat (Thorsen *et al.* 1995) and the southern Norwegian coast (Thorsen & Dale 1997) have confirmed these fluctuations and suggest *Gymnodinium nolleri* to be a regional warmer-water indicator in open coastal waters off southern Scandinavia, although this interpretation has been questioned (Harland & Nordberg 2011). While other environmental factors may also be at play, the significant fluctuations in *Gymnodinium nolleri* over the past several thousand years (Harland & Nordberg 2011) demonstrate the effects of environmental change on the cyst record during historic times but before major anthropogenic impact on the marine environment.

Inner Oslofjord, Norway: a eutrophication signal

Studies from the inner Oslofjord are useful for assessing anthropogenic impact because of the known history of population growth in Oslo and its effects on nutrient loading. This started in the mid-nineteenth century with a rise in population growth along with the country's first large industrial development, and then with the introduction of the water closet in the early twentieth century and its widespread use after *c.* 1930 (Dale & Fjellså 1994; Dale *et al.* 1999). Eutrophication is defined as an increase in the rate of supply of organic matter in an ecosystem (Nixon 1995). It results from increased nutrient loading leading to excessive growth of algae and hence an increase in decomposing organic matter, which depletes oxygen levels in the water body. The inner Oslofjord has restricted marine circulation and water exchange with the Skagerrak, and is therefore particularly sensitive to cultural eutrophication.

Four cores obtained from the inner Oslofjord and spanning the interval of population growth and industrial development from the mid-nineteenth century to present have yielded a clear eutrophication signal in the dinocyst record, especially since three of these cores record conditions before human population increase and so provide a background signal before eutrophication. From about the sixteenth century to the mid-nineteenth century, relatively low cyst concentrations with a co-dominance of the cysts of *Protoceratium reticulatum* (as *Operculodinium centrocarpum* in Dale *et al.* 1999) and *Pentapharsodinium dalei* (as *Peridinium faeroense* in Dale *et al.* 1999) and with few specimens of *Lingulodinium machaerophorum* (<5%) represent pre-industrial assemblages and are similar to assemblages found in the unpolluted outer Oslofjord and similar unpolluted environments along the Norwegian coast today. In cores where the signal is best represented, the first signs of eutrophication are marked by an increase in cyst concentration involving first the opportunistic species *Protoceratium reticulatum* in response to environmental disturbance, and then *Lingulodinium machaerophorum* as an indicator of elevated nutrients. The response of *Lingulodinium machaerophorum* to artificial nutrient input can be explained by the fact that its abundance under natural conditions would be held in check by limited nutrient levels during the summer. In contrast, *Pentapharsodinium dalei* appears to be largely unaffected by eutrophication because it blooms in spring in the Oslofjord, at a time when the environment is not naturally limited in nutrients. Hence, a rise in *Lingulodinium machaerophorum* along with a general rise in cyst production can be recognized as the primary dinocyst signal for cultural eutrophication, with a clear initial response occurring in the mid- to late nineteenth century that strengthened during the early to mid-twentieth century, in step with historical records of population growth in Oslo. The youngest part of the cyst record shows a reversal of these trends back to pre-industrial levels, reflecting improvements in sewage treatment during the 1980s and 1990s (Dale & Fjellså 1994; Dale *et al.* 1999; Fig. 3).

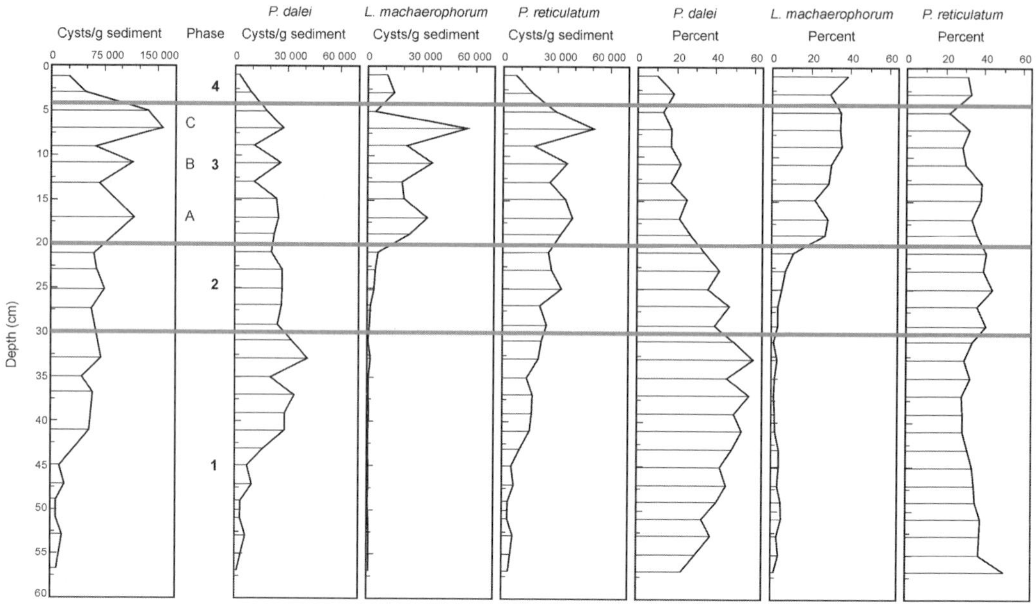

Fig. 3. Dinocyst record from the inner Oslofjord, Norway, showing total cyst concentrations, and the concentrations and relative abundances of the three most important species, *Pentapharsodinium dalei*, *Lingulodinium machaerophorum* and *Protoceratium reticulatum* (*Operculodinium centrocarpum* in Dale *et al.* 1999). This record primarily illustrates the effects of cultural eutrophication. Four distinct phases of development are shown: Phase 1, sixteenth century up to about the mid-nineteenth century; Phase 2, about the mid- to late nineteenth century; Phase 3, *c.* 1900 to 1980, in which A, B and C refer to distinct pulses of increase in cyst concentration; Phase 4, 1980s to 1985 (after Dale *et al.* 1999, fig. 3).

Southwestern Norway: a pollution signal

The restricted fjord Nordåsvannet had been receiving untreated sewage from nearby Bergen during a substantial growth in population that began in the early nineteenth century. By the 1950s, this had reached *c.* 30 000 person equivalents (Thorsen & Dale 1997). The cyst signal at Nordåsvannet shows a peak in *Protoceratium reticulatum* cysts (as *Operculodinium centrocarpum* in Thorsen & Dale 1997), which probably reflects transitional conditions associated with the population expansion of Bergen in the mid-nineteenth century. This is followed by a peak in cysts of the cosmopolitan heterotrophic dinoflagellate *Protoperidinium conicum* *c.*1930, and then a possible increase in percentages of *Lingulodinium machaerophorum*. An increase in cyst concentrations in the 1980s together with an uppermost peak in *Protoceratium reticulatum* cysts and declining numbers of *Protoperidinium conicum* may be the response to a shifting of the sewage discharge away from Nordåsvannet in the 1980s. The anthropogenic signal in Nordåsvannet, as with the inner Oslofjord, seems to begin with elevated levels of *Protoceratium reticulatum*, but is then replaced by the heterotroph *Protoperidinium conicum*, which seems a response to increased particulate matter and/or nutrient levels, both of which could result in reduced light levels and would favour the growth of a cosmopolitan heterotroph. This appears to represent primarily a general pollution response rather than a specific cultural eutrophication signal, as characterized by a rise in *Lingulodinium machaerophorum* (Thorsen & Dale 1997).

Southern Norway: a pollution signal?

The cyst record of a southern Norwegian fjord, Frierfjord, reveals a similar response to human impact (Sætre *et al.* 1997). Frierfjord is one of Norway's most polluted fjords, with both industrial waste and sewage effluent characterizing the changing environment. Industrial pollution began in the sixteenth century with the establishment of a sawmill, and during the nineteenth and twentieth centuries the area developed into an industrial centre that discharged increasing amounts of industrial effluent into Frierfjord along with untreated sewage from the expanding population. The industrial discharge consisted primarily of suspended material including wood fibre and nutrients (P and N) from the late eighteenth century to the 1970s, with heavy metals and toxins being introduced from the 1900s and 1910s along with suspended material. In the mid-1970s, measures were introduced to reduce industrial discharge.

For control purposes, the cyst record at Frierfjord was compared with that from the much less polluted Brevikfjord nearby. Frierfjord is a freshwater-influenced fjord with a 2–8-m-thick brackish surface layer of *c.* 0.5–10 psu, below which salinities increase rapidly to *c.* 22 psu. The cyst signal (Fig. 4) first registers changes from about the 1830s, with reduced cyst concentrations (possibly a combination of increased sedimentation rates as well as reduced cyst production) and an increase in heterotrophic species, particularly *Protoperidinium conicum*. Elevated levels of *Protoceratium reticulatum* cysts (as *Operculodinium centrocarpum* in Sætre *et al.* 1997) mark the onset of major environmental disturbance. *Pentapharsodinium dalei* cysts (as *Protoperidinium faeroense* cysts in Sætre *et al.* 1997) show a decline with increasing pollution, suggesting sensitivity to pollution. Surprisingly, and in contrast with other fjords (e.g. Dale & Fjellså 1994; Dale *et al.* 1999; Fig. 3), the autotrophic *Lingulodinium machaerophorum* does not increase substantially in relative abundance with increasing pollution and this is explained by the negative influence of decreasing light levels (and possible chemical toxicity) counterbalancing the benefits of increased nutrients. The strongest signal of industrial pollution therefore seems to be the increased relative abundance of *Protoperidinium conicum* cysts, presumably in response to some combination of decreasing light levels and increasing prey for this heterotrophic dinoflagellate, together with strongly reduced cyst concentrations implying at least some decline in cyst production. From the mid-1970s an increase in cyst concentration and decrease in heterotrophic species seems to reflect an improvement in water quality at Frierfjord (Sætre *et al.* 1997; Fig. 4). While industrial pollution and attendant chemical toxicity may have been responsible for these changes in the dinocyst record (Sætre *et al.* 1997), it is not possible to rule out cultural eutrophication as an important driver, with reduced light penetration caused by increased diatom production explaining the drop in autotrophic cyst concentrations (Thorsen & Dale 1997; Dale 2009).

Gullmar Fjord, Sweden: questionable correlation with cultural eutrophication

Gullmar Fjord on the west coast of Sweden has comprehensive hydrographic records, and a high-resolution study (*c.* 1.5 years per sample) by Harland *et al.* (2006) allows the dinoflagellate cyst response to be compared in detail to both climatic and anthropogenic changes over the past 85 years. Gullmar Fjord has surface salinities of 24–27 psu, with more saline waters in excess of 32 psu at depths below *c.* 15–20 m. Although this fjord has been subjected in the past to considerable industrial pollution and human effluent, the practice was largely stopped during the late 1960s to mid-1970s over concerns of eutrophication. Surprisingly, however, eutrophication has persisted since the 1970s and remains a major concern. The

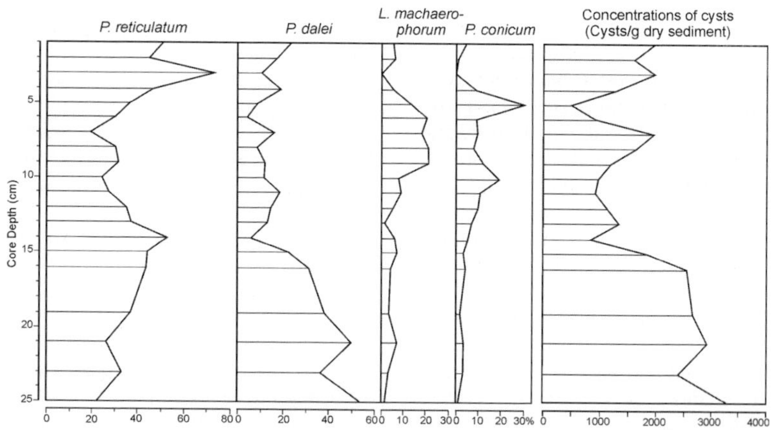

Fig. 4. Dinocyst record from Frierfjord, near Bergen, Norway, showing changes in relative abundance of the four most important species, *Protoceratium reticulatum* (*Operculodinium centrocarpum* in Sætre *et al.* 1997), *Pentapharsodinium dalei*, *Lingulodinium machaerophorum* and *Protoperidinium conicum* (*Selenopemphix quanta*), and changes in cyst concentration. This record illustrates the effects of industrial pollution and cultural eutrophication. Depths below 15 cm, before *c.* 1836; 14–4 cm, *c.* 1836–1974; 3–1 cm, 1974–1991 (after Sætre *et al.* 1997, figs 6 & 7).

dinoflagellate cyst record shows a significant change at *c.* 1969–1970, although this is perhaps better explained as a response to a shift in the North Atlantic oscillation from a negative phase to the present positive phase, rather than by any anthropogenic changes in the fjord itself. *Lingulodinium machaerophorum* declines after *c.* 1970, partly coincident with the curtailment of activity of a local sulphite pulp mill, a fish-canning activity, and cessation of untreated sewage input from water closets (Harland *et al.* 2010). Yet, eutrophication continues in Gullmar Fjord, questioning any simple relationship between *Lingulodinium machaerophorum* abundance and eutrophication (Harland *et al.* 2006, 2010). Dinoflagellate behaviour is complex. For example, the motile stage has the ability to swim vertically within the water column to access depths favourable for nutrients, prey and other requirements. These complexities must be taken into consideration when interpreting the cyst record, particularly in the stratified water masses that characterize most fjords.

Adriatic Sea, Italy: cultural eutrophication signal

The northwestern Adriatic Sea just off the Po River delta is a suitable location for studying cultural eutrophication because of the restricted circulation in this part of the Adriatic Sea and the opportunity to assess the effects of agricultural runoff on the marine environment (Sangiorgi & Donders 2004). The Po River delta was subject to wetland reclamation between the end of the nineteenth century and the 1970s. A study by Sangiorgi & Donders (2004) combined pollen and dinocyst records from a core spanning 1830–1990 off the Po River delta to compare marine and terrestrial responses to this wetland reclamation. The pollen record shows high nutrient-indicating taxa from *c.* 1910, although land disturbance is reflected from *c.* 1890. The dinocyst records shows a progressive increase in the heterotrophic/autotrophic ratio from *c.* 1910, mainly owing to increased relative abundances of *Brigantedinium* spp. and *Selenopemphix quanta*, and elevated numbers of *Lingulodinium machaerophorum* from 1930 onwards. These signals together suggest an increase in eutrophication, presumably in response to the introduction of modern farming methods and increasing industrialization and urbanization. The reversal of the heterotrophic/autotrophic ratio in the highest part of the core may correspond to a documented reduction in pollutants discharging from the Po River, and suggests that the heterotrophic/autotrophic cyst ratio is a valid proxy for assessing the trophic state of the area (Sangiorgi & Donders 2004).

Massachusetts, USA: cultural eutrophication and pollution signals

Sediment cores from two nearby coastal sites, New Bedford Harbor and Apponagansett Bay, in Massachusetts, USA, were examined by Pospelova *et al.* (2002) to assess human impacts over the past 500 years. For New Bedford Harbor these included agriculture (1676–1780), whaling (1750–1900), a textile industry (1880–1940), post-textile (1940–present) industries including electronics, and a hurricane barrier constructed in 1964 that altered current patterns. These developments coincided with a dramatic population growth at the end of the eighteenth century. Apponagansett Bay has no history of industrial pollution and lower levels of nutrient input, making it a good control site.

At New Bedford Harbor, species richness initially increased with nutrient levels during the seventeenth to nineteenth centuries, but declined during the twentieth century when this site was subjected both to toxic pollution and heavy nutrient loading. This decline was accompanied by wide fluctuations in cyst production, with abrupt increases in certain taxa including *Spiniferites* spp. (a group tolerant of extreme environmental conditions) in the 1940s. The greatest changes in the dinocyst record occurred at New Bedford Harbor during the twentieth century with the increase in industrial pollution. At the control site of Apponagansett Bay, the dinocyst response was an increase in species richness, coincident with increasing nutrient levels, during the twentieth century. Pospelova *et al.* (2002) noted that the responses of indicator species to cultural eutrophication and pollution were different from those reported for Norwegian fjords, owing to the mixing of water masses in shallow estuaries. For example, *Lingulodinium machaerophorum* and the cysts of *Protoperidinium conicum* (as *Selenopemphix quanta*) tend to have responded negatively to eutrophication and inorganic pollution, respectively, unlike records from the Norwegian fjords.

St. Lawrence Estuary, Canada: cultural eutrophication signal

Historical measurements of dissolved oxygen in the Lower St. Lawrence Estuary have revealed a decrease of nearly 50% over the past 70 years. This hypoxia is attributed to rising eutrophication linked to increasing land clearance and industrial development, especially in the paper mill industry, which peaked in the second half of the twentieth century (Thibodeau *et al.* 2006). Two cores 9 km apart were studied by Thibodeau *et al.* (2006) for dinocysts, benthic foraminifera and geochemistry to explore the relationship between surface water productivity and dissolved oxygen levels over the past several decades.

The dinocyst record at both sites shows an increase in cyst concentration and flux since the 1960s, and especially from *c.* 1980, and this is attributed to an increase in surface-water productivity. Slight differences in assemblage compositions in the upper parts of both cores seem related to localized changes in hydrography. The rise in dinocyst concentration and flux is mirrored by increased organic matter flux to the sediment and a proliferation of benthic foraminifera. These relationships therefore support the hypothesis that the progressive depletion of dissolved oxygen in the bottom waters of the lower St. Lawrence Estuary is at least partly related to increased primary productivity in the surface waters (Thibodeau *et al.* 2006).

Tokyo Bay, Japan: cultural eutrophication and pollution signals

Studies conducted by Matsuoka (1999, 2001) on two sediment cores from Yokohama Port, Toyko Bay (stations 1 and 2), cover the past *c.* 150 years, over which time Yokohama Port has become one of the most polluted and eutrophicated environments anywhere. Eutrophication in Toyko Bay began in the early 1930s, although the highest concentrations of nutrients are recorded between 1960 and 1965. Increased organic carbon since the 1900s, with highest concentrations in the 1950s and 1960s, reflects the impacts of urbanization and industrialization. Mirroring these results, such heavy metals as Zn, Cu and Pb have gradually accumulated since the 1920s, with the highest concentrations recorded in the 1950s and 1960s.

Of the two cores examined by Matsuoka (1999, 2001), the core at station 1 showed total cyst concentrations at approximately constant levels, although with erratic fluctuations from *c.* 1970, and with heterotrophic dinocyst concentrations somewhat increased after the 1950s. The core at station 2 showed, with the exception of high concentrations in the uppermost sample, a general decline in concentrations from the 1930s onwards, mostly caused by the decline in autotrophs, with heterotrophic cyst concentrations remaining fairly constant. If cyst concentrations in this case are related to productivity, then no significant decline in productivity has taken place during the later phase of heavy pollution and eutrophication, and indeed there has been a moderate increase in heterotrophic cyst production at station 1. This contrasts with the results from Frierfjord, Norway, where heavy pollution was linked to reduced cyst production (Sætre *et al.* 1997). Matsuoka (2001) noted that in Tokyo Bay, heavy pollution can actually lead to increased phytoplankton productivity (especially diatoms) on which the heterotrophic dinoflagellates feed. This could then account for an increase in their cysts in the core at station 1.

Discussion

The foregoing examples do not represent a comprehensive review of the literature (see Dale 2009 for a more complete appraisal), but they show the potential of the cyst record to capture anthropogenic impacts where these are relatively extreme or localized. These examples also highlight the sometimes challenging task of untangling 'natural' environmental signals, such as the North Atlantic oscillation, from human-induced ones.

Signals vary depending on whether the impact is caused by eutrophication, often characterized by an increase in *Lingulodinium machaerophorum* (in Scandinavia) and in cyst production, or by industrial pollution causing an increase in the heterotrophic/autotrophic cyst ratio simply by limiting the numbers of autotrophic cysts, that is without necessarily an associated increase in cyst production. It is also evident that cyst signals for pollution and eutrophication may vary depending on the local hydrographic setting.

The cyst record of human impact in freshwater systems is less well documented than for marginal marine environments, in part because of the much lower diversity of cysts in freshwater environments, their generally less distinctive morphologies, and the ubiquitous availability of other well-tested proxies for environmental change such as pollen, diatoms and thecamebians. Nonetheless, recent studies of the Great Lakes region of Canada (e.g. McCarthy *et al.* 2011) and elsewhere (Mertens *et al.* 2012) have documented the sensitivity of the freshwater cyst record to eutrophication.

The presented examples, and others (Dale 2009), show that the cyst signal for anthropogenic impact is related to local phenomena, such as population growth, industrialization, land clearance and reclamation, and modern agricultural practices. These began as early as the mid-nineteenth century in the examples given. It remains to be seen whether dinoflagellate cyst studies can contribute to the identification of anthropogenically-induced chronostratigraphic events on the global scale.

Foraminifera

Introduction

Foraminifera (see Scott *et al.* 2001; Sen Gupta 2002; Murray 2006 for further reading) are unicellular protozoans. Both planktonic and benthic species exist, and have colonized virtually every marine environment over a wide range of latitudes, depths and salinity gradients. Although naked forms occur, the eukaryotic cell is enclosed by a test in most species, which is typically in the 50–500 μm size range. Depending on test composition and structure, two morphological groups are recognized. Agglutinated species make up their test from detrital grains collected in the environment, which are glued together with organic cement. Calcareous species secrete calcite crystals that are arranged either randomly (white, opaque 'porcellaneous' wall structure) or following a defined pattern (translucent, glassy 'hyaline' wall structure). As they accumulate on the seafloor, agglutinated and calcareous tests produce an abundant and diverse fossil record dating back as far as the Cambrian. Owing to rapid evolutionary changes through geological time, fossil foraminifera are fundamental to biostratigraphy. They served as a basis for concepts founded by eighteenth-century French micropalaeontologist Alcide d'Orbigny, and have been routinely used since then to correlate rock strata, especially for petroleum exploration purposes (Cushman 1928). Since the more recent development of radiometric dating techniques, foraminifera have also been increasingly used to document Quaternary environmental changes, based on knowledge of present-day living species (Haslett 2002). Benthic foraminifera, which prevail in nearshore areas, are sensitive bio-indicators (Schönfeld *et al.* 2012) because they show a wide variety of ecological preferences regarding trophic strategies and habitat selection, and because they allow both long-term water quality decline and episodic degradation events to be identified, due to their short life cycle. Therefore, many case studies have described changes in faunal composition and the diversity of foraminiferal assemblages in response to human-induced environmental stressors, and this offers an opportunity to improve the micropalaeontological characterization of Anthropocene sediments in nearshore areas.

Eutrophication-induced hypoxia

Gooday *et al.* (2009) have already published an extensive review of the historical records of eutrophication-induced hypoxia based on several proxies, including benthic foraminifera. Considerable attention has been paid to the Louisiana shelf (Blackwelder *et al.* 1996; Sen Gupta *et al.* 1996; Osterman *et al.* 2005, 2008, 2009; Platon *et al.* 2005; Strauss *et al.* 2012), where natural stratification of the water column exacerbates the effects of human-induced hypoxia. Several qualitative foraminiferal indices have been developed in this area, including the A–E index (*Ammonia*:*Elphidium* ratio, Sen Gupta *et al.* 1996), the PEB index (sum of percentages of *Protononion*, *Epistominella* and *Buliminella*, Osterman *et al.* 2005, 2008, 2009), and more recently the A–P index (agglutinated:porcellaneous ratio, Platon *et al.* 2005). The indices provided globally consistent results, but not the cause of changes in faunal composition and diversity. Instead, the rapid disappearance of the porcellaneous species *Pyrgo nasutus* in this area provides a more direct stratigraphic marker (Blackwelder *et al.* 1996). As with the A–E index, stress-tolerant species of the genus *Ammonia* also characterize post-impact sediments in other areas, including Chesapeake Bay where deformities in tests are reported (Karlsen *et al.* 2000), Long Island Sound (Thomas *et al.* 2004), Osaka Bay (Japan; Tsujimoto *et al.* 2008) and the Pearl River Estuary (China; Ye *et al.* 2012). Elongate species of the order Buliminida provide another useful marker of eutrophication events in the sedimentary record, including in the Po Delta (*Hopkinsina* and *Bolivina*, Barmawidjaja *et al.* 1995), St. Lawrence Estuary (*Brizalina* and *Bulimina*, Thibodeau *et al.* 2006) and the Skagerrak silled basins (*Stainforthia*, Alve 1996, 2000; Alve *et al.* 2009). In the Skagerrak records, a shift from calcareous to agglutinated-dominated assemblages was also recorded. Although oxygen depletion may result from natural processes, foraminifera-based records of eutrophication indicate a dramatic increase in the frequency and intensity of bottom-water hypoxia events since the mid-twentieth century, which coincides with increased use of N and P-based fertilizers (Blackwelder *et al.* 1996).

Heavy metal and organic pollution

The impact of industrial activities on foraminiferal assemblages has been extensively documented in Spain, including the Bilbao (Cearreta *et al.* 2000, 2002), Huelva (Ruiz *et al.* 2004, 2009) and Suances (Irabien *et al.* 2008) estuaries. Exposure to increased concentrations of heavy metals coincided with a complete defaunation of sediments. Restoration plans applied in the mid-1980s allowed limited post-impact recovery in the Suances estuary. A similar defaunation and recovery trend is documented in the Bosphorus Strait (Meriç *et al.* 2007), but low-resolution sampling and inaccurate dating limits the comparison of records. Exposure to environmental stress, especially heavy metal pollution, induces morphological abnormalities in foraminiferal tests. While many studies have explored the relationships between pollution and test deformation based on both *in situ* and laboratory studies (e.g. Samir 2000), very few have applied the proxy to fossil records. Deformed *Melonis barleeanus* appeared in sediments from Kangerlussuaq Fjord, western Greenland (Elberling *et al.* 2003) in response to increased Zn and Pb contamination from submarine tailing disposal of mining wastes. In New Bedford Harbour (Nova Scotia, eastern Canada), the appearance of deformed *Haynesina orbiculare* was correlated with high concentrations of PCBs in the sediments (Scott *et al.* 2005). It also corresponded to a shift from calcareous *Elphidium* to agglutinated *Textularia* and *Ammobaculites*-dominated assemblage. Studies have also documented the impact of organic pollution originating from pulp and saw mill industries in Drammensfjord, Norway (Alve 1991), of sewage outfall in Santa Monica Bay, USA (McGann *et al.* 2003), and in Manukau Harbour, New Zealand (Matthews *et al.* 2005), and from aquaculture in Bliss Harbour, Canada (Scott *et al.* 1995). By contrast with heavy metal pollution, post-impact sediments in this case show a very consistent foraminiferal content, with more abundant buliminids (*Bulimina*, *Stainforthia*, *Virgulopsis*) and agglutinated species (*Eggerella*, *Reophax*, *Spiroplectammina*, *Trochammina*). Likewise, the buliminid *Saidovina karreriana* provides a stratigraphic marker of organic matter enrichment of the Tagus Prodelta (Bartels-Jónsdóttir *et al.* 2009). The timing of changes related to both heavy metal and organic pollution recorded by foraminifera is variable, but coincides with either the onset of the Industrial Revolution or the mid-twentieth century.

Water management practices

Water management practices modify the amount of freshwater, sediment and nutrients transported by rivers to nearshore environments. The foraminiferal content of post-disturbance sediments is difficult to compare, because of the variable effects of such practices. In Japan, the agglutinated species *Trochammina hadai* characterizes the impact of constructing dykes in Isahaya Bay (Akimoto *et al.* 2004) and Nakaumi Lake (Katsuki *et al.* 2008), as well as widening the Kugushi Lake channel (Nomura & Kawano 2011). Greater confinement and reduced exchange with the open ocean increased nutrient concentration and promoted eutrophication-induced hypoxia, along with salinity changes, and the consequent development of the stress-tolerant *T. hadai*. Following similar processes, oxygen depletion also explains faunal changes that coincided with heavy metal pollution in the Bilbao Estuary (Cearreta *et al.* 2002). The deleterious effects of industrial effluents were enhanced by artificial confinement resulting from dyking and dredging, which turned the natural estuaries into straight, narrow tidal channels with poor renewal of water masses. River channel diversion provoked a drastic salinity decrease in Valo Grande tidal inlet, Brazil (de Mahiques *et al.* 2009), in which a diverse calcareous *Ammonia* and *Elphidium* assemblage was sharply replaced by a single agglutinated *Miliammina* species. River damming of the Sele River (Italy; Vallefuoco *et al.* 2012) was evidenced in the foraminiferal record by the unprecedented appearance of a *Bulimina* species in the Gulf of Salerno. The most complex changes related to water management practices have been described in Florida Bay, southeastern USA. Despite natural salinity fluctuations recorded as cyclic transitions between *Ammonia*–*Elphidium* to *Miliolinella* (Brewster-Wingard & Ishman 1999), a persistent shift to a *Quinqueloculina*-dominated assemblage (Ishman 2000) occurred in response to dyking. Increased retention of organic matter in the closed bay also favoured the development of the buliminid genus *Bolivina* (Cheng *et al.* 2012). The timing of changes related to water management practices varies, but is not older than 1850 (de Mahiques *et al.* 2009).

Introduction of non-indigenous species

In nearshore environments, non-indigenous species are intentionally or accidentally introduced for aquaculture purposes, and via ship ballast waters or hull fouling. A few examples of the introduction of non-indigenous benthic foraminiferal species have been reported, but only two of these have been dated to historical times. Grenfell *et al.* (2007) found specimens of stress-tolerant *Elphidium vellai* in the uppermost 20 cm of sediments under oyster farms of Mahurangi Harbour, New Zealand, which corresponds to a date of introduction between the 1950s and 1980s. The best documented case is that of *Trochammina hadai* (see above), which is a native of the northwestern Pacific and especially Japan (Akimoto *et al.* 2004; Katsuki *et al.* 2008; Nomura & Kawano 2011). The species was identified in sub-surface sediments of San Francisco Bay (McGann 2008) and Padilla Bay, western USA (McGann *et al.* 2012). After an initial expansion followed by a setback, the species now dominates the surface assemblage in the latter, with a relative abundance greater than 80%. In the absence of an exploitable chronological model, the authors postulated that *T. hadai* arrived in Padilla Bay around the 1930s. Indirectly, the introduction of the cordgrass *Spartina alterniflora* in New Zealand estuaries from the 1910s to 1960s triggered the expansion of agglutinated species commonly found in salt marshes. Based on the occurrence of *E. vellai* and *T. hadai* in sub-recent sediments, non-indigenous benthic foraminifera are not yet a reliable stratigraphic marker on a global scale. However, such species might become more frequent if maritime transportation exchanges intensify, and they would therefore help to characterize Anthropocene sediment layers better in the future.

Land-use and land-cover changes

Land-use and land-cover (LULC) changes are provoked by human activities on continental surfaces such as deforestation, agriculture, or stone extraction. Along with natural processes (topography, climate, pedology), LULC changes control soil erosion and consequently the amount of sediments that settle in nearshore environments. In contrast to the previous categories of anthropogenic disturbance, which induced change during the last 200 years, foraminiferal records of LULC changes point to much earlier impacts. Development of agriculture in the Fertile Crescent during the Bronze Age triggered a pulse of muddy sedimentation around 3700 years BP in Iskenderun Bay, Turkey. Benthic foraminiferal diversity increased, but no major change in faunal composition was evident in this case (Spezzaferri *et al.* 2000). In response to deforestation, increased fine-grained sedimentation in Loch Etive, Scotland (*c.* 1000 years BP; Nørgaard-Pedersen *et al.* 2006), and in several estuaries around Auckland city, New Zealand (*c.* 1950; Hayward *et al.* 2004, 2006, 2008; Grenfell *et al.* 2007), was recorded by a shift from calcareous to agglutinated-dominated assemblages. In the latter, a drastic decline in salinity resulting from increased freshwater runoff was attributed by the authors to the urban development of impervious surfaces such as roofs and made ground. Post-impact sediments were also characterized by more abundant *Bulimina* species off Guadiana Estuary (*c.* 1500–1200 years BP; Mendes *et al.* 2010) and in the central Adriatic Sea (1345 years BP; Oldfield *et al.* 2003). An inverse trend was described by Baptista Neto *et al.* (2012) in Jurujuba Sound, Brazil, where *Ammonia* replaced *Buliminella* in response to greater soil erosion. In this case, however, exposure to increased heavy metal concentrations from untreated sewage outfall may account for this ecological transition. LULC changes have also increased the amount of sediments deposited in coral reef ecosystems, where symbiont-bearing species are a distinctive feature of benthic foraminiferal assemblages. For example, in the Great Barrier Reef, Australia (Uthicke *et al.* 2012), higher concentrations of suspended matter in the water column allowed the expansion of three low light-tolerant species of symbiont-bearing foraminifera.

Discussion

Benthic foraminifera are sensitive bio-indicators that successfully document ecological changes in response to anthropogenic disturbances, including eutrophication-induced hypoxia, heavy metal and organic pollution, water management practices, introduction of non-indigenous species, and LULC changes. Although not uniform at a global scale, the foraminiferal content of post-impact sediments in nearshore environments is consistent. Regardless of the category of environmental stressors and of the timing of change, some striking similarities between the records cited above are found (Fig. 5), including the following:

(1) a shift from calcareous to agglutinated-dominated assemblages (Alve 1996; Hayward *et al.* 2004, 2006, 2008; Nørgaard-Pedersen *et al.* 2006; Grenfell *et al.* 2007; Alve *et al.* 2009);

(2) the expansion of species of the order Buliminida (Alve 1991, 2000; Barmawidjaja *et al.* 1995; Elberling *et al.* 2003; McGann *et al.* 2003; Oldfield *et al.* 2003; Matthews *et al.* 2005; Thibodeau *et al.* 2006; Alve *et al.* 2009; Bartels-Jónsdóttir *et al.* 2009; Mendes *et al.* 2010; Vallefuoco *et al.* 2012);

(3) the expansion of species of the genus *Ammonia* (Akimoto *et al.* 2004; Thomas *et al.* 2004; Tsujimoto *et al.* 2008; Nomura & Kawano 2011; Baptista Neto *et al.* 2012; Ye *et al.* 2012).

These similarities are fully consistent with ecological studies, which often describe the above-mentioned benthic foraminifera as stress-tolerant species, especially with respect to oxygen depletion or organic matter enrichment. With the exception of a few cases (especially Barmawidjaja *et al.* 1995), most changes in faunal composition or diversity are sharp, and coincide with lithological or chemical boundaries in the sedimentary records. This review suggests that benthic foraminifera could be used as an effective biostratigraphic marker of the early Anthropocene transition, as they leave a distinguishable imprint of anthropogenic disturbances in marine sediments.

Ostracods

Ostracods are crustaceans that secrete calcareous carapaces, are generally between 500 and 1500 μm in length, and occupy all aquatic environments from the deep sea to inland freshwater ponds and wetland to terrestrial environments. As carapaces are preserved after death, they provide a record of change through time and form ideal proxies in palaeoenvironmental modelling (de Deckker *et al.* 1988). In addition to assemblage and population structure, environmental change can be determined by chemical analysis of carapaces because, during ecdysis, information regarding their immediate environment (e.g. Na, Al, K, Fe, S, P and traces of Cr, Mn, Cu, Zn, As and Sr) is preserved within the crystal lattice. Ostracod populations from the coastal zone (e.g. Schornikov 2000*a*) through to freshwater systems (e.g. Mezquita *et al.* 1996;

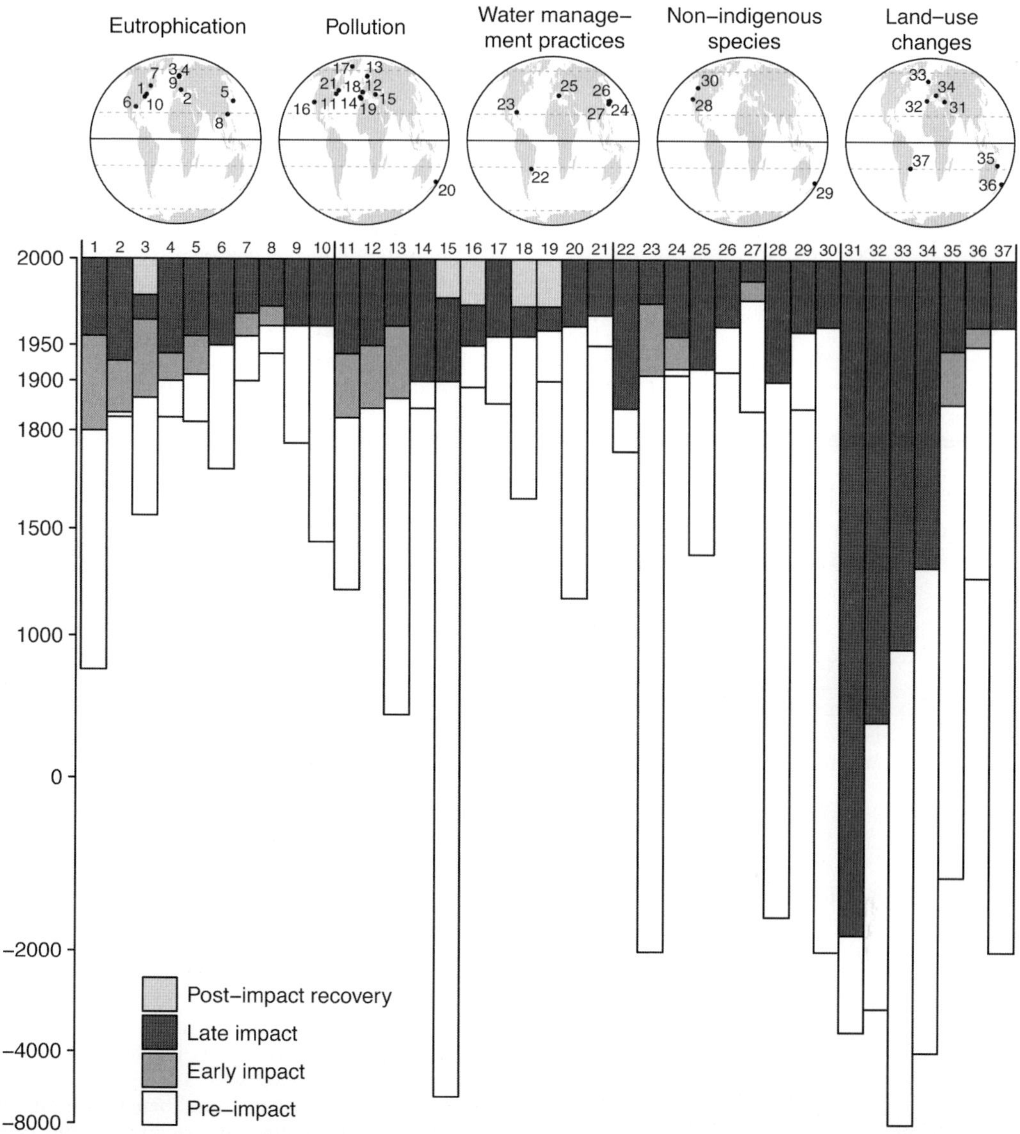

Fig. 5. Relationship between changes in benthic foraminiferal assemblages and human drivers through time. Note the non-linear timescale. (1) Long Island Sound, Thomas *et al.* 2004; (2) Po Delta, Barmawidjaja *et al.* 1995; (3) Frierfjord, Alve 2000; (4) Oslofjord, Alve *et al.* 2009; (5) Osaka Bay, Tsujimoto *et al.* 2008; (6) Louisiana Shelf, Blackwelder *et al.* 1996; Sen Gupta *et al.* 1996; Osterman *et al.* 2005, 2008, 2009; Platon *et al.* 2005; Strauss *et al.* 2012; (7) St. Lawrence Estuary, Thibodeau *et al.* 2006; (8) Pearl River Estuary, Ye *et al.* 2012; (9) Skagerrak Strait, Alve 1996; (10) Chesapeake Bay, Karlsen *et al.* 2000; (11) New Bedford Harbor, Scott *et al.* 2005; (12) Bilbao Estuary, Cearreta *et al.* 2002; (13) Drammensfjord, Alve 1991; (14) Tagus Delta, Bartels-Jónsdóttir *et al.* 2009; (15) Bosphorus Strait, Meriç *et al.* 2007; (16) Santa Monica Bay, McGann *et al.* 2003; (17) Kangerlussuaq Fjord, Elberling *et al.* 2003; (18) Suances Estuary, Irabien *et al.* 2008; (19) Huelva Estuary, Ruiz *et al.* 2004, 2009; (20) Mahurangi Harbour, Matthews *et al.* 2005; (21) Bliss Harbour, Scott *et al.* 1995; (22) Valo Grande Channel, de Mahiques *et al.* 2009; (23) Florida Bay, Brewster-Wingard & Ishman 1999; Ishman 2000; Cheng *et al.* 2012; (24) Nakaumi Lake, Katsuki *et al.* 2008; (25) Gulf of Salerno, Vallefuoco *et al.* 2012; (26) Kugushi Lake, Nomura & Kawano 2011; (27) Isahaya Bay, Akimoto *et al.* 2004; (28) San Francisco Bay, McGann 2008; (29) Auckland region, Grenfell *et al.* 2007; (30) Padilla Bay, McGann *et al.* 2012; (31) Iskenderun Bay, Spezzaferri *et al.* 2000; (32) Guadiana Estuary, Mendes *et al.* 2010; (33) Loch Etive, Nørgaard-Pedersen *et al.* 2006; (34) Adriatic Sea, Oldfield *et al.* 2003; (35) Great Barrier Reef, Uthicke *et al.* 2012; (36) Auckland region, Grenfell *et al.* 2007; Hayward *et al.* 2004, 2006, 2008; (37) Jurujuba Sound, Baptista Neto *et al.* 2012.

Külköylüoğlu *et al.* 2007) have suffered human impact in terms of water pollution by industrial and agricultural processes, sewage and concomitant eutrophication.

Eutrophication is caused by nutrient enrichment resulting in the increased production of particulate organic matter that may lead to reduced oxygenation of the water (hypoxia) and, in extreme cases, may be sufficient to induce anoxia (e.g. Gooday *et al.* 2009). Ostracods are usually intolerant of hypoxia and respond by a reduction in diversity and richness, and, in some cases, populations become monospecific (e.g. Boomer 2002; Frenzel & Boomer 2005). There are a few exceptions, such as *Cyprideis torosa* (Vilela *et al.* 2003) and, perhaps, *Heterocyprideis sorbyana* and *Candona neglecta*, which survive acute hypoxia by using anaerobic metabolic pathways to detoxify sulphide (Jahn *et al.* 1996; Modig & Olafsson 1998).

Industrial impacts

Industrial pollution and the need to monitor and mitigate its effect on human health and natural resources has led to the realization that ostracods can be used as bio-indicators to assess the impact of toxins, heavy metals etc. on freshwater and marginal marine ecosystems. Laboratory experimentation shows that freshwater ostracods are particularly useful in measuring the toxicity of industrial pollutants (Khangarot *et al.* 1983: Bodergat *et al.* 1998; Sánchez-Bayo 2006), heavy metals (Khangarot & Ray 1987; Bergin *et al.* 2006), sewage sludges (Hamdi *et al.* 2006; Oleszczuk 2007, 2008) and other human activity. Sánchez-Bayo & Goka (2006), for example, showed that organic pollutants used in insecticides, toiletries and antifouling paints (imidacloprid, a neonicotinoid insecticide, and zinc pyrithione) were acutely toxic to *Ilyocypris dentifera*, *Cypridopsis vidua* and *Cyretta seurati*.

The uptake of metals via the food chain by the freshwater ostracod *Chrissia halyi* is rapid (Prasuna Solomon *et al.* 2012). Significant amounts of toxins accumulated after just four days and the mortality rate increased with the concentration of lead nitrate and exposure time. However, the tolerance limits of ostracods to toxicity and the concentration of pollutants required to cause mortality or mutation (such as dwarfism) is known for only a very few species studied under laboratory conditions. Nevertheless, they prove their potential in providing temporal information.

The reaction of *Stenocypris major* to a number of elements associated with anthropogenic pollution (Cu, Cd, Zn, Pb, Ni, Fe, Al and Mn) and the bioconcentration of these metals in its body were measured under laboratory conditions (Shuhaimi-Othman *et al.* 2011). It succumbed most readily to the toxicity of Cd followed by Cu, Fe, Mn, Pb, Zn, Al and Ni, and bioconcentrations of Cu and Cd were high. The toxicity of Cu on *Cypris subglobosa* changed with pH value, with mortality rates being higher in the pH range 5.5–6.5 than at 8.5 and above (Khangarot & Ray 1987). This indicates that there is a relationship between acidification of freshwater bodies (as a result of contamination by acids and dissolved metals, such as Cu and Zn) and mortality. The species was most sensitive to the toxicity of Os, and linear regression analysis indicated a significant relationship between the biotic immobilization of this ostracod and electronegativity, ionization potential, solubility product of metal sulphides and other physicochemical factors (Khangarot & Das 2009). The value of the pollution-sensitive *Heterocypris incongruens* as a biomonitor was proved when ecotoxicological laboratory tests were evaluated by monitoring the relationship between growth inhibition (at sublethal pollution levels), immobilization and mortality in contaminated sediments from various rivers in Belgium and Canada (Chial & Persoone 2002*a*, *b*, *c*; Chial *et al.* 2003). Specimens from urban and industrial sites demonstrated that this species is particularly sensitive to the toxicity of PAHs (Manzo *et al.* 2008) and hydrophobic compounds (Hamdi *et al.* 2006; Oleszczuk 2008).

The effects of industrial activities on ostracods can be observed in many parts of the world, and a number of studies in Japan have resulted in a better understanding of the relationship between ostracod populations and anthropogenetic influences (e.g. Yasuhara & Yamazaki 2005). In the inner part of Osaka Bay, the total abundance of ostracods decreased between 1910 and 1920 (Yasuhara *et al.* 2007). However, eutrophication appears to have accelarated through the 1950s and 1960s and acute hypoxia was reached during the 1970s. Although there was a major decline in diversity, some ostracods faired better than others, the more tolerant being *Bicornucythere bisanensis*, *Cytheromorpha acupunctata* and *Loxoconcha viva*. Reduced ostracod population densities were also observed in Hiroshima Bay, commencing in 1940, caused by the rapid increase in heavy metal concentrations (Cu, Zn, Pb) related to the increased industrialization that began during the Second World War (e.g. Bodergat & Ikeya 1988; Bodergat *et al.* 1997; Yasuhara *et al.* 2003, 2007). Ostracodal response to pollutants varied, so some species (e.g. *Callistocythere alata*) proved to be very sensitive, whereas others (such as *Bicornucythere bisanensis* and *Cytheromorpha acupunctata*) were more resistant. Even species resistant to pollutants may show a change in population structure, as in Ise Bay and Mikawa Bay, where there is a relationship between low ostracod productivity and high concentrations of Zn, Pb, Cr or Cu. Evidence suggests that pollution may also result in morphological change, as in Aburatsubo Cove, Japan, where *Bicornucythere bisanensis* exhibited a mutated carapace ornamentation in modern assemblages that was not observed in pre-industrial specimens (Abe 1983).

Ostracod abundance and diversity is related to industrial sewage and urban effluents, although there is local variability (Ruiz *et al.* 2005), as can be seen in examples in southern Europe. In Spain, for example, the absence of ostracods within the Gernika Estuary since *c.* 1940 can be related to the high levels of heavy metals, principally As and Pb (Pascual *et al.* 2002), whereas in the Tinto-Odiel Estuary, metals such as Pb, Cu, Cr and Zn derived from mining, industrial waste (from the production of fertilizers, hydrocarbons and chemicals) and the introduction of urban effluents have had the same effect in several parts of the estuary (Ruiz *et al.* 2004). Further east, the industrialized inner part of the Gulf of Izmir, Western Turkey, is contaminated by numerous heavy metals (e.g. Cu, Zn, Ni, Pb, Cr, Hg and Cd) from such manufacturing activities as iron, paper, antifouling paints, chlorine-alkali plants, cement factories, beer industries and shipping, together with many domestic discharge outlets (Bergin *et al.* 2006). Of the 27 ostracod species recorded in the middle and outer parts of the Gulf of Izmir, only four species tolerated the highly polluted inner part of the Gulf: *Carinocythereis carinata*, *Bosquetina dentata*, *Xestoleberis depressa* and *Loxoconcha rhomboidea*. The negative response to heavy metals such as copper (Lenihan *et al.* 2003) and zinc (Watzin & Roscigno 1997) under laboratory conditions can be related to populations affected by mining pollution in northern Chile (Lee & Correa 2005).

In order to monitor the effects of industry on ostracod populations during the Anthropocene transition, temporally controlled successions are required. The study of Chesapeake Bay by Cronin & Vann (2003) is a good example. Between the fourteenth and seventeenth centuries, when ostracod populations were controlled by natural conditions, *Loxoconcha* sp., *Cytheromorpha newportensis* and *Actinocythereis captionis* formed characteristic assemblages in well-oxygenated waters and *Cytheromorpha curta* thrived in conditions of low dissolved oxygen. The first signs of anthopogenic effects occurred during the eighteenth and nineteenth centuries, when increases in the proportion of *Leptocythere nikraveshae* and *C. curta* coincided with a period of increased organic C, N and S, progressive oxygen depletion and higher turbidity and sediment flux caused by extensive land clearance for agriculture and timber production. However, the main phase of anthropogenic change appears to have taken place between 1920 and 1960, when *P. brachyforma* became abundant, *C. newportensis* and *Loxoconcha* sp. decreased and *A. captionis* disappeared. Between 1934 and 1948, oxygen depletion became an issue in Chesapeake Bay and the frequency of severe hypoxia increased. After 1960, *C. curta* increased in dominance and *Paracytheroma repexa* appeared, although, in general, ostracods virtually disappeared, reflecting the acute hypoxia and a general decline in water quality associated with anthropogenic activities.

Within Alpine lakes, anthropogenic eutrophication began in many lakes in the twentieth century due to industrial impacts, engineering works and deforestation. Eutrophication affected lacustrine ostracod communities such as *Cytherissa lacustris*, *Limnocythere inopinata* and *Fabaeformiscandona protzi*, and particularly the rarer components such as *Leucocythere mirabilis*, which is locally extirpated and found dead in a number of lakes (Löffler 1969, 1972, 1983; Danielopol *et al.* 1985, 1988, 1990; Scharf 1998; Geiger 1990*a*, *b*, 1993; Filippi *et al.* 1999). In Lake Sevan, Armenia, eutrophication has similarly reduced the richness of the living ostracod community (Wilkinson *et al.* 2005; Wilkinson & Gulakyan 2010) and the increased dominance of environmentally tolerant taxa. Several species recorded prior to 1950 were not found living in the lake in 2005, including *Cytherissa lacustris* and *Ilyocypris gibba*, while *Cyclocypris ovum* is very rare, perhaps mirroring the decline in macrophytes (Fig. 6).

Oil pollution

Nearshore marine and brackish water ostracods are highly sensitive to oil pollution, mortality being high in contaminated surface sediments, as demonstrated by laboratory and field experiments (Liljenstroem *et al.* 1987; Ruiz *et al.* 2005). The effect of hydrocarbons on ostracod communities was demonstrated by Schornikov (2000b), who briefly outlined the extreme impact of anthropogenic pollutants on the marginal marine ostracods of the Vladivostok Port area, Russia. Living specimens were absent from the heavily polluted, nearshore environments as well as the main entrance channels of the harbour, where the bottom sediments are covered by an oil film that also prevents algal growth.

During the 1991 Gulf War, vast quantities of Kuwaiti crude oil poured into the Gulf and, together with other by-products associated with burning oil, resulted in severe pollution and a deleterious impact on the ecosystem. Plumes of burning oil also had the effect of reducing mean sea temperatures by as much as 6.9 °C and significant hypoxia caused major faunal decline and increased microbial activity. Mostafawi (2001) summarized the extent

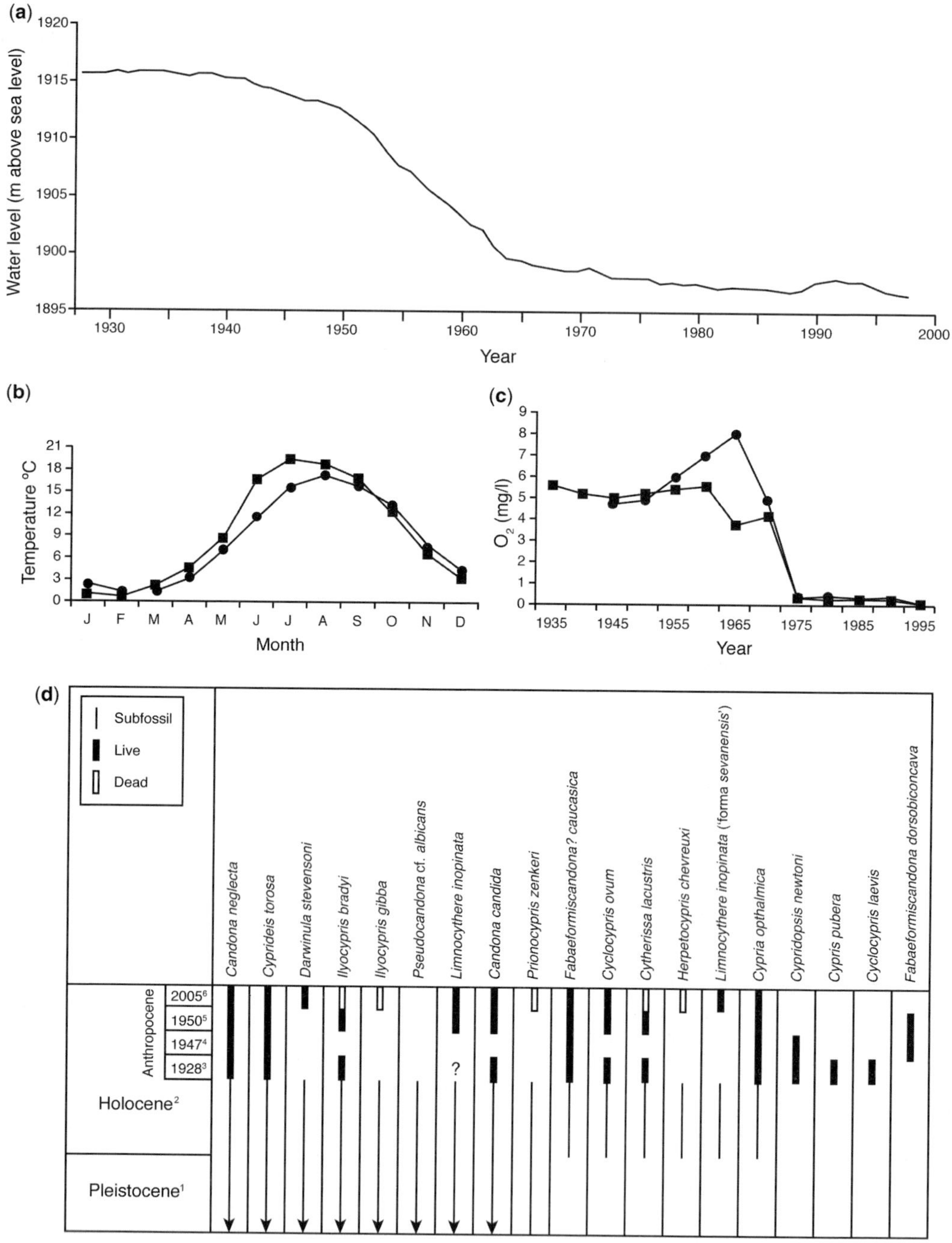

Fig. 6. Anthropogenic change in Lake Sevan, Armenia: (**a**) water level change due to extraction for irrigation and hydroelectricity generation; (**b**) average surface water temperature in Great Sevan (dashed line 1927–1933; solid line 1982–1985) indicating warmer waters in spring and summer; (**c**) the collapse of oxygen concentration in the bottom waters of Little Sevan (dashed line) and Great Sevan (solid line) during the late twentieth century (data after Hovhannissian 1994, 1996); (**d**) Ostracod populations from the Pleistocene to present day ([1]Bubikyan 1984; [2]Wilkinson *et al.* 2005; [3]Bronstein 1928; [4]Bronstein 1947; [5]Friedman 1950; [6]Wilkinson & Gulakyan 2010 (after Wilkinson & Gulakyan 2010). 'Anthropocene' is shown informally to indicate that part of the lake's history during which human activity has resulted in major changes to the ecosystem.

of the contamination, the greatest pollution being found in the halophyte marsh/algal complexes and mudflats at the head of sheltered bays, where oil had penetrated into the sediment substrate and polycyclic aromatic components were greatly increased. Comparison of the ostracod population pre- (Paik 1977) and post- (Mostafawi 2001) conflict prove a clear deterioration of the ostracod assemblage associated with the heavily contaminated surface sediments. Ostracods with soft parts were very rare after 1991, and the closed carapaces of both herbivores and detrivores (e.g. *Keijella karwarensis*, *Krithe kroemmelbeini*, *Neomonoceratina iniqua* and *Propontocypris bengalensis*) recovered from the uppermost surface sediments contained particles of bitumen.

Sewage

The discharge of untreated or poorly treated urban effluent and varying concentrations of biological or chemical pollutants into inshore brackish and marine ecosystems results in eutrophication and hypoxia. Some ostracod species are particularly sensitive to high levels of pollution, whereas others can tolerate a broader spectrum of conditions. Discharge of untreated domestic effluent causes low diversity, sometimes monospecific, populations, as is the case in Nador Lagoon, Morrocco, where the opportunistic, hypoxic-tolerant species *Palmoconcha turbida* is the most successful (Ruiz *et al.* 2005; Bodergat *et al.* 1998). However, ostracods recover rapidly away from the effluent. Near Wellington, New Zealand, only *Loxoconcha parvifoveata* survived in the area adjacent to a sewage discharge, but just 200 m away, *Xestoleberis atra*, *L. punctata* and *L. parvifoleata* were present, and diversity rose to eight species just 550 m from the discharge (Eager 1999). A similar relationship between ostracods and sewage pollution was found at Tarawa Atoll, Republic of Kiribati, where *Touroconcha marcida* was found to be the most tolerant of sewage (Eager 2000).

Sewage-contaminated water discharged into freshwater environments usually results in a decrease in oxygen content and an increase in the concentration of toxic substances, resulting in a loss of faunal diversity. Ostracod faunas become biased towards the more tolerant species and there is some evidence that certain taxa can be used as indices for the degree of contamination, thereby having potential in the recognition of the lower boundary of the Anthropocene. In the Magre River, Spain (Mezquita *et al.* 1999), *Herpetocypris brevicaudata*, *Cypridopsis vidua*, *Sarscypridopsis lanzarotensis* and *Ilyocypris bradyi* occurred where water quality was clean to moderately polluted (b-mesosaprobic to oligosaprobic). However, *Herpetocypris intermedia*, *Heterocypris incongruens*, *H. salina* and sometimes *Eucypris virens* were more frequent in high to moderate pollution (a- to b-mesosaprobic waters) and *Trajancypris clavata* was found to be one of the more tolerant species, being found in oligosaprobic–mesosaprobic waters. Ostracods were totally absent at highly polluted sites due to acute hypoxia and high concentrations of toxins such as ammonia, a situation that has been observed in several studies (e.g. Bromley & Por 1975; Rosenfeld & Ortal 1983). The Harod Valley, northern Israel, is an anthropogenic fluviatile system in which much of the water is recycled for fish farming, irrigation systems and sewage flushing (Rosenfeld *et al.* 2000). High concentrations of pollutants, including ammonia and phosphates, have been recorded at some localities. Ostracod diversity and density vary with pollution levels and respond quickly to change; *Ilyocypric bradyi*, *Candona neglecta*, *Cypridopsis aculeata* and *Cyprideis torosa* are restricted to clean to oligosaprobic waters, whereas *H. salina* dominates in moderate pollution (oligosaprobic to mesosaprobic) and *H. incongruens* dominates in higher levels of contamination (oligosaprobic–polysaprobic). Similarly, at Ouseburn, northeastern England, pollutants include sewage, agricultural slurry and de-icer runoff from the local airport, and more than 90% of the total population were pollution-tolerant (*Cypria ophthalmica*, *Candona candida*, *Cyclocypris ovum*, *Ilyocypris bradyi* and *Candona neglecta*), although *Ilyocypris bradyi* and *Candona neglecta* occurred in less polluted localities (Boomer & Attwood 2007).

Shell chemistry is useful to recognize anthropogenic effects, although data relating to carapace chemistry and pollution are scarce and more information is required. Investigations have been carried out on the geochemical composition of carapaces of, for example, *Leptocythere psammophyla* (Bodergat 1978*a*, *b*; Bodergat *et al.* 1993) and *Cyprideis torosa* (Carbonel & Tölderer-Farmer 1988; Ruiz *et al.* 2005) in order to determine relationships to environmental variables. Near Marseille, southern France, Ce was concentrated in the carapace of *Aurila speyeri* close to a large sewage outfall (Bodergat 1978*a*; Ruiz *et al.* 2005). In the Camargue, southern France, chlorides and Fe rise in the carapace chemistry of *C. torosa* with increasing industrial sewage (Bodergat *et al.* 1991). Variations in the proportion of Mn, Fe and rare-earth elements were also observed in the carapace of *Cyprideis* spp. from polluted lagoons of America and Europe (Palacios-Fest *et al.* 2003)

Agricultural impacts

It is difficult to determine the extent to which agriculture has affected ostracod populations in the coastal zone, as there are very few data. However, one example is the Louisiana coastline, USA, where agricultural use of fertilizers throughout the twentieth century resulted in nutrient-rich discharge of the Mississippi, causing seasonal hypoxia and a reduction in the diversity of ostracods (Alvarez Zarikian *et al.* 2000; Rabalais *et al.* 2002; Ruiz *et al.* 2005). Hypoxia-tolerant species of *Loxoconcha* were abundant throughout the early twentieth century, eventually reaching flood proportions. Fish farming is another source of pollution and, although carried out on a relatively small scale, organic nutrients (proteins, carbohydrates, lipids and so on) accumulate in bottom sediments, causing dysaerobic and anoxic conditions (Henderson *et al.* 1997). As fish farming will continue to increase as human population rises, it may become a major problem globally as it already has locally, for example, in the Western Mediterranean (Mazzola *et al.* 1999) and eastern Baltic Sea (Pallo *et al.* 1998).

Ostracod biodiversity in the freshwater environments of Winsconsin is also affected by land use, agriculture being the most significant driver. Candonids such as *Candona ohioensis* and *Candona crogmaniana* may be more tolerant of human impacts (Allen & Dodson 2011), although a decrease in *C. ohioensis* in Illinois may be caused by seasonal anoxia driven by agricultural land use and lake development (Curry & Filippelli 2010).

One of the more important influences on ostracod populations is agriculturally induced salinity change. Perhaps the best documented example is the Aral Sea, where major water extraction for irrigation, particularly since 1960, has resulted in increased salinity and a change in water chemistry (e.g. Zenkevitch 1963; Shnitnikov 1969; Kotlyakov 1991; Boomer 1993*a*, 2013; Boomer *et al.* 1996). The change in the ostracod fauna has been profound. Boomer (1993*b*) and Aladin (1993) demonstrated that 11 species occupied the Aral Sea prior to the extraction of water. Between 1974 and 1976 salinity rose to 14 psu and hyperosmotic ostracods disappeared, between 1985 and 1987 salinity rose to 24 psu, and the brackish water species disappeared, and by 1993 salinity had reached 28–30 psu. At these salinities, 10 of the 11 species had disappeared entirely and just *Cyprideis torosa* survived.

Another example where salinity is a cause of concern is the Wheatbelt of SW Australia. Indigenous upland vegetation comprising perennial woodlands and shrublands has been extensively cleared for agriculture. As a result, freshwater wetlands have become, or are becoming, saline, and secondary salinization is also a problem as a result of increased stream and wetland salinity, increased stream flow, more prolonged periods of wetland inundation, and altered water chemistry (Pinder *et al.* 2005). Of the ostracods, halophiles, such as species of *Diacypris*, *Mytilocypris*, *Australocypris*, *Reticypris* and *Cyprinotus*, have become diverse, but freshwater ostracods with a low tolerance to rising salinity have dropped from five to 10 species per wetland at salinities $<2 g\ l^{-1}$, to fewer than seven species at $2-7\ g\ l^{-1}$ and fewer than three species at higher salinities. The salinity 'tipping point' where the richness of freshwater species declines and halophilic species increases is 2–3.2 $g\ l^{-1}$, apparently due to a change in ionic composition of the water (Radke *et al.* 2003; Pinder *et al.* 2005).

Salinity (its concentration, variability and ionic composition) is a major factor controlling the distribution and reproduction of ostracods (e.g. Forester & Brouwers 1985; Horne & Boomer 2000; Ruiz *et al.* 2000; Boomer & Eisenhauer 2002; and references therein). Assemblages are controlled by hydrochemical change, where thalassic water (rich in Na and Cl ions) passes into athalassic water (rich in Ca_2 and HCO_3 ions) below salinities of *c.* 10% (Forester & Brouwers 1985). Assemblage composition, abundance and ecophenotypes also change at this 'tipping point' (e.g. Anadón *et al.* 2002). Ionic concentration controls the abundance of several species, their presence or absence being primarily controlled by salinity (Boomer & Eisenhauer 2002). Salinity has also been attributed to changes in ostracods morphology such as dimensions (Neale 1988; Anadón *et al.* 2002), calcification (for example, *Leptocythere psammophila*, *L. lacertosa* and *Limnocythere floridensis*; Hartmann & Kühl 1978; Kühl 1980) and sieve pore morphology (Rosenfeld & Vesper 1977). Mg/Ca ratios have also been related to ornamentation (Hartmann & Kühl 1978; Carbonel 1982).

Anthropogenic dispersal

Methods usually postulated to account for the spatial distribution of fossil ostracods include such vectors as avian (e.g. Proctor 1964; Proctor *et al.* 1967; De Deckker 1977) and fish (Kornicker & Sohn 1971; Victor *et al.* 1979) migration, and hydrological dispersal by the drift of eggs and carapaces (e.g. Williams & Hynes 1976; Dance & Hynes 1979; Victor *et al.* 1981; Fryer 1981; Fryer & Joyce 1981). However, evidence is accumulating that anthropogenic dispersal is also a factor and a possible characteristic of the Anthropocene.

There are few well-documented instances of anthropogenic introduction of 'alien' ostracods into a new region, but one such example is the myodocope *Eusarsiella zostericola*. This appeared in British coastal waters when the American oyster *Crassostrea virginica* was introduced into the Thames estuary during the early twentieth century for human consumption, and has been recorded in the Blackwater Estuary, Essex, and the River Medway, Kent (Kornicker 1975; Bamber 1987). It later spread from SE England across the North Sea to colonize Oosterschelde near Zierikzee, The Netherlands (Faasse 2013). Risiculture also appears to have been a significant cause of anthropogenic dispersal of ostracods in SE Asia and in between Asia and Europe (Ghetti 1973; Victor & Fernando 1982; McKenzie & Moroni 1986; Petkovski & Meisch 1996). It would appear that ostracods or their eggs have been introduced to paddy fields with seeds or seedlings.

The introduction of ostracods into a region as a result of anthropogenic activities has been surmised after a study comparing the spatial distribution of Holocene fossil ostracods with that of the present day. Witte (1993), for example, pointed out that the living ostracod population of Senegal, West Africa, included *Kotoracythere inconspicua*, *Tanella gracilis* and *Keijia demissa*, but these same species had a Holocene fossil record restricted to the Indo-West Pacific. Their modern distribution is now broadly circumtropical, although within that province they have a very patchy distribution (and are absent from the Eastern Pacific). No other representatives of these genera are found in the Atlantic, and intraspecific variation and polymorphism for these species are higher in the area of their fossil record than the Atlantic (e.g. Witte & Van Harten 1991). Witte (1993) concluded that their centre of origin, the Indo-West Pacific, could be defined by their fossil record and that their modern distribution was the result of dispersal by shipping. *Loxoconcha parvifoveata*, a species that is restricted to just two known localities near ports in New Zealand, also appears to have been dispersed passively from Australia by shipping (Witte 1994; Eager 1999).

Martens *et al.* (2008) pointed out that, in the freshwater realm, very few species have a cosmopolitan distribution and suggested that many species that are found on several (or all) continents (e.g. *Heterocypris incongruens*, *Eucypris virens*, *Cypridopsis vidua*) may have been introduced by human activity. One clear example where freshwater ostracod dispersal can be assigned to anthropogenic causes is that of *Eucypris virens*. The species is found throughout the Holarctic (e.g. Martens 1998; Meisch 2000), but has also been reported from Australia (De Deckker 1981; Davis & Christidis 1997) and New Zealand (Barclay 1966). Koenders *et al.* (2012) analysed DNA sequence data from Western Australian specimens in order to determine whether allopatric speciation had led to the Australian *E. virens* or whether they had been introduced. Phylogenetic reconstructions, genetic networks and estimates of genetic distances proved that *E. virens* of Western Australia were very closely related to European specimens and it was concluded that this species had been anthropogenically introduced into Western Australia, probably from western Europe, and had not yet diverged in Australia.

The distribution of endemic species, with a fossil record, clearly pre-dates human influence; for example, *Fabaeformiscandona siliquosa*, *Eucypris anglica* and *Psychrodromus robertsoni*, which are endemic to the British Isles, probably represent a relict assemblage after glacial decay. However, it is often difficult to determine whether the distribution of an ostracod can be related to human vectors, particularly where taxa are rare, apparently endemic to a few localities, restricted to unusual habitats and/or with no fossil record. Live *Terresticythere*, for example, are known from the Sea of Japan (Schornikov 1980; Hiruta *et al.* 2007) and two British intertidal locations on the southern English coast (Horne *et al.* 2004). This strange geographical distribution begs the question – do these isolated occurrences represent lineages of endemic taxa or, if not, what is the cause of such an unusual distribution? Might dispersal be related to the flight path of migrating aquatic birds or does anthropogenic activity play a part in their distribution? A further complication is that two dead specimens of *Terrestricythere ivanovae* were found in Lac du Bouchet in the Massif Central, France, far from the coast and were presumably introduced into the lake accidentally (Scharf & Keyser 1991).

Discussion

The nearshore coastal strip and freshwater systems have suffered from the introduction of a variety of pollutants, which affect different trophic levels of the ecosystems and, in turn, the morphology, population structure, densities and diversities of the ostracod communities, colonization by 'exotic' species, and so on. The abundance and distribution of key species and their carapace chemistry mean that ostracods can be used to monitor anthropogenic environmentally stressed regions through time (e.g. Malard *et al.* 1996; Mosslacher 2000; Anadón *et al.* 2002; Boomer & Eisenhauer 2002; Ruiz *et al.* 2005). Anthropogenic eutrophication is the principal cause of change within the ostracod community. Although there are earlier historical examples (e.g. Nixon 1995; Clarke *et al.* 2003, 2006; Cronin & Vann 2003), anthropogenic eutrophication has become widespread and more profound only since the middle of the twentieth century (e.g Justíc *et al.* 1987; Ikeya 1995; Alvarez Zarikian *et al.* 2000; Rabalais *et al.* 2002; Cronin & Vann 2003; Conley *et al.* 2007; Cronin *et al.* 2013). If anthropogenic pollution is chosen as a method of defining the base of the Anthropocene, then ostracods have the potential to play a useful role.

Microfloral and microfaunal discussion

Limitations

There can be little doubt that anthropogenic activity has affected ecosystems on a global scale, in fact it is now difficult to find an entirely pristine example (e.g. Halpern *et al.* 2008). In order to recognize the Anthropocene, an episode defined by intense human activity, we have attempted to demonstrate the way in which the biotic response can be used to recognize anthropogenically induced changes of marginal marine and freshwater ecosystems, the timing of such change and its spatial variability. It must be borne in mind, however, that there are large spatial gaps in available datasets, which have a tendency to skew results; parts of Europe, America, Australia and Japan have been studied in some detail, but much of South America, Africa and southeastern Asia remains understudied.

Accurate dating is required to differentiate local historical anthropogenic change from that which might be used to determine the base of the Anthropocene. However, even data related to better known regions are limited in respect to the resolution of dating methods. Without high-resolution temporal data, it is difficult to determine on a global scale when the influence of anthropogenic activities began to be measurable within microfossil assemblages. This poses a significant problem in determining the base of the Anthropocene. For example, radiocarbon (^{14}C) dating has an error of several decades and lacks an effective resolution for the base of the Anthropocene, if its base is within the last few centuries. However, the short half-life of ^{210}Pb means that it may be useful for dating microfaunas and microfloras within the last 200 years and ^{137}Cs can be used for the period post-dating nuclear testing. Other dating methods including tephrochronology, varves, archaeological artefacts or documented historical collections also have limitations. Biostratigraphy is useful where there is an expanded succession and little reworking, although resolution is unlikely to be better than centennial, let alone decadal, as is required to identify the Anthropocene.

From the little information available, it appears that there is a lag between the onset of human activity and its influence on microbiotas. The Industrial Revolution, one of the suggested markers for the Anthropocene (Zalasiewiscz *et al.* 2008), did not immediately result in biotic change on a large scale. Its earliest manifestation can be found in Europe and the USA where local changes can be observed, but industrialization did not affect Japan until the second half of the twentieth century and in many areas it did not become profound and widespread until increased industrialization during the Second World War, from *c.* 1940 onwards. The transgressive nature of the lower boundary, however defined, is not easily overcome.

Methodology

The above discussion of the major microfossil groups outlines the biotic response to anthropogenic pollution of various types. It is apparent that the principal method by which the impact of human activity on ecosystems can be identified is by a reduction in both biodiversity and abundance within the benthic microfloral and microfaunal communities, together with the replacement of pollution-sensitive species with taxa tolerant of eutrophication, hypoxia, metal pollution, water acidification or salinity change.

In the coastal zone, opportunistic and pollution-tolerant benthic foraminifera and ostracods often take over environmentally stressed communities and, providing that waters are oxygenated to some degree, out-compete the indigenous populations and may reach flood proportions (Alvarez Zarikian *et al.* 2000; Cronin & Vann 2003; Irizuki *et al.* 2003; Yasuhara *et al.* 2003, 2007, 2012). Planktonic associations also show change related to anthropogenic environmental stress. Planktonic diatom communities tend to increase in diversity and abundance in response to eutrophication (Clarke *et al.* 2006; Katsuki *et al.* 2008; Tuovinen *et al.* 2010), a notable example being in Chesapeake Bay after the land clearance *c.* 1700 (Cooper 1995). Heterotrophic dinoflagellates tend to replace autotrophs with increased nutrient levels, a good example being the eutrophication of the Adriatic coastal zone after *c.* 1900 (Sangiorgi & Donders 2004).

Freshwater lakes are particularly susceptible to eutrophication related to human activity. Such eutrophication affects all microfloral and microfaunal groups, and thus the entire food web, as exemplified by Lake Sevan, Armenia. Between 1948 and the mid-1970s, the diversity of deep-water cladocerans and ostracods collapsed. Macrophytes decreased from more than 20 common species to one (although 13 rare species occur) with *Chara, Potamogeton* and *Zannichelia* almost disappearing. The biomass of gammarids sharply declined. Between 1972 and 1980, the population of zooplankton changed from being dominated by copepods and cladocerans to one dominated by smaller zooplankton such as rotifers. The zoobenthos, which was dominated by gammarids, chironomids and oligochaetes prior to 1948, declined, mirroring the reduction of macrophytes (Friedman 1950;

Markosyan 1959; Meshkova 1976; Gambaryan 1979; Legovich 1979; Ostrovsky 1983; Kurashov & Belyakov 1987; Simonyan 1988, 1991; Hovhannissian 1994; Wilkinson & Gulakyan 2010).

Deformation of foramininiferal tests and ostracod carapaces has been recorded in association with anthropogenic environmental stress, such as the reaction of *Melonis barleeanus* to heavy metals pollution (Elberling *et al.* 2003; Yasuhara *et al.* 2012) and to polychlorinated biphenyls (Scott *et al.* 2005), and of *Ammonia* to hypoxia/anoxia (Karlsen *et al.* 2000). Ostracods are particularly susceptible to salinity change causing mutated sieve pores (Rosenfeld & Vesper 1977) and deformed *Bicornucythere bisanensis* (Abe 1983).

In some areas, the appearance of accidentally introduced, or 'exotic', species via shipping or fish farming can also indicate the human 'footprint'. Anthropogenic dispersal of microfloras and faunas can be recognized from the Bronze Age through to more recent times: the introduction of dinoflagellates into the Black Sea *c.* 1000 BC, possibly associated with early Greek exploration (Marret *et al.* 2009); the Japanese foraminifer *Trochammina hadai* into the western USA *c.* 100 years ago (McGann 2008; McGann *et al.* 2012) and *Elphidium vellai* into New Zealand (Grenfell *et al.* 2007); and the ostracod *Eusarsiella zostericola* into British waters (Kornicker 1975) as a result of oyster farming. The introduction of exotic species along the coastal zone of eastern Australia is also believed to have been caused by the release of water in the ballast of shipping during the late nineteenth to twentieth centuries (McMinn *et al.* 1997; Saunders *et al.* 2007; Taffs *et al.* 2008; Saunders & Taffs 2009). It must be emphasized, however, that reductions in diversity, morphological mutation and appearance of 'exotic' species can occur naturally (although usually over long time intervals), so care must be exercised in using these indicators.

A date for the Anthropocene

The date of the start of anthropogenic deterioration of the environment and hence the base of the Anthropocene is difficult to identify

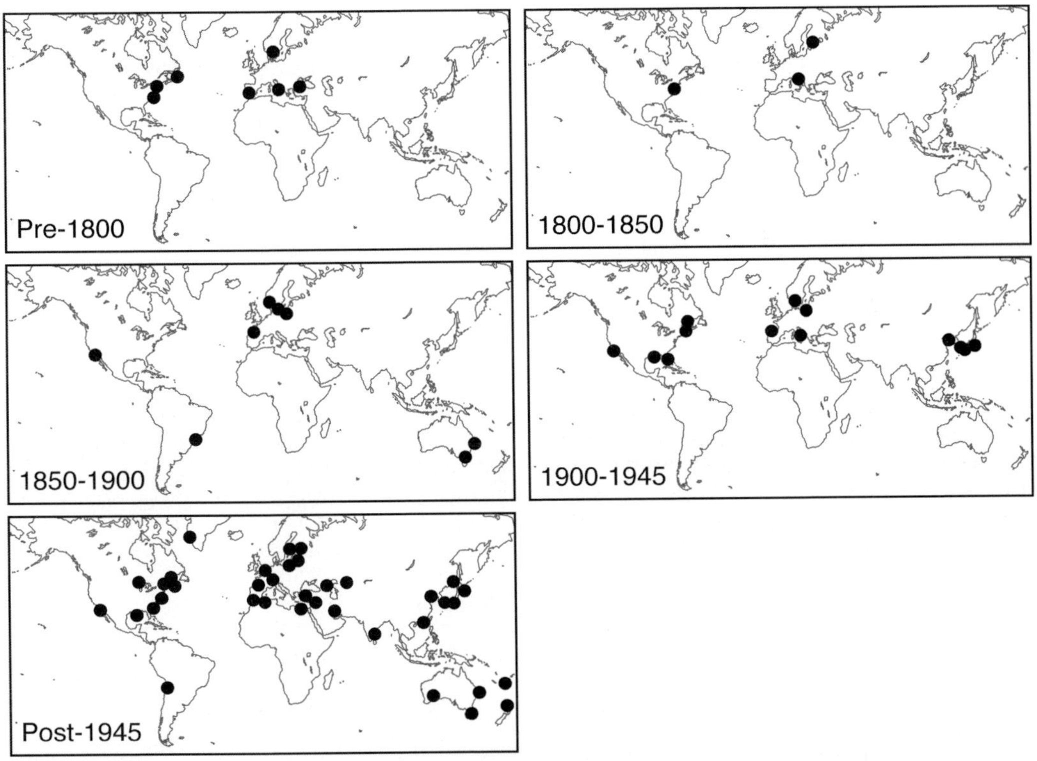

Fig 7. Locations where chronologically constrained microbiotic data indicate the influence of anthropogenic activity (related to industrialization, agriculture, chemical and hydrocarbon pollution, introduction of exotic organisms, and so on). Profound and widespread pollution after *c.* 1940 is demonstrated, although the information is skewed to some extent due to bias in temporal and spatial datasets. Locations shown are discussed herein with additional information by Yasuhara *et al.* (2012).

micropalaeontologically. It varies in intensity over a period of time and must be considered transitional (Fig. 7). Local change can be recognized in prehistoric times, for example, deforestation around Lake Ohrid 5000 years ago (Wagner *et al.* 2009); as a result of mining in the Tinto Estuary, Spain, in *c.* 1500 BC (Ruiz *et al.* 2009); and as a result of the late Neolithic and Bronze Age occupation of Apulia, Italy (Caldara & Simone 2005). Ostracod populations from the coastal strip of Portugal, Spain and Malta and Lake Qarun, Egypt, preserve the effects of anthropogenic pollution, deforestation and agriculture from as early as Roman times (Almeida *et al.* 2009; Keatings *et al.* 2010; Currás *et al.* 2012; Marriner *et al.* 2012). By *c.* 1550 AD, eutrophication in parts of the Norwegian coastal zone (e.g. Frierfjord) was related to the timber industry (Alve 2000; Dale 2000), and the coastal area of San André, Portugal, was affected by human activity some 400 years ago (Cearreta *et al.* 2003). During the eighteenth and nineteenth centuries, anthropogenic change began in Europe and America on a larger scale. Examples include the Baltic Sea, Sweden (Andrén 1999; Andrén *et al.* 1999), along the Norwegian coastal zone (e.g. Alve 1991; Dale & Fjellsa 1994; Dale 2000; Alve *et al.* 2009) and Chesapeake Bay, USA (Brush & Hilgartner 2000; Cronin & Vann 2003; Willard & Cronin 2007; Brush 2009). However, anthropogenic pollution and eutrophication did not reach Asia until the twentieth century (e.g. Yasuhara *et al.* 2007; Katsuki *et al.* 2008; Tsujimoto *et al.* 2008). This difference is a consequence of temporal and spatial differences in the commencement and intensity of industrialization, deforestation for agricultural purposes, increased use of fertilizers, intensity of fish farming, population growth, sewage systems, and so on.

Despite the fact that anthropogenic activity can be recognized throughout pre-history and historic times, these instances are all at a local scale. However, an acceleration of change on a global scale can certainly be recognized after *c.* 1945, with increased industrialization related to the Second World War and increased use of fertilizers necessary to increase food production for the post-war population growth (Fig. 7). Since that time, anthropogenic-induced population change has been reported in remote parts of the world, including Greenland (Elberling *et al.* 2003) and New Caledonia, SW Pacific (Debenay & Fernandez 2009).

Conclusions

Few parts of the world can be considered pristine, and the effect of human activity is clearly reflected in changes to microbiotic communities. As a consequence, these organisms have the potential to define the base of the Anthropocene via the anthropogenic signature they preserve, although there are limitations due to gaps in the data and both temporal and spatial differences in the degree of human impact. Caution is necessary when using microfloras and microfaunas as there are several possible boundaries that may be chosen for the base, and an argument can be made that it is better to use indications of the drivers of the Anthropocene, rather than the responders. There is also a question as to whether its base should be placed at the first, prehistoric, signs of human activity, at the nineteenth- and twentieth-century land clearance and rise in agriculture in Europe and the USA, or at the more modern change caused by global industrialization and population growth.

There is sufficient biotic (diatom, foraminiferal and ostracodal) evidence to suggest that the best position for its base is at *c.* 1940–1945 AD, although the scant data from the dinoflagellate cysts are insufficient to identify a base with confidence. This was a period of intense industrialization and release of pollutants, toxins, chemicals and hydrocarbons; increased agricultural output, use of fertilizers, fish farming and salinization; water extraction for energy, irrigation and consumption; and increased population and sewage output (e.g. Steffen *et al.* 2011). This contamination has had the combined effect of causing biotic change on a global scale in both marine and freshwater ecosystems. Profound changes to microfloras and microfaunas may individually occur on a local scale (an individual lake or embayment of the coastal zone) or regional scale, but, seen as a whole, these local effects compound to result in spatially widespread anthropogenic change that can be interpreted as being characteristic of the Anthropocene.

IPW publishes with the permission of the Director of the British Geological Survey (N.E.R.C.). MJH acknowledges support from a Natural Sciences and Engineering Research Council of Canada Discovery Grant. We are grateful to the anonymous reviewers for their helpful comments.

References

Abe, K. 1983. Population structure of *Keijella bisanensis* (Okubo) (Ostracoda, Crustacea) – an inquiry into how far the population structure will be preserved in the fossil record. *Journal of the Faculty of Science, University of Tokyo, Series II, Geology, Mineralogy, Geography and Geophysics*, **20**, 443–488.

Adl, S. M., Simpson, A. G. B. et al. 2012. The revised classification of eukaryotes. *Journal of Eukaryotic Microbiology*, **59**, 429–493.

Akimoto, K., Nakahara, K., Kondo, H., Ishiga, H. & Dozen, K. 2004. Environmental reconstruction based on heavy metals, diatoms and benthic foraminifers in the Isahaya reclamation area, Nagasaki, Japan. *Environmental Micropaleontology, Microbiology and Meiobenthology*, **1**, 83–104.

Aladin, N. V. 1993. Salinity tolerance, morphology and physiology of the osmoregulation organs in Ostracoda with special reference to Ostracoda from the Aral Sea. *In*: McKenzie, K. G. & Jones, P. J. (eds) *Ostracoda in Earth and Life Sciences*. Balkema, Rotterdam, 387–403.

Allen, P. E. & Dodson, S. I. 2011. Land use and ostracod community structure. *Hydrobiologia*, **668**, 203–219.

Almeida, I. M., Andrade, C., Freitas, M. C., Moreno, J., Cabral, M. C., Craveiro, S. C. & Marques, F. M. S. F. 2009. Holocene palaeoenvironmental evolution of the Lisbon Downtown area as recorded in the Esteiro da Baixa sediments – first results. *Journal of Coastal Research*, **56**, 574–578.

Alvarez Zarikian, C. A., Blackwelder, P. L., Hood, T., Nelsen, T. A. & Featherstone, C. 2000. Ostracods as indicators of natural and anthropogenically induced changes in coastal marine environments. *In*: *Coasts at the Millennium. Proceedings of the 17th International Conference of the Coastal Society*, 9–12 July, Portland, OR, USA, 896–905.

Alve, E. 1991. Foraminifera, climate change, and pollution: a study of late Holocene sediments in Drammensfjord, southeast Norway. *The Holocene*, **1**, 243–261.

Alve, E. 1996. Benthic foraminiferal evidence of environmental change in the Skagerrak over the past six decades. *Norges Geologiske Undersøkelse Bulletin*, **430**, 85–93.

Alve, E. 2000. Environmental Stratigraphy – a case study reconstructing bottom water oxygen conditions in Frierfjord, Norway, over the past five centuries. *In*: Martin, R. (ed.) *Environmental Micropaleontology – Topics in Geobiology*. Kluwer Academic/Plenum Publishers, New York, **15**, 323–350.

Alve, E., Lepland, A., Magnusson, J. & Backer-Owe, K. 2009. Monitoring strategies for re-establishment of ecological reference conditions: possibilities and limitations. *Marine Pollution Bulletin*, **59**, 297–310.

Anadón, T., Gliozzi, E. & Manzini, I. 2002. Paleoenvironmental reconstruction of marginal marine environments from combined paleoecological and geochemical analysis on ostracods. *In*: Holmes, J. A. & Chivas, A. R. (eds) *The Ostracoda: Applications in Quaternary Research*. American Geophysical Union, Washington, DC, Geophysical Monographs, **131**, 227–247.

Andrén, E. 1999. Changes in the composition of the diatom flora during the last century indicate increased eutrophication of the Oder Estuary, south-western Baltic Sea. *Estuarine, Coastal and Shelf Science*, **48**, 665–676.

Andrén, E., Shimmield, G. & Brand, T. 1999. Environmental changes of the last three centuries indicated by siliceous microfossil records from the southwestern Baltic Sea. *Holocene*, **9**, 25–38.

Bamber, R. N. 1987. Some aspects of the biology of the North American ostracod *Sarsiella zostericola* Cushman in the vicinity of a British power station. *Journal of Micropalaeontology*, **6**, 57–62.

Baptista Neto, J. A., Barreto, C. F., Martins da Silva, M. A., Smith, B. J., McAllister, J. J. & Vilela, C. G. 2012. Nearshore sedimentation as a record of landuse change and erosion: Jurujuba Sound, Niterói, SE Brazil. *Ocean & Coastal Management*, **19**, 153–177.

Barclay, M. H. 1966. An ecological study of a temporary pond near Auckland, New Zealand. *Australian Journal of Marine and Freshwater Research*, **17**, 239–258.

Barmawidjaja, D. M., van der Zwaan, G. J., Jorissen, F. J. & Puskaric, S. 1995. 150 years of eutrophication in the nothern Adriatic Sea: evidence from a benthic foraminiferal record. *Marine Geology*, **122**, 367–384.

Barnosky, A. D., Matzke, N. *et al.* 2011. Has the Earth's sixth mass extinction already arrived? *Nature*, **471**, 51–57.

Bartels-Jónsdóttir, H. B., Voelker, A. H. L., Knudsen, K. L. & Abrantes, F. 2009. Twentieth-century warming and hydrographical changes in the Tagus Prodelta, eastern North Atlantic. *The Holocene*, **19**, 369–381.

Battarbee, R. W. 1984. Diatom analysis and the acidification of lakes. *Philosophical Transactions of the Royal Society of London B*, **305**, 451–477.

Battarbee, R. W. & Charles, D. 1987. The use of diatom assemblages in lake sediments as a means of assessing the timing, trends and causes of lake acidification. *Progress in Physical Geography*, **11**, 552–580.

Battarbee, R. W., Jones, V. J., Flower, R. J., Cameron, N. G., Bennion, H., Carvalho, L. & Juggens, S. 2001. Diatoms. *In*: Smol, J. P., Birks, H. J. B. & Last, W. M. (eds) *Tracking Environmental Change Using Lake Sediments, Volume 3, Terrestrial, Algal, and Siliceous Indicators*. Kluwer Academic Publishers, Dordrecht, 155–202.

Battarbee, R. W., Morley, D., Bennion, H., Simpson, G. L., Hughes, M. & Bauere, V. 2011. A palaeolimnological meta-database for assessing the ecological status of lakes. *Journal of Paleolimnology*, **45**, 405–414.

Bennion, H., Fluin, J. & Simpson, G. L. 2004. Assessing eutrophication and reference conditions for Scottish freshwater lochs using subfossil diatoms. *Journal of Applied Ecology*, **41**, 124–138.

Bennion, H., Sayer, C. D., Tibby, J. & Carrick, H. 2010. Diatoms as indicators of environmental change in shallow lakes. *In*: Smol, J. P. & Stoermer, E. F. (eds) *The Diatoms: Applications for the Environmental and Earth Sciences*, 2nd edn. Cambridge University Press, Cambridge, 152–173.

Bergin, F., Kucuksezgin, F., Uluturhan, E., Barut, I. F., Meric, E., Avsar, N. & Nazik, A. 2006. The response of benthic foraminifera and ostracoda to heavy metal pollution in Gulf of Izmir (Eastern Aegean Sea). *Estuarine, Coastal and Shelf Science*, **66**, 368–386.

Birks, H. J. B., Jones, V. J. & Rose, N. L. 2004. Recent environmental change and atmospheric contamination on Svalbard as recorded in lake sediments – synthesis and general conclusions. *Journal of Paleolimnology*, **31**, 531–546.

Bjerring, R., Bradshaw, E. G., Amsinck, S. L., Johansson, L. S., Odgaard, B. V., Nielsen, A. B. & Jeppesen, E. 2008. Inferring recent changes in the ecological state of 21 Danish candidate reference lakes (EU Water

Framework Directive) using palaeolimnology. *Journal of Applied Ecology*, **45**, 1566–1575.

Blackwelder, P., Hood, T., Alvarez-Zarikian, C., Nelsen, T. A. & McKee, B. 1996. Benthic Foraminifera from the NECOP study area impacted by the Mississippi River plume and seasonal hypoxia. *Quaternary International*, **31**, 19–36.

Bodergat, A.-M. 1978*a*. Un marqueur instantané de la pollution en cérium dans la zone de Cortiou (rade sud de Marseille, Bouches-du-Rhône): *Aurila speyeri* Brady, 1868 (Ostracoda, Podocopida). *Documents des laboratoires de géologie de la Faculté des sciences de Lyon*, **4**, 71–85.

Bodergat, A.-M. 1978*b*. L'intensité lumineuse, son influence sur la teneur en phosphore des carapaces d'ostracodes. *Geobios*, **11**, 715–735.

Bodergat, A.-M. & Ikeya, N. 1988. Distribution of recent ostracoda in Ise and Mikawa Bays, Pacific coast of central Japan. *In*: Hanai, T., Ikeya, N. & Ishizaki, K. (eds) *Evolutionary Biology of Ostracoda, Proceedings of the Ninth International Symposium on Ostracoda*, Tokyo, 413–428.

Bodergat, A.-M., Rio, M. & Andréani, A. M. 1991. Composition chimique et ornementation de Cyprideis torosa (Crustacea, Ostracoda) dans le domaine paralique. *Oceanologica Acta*, **14**, 505–514.

Bodergat, A.-M., Carbonel, G., Rio, M. & Keyser, D. 1993. Chemical composition of *Leptocythere psammophyla* (Crustacea: Ostracoda) as influenced by winter metabolism and summer supplies. *Marine Biology*, **117**, 53–62.

Bodergat, A.-M., Rio, M. & Ikeya, N. 1997. Tide v. eutrophication. Impact on ostracod population structure of Mikawa Bay (Japan). *Revue de Micropaléontologie*, **40**, 3–13.

Bodergat, A.-M., Ikeya, N. & Irzi, Z. 1998. Domestic and industrial pollution: use of ostracods (Crustacea) as sentinels in the marine coastal environment. *Journal de Recherche Océanographique*, **23**, 139–144.

Boomer, I. 1993*a*. Sub-fossil Ostracoda and the death of the Aral Sea. *Geology Today*, **9**, 18–22.

Boomer, I. 1993*b*. Palaeoenvironmental indicators from Late Holocene and contemporary ostracoda of the Aral Sea. *Palaeogeography, Palaeoclimatology, Palaeoecology*, **103**, 141–153.

Boomer, I. 2002. Environmental applications of marine and freshwater Ostracoda. *In*: Haslett, S. K. (ed.) *Quaternary Environmental Micropalaeontology*. Arnold, London, 115–138.

Boomer, I. 2013. Ostracoda as indicators of climatic and human-influenced changes in the Late Quaternary of the Ponto-Caspian Region (Aral, Caspian and Black Seas). *In*: Horne, D. J., Holmes, J., Rodriguez-Lazaro, J. & Viehberg, F. A. (eds) *Ostracoda as Proxies for Quaternary Climate Change*. Elsevier, The Netherlands, Developments in Quaternary Science, **17**, 205–216.

Boomer, I. & Attwood, F. 2007. Ostracods as freshwater pollution indicators: a case study from the Ouseburn, a polluted urban catchment (Tyneside, NE England). *Journal of Micropalaeontology*, **26**, 117–125.

Boomer, I. & Eisenhauer, G. 2002. Ostracod faunas as paleoenvironmental indicators in marginal marine environments. *In*: Holmes, J. A. & Chivas, A. R. (eds) *The Ostracoda: Applications in Quaternary Research*. American Geophysical Union, Washington, DC, Geophysical Monographs, **131**, 135–149.

Boomer, I., Whatley, R. & Aladin, N. V. 1996. Aral Sea Ostracoda as environmental indicators. *Lethaia*, **28**, 77–85.

Bradbury, J. P. 2000. Limnologic history of Lago de Patzcuaro, Michoacan, Mexico for the past 48,000 years: impacts of climate and man. *Palaeogeography, Palaeoclimatology, Palaeoecology*, **163**, 69–95.

Bradshaw, E. G., Rasmussen, P., Nielsen, H. & Anderson, N. J. 2005. Mid- to late-Holocene land-use change and lake development at Dallund Sø, Denmark: trends in lake primary production as reflected by algal and macrophyte remains. *The Holocene*, **15**, 1130–1142.

Brewster-Wingard, G. L. & Ishman, S. E. 1999. Historical trends in salinity and substrate in central Florida Bay: a paleoecological reconstruction using modern analogue data. *Estuaries*, **22**, 369–383.

Bromley, H. J. & Por, F. D. 1975. The metazoan fauna of a sewage carrying wadi, Nahal Soreq (Judean Hills, Israel). *Freshwater Biology*, **5**, 121–133.

Bronstein, Z. S. 1928. Contribution to the identification of the ostracod fauna of the Caucasus and Persia. *Works of the North-Caucasus Hydrobiological Station*, **2**, 67–119 (in Russian, with German summary).

Bronstein, Z. S. 1947. Faune de l'URSS. Crustacés, volume 2, numéro 1: Ostracodes des eaux douces. *Zoologicheskiy Institut Akademii Nauk SSSR, new series*, **31**, 1–339 (English translation, Freshwater Ostracoda, 1988, Oconian Press, New Delhi, for the US Department of Commerce, Springfield, Virginia).

Brush, G. S. 2009. Historical land use, nitrogen, and coastal eutrophication: a paleoecological perspective *Estuaries and Coasts*, **32**, 18–28.

Brush, G. S. & Hilgartner, W. B. 2000. Paleoecology of submerged macrophytes in the upper Chesapeake Bay. *Ecological Monographs*, **70**, 645–667.

Bubikyan, S. A. 1984. Recent Ostracoda from the bottom beds of the south-western part of Lake Sevan. *Izvestiya Akadeii Nauki Armyanskoy SSR, Naukii o Zemlye*, **37**, 10–24 (in Russian).

Caldara, M. & Simone, O. 2005. Coastal changes in the eastern Tavoliere Plain (Apulia, Italy) during the Late Holocene: natural or anthropic? *Quaternary Science Reviews*, **24**, 2137–2145.

Carbonel, P. 1982. Les Ostracodes, traceurs des variations hydrologiques dans des systèmes de transition eaux douces – eaux salées. *Mémoires de la Société Géologique de France, N. S.*, **144**, 117–128.

Carbonel, P. & Tölderer-Farmer, M. 1988. The ostracod carapace as a hydrochemical source of information at water/sediment interface. *In*: Hanai, T., Ikeya, N. & Ishizaki, K. (eds) *Evolutionary Biology of Ostracoda: Its Fundamental and Applications. Proceedings of the Ninth International Symposium on Ostracoda*, Shizouka, Japan. Kodansha, Tokyo and Elsevier, Tokyo, Developments in Palaeontology and Stratigraphy, **11**, 341–351.

Casas-Monroy, O., Roy, S. & Rochon, A. 2011. Ballast sediment-mediated transport of non-indigenous species of dinoflagellates on the east coast of Canada. *Aquatic Invasions*, **6**, 231–248.

Cearreta, A., Irabien, M. J., Leorri, E., Yusta, I., Croudace, I. W. & Cundy, A. B. 2000. Recent anthropogenic impacts on the Bilbao Estuary, northern Spain: geochemical and microfaunal evidence. *Estuarine, Coastal and Shelf Science*, **50**, 571–592.

Cearreta, A., Irabien, M. J., Leorri, E., Yusta, I., Quintanilla, A. & Zabaleta, A. 2002. Environmental transformation of the Bilbao estuary, N. Spain: microfaunal and geochemical proxies in the recent sedimentary record. *Marine Pollution Bulletin*, **44**, 487–503.

Cearreta, A., Cachão, M., Cabral, M. C., Bao, R. & Ramalho, M. de J. 2003. Lateglacial and Holocene environmental changes in Portuguese coastal lagoons 2: microfossil multiproxy reconstruction of the Santo André coastal area. *The Holocene*, **13**, 447–458.

Cheng, J., Collins, L. S. & Holmes, C. 2012. Four thousand years of habitat change in Florida Bay, as indicated by benthic Foraminifera. *Journal of Foraminiferal Research*, **42**, 3–17.

Chial, Z. B. & Persoone, G. 2002*a*. Cyst-based toxicity tests XII – development of a short chronic sediment toxicity test with the Ostracod Crustacean *Heterocypris incongruens*: selection of test parameters. *Environmental Toxicology*, **17**, 520–527.

Chial, Z. B. & Persoone, G. 2002*b*. Cyst-based toxicity tests XIII – development of a short chronic sediment toxicity test with the Ostracod Crustacean *Heterocypris incongruens*: methodology and precision. *Environmental Toxicology*, **17**, 528–532.

Chial, Z. B. & Persoone, G. 2002*c*. Cyst-based toxicity tests XIV – application of the ostracod solid phase microbiotest for toxicity monitoring of river sediments in Flanders (Belgium). *Environmental Toxicology*, **17**, 533–537.

Chial, Z. B., Persoone, G. & Blaise, C. 2003. Cyst-based toxicity tests XVI – sensitivity comparison of the solid phase *Heterocypris incongruens* microbiotest with the *Hyalella azteca* and *Chironomus riparius* contact assays on freshwater sediments from Peninsula Harbour (Ontario, Canada). *Chemosphere*, **52**, 95–101.

Clarke, A., Juggins, S. & Conley, D. 2003. A 150-year reconstruction of the history of coastal eutrophication in Riskilde Fjord, Denmark. *Marine Pollution Bulletin*, **46**, 1615–1629.

Clarke, A. L., Weckström, K. *et al.* 2006. Long-term trends in eutrophication and nutrients in the coastal zone. *Limnology and Oceanography*, **51**, 385–397.

Conley, D. J., Carstensen, J., Ærtebjerg, G., Christiansen, P. B., Dalsgaard, T., Hansen, J. L. S. & Josefson, A. B. 2007. Long-term changes and impacts of hypoxia in Danish coastal waters. *Ecological Applications*, **17**(Suppl), 165–184.

Cooper, S. R. 1995. Chesapeake Bay watershed historical land use: impact on water quality and diatom communities. *Ecological Applications*, **5**, 703–723.

Cronin, T. M. & Vann, C. D. 2003. The sedimentary record of climate and anthropogenic influence on the Patuxent Estuary and Chesapeake Bay ecosystems. *Estuaries*, **26**, 196–209.

Cronin, T. M., Wingard, G. L., Dwyer, G. S., Swart, P. K., Willard, D. A. & Albietz, J. 2013. Climate variability during the Medieval Climate Anomaly and Little Ice Age based on ostracode faunas and shell geochemistry from Biscayne Bay, Florida. *In*: Horne, D. J., Holmes, J., Rodriguez-Lazaro, J. & Viehberg, F. A. (eds) *Ostracoda as Proxies for Quaternary Climate Change*. Elsevier, The Netherlands, Developments in Quaternary Science, **17**, 241–262.

Crutzen, P. J. 2002. Geology of mankind. *Nature*, **415**, 23.

Currás, A., Zamora, L. *et al.* 2012. Climate change and human impact in central Spain during Roman times: high-resolution multi-proxy analysis of a tufa lake record (Somolinos, 1280 m asl). *Catena*, **89**, 31–53.

Curry, B. B. & Filippelli, G. M. 2010. Episodes of low dissolved oxygen indicated by ostracodes and sediment geochemistry at Crustal Lake, Illinois, USA. *Limnology and Oceanography*, **55**, 2403–2423.

Cushman, J. A. 1928. *Foraminifera, Their Classification and Economic Use*. Cushman Laboratory for Foraminiferal Research, Sharon, Massachusetts, Special Publications, **1**, 1–401.

Dale, B. 1996. Dinoflagellate cyst ecology: modeling and geological applications. *In*: Jansonius, J. & McGregor, D. C. (eds) *Palynology: Principles and Applications*. American Association of Stratigraphic Palynologists Foundation, Dallas, TX, **3**, 1249–1275.

Dale, B. 2000. Dinoflagellate cysts as indicators of cultural eutrophication and industrial pollution in coastal sediments. *In*: Martin, R. E. (ed.) *Environmental Micropaleontology*. Kluwer Academic/Plenum Publishing, New York, NY, Topics in Geobiology, **15**, 305–321.

Dale, B. 2001. Marine dinoflagellate cysts as indicators of eutrophication and industrial pollution: a discussion. *The Science of the Total Environment*, **264**, 235–240.

Dale, B. 2009. Eutrophication signals in the sedimentary record of dinoflagellate cysts in coastal waters. *Journal of Sea Research*, **61**, 103–113.

Dale, B. & Dale, A. L. 2002. Environmental applications of dinoflagellate cysts and acritarchs. *In*: Haslett, S. K. (ed.) *Quaternary Environmental Micropalaeontology*. Arnold, London/Oxford University Press, New York, NY, 207–240.

Dale, B. & Fjellså, A. 1994. Dinoflagellate cysts as paleoproductivity indicators: state of the art, potential, and limits. *In*: Zahn, R., Pedersen, T. F., Kaminski, M. A. & Labeyrie, L. (eds) *Carbon Cycling in the Glacial Ocean: Constraints on the Ocean's Role in Global Change*. NATO ASI Series, Springer-Verlag, Berlin, **I, 17**, 521–537.

Dale, B. & Nordberg, K. 1993. Possible environmental factors regulating prehistoric and historic 'blooms' of the toxic dinoflagellate *Gymnodinium catenatum* in the Kattegat–Skagerrak region of Scandinavia. *In*: Smayda, T. J. & Shimizu, Y. (eds) *Toxic Phytoplankton Blooms in the Sea*. Elsevier Science, Amsterdam, 53–57.

Dale, B., Madsen, A., Nordberg, K. & Thorsen, T. A. 1993. Evidence for prehistoric and historic 'blooms' of the toxic dinoflagellate *Gymnodinium catenatum* in the Kattegat–Skagerrak region of Scandinavia. *In*: Smayda, T. J. & Shimizu, Y. (eds) *Toxic Phytoplankton Blooms in the Sea*. Elsevier Science, Amsterdam, 47–52.

Dale, B., Thorsen, T. A. & Fjellså, A. 1999. Dinoflagellate cysts as indicators of cultural eutrophication in the

Oslofjord, Norway. *Estuarine, Coastal and Shelf Science*, **48**, 371–382.

DANCE, K. W. & HYNES, H. B. N. 1979. A continuous study of drift in adjacent intermittent and permanent streams. *Archives für Hydrobiologie*, **87**, 253–261.

DANIELOPOL, D. L., GEIGER, W., TÖLDERER-FARMER, M., ORELLANA, C. P. & TERRAT, M. N. 1985. The Ostracoda of Mondsee: spatial and temporal changes during the last 50 years. *In*: DANIELOPOL, D. L., SCHMIDT, R. & SCHULTZ, E. (eds) *Contributions to the Paleolimnology of the Trumer Lakes (Salzburg) and the Lakes Mondsee, Attersee and Traunsee (Upper Austria)*. Limnologisches Institut der Österreichischen Akademie der Wissenschaften, Mondsee, 99–121.

DANIELOPOL, D. L., GEIGER, W., TÖLDERER-FARMER, M., ORELLANA, C. P. & TERRAT, M.-N. 1988. In search of Cypris and Cythere. A report of the evolutionary ecological project on limnic Ostracoda from the Mondsee (Austria). *In*: HANAI, T., IKEYA, N. & ISHIZAKI, K. (eds) *Evolutionary Biology of Ostracoda*. Kodansha/Elsevier, Tokyo, 485–500.

DANIELOPOL, D. L., OLTEANU, R., LÖFFLER, H. & CARBONEL, P. 1990. Present and past geographical–ecological distribution of Cytherissa (Ostracoda, Cytherideidae). *In*: DANIELOPOL, D. L., CARBONEL, P. & COLIN, J.-P. (eds) *Cytherissa (Ostracoda)—The Drosophila of Paleolimnology. Bulletin de l'Institut de Géologie du Bassin d'Aquitaine*, **47–48**, 97–118.

DAVIS, J. & CHRISTIDIS, F. 1997. *A Guide to Wetland Invertebrates of Southwestern Australia*. Western Australian Museum, Perth.

D'COSTA, D. M., EDNEY, P., KERSHAW, A. P. & DE DECKKER, P. 1989. Late Quaternary palaeoecology of Tower Hill, Victoria, Australia. *Journal of Biogeography*, **16**, 461–482.

DEBENAY, J. P. & FERNANDEZ, J. M. 2009. Benthic foraminifera records of complex anthropogenic environmental changes combined with geochemical data in a tropical bay of New Caledonia (SW Pacific). *Marine Pollution Bulletin*, **59**, 311–322.

DE DECKKER, P. 1977. The distribution of the 'giant' ostracods (Family Cyprididae Baird, 1845) endemic to Australia. *In*: LÖFFLER, H. & DANIELOPOL, D. (eds) *Aspects of the Ecology and Zoogeography of Recent and Fossil Ostracoda*. Dr. W. Junk, The Hague, 285–294.

DE DECKKER, P. 1981. Ostracoda from Australian inland waters – notes on taxonomy and ecology. *Transactions of the Royal Society of Victoria*, **93**, 43–85.

DE DECKER, P., COLIN, J.-P. & PEYPOUQUET, J.-P. (eds) 1988. *Ostracoda in the Earth Sciences*. Elsevier, Amsterdam.

DE MAHIQUES, M. M., BURONE, L. *ET AL.* 2009. Anthropogenic influences in a lagoonal environment: a multiproxy approach at the Valo Grande mouth, Cananéia-guape system (SE Brazil). *Brazilian Journal of Oceanography*, **57**, 325–337.

DIGERFELDT, G. & HÅKANSSON, H. 1993. The Holocene palaeolimnology of Lake Sämbosjön, southwestern Sweden. *Journal of Paleolimnology*, **8**, 189–210.

EAGER, S. H. 1999. Distribution of Ostracoda around a coastal sewer outfall: a case study from Wellington, New Zealand. *Journal of the Royal Society of New Zealand*, **29**, 257–264.

EAGER, S. H. 2000. Ostracoda in detection of sewage discharge on a Pacific Atoll. *In*: MARTIN, R. E. (ed.) *Environmental Micropaleontology: The Application of Microfossils to Environmental Geology*. Kluwer Academic, New York, NY, Topics in Geobiology, **15**, 151–167.

EKDAHL, E. J., TERANES, J. L., WITTKOP, C. A., STOERMER, E. F., REAVIE, E. D. & SMOL, J. P. 2007. Diatom assemblage response to Iroquoian and Euro-Canadian eutrophication of Crawford Lake, Ontario, Canada. *Journal of Paleolimnology*, **37**, 233–246.

ELBERLING, B., KNUDSEN, K. L., KRISTENSEN, P. H. & ASMUND, G. 2003. Applying foraminiferal stratigraphy as a biomarker for heavy metal contamination and mining impact in a fiord in West Greenland. *Marine Environmental Research*, **55**, 235–256.

ELNER, J. K. & HAPPEY-WOOD, C. M. 1980. The history of two linked but contrasting lakes in North Wales from a study of pollen, diatoms and chemistry in sediment cores. *Journal of Ecology*, **68**, 95–121.

EUROPEAN UNION 2000. Directive 2000/60/EC of the European Parliament and of the Council of 23 October 2000 on establishing a framework for community action in the field of water policy. *Official Journal of the European Communities*, **L327**, 1–72.

FAASSE, M. 2013. The North American ostracod Eusarsiella zostericola (Cushman, 1906) arrives in mainland Europe. *BioInvasions Records*, **2**, 47–50.

FENSOME, R. A., TAYLOR, F. J. R., NORRIS, G., SARJEANT, W. A. S., WHARTON, D. I. & WILLIAMS, G. L. 1993. A classification of living and fossil dinoflagellates. *Micropaleontology* (Special Publication), **7**, 1–351.

FILIPPI, M. L., LAMBERT, P., HUNZIKER, J., KÜBLER, B. & BERNASCONI, S. 1999. Climatic and anthropogenic influence on the stable isotope record from bulk carbonates and ostracodes in Lake Neuchâtel, Switzerland, during the last two millennia. *Journal of Paleolimnology*, **21**, 19–34.

FORESTER, R. M. & BROUWERS, E. M. 1985. Hydrochemical parameters governing the occurrence of estuarine and marginal estuarine ostracodes: an example from south-central Alaska. *Journal of Paleontology*, **59**, 344–369.

FRENZEL, P. & BOOMER, I. 2005. The use of ostracods from marginal marine, brackish waters as bioindicators of modern and Quaternary environmental change. *Palaeogeography, Palaeoclimatology, Palaeoecology*, **225**, 68–92.

FRIEDMAN, G. M. 1950. Benthos of Lake Sevan. *Works of the Sevan Hydrobiological Station*, **11**, 7–93.

FRITZ, S. C. 1989. Lake development and limnological response to prehistoric and historic land-use in Diss, Norfolk, U.K. *Journal of Ecology*, **77**, 182–202.

FRYER, G. 1981. The copepod *Salmincola edwardsii* as a parasite of Salvelinus alpinus in Britain, and a consideration of the so-called relict fauna of Ennerdale Water. *Journal of Zoology*, **193**, 253–268.

FRYER, G. & JOYCE, A. 1981. The distribution of some freshwater copepods and its bearing on the history of the fauna and flora of the British Isles. *Journal of Biogeography*, **8**, 253–268.

GAMBARYAN, M. G. 1979. Temperature regime of Lake Sevan. *Trudy Sevanskoi Gidrobiologicheskoi Stantsii*, **17**, 123–129 (in Russian).

GEIGER, W. 1990*a*. The role of oxygen in the disturbance and recovery of the *Cytherissa lacustris* population of Mondsee (Austria). *In*: DANIELOPOL, D. L., CARBONEL, P. & COLIN, J. P. (eds) *Cytherissa – The Drosophila of Paleolimnology. Bulletin de l'Institut de Géologie du Bassin d'Aquitaine*, **47**, 167–189.

GEIGER, W. 1990*b*. Field and laboratory studies on the life cycle of *Cytherissa lacustris* (Sars) (Crustacea, Ostracoda) with special emphasis on the role of temperature. *In*: DANIELOPOL, D. L., CARBONEL, P. & COLIN, J. P. (eds) *Cytherissa – The Drosophila of Paleolimnology. Bulletin de l'Institut de Géologie du Bassin d'Aquitaine*, **47**, 191–208.

GEIGER, W. 1993. *Cytherissa lacustris* (Ostracoda, Crustacea): its use in detecting and reconstructing environmental changes at the sediment–water interface. *Verhandlungen der Internationalen Vereinigung für theoretische und angewandte Limnologie*, **25**, 1102–1107.

GELL, P. A., BARKER, P. A., DE DECKKER, P., LAST, W. M. & JELICIC, L. 1994. The Holocene history of West Basin Lake, Victoria, Australia; chemical changes based on fossil biota and sediment mineralogy. *Journal of Paleolimnology*, **12**, 235–258.

GELL, P. A., TIBBY, J. *ET AL*. 2005*a*. Accessing limnological change and variability using fossil diatom assemblages, south-east Australia. *River Research and Applications*, **21**, 257–269.

GELL, P. A., BULPIN, S., WALLBRINK, P., BICKFORD, S. & HANCOCK, G. 2005*b*. Tareena Billabong – a palaeolimnological history of an ever-changing wetland, Chowilla Floodplain, lower Murray–Darling Basin, Australia. *Marine and Freshwater Research*, **56**, 441–456.

GHETTI, P. F. 1973. L'ostracodofauna di alcune risaei turche e persiane. *Ateneo Parmense, Sezione II, Acta Naturalia*, **8**, 117–147.

GOODAY, A. J., JORISSEN, F. *ET AL*. 2009. Historical records of coastal eutrophication-induced hypoxia. *Biogeosciences*, **6**, 1707–1745.

GOTTSCHALK, S. 2011. EU reference conditions in Swedish lakes identified with diatoms as palaeoindicators – a review. *Boreal Environment Research*, **16**, 473–494.

GRENFELL, H. R., HAYWARD, B. W. & HORROCKS, M. 2007. Foraminiferal record of ecological impact of deforestation and oyster farms, Mahurangi Harbour, New Zealand. *Marine and Freshwater Research*, **58**, 475–491.

HABERLE, S. G., TIBBY, J., DIMITRIADIS, S. & HEIJNIS, H. 2006. The impact of European occupation on terrestrial and aquatic ecosystem dynamics in an Australian tropical rainforest. *Journal of Ecology*, **94**, 987–1002.

HÅKANSSON, H. & REGNÉLL, J. 1993. Diatom succession related to land use change during the last 6000 years: a study of a small eutrophic lake in southern Sweden. *Journal of Paleolimnology*, **8**, 49–69.

HALPERN, B. S., WALBRIDGE, S. *ET AL*. 2008. A global map of human impact on marine ecosystems. *Science*, **319**, 948–952.

HAMDI, H., MANUSADŽIANAS, L., AOYAMA, I. & JEDIDI, N. 2006. Effects of anthracene, pyrene and benzo[a]pyrene spiking and sewage sludge compost amendment on soil ecotoxicity during a bioremediation process. *Chemosphere*, **65**, 1153–1162.

HARLAND, R. & NORDBERG, K. 2011. The identification, occurrence and importance of microreticulate dinoflagellate cysts in the latest Holocene sediments of the Skagerrak and Kattegat, west coast of Sweden. *Review of Palaeobotany and Palynology*, **164**, 84–92.

HARLAND, R., NORDBERG, K. & FILIPSSON, H. L. 2006. Dinoflagellate cysts and hydrographical change in Gullmar Fjord, west coast of Sweden. *Science of the Total Environment*, **355**, 204–231.

HARLAND, R., NORDBERG, K. & FILIPSSON, H. L. 2010. A major change in the dinoflagellate cyst flora of Gullmar Fjord, Sweden, at around 1969/1970 and its possible explanation. *In*: HOWE, J. A., AUSTIN, W. E. N., FORWICK, M. & PAETZEL, M. (eds) *Fjord Systems and Archives*. Geological Society, London, Special Publications, **344**, 75–82.

HARTMANN, G. & KÜHL, C. 1978. Zur Variabilität der Oberflächenornamente der Schlenlebender Ostracoden-Populationen. *Mitteilungen aus dem Hamburgischen Museum und Institut*, **75**, 221–223.

HASLETT, S. K. (ed.) 2002. *Quaternary Environmental Micropalaeontology*. Arnold, London.

HAYWARD, B. W., GRENFELL, H. R. *ET AL*. 2004. Foraminiferal record of human impact on intertidal estuarine environments in New Zealand's largest city. *Marine Micropaleontology*, **53**, 37–66.

HAYWARD, B. W., GRENFELL, H. R., SABAA, A. T., MORLEY, M. S. & HORROCKS, M. 2006. Effect and timing of increased freshwater runoff into sheltered harbor environments around Auckland City, New Zealand. *Estuaries and Coasts*, **29**(2), 165–182.

HAYWARD, B. W., GRENFELL, H. R., SABAA, A. T. & MORLEY, M. S. 2008. Ecological impact of the introduction to New Zealand of Asian date mussels and cordgrass – the foraminiferal, ostracod and molluscan record. *Estuaries and Coasts*, **31**, 941–959.

HEAD, M. J. 1996. Modern dinoflagellate cysts and their biological affinities. *In*: JANSONIUS, J. & MCGREGOR, D. C. (eds) *Palynology: Principles and Applications*. American Association of Stratigraphic Palynologists Foundation, Dallas, TX, **3**, 1197–1248.

HENDERSON, R. J., FORREST, D. A. M., BLACK, K. D. & PARK, M. T. 1997. The lipid composition of sealoch sediments underlying salmon cages. *Aquaculture*, **158**, 69–83.

HIRUTA, S. F., HIRUTA, S. I. & MAWATARI, S. F. 2007. A new, interstitial species of Terresticythere (Crustacea: Ostracoda) and its microdistribution at Onto Beach, northeastern Sea of Japan. *Hydrobiologia*, **585**, 43–56.

HOLMGREN, S. U., BIGLER, C., INGÓLFSEN, O. & WOLFE, A. P. 2010. The Holocene–Anthropocene transition in lakes of western Spitzbergen, Svalbard (Norwegian High Arctic): climate change and nitrogen deposition. *Journal of Palaeolimnology*, **43**, 393–412.

HORNE, D. J. & BOOMER, I. 2000. The role of Ostracoda in saltmarsh meiofaunal communities. *In*: SHERWOOD, B. R., GARDINER, B. G. & HARRIS, T. (eds) *British Saltmarshes*. Forrest Text, Cardigan, for the Linnean Society of London, 182–202.

HORNE, D. J., SMITH, R. J., WHITTAKER, J. E. & MURRAY, J. W. 2004. The first British record and a new species of the superfamily Terrestricytheroidea (Crustacea, Ostracoda): morphology, ontogeny, lifestyle and phylogeny. *Zoological Journal of the Linnean Society*, **142**, 253–288.

HOVHANNISSIAN, R. H. 1994. *Lake Sevan, Yesterday, Today*. Armenia National Academy of Science, Erevan (in Russian, with extended summaries in English and Armenian).

HOVHANNISSIAN, R. H. 1996. Evolution of eutrophication processes of Lake Sevan and how to control them. *In*: *Lake Sevan: Problems and Strategies of Action. Proceeding of the International Conference*, Sevan, Armenia, 13–16 October. Yeravan, 81–85.

HÜBENER, T., ADLER, S. A., WERNER, P., SCHULT, M., ERLENKEUSER, H., MEYER, H. & BAHNWART, M. 2009. A multi-proxy paleolimnological reconstruction of trophic state reference conditions for stratified carbonate-rich lakes in northern Germany. *Hydrobiologia*, **631**, 303–327.

IKEYA, N. 1995. *Ostracoda in Sediment Cores from Yokohama Port. Changes in Marine Organisms and Environments at Yokohama Port: Research on Fossils in Sediment Core Samples*. Yokohama Environmental Research Institute, Report no. **116**.

IRABIEN, M. J., CEARRETA, A., LEORRI, E., GÓMEZ, J. & VIGURI, J. 2008. A 130 year record of pollution in the Suances estuary (southern Bay of Biscay): implications for environmental management. *Marine Pollution Bulletin*, **56**, 1719–1727.

IRIZUKI, T., NAKAMURA, Y., TAKAYASU, K. & SAKAI, S. 2003. Faunal changes of Ostracoda (Crustacea) in Lake Nakaumi, southwestern Japan, during the last 40 years. *Geoscience Reports of Shimane University*, **22**, 149–160 (in Japanese with English abstract).

ISHMAN, S. E. 2000. Benthic foraminiferal distributions in South Florida – analogues to historical changes. *In*: MARTIN, R. (ed.) *Environmental Micropaleontology*. Kluwer Academic/Plenum Publishers, New York, Topics in Geobiology, **15**, 371–384.

JAHN, A., GAMENICK, I. & THEEDE, H. 1996. Physiological adaptations of *Cyprideis torosa* (Crustacea, Ostracoda) to hydrogen sulphide. *Marine Ecology Progress Series*, **142**, 215–223.

JONES, V. J., STEVENSON, A. C. & BATTARBEE, R. W. 1989. Acidification of lakes in Galloway, south-west Scotland: a diatom and pollen study of the post-glacial history of the Round Loch of Glenhead. *Journal of Ecology*, **77**, 1–23.

JUSTÍC, D., LEGOVÍC, T. & ROTTINI-SANDRI, L. 1987. Trends in oxygen content 1911–1984 and occurrence of benthic mortality in the northern Adriatic Sea. *Estuarine, Coastal and Shelf Science*, **25**, 435–445.

KARLSEN, A. W., CRONIN, T. M., ISHMAN, S. E., WILLARD, D. A., KERHIN, R., HOLMES, C. W. & MAROT, M. 2000. Historical trends in Chesapeake Bay dissolved oxygen based on benthic Foraminifera from sediment cores. *Estuaries*, **24**, 488–508.

KATSUKI, K., MIYAMOTO, Y. *ET AL*. 2008. Eutrophication-induced changes in Lake Nakaumi, southwest Japan. *Journal of Paleolimnology*, **40**, 1115–1125.

KEATINGS, K., HOLMES, J., FLOWER, R., HORNE, D., WHITTAKER, J. E. & ABU-ZIED, R. H. 2010. Ostracods and the Holocene palaeolimnology of Lake Qarun, with special reference to past human–environment interactions in the Faiyum (Egypt). *Hydrobiologia*, **654**, 155–176.

KEATLEY, B. E., BENNET, E. M., MACDONALD, G. K., TARANU, Z. E. & GREGORY-EAVES, I. 2011. Land-use legacies are important determinants of lake eutrophication in the Anthropocene. *PLoS ONE*, **6**, e15913, http://dx.doi.org/10.1371/journal.pone.0015913

KHANGAROT, B. S. & DAS, S. 2009. Acute toxicity of metals and reference toxicants to a freshwater ostracod, Cypris subglobosa Sowerby, 1840 and correlation to EC50 values of other test models. *Journal of Hazardous Materials*, **172**, 641–649.

KHANGAROT, B. S. & RAY, P. K. 1987. Response of a freshwater ostracod (*Cypris subglobosa* SOWERBY) exposed to copper at different pH levels. *Acta Hydrochimica et Hydrobiologica*, **15**, 553–558.

KHANGAROT, B. S., RAO, P. S. & SHEKHAWAT, S. S. 1983. Acute toxicity of phenol, pentachlorophenol and sodium pentachlorophenate on a freshwater ostracod, *Cypris subglobosa*. *Acta Hydrochimica et Hydrobiologica*, **11**, 457–465.

KOENDERS, A., MARTENS, K., HALSE, S. & SCHÖN, I. 2012. Cryptic species of the *Eucypris virens* species complex (Ostracoda, Crustacea) from Europe have invaded Western Australia. *Biological Invasions*, **14**, 2187–2201.

KORNICKER, L. S. 1975. Spread of ostracodes to exotic environs on transplanted oysters. *Bulletin of American Paleontology*, **65**, 129–139.

KORNICKER, L. S. & SOHN, I. G. 1971. Viability of ostracode eggs egested by fish and effect of digestive fluids on ostracode shells – ecologic and paleontologic implications. *Bulletin du Centre de recherches Pau SNPA*, **5**(Suppl), 125–135.

KOTLYAKOV, V. M. 1991. The Aral Sea Basin: a critical environmental zone. *Environment*, **33**, 36–38.

KREMER, H. & CROSSLAND, C. 2002. Coastal change and the 'Anthropocene'; past and future perspectives of the IGBP-LOICZ project. *In*: HOFIUS, K. (ed.) *Low-lying Coastal Areas; Hydrology and Integrated Coastal Zone Management*. IHP/OHP Berichte, Deutsches IHP/HWRP-Nationalkomitee Koblenz, Federal Republic of Germany, **13**, 3–19.

KÜHL, C. 1980. Die Variabilität von Leptocythere psammophila Guillaume, 1976: Schalenabmessungen und Schalenstrukturen (Crust.: Ostracoda: Cytheridae). *Verhandlungen des Naturwissenschaftlichen Vereins in Hamburg*, **23**, 275–301.

KÜLKÖYLÜOĞLU, O., DÜGEL, M. & KILIÇ, M. 2007. Ecological requirements of Ostracoda (Crustacea) in a heavily polluted shallow lake, Lake Yeniçağa (Bolu, Turkey). *Hydrobiologia*, **585**, 119–133.

KURASHOV, E. A. & BELYAKOV, V. P. 1987. Role of meiofauna in benthic communities of various types in the Lattgalian lakes. *Hydrobiological Journal*, **23**, 46–50.

LEAHY, P. J., TIBBY, J., KERSHAW, A. P. & HEIJNIS, H. 2005. A palaeoecological reconstruction of the impact of European settlement on Bolin Billabong, Yarra River floodplain. *River Research and Applications*, **21**, 131–149.

LEE, M. R. & CORREA, J. A. 2005. Effects of copper mine tailings disposal on littoral meiofaunal assemblages in the Atacama region of northern Chile. *Marine Environment Research*, **59**, 1–18.

Legovich, N. A. 1979. 'Blooms' in the water of Lake Sevan (Observations 1964–1972). *Materials of Sevan Hydrobiological Station*, **17**, 51–74 (in Russian).

Leira, M., Jordan, P., Taylor, D., Dalton, C., Bennion, H., Rose, N. & Irvine, K. 2006. Assessing the ecological status of candidate reference lakes in Ireland using palaeolimnology. *Journal of Applied Ecology*, **43**, 816–827.

Lenihan, H. S., Peterson, C. H. *et al.* 2003. Variation in marine benthic community composition allows discrimination of multiple stressors. *Marine Ecology Progress Series*, **261**, 63–73.

Liljenstroem, S., Widbom, B. & Mattson, J. 1987. *Effects of Two Oil Refinery Effluents on Benthic Meiofauna in Mesocosms*. Swedish Environmental Research Institute, Karlstrona.

Löffler, H. 1969. Recent and subfossil distribution of *Cytherissa lacustris* (Ostracoda) in Lake Constance. *Verhandlungen des Internationalen Verein Limnologie*, **17**, 240–251.

Löffler, H. 1972. The distribution of subfossil ostracods and diatoms in pre-alpine lakes. *Verhandlungen des Internationalen Verein Limnologie*, **18**, 1039–1050.

Löffler, H. 1983. Aspects of the history and evolution of Alpine lakes in Austria. *Hydrobiologia*, **100**, 143–152.

Malard, F., Pienet, S. & Gibert, J. 1996. The use of invertebrates in ground water monitoring: a rising research field. *Ground Water Monitoring and Remediation*, **16**, 103–113.

Manzo, S., De Nicola, F., De Luca Picione, F., Maisto, G. & Alfani, A. 2008. Assessment of the effects of soil PAH accumulation by a battery of ecotoxicological tests. *Chemosphere*, **71**, 1937–1944.

Markosyan, A. G. 1959. Benthos production in Lake Sevan. *In*: *Trudy VI Soveshchaniya po Problemam Biologii Vnutrennikh Vod.* Akademii Nauk SSSR, Moscow, Leningrad, 139–145 (in Russian).

Marret, F., Mudie, P., Aksu, A. & Hiscott, R. N. 2009. A Holocene dinocyst record of a two-step transformation of the Neoeuxinian brackish water lake into the Black Sea. *Quaternary International*, **197**, 72–86.

Marriner, N., Gambin, T., Djamali, M., Morhange, C. & Spiteri, M. 2012. Geoarchaeology of the Burmarrad ria and early Holocene human impacts in western Malta. *Palaeogeography, Palaeoclimatology, Palaeoecology*, **339–341**, 52–65.

Martens, K. 1998. Diversity and endemicity of Recent nonmarine ostracods (Crustacea, Ostracoda) from Africa and South America: a faunal comparison. *Verhandlungen der Internationalen Vereinigung für Limnologie*, **26**, 2093–2097.

Martens, K., Schön, I., Meisch, C. & Horne, D. J. 2008. Global diversity of ostracods (Ostracoda, Crustacea) in freshwater. *Hydrobiologia*, **595**, 185–193.

Matsuoka, K. 1999. Eutrophication process recorded in dinoflagellate cyst assemblages – a case of Yokohama Port, Tokyo Bay, Japan. *The Science of the Total Environment*, **231**, 17–35.

Matsuoka, K. 2001. Further evidence for a marine dinoflagellate cyst as an indicator of eutrophication in Yokohama Port, Tokyo Bay, Japan: comments on a discussion by B. Dale. *The Science of the Total Environment*, **264**, 221–233.

Matthews, A., Grenfell, H. R., Hayward, B. W. & Horrocks, M. 2005. Foraminiferal record of sewage outfall impacts on the inner Manukau Harbour, Auckland, New Zealand. *New Zealand Journal of Marine and Freshwater Research*, **39**, 193–215.

Mazzola, A., Mirto, S. & Danovaro, R. 1999. Initial fish-farm impact on meiofaunal assemblages in coastal sediments of the Western Mediterranean. *Marine Pollution Bulletin*, **38**, 1126–1133.

McCarthy, F. M. G., Mertens, K. N. *et al.* 2011. Resting cysts of freshwater dinoflagellates in southeastern Georgian Bay (Lake Huron) as proxies of cultural eutrophication. *Review of Palaeobotany and Palynology*, **166**, 46–62.

McGann, M. 2008. High-resolution foraminiferal, isotopic, and trace element records from Holocene estuarine deposits of San Francisco Bay, California. *Journal of Coastal Research*, **245**, 1092–1109.

McGann, M., Alexander, C. R. & Bay, S. M. 2003. Response of benthic foraminifers to sewage discharge and remediation in Santa Monica Bay, California. *Marine Environmental Research*, **56**, 299–342.

McGann, M., Grossman, E. E., Takesue, R. K., Penttila, D., Walsh, J. P. & Corbett, R. 2012. Arrival and expansion of the invasive foraminifera Trochammina hadai Uchio in Padilla Bay, Washington. *Northwest Science*, **86**, 9–26.

McKenzie, K. G. & Moroni, A. 1986. Man as an agent of crustacean passive dispersal via useful plants – exemplified by Ostracoda Ospiti esteri of the Italian ricefields ecosystem – and implications arising therefrom. *Journal of Crustacean Biology*, **6**, 181–198.

McMinn, A., Hallegraeff, G. M., Thomson, P., Jenkinson, A. V. & Heijnis, H. 1997. Cyst and radionucleotide evidence for the recent introduction of the toxic dinoflagellate *Gymnodinium catenatum* into Tasmanian waters. *Marine Ecology Progress Series*, **161**, 165–172.

McWethy, D. B., Whitlock, C. *et al.* 2010. Rapid landscape transformation in South Island, New Zealand following initial Polynesian settlement. *Proceedings of the National Academy of Sciences*, **107**, 21343–21348.

Meisch, C. 2000. Freshwater Ostracoda of Western and Central Europe. *In*: *Süsswasserfauna von Mitteleuropa*. Spektrum Akademischer Verlag, Heidelberg, **8/3**, 522.

Mendes, I., Rosa, F., Dias, J. A., Schönfeld, J., Ferreira, Ó. & Pinheiro, J. 2010. Inner shelf paleoenvironmental evolution as a function of land–ocean interactions in the vicinity of the Guadiana River, SW Iberia. *Quaternary International*, **221**, 58–67.

Meriç, E., Görmüş, M. & Avşar, N. 2007. Holocene geologic history of the Golden Horn (Istanbul, NW Turkey) based on foraminiferal data. *Journal of Asian Earth Sciences*, **30**, 353–363.

Merritts, D., Walter, R. *et al.* 2011. Anthropocene streams and base-level controls from historic dams in the unglaciated mid-Atlantic region, USA. *Philosophical Transactions of the Royal Society, A*, **369**, 976–1009.

Mertens, K. N., Rengefors, K., Moestrup, Ø. & Ellegaard, M. 2012. A review of recent freshwater dinoflagellate cysts: taxonomy, phylogeny, ecology and palaeocology. *Phycologia*, **51**, 612–619.

MESHKOVA, T. M. 1976. Eutrophication of Lake Sevan. *Biological Journal of Armenia*, **29**, 14–22 (in Russian).

MEYBECK, M. & VÖRÖSMARTY, C. 2005. Fluvial filtering of land-to-ocean fluxes; from natural Holocene variations to Anthropocene. *In*: DE MARSILY, G. (ed.) *Eaux Continentales*. Comptes Rendus Academie des Sciences Geoscience, Elsevier, Paris, **337**, 107–123.

MEZQUITA, F., SANZ-BRAU, A. & MIRACLE, M. R. 1996. New data on freshwater ostracod assemblages (Crustacea, Ostracoda) from Cuenca (Central Spain). *Bulletin de la Société des naturalistes Luxembourgeois*, **97**, 239–247.

MEZQUITA, F., HERNÁNDEZ, R. & RUEDA, J. 1999. Ecology and distribution of ostracods in a polluted Mediterranean river. *Palaeogeography, Palaeoclimatology, Palaeoecology*, **148**, 87–103.

MODIG, H. & OLAFSSON, E. 1998. Responses of Baltic invertebrates to hypoxic events. *Journal of Experimental Marine Biology and Ecology*, **229**, 133–148.

MORGAN, P., APLET, G. H., HAUFLER, J. B., HUMPHRIES, H. C., MOORE, M. M. & WILSON, W. D. 1994. Historical range of variability: a useful tool for evaluating ecosystem change. *Journal of Sustainable Forestry*, **2**, 87–111.

MOSSLACHER, F. 2000. Sensitivity of groundwater and surface water crustaceans to chemical pollutants and hypoxia: implications for pollution management. *Archiv für Hydrobiologie*, **149**, 51–66.

MOSTAFAWI, N. 2001. How severely was the Persian Gulf affected by oil spills following the 1991 Gulf War. *Environmental Geology*, **40**, 1185–1191.

MURRAY, J. W. (ed.) 2006. *Ecology and Applications of Benthic Foraminifera*. Cambridge University Press, Cambridge.

NEALE, J. W. 1988. Ostracods and palaeosalinity reconstruction. *In*: DE DECKKER, P., COLIN, J. P. & PEYPOUQUET, J. P. (eds) *Ostracoda in the Earth Sciences*. Elsevier, Amsterdam, 125–157.

NIXON, S. W. 1995. Coastal marine eutrophication: a definition, social causes, and future concerns. *Ophelia*, **41**, 199–219.

NOMURA, R. & KAWANO, S. 2011. Foraminiferal assemblages response to anthropogenic influence and parallel to decadal sea-level changes over the last 70 years in Lake Kugushi, Fukui Prefecture, southwest Japan. *Quaternary International*, **230**, 44–56.

NØRGAARD-PEDERSEN, N., AUSTIN, W. E. N., HOWE, J. A. & SHIMMIELD, T. 2006. The Holocene record of Loch Etive, western Scotland: influence of catchment and relative sea level changes. *Marine Geology*, **228**, 55–71.

OLDFIELD, F., ASIOLI, A. *ET AL*. 2003. A high resolution late Holocene palaeo environmental record from the central Adriatic Sea. *Quaternary Science Reviews*, **22**, 319–342.

OLESZCZUK, P. 2007. The evaluation of sewage sludge and compost toxicity to *Heterocypris incongruens* in relation to inorganic and organic contaminants content. *Environmental Toxicology*, **22**, 587–596.

OLESZCZUK, P. 2008. *Heterocypris incongruens* as a tool to estimate sewage sludge toxicity. *Environmental Toxicology and Chemistry*, **27**, 864–872.

ORTEGA, B., VÁZQUEZ, G., CABALLERO, M., ISRADE, I., LOZANO-GARCÍA, S., SCHAAF, P. & TORRES, E. 2010. Late Pleistocene: Holocene record of environmental changes in Lake Zirahuen, Central Mexico. *Journal of Paleolimnology*, **44**, 745–760.

OSTERMAN, L. E., POORE, R. Z., SWARZENSKI, P. W. & TURNER, R. E. 2005. Reconstructing a 180 yr record of natural and anthropogenic induced low-oxygen conditions from Louisiana continental shelf sediments. *Geology*, **33**, 329–332.

OSTERMAN, L. E., POORE, R. Z. & SWARZENSKI, P. W. 2008. The last 1000 years of natural and anthropogenic low-oxygen bottom-water on the Louisiana shelf, Gulf of Mexico. *Marine Micropaleontology*, **66**, 291–303.

OSTERMAN, L. E., POORE, R. Z., SWARZENSKI, P. W., SENN, D. B. & DIMARCO, S. F. 2009. The 20th-century development and expansion of Louisiana shelf hypoxia, Gulf of Mexico. *Geo-Marine Letters*, **29**, 405–414.

OSTROVSKY, I. S. 1983. *Productivity of the Common Zoobenthos Species and their Role in Lake Sevan Ecosystem*. University of Moscow, Moscow (in Russian).

PAIK, H. K. 1977. Regionale Untersuchungen zur Verteilung der Ostracoden im Persischen Golf und im Golf von Oman. *'Meteor' Forschungs-Ergebnisse, C*, **28**, 37–76.

PALACIOS-FEST, M. R., PARK, L. E., GONZÁLEZ-PORTA, J., PALACIOS-FEST, M. R. & DIX, G. R. 2003. Química de conchas de ostrácodos: una alternativa para medir la contaminación por metales en sistemas acuáticos?. *Revista Mexicana de Ciencias Geológicas UNAM Mexico*, **20**, 139–153.

PALLO, P., WIDBOM, B. & OLAFSSON, E. 1998. A quantitative survey of the benthic meiofauna in the Gulf of Riga (eastern Baltic Sea), with special reference to the structure of nematode assemblage. *Ophelia*, **49**, 117–139.

PASCUAL, A., RODRIGUEZ-LAZARO, J., WEBER, O. & JOUANNEAU, J. M. 2002. Late Holocene pollution in the Geika Estuary (Southern Bay of Biscay) evidenced by the study of foraminifera and Ostracoda. *Hydrobiologia*, **475/476**, 477–491.

PERREN, B. P., DOUGLAS, M. S. V. & ANDERSON, N. J. 2009. Diatoms reveal complex spatial and temporal patterns of recent limnological change in West Greenland. *Journal of Paleolimnology*, **42**, 233–247.

PETKOVSKI, T. & MEISCH, C. 1996. Species of the genus *Stenocypris* Sars, 1889 from the ricefields of Macedonia (Crustacea, Ostracoda). *Travaux scientifiques du Musée national d'histoire naturelle de Luxembourg*, **23**, 57–85.

PINDER, A. M., HALSE, S. A., MCRAE, J. M. & SHIEL, R. J. 2005. Occurrence of aquatic invertebrates of the wheatbelt region of Western Australia in relation to salinity. *Hydrobiologia*, **543**, 1–24.

PLATON, E., SEN GUPTA, B. K., RABALAIS, N. N. & TURNER, R. E. 2005. Effect of seasonal hypoxia on the benthic foraminiferal community of the Louisiana inner continental shelf: the 20th century record. *Marine Micropaleontology*, **54**, 263–283.

POSPELOVA, V., CHMURA, G. L., BOOTHMAN, W. S. & LATIMER, J. S. 2002. Dinoflagellate cyst records and human disturbance in two neighboring estuaries, New Bedford Harbor and Apponagansett Bay,

Massachusetts (USA). *The Science of the Total Environment*, **298**, 81–102.

PRASUNA SOLOMON, G., PAVANI, V. & NAGESWARA, R. A. 2012. Food chain bioaccumulation of lead in *Chrissia halyi* (Ferguson 1969) using cladophora as feed. *European Journal of Experimental Biology*, **2**, 83–87.

PROCTOR, V. W. 1964. Viability of crustacean eggs recovered from ducks. *Ecology*, **45**, 656–658.

PROCTOR, V. W., MALONE, C. R. & DE VLAMING, V. L. 1967. Dispersal of aquatic organisms: viability of disseminules recovered from the intestinal tract of captive killdeer. *Ecology*, **48**, 672–676.

RABALAIS, N. N., TURNER, R. E. & WISEMAN, W. J. 2002. Gulf of Mexico, a. k.a. 'The Dead Zone'. *Annual Review of Ecology and Systematics*, **33**, 235–263.

RADKE, L. C., JUGGINS, S., HALSE, S. A., DE DECKKER, P. & FINSTON, T. 2003. Chemical diversity in southeastern Australian saline lakes II: biotic implications. *Marine and Freshwater Research*, **54**, 895–912.

RÄSÄNEN, J., KAUPPILA, T. & SALONOEN, V.-P. 2006. Sediment-based investigation of naturally or historically eutrophic lakes – implications for lake management. *Journal of Environmental Management*, **79**, 253–265.

REID, M. A., FLUIN, J., OGDEN, R. W., TIBBY, J. C. & KERSHAW, A. P. 2002. Long-term perspectives on human impacts on floodplain-river ecosystems, Murray-Darling Basin, Australia. *Verhandlungen der Internationalen Vereinigung für Theoretische und Angewandte Limnologie*, **28**, 710–716.

REID, M. A., SAYER, C. D., KERSHAW, A. P. & HEIJNIS, H. 2007. Palaeolimnological evidence for submerged plant loss in a floodplain lake associated with accelerated catchment soil erosion (Murray River, Australia). *Journal of Paleolimnology*, **38**, 191–208.

RENBERG, I., KORSMAN, T. & BIRKS, H. J. B. 1993*a*. Prehistoric increases in the pH of acid-sensitive Swedish lakes caused by land-use changes. *Nature*, **362**, 824–826.

RENBERG, I., KORSMAN, T. & ANDERSON, N. J. 1993*b*. A temporal perspective of lake acidification in Sweden. *Ambio*, **22**, 264–271.

ROSENFELD, A. & ORTAL, R. 1983. Ostracodes as indicators of water pollution in Nahal Harod, Northern Israel. *In*: MADDOCKS, R. F. (ed.) *Applications of Ostracoda*. University of Houston, Geosciences, Houston, TX, 229–237.

ROSENFELD, A. & VESPER, B. 1977. The variability of the sieve-pores in Recent and fossil species of *Cyprideis torosa* (Jones, 1850) as an indicator for salinity and palaeosalinity. *In*: LÖFFLER, H. & DANIELOPOL, D. (eds) *Aspects of Ecology and Zoogeography of Recent and Fossil Ostracoda. Sixth International Ostracod Symposium, Saalfelden*, Kluwer Academic Publishers, Dordrecht, 55–67.

ROSENFELD, A., ORTAL, R. & HONIGSTEIN, A. 2000. Ostracodes as indicators of river pollution in Northern Israel. *In*: MARTIN, R. E. (ed.) *Environmental Micropaleontology: The Application of Microfossils to Environmental Geology*. Kluwer Academic, New York, NY, 167–180.

RUIZ, F., GONZALEZ-REGALADO, M. L., BACETA, J. I. & MUNOZ, J. M. 2000. Comparative ecological analysis of the ostracod faunas from low- and high-polluted southwestern Spanish estuaries: a multivariate approach. *Marine Micropaleontology*, **40**, 345–376.

RUIZ, F., GONZÁLEZ-REGALADO, M. L., BORREGO, J., ABAD, M. & PENDÓN, J. G. 2004. Ostracoda and foraminifera as short-term tracers of environmental changes in very polluted areas: the Odiel Estuary (SW Spain). *Environmental Pollution*, **129**, 49–61.

RUIZ, F., ABAD, M., BODERGAT, A. M., CARBONEL, P., RODRIGUEZ-LÁZARO, J. & YASUHARA, M. 2005. Marine and brackish-water ostracods as sentinels of anthropogenic impacts. *Earth-Science Reviews*, **72**, 89–111.

RUIZ, F., BORREGO, J., GONZÁLEZ-REGALADO, M. L., LÓPEZ-GONZÁLEZ, N., CARRO, B. & ABAD, M. 2009. Interaction between sedimentary processes, historical pollution and microfauna in the Tinto Estuary (SW Spain). *Environmental Geology*, **58**, 779–783.

SÆTRE, M. M. L., DALE, B., ABDULLAH, M. I. & SÆTRE, G.-P. 1997. Dinoflagellate cysts as potential indicators of industrial pollution in a Norwegian fjord. *Marine Environmental Research*, **44**, 167–189.

SAMIR, A. 2000. The response of benthic foraminifera and ostracods to various pollution sources: a study from two lagoons in Egypt. *Journal of Foraminiferal Research*, **30**, 83–98.

SÁNCHEZ-BAYO, F. 2006. Comparative acute toxicity of organic pollutants and reference values for crustaceans. *I. Branchiopoda, Copepoda and Ostracod. Environmental Pollution*, **139**, 385–420.

SÁNCHEZ-BAYO, F. & GOKA, K. 2006. Influence of light in acute toxicity bioassays of imidacloprid and zinc pyrithione to zooplankton crustaceans. *Aquatic Toxicology*, **78**, 262–271.

SANGIORGI, S. & DONDERS, T. H. 2004. Reconstructing 150 years of eutrophication in the north-west Adriatic Sea (Italy) using dinoflagellate cysts, pollen and spores. *Estuarine and Coastal Shelf Sciences*, **60**, 69–79.

SAUNDERS, K. M. & TAFFS, K. H. 2009. Palaeoecology: a tool to improve the management of Australian estuaries. *Journal of Environmental Management*, **90**, 2730–2736.

SAUNDERS, K. M., MCMINN, A., ROBERTS, D., HODGSON, D. A. & HEIJNIS, H. 2007. Recent human-induced salinity changes in Ramsar-listed Orielton Lagoon, southeast Tasmania, Australia: a new approach for coastal lagoon conservation and management. *Aquatic Conservation*, **17**, 51–70.

SCHARF, B. W. 1998. Eutrophication history of Lake Arendsee (Germany). *Palaeogeography, Palaeoclimatology, Palaeoecology*, **140**, 85–96.

SCHARF, B. W. & KEYSER, D. 1991. Living and subfossil Ostracoda (Crustacea) from Lac du Bouchet (France, Auvergne). *Documents du Centre d'Études et de Recherches sur les Lacs, Anciens Lacs et Tourbières du Massif-Central (C.E.R.L.A.T.), Mémoire*, **2**, 387–391.

SCHÖNFELD, J., ALVE, E., GESLIN, E., JORISSEN, F., KORSUN, S. & SPEZZAFERRI, S. 2012. The FOBIMO (FOraminiferal BIo-MOnitoring) initiative – Towards a standardised protocol for soft-bottom benthic foraminiferal monitoring studies. *Marine Micropaleontology*, **94–95**, 1–13.

Schornikov, E. I. 1980. Ostracodes in terrestrial biotopes. *Russkii Zoologicheskii Zhurnal*, **59**, 1306–1319 (in Russian).

Schornikov, E. I. 2000*a*. Ostracod fauna from the East Sea coast of Korea and their distribution – preliminary study on Ostracoda as an indicator of water pollution. *Journal of the Geological Society of Korea*, **36**, 435–472.

Schornikov, E. I. 2000*b*. Ostracoda as indicators of conditions and dynamics of water ecosystems. *In*: Martin, R. E. (ed.) *Environmental Micropalaeontology*. Kluwer, New York, NY, 181–187.

Scott, D. B., Schafer, C. T., Honig, C. & Younger, D. C. 1995. Temporal variations of benthic foraminiferal assemblages under or near aquaculture operations: documentation of impact history. *Journal of Foraminiferal Research*, **25**, 224–235.

Scott, D. B., Medioli, F. S. & Schafer, C. T. (eds) 2001. *Monitoring in Coastal Environments Using Foraminifera and Thecamoebian Indicators*. Cambridge University Press, Cambridge.

Scott, D. B., Tobin, R. *et al.* 2005. Pollution monitoring in two North American estuaries: historical reconstructions using benthic Foraminifera. *Journal of Foraminiferal Research*, **35**, 65–82.

Sen Gupta, B. K. 2002. *Modern Foraminifera*. Kluwer Academic Publishers, Dordrecht.

Sen Gupta, B. K., Turner, R. E. & Rabalais, N. N. 1996. Seasonal oxygen depletion in continental-shelf waters of Louisiana: historical record of benthic foraminifers. *Geology*, **24**, 227–230.

Shnitnikov, A. V. 1969. On the Holocene history of the Aral. *Mitteilungen Internationale Vereinigung für Theoretische und Angewandte Limnologie*, **17**, 400–413.

Shuhaimi-Othman, M., Yakub, N., Ramle, N.-A. & Abas, A. 2011. Toxicity of metals to a freshwater ostracod: *Stenocypris major*. *Journal of Toxicology*, **2011**, 136104, http://dx.doi.org/10.1155/2011/136104

Simonyan, A. A. 1988. *Zooplankton in changing conditions of a water body (case study: Lake Sevan)*. PhD thesis, Leningrad (in Russian).

Simonyan, A. A. 1991. *Zooplankton of the Lake Sevan*. Izd-vo Akademii nauk Armenii, Yerevan (in Russian).

Smol, J. P. & Stoemer, E. F. 2010. *The Diatoms: Applications for the Environmental and Earth Sciences*, 2nd edn. Cambridge University Press, Cambridge.

Smol, J. P., Wolfe, A. P. *et al.* 2005. Climate-driven regime shifts in the biological communities of Arctic lakes. *Proceedings of the National Acadamy of Sciences*, **102**, 4397–4402.

Spezzaferri, S., Basso, D. & Koral, H. 2000. Holocene palaeoceanographic evolution of the Iskenderun Bay, South-Eastern Turkey, as a response to river mouth diversions and human impact. *Mediterranean Marine Science*, **1**, 19–43.

Stoppani, A. 1873. *Corsa di Geologia*. Bernardoni and Brigola, Milan.

Strauss, J., Grossman, E. L., Carlin, J. A. & Dellapenna, T. M. 2012. 100 years of benthic foraminiferal history on the inner Texas shelf inferred from fauna and stable isotopes: preliminary results from two cores. *Continental Shelf Research*, **38**, 89–97.

Syvitski, J. P. M. & Kettner, A. 2011. Sediment flux and the Anthropocene. *Philosophical Transactions of the Royal Society, A*, **369**, 957–975.

Steffen, W., Persson, Å. *et al.* 2011. The Anthropocene: from global change to planetary stewardship. *Ambio*, **40**, 739–761.

Taffs, K. H., Farago, L. J., Heijnis, H. & Jacobsen, G. 2008. A diatom-based Holocene record of human impact from a coastal environment: Tuckean Swamp, eastern Australia. *Journal of Paleolimnology*, **39**, 71–82.

Taylor, F. J. R. & Pollingher, U. 1987. Ecology of dinoflagellates. *In*: Taylor, F. J. R. (ed.) *The Biology of Dinoflagellates*. Blackwell Scientific, Oxford, London, 398–529.

Thibodeau, B., de Vernal, A. & Mucci, A. 2006. Recent eutrophication and consequent hypoxia in the bottom waters of the Lower St. Lawrence Estuary: micropaleontological and geochemical evidence. *Marine Geology*, **231**, 37–50.

Thorsen, T. A. & Dale, B. 1997. Dinoflagellate cysts as indicators of pollution and past climate in a Norwegian fjord. *The Holocene*, **7**, 433–446.

Thorsen, T. A., Dale, B. & Nordberg, K. 1995. 'Blooms' of the toxic dinoflagellate *Gymnodinium catenatum* as evidence of climatic fluctuations in the late Holocene of southwestern Scandinavia. *The Holocene*, **5**, 435–446.

Tibby, J. & Haberle, S. G. 2007. A late glacial to present diatom record from Lake Euramoo, wet tropics of Queensland, Australia. *Palaeogeography, Palaeoclimatology, Palaeoecology*, **251**, 46–56.

Tibby, J., Kershaw, A. P., Builth, H., Philibert, A. & White, C. 2006. Environmental change and variability in south-western Victoria: changing constraints and opportunities for occupation and land use. *In*: David, B., Barker, B. & McNiven, I. J. (eds) *The Social Archaeology of Australian Indigenous Societies*. Aboriginal Studies Press, Canberra, 254–269.

Thomas, E., Abramson, I., Varekamp, J. C. & ten Brink, M. R. B. 2004. Eutrophication of Long Island Sound as traced by benthic foraminifera. *In*: *Proceedings of the 6th Biennual Long Island Sound Meeting*. Groton, CT, October 2002, 87–91.

Tsujimoto, A., Yasuhara, M., Nomura, R., Yamazaki, H., Sampei, Y., Hirose, K. & Yoshikawa, S. 2008. Development of modern benthic ecosystems in eutrophic coastal oceans: the foraminiferal record over the last 200 years, Osaka Bay, Japan. *Marine Micropaleontology*, **69**, 225–239.

Tuovinen, N., Weckström, K. & Virtasalo, J. 2010. Assessment of recent eutrophication and climate influence in the Archipelago Sea based on the subfossil diatom record. *Journal of Paleolimnology*, **44**, 95–108.

Uthicke, S., Patel, F. & Ditchburn, R. 2012. Elevated land runoff after European settlement perturbs persistent foraminiferal assemblages on the Great Barrier Reef. *Ecology*, **93**, 111–121.

Vallefuoco, M., Lirer, F., Ferraro, L., Pelosi, N., Capotondi, L., Sprovieri, M. & Incarbona, A. 2012. Climatic variability and anthropogenic signatures in the Gulf of Salerno (southern-eastern

Tyrrhenian Sea) during the last half millennium. *Rendiconti Lincei: Scienze Fisiche e Naturali*, **23**, 13–23.

VICTOR, R. & FERNANDO, C. H. 1982. Distribution of freshwater Ostracoda (Crustacea) in Southeast Asia. *Journal of Biogeography*, **9**, 281–288.

VICTOR, R., CHAN, G. L. & FERNANDO, C. H. 1979. Notes on the recovery of live ostracods from the gut of the white sucker (*Catostomus commersoni* Lacepede) 1808, (Pisces: Catostomidae). *Canadian Journal of Zoology*, **57**, 1745–1747.

VICTOR, R., DANCE, K. W. & HYNES, H. B. N. 1981. Drift of ostracod crustaceans in adjacent intermittent and permanent streams. *Hydrobiologia*, **80**, 219–223.

VILELA, C. G., SANJINES, A. E. S., GHISELLI, R. O., MENDONCA, J. G., BAPTISTA, J. A. & BARBOSA, C. F. 2003. Search for bioindicators of pollution in the Guanabara Bay: integrations of ecologic patterns. *Anuário do instituto de geociências – UFRJ*, **26**, 22–33.

WAGNER, B., LOTTER, A. F. ET AL. 2009. A 40,000-year record of environmental change from ancient Lake Ohrid (Albania and Macedonia). *Journal of Paleolimnology*, **41**, 407–430.

WATZIN, M. C. & ROSCIGNO, P. R. 1997. The effects of zinc contamination on the recruitment and early survival of benthic invertebrates in an estuary. *Bulletin of Marine Pollution*, **34**, 443–455.

WILKINSON, I. P. & GULAKYAN, S. Z. 2010. Holocene to Recent Ostracoda of Lake Sevan, Armenia: biodiversity and ecological controls. *Stratigraphy*, **7**, 301–315.

WILKINSON, I. P., BUBIKYAN, S. A. & GULAKYAN, S. Z. 2005. The impact of late Holocene environmental change on lacustrine Ostracoda in Armenia. *Palaeoecology, Palaeogeography, Palaeoclimatology*, **225**, 187–202.

WILLARD, D. A. & CRONIN, T. M. 2007. Paleoecology and ecosystem restoration: case studies from Chesapeake Bay and the Florida Everglades. *Frontiers in Ecology and the Environment*, **5**, 491–498.

WILLIAMS, D. D. & HYNES, H. B. N. 1976. The recolonization mechanisms of stream benthos. *Oikos*, **27**, 265–272.

WILSON, E. O. 2002. *The Future of Life*. Alfred A. Knopf, Random House, New York, NY.

WITTE, L. 1993. Pacific ostracods on West African beaches; a case of anthropogenic faunal contamination by shipping. *In*: WITTE, L. J. (ed.) *Taxonomy and Origin of Modern West African Shallow Marine Ostracoda. Part 2. Biogeography*. Proefschrift Vrije Universiteit Amsterdam, 145–163.

WITTE, L. J. 1994. Man-made dispersal of microbenthos. *Marine Micropaleontology*, **23**, 87–88.

WITTE, L. & VAN HARTEN, D. 1991. Polymorphism, biogeography and systematics of *Kotoracythere inconspicua* (Brady, 1880) (Ostracoda: Pectocytheridae). *Journal of Biogeography*, **18**, 427–436.

WOLFE, A. P., BARON, J. S. & CORNETT, R. J. 2001. Anthropogenic nitrogen deposition induced rapid ecological changes in alpine lakes of the Colorado Front Range (USA). *Journal of Paleolimnology*, **25**, 1–17.

WOLFE, A. P., VAN GORP, A. C. & BARON, J. S. 2003. Recent ecological and biogeochemical changes in alpine lakes of Rocky Mountain National Park (Colorado, USA): a response to anthropogenic nitrogen deposition. *Geobiology*, **1**, 153–168.

WOLFE, A. P., HOBBS, W. O. ET AL. 2013. Stratigraphic expressions of the Holocene–Anthropocene transition revealed in sediments from remote lakes. *Earth-Science Reviews*, **116**, 17–34.

YASUHARA, M. & YAMAZAKI, H. 2005. The impact of 150 years of anthropogenic pollution on the shallow marine ostracode fauna, Osaka Bay, Japan. *Marine Micropaleontology*, **55**, 63–74.

YASUHARA, M., YAMAZAKI, H., IRIZUKI, T. & YOSHIKAWA, S. 2003. Temporal changes in ostracod assemblages and anthropogenic pollution during the last 100 years, in sediment cores from Hiroshima Bay, Japan. *The Holocene*, **13**, 527–536.

YASUHARA, M., YAMAZAKI, H., TSUJIMOTO, A. & HOROSE, K. 2007. The effect of long-term spatiotemporal variations in urbanization-induced eutrophication on a benthic ecosystem, Osaka Bay, Japan. *Limnology and Oceanography*, **52**, 1633–1644.

YASUHARA, M., HUNT, G., BREITBURG, D., TSUJIMOTO, A. & KATSUKI, K. 2012. Human-induced marine ecological degradation: micropaleontological perspectives. *Ecology and Evolution*, **2**, 3242–3268.

YE, F., HUANG, X., ZHANG, X., ZHANG, D., ZENG, Y. & TIAN, L. 2012. Recent oxygen depletion in the Pearl River Estuary, South China: geochemical and microfaunal evidence. *Journal of Oceanography*, **68**, 387–400.

ZALASIEWICZ, J., WILLIAMS, M. ET AL. 2008. Are we now living in the Anthropocene?. *GSA Today*, **18**, 4–8.

ZALASIEWICZ, J., WILLIAMS, M., STEFFEN, W. & CRUTZEN, P. 2010. The New World of the Anthropocene. *Environmental Science & Technology*, **44**, 2228–2231.

ZALASIEWICZ, J., WILLIAMS, M. ET AL. 2011. Stratigraphy of the Anthropocene. *Philosophical Transactions of the Royal Society A*, **369**, 1036–1055.

ZENKEVITCH, L. 1963. *Biology of the Seas of the U.S.S.R.* Allen and Unwin, London (translation).

Assessing the Anthropocene with geochemical methods

AGNIESZKA GAŁUSZKA[1]*, ZDZISŁAW M. MIGASZEWSKI[1] & JAN ZALASIEWICZ[2]

[1]*Geochemistry and the Environment Division, Institute of Chemistry, Jan Kochanowski University, 15G Świętokrzyska St, 25-406 Kielce, Poland*

[2]*Department of Geology, University of Leicester, Leicester LE 1 7RH, UK*

**Corresponding author (e-mail: aggie@ujk.edu.pl)*

Abstract: Anthropogenic chemical contamination is one of the most evident signals of human influence on the environment. The large amounts of industrially produced pollutants that have been introduced, over decades, into air, soil and water have caused modifications to natural elemental cycling. Anthropogenic contamination usually leads to enrichment in many elements, particularly in industrial areas. Thus, certain elements and their isotopes can be used as geochemical tracers of anthropogenic impact. Some human-induced changes in the environment may be regarded as a secondary effect of pollution, such as acidification, which causes increased geochemical mobility of several trace elements in surficial deposits. Methods used by geochemists to assess the scale of anthropogenic influence on the environment include calculations of anthropogenic influence on the environment via enrichment and contamination factors, geoaccumulation index and pollution load index. The use of geochemical background levels for delineating between natural and anthropogenic pollution is important. A historical perspective of anthropogenic contamination, allied with isotopic and geochemical signatures in dated sediment cores, may be applied to help define the Anthropocene.

Human influence on the environment has been complex, encompassing different environmental compartments (air, water, soil) and scales of changes (local, regional, global). The Earth is a dynamic system, and geochemical dynamics are best illustrated by consideration of element cycling. Although anthropogenic pollution sources have contributed to chemical changes in the atmosphere, hydrosphere, pedosphere and biosphere, their impact on the total environment is best preserved in sediments, which are often referred to as archives of contamination (Choi & Wania 2011). Significant evidence of a distinctive Anthropocene signal can be found in recent sedimentary deposits that have been impacted by anthropogenic activity, and dated sediment cores can be used to reconstruct the chronology of contamination (Buckley *et al.* 1995).

In assessing the Anthropocene it is important to quantify the human influence on the environment. The results of this influence, recorded in sediments, can be evaluated using different approaches and criteria depending on the nature of the signal: mechanical, biological or chemical (Fig. 1) (Zalasiewicz *et al.* 2011*a*, *b*; Price *et al.* 2011; Vane *et al.* 2011; Schrag 2012).

Geochemical methodology may help discriminate natural and anthropogenic sources of a given substance in the environment. Some of the methods and concepts (e.g. anomaly-background concentrations, geochemical calculations and geochemical tracers) traditionally used in geochemical exploration and further adapted to the needs of environmental geochemistry have found application in the study of anthropogenic contamination (Reimann & de Caritat 2005; Reimann *et al.* 2005; Gałuszka 2007; Gałuszka & Migaszewski 2011). These methods may be extended in assessment of the Anthropocene.

Anthropogenic anomalies and geochemical background concept

Geochemical anomalies can be defined as 'abnormal concentrations of elements in earth materials compared with background levels' (Allaby & Allaby 1999). Anomalous concentrations of elements are both of natural origin (e.g. via ore-forming processes, weathering of ore bodies or mineralized rock formations, sorption by mineral phases, volcanic and hydrothermal activity) and of anthropogenic origin (McQueen 2009). Of the different types of geochemical anomalies (Fig. 2), the positive anthropogenic anomalies of regional scale may provide a useful marker for the Anthropocene (Selinus & Esbensen 1995).

Identification of geochemical anomalies requires knowledge of the 'normal' or 'average' distribution of elements in the environment, which is usually referred to as a 'geochemical background'. Lower and upper limits of the geochemical background

From: WATERS, C. N., ZALASIEWICZ, J. A., WILLIAMS, M., ELLIS, M. A. & SNELLING, A. M. (eds) 2014. *A Stratigraphical Basis for the Anthropocene*. Geological Society, London, Special Publications, **395**, 221–238.
First published online October 24, 2013, http://dx.doi.org/10.1144/SP395.5

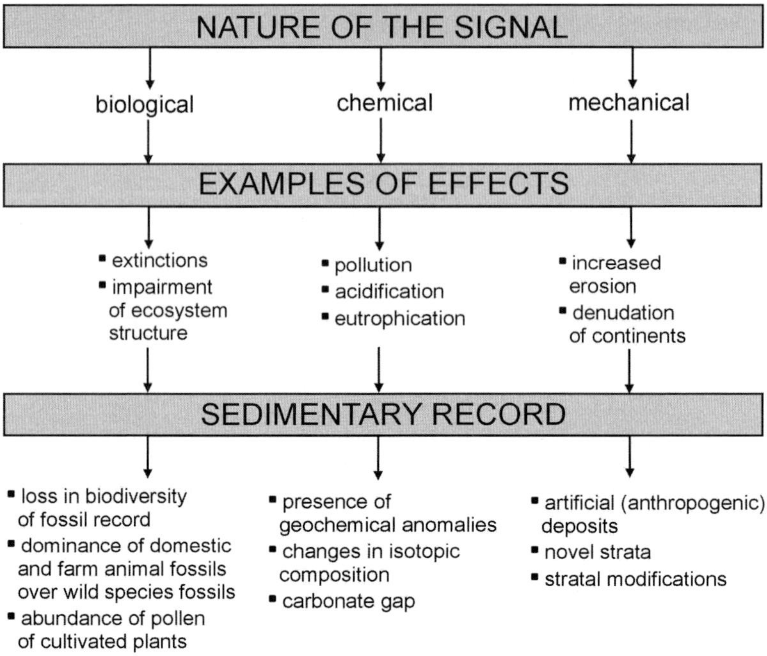

Fig. 1. Types of human-induced changes, their possible effects and sedimentary record.

range are considered thresholds of negative (depletion in an element) and positive (enrichment in an element) anomalies (Fig. 3). Historically, the search for positive anomalies has been essential to mineral exploration and, in the 1960s, geochemists established several methodologies for separating anomalies from the geochemical background (Hawkes & Webb 1962; Miesch 1967). Many

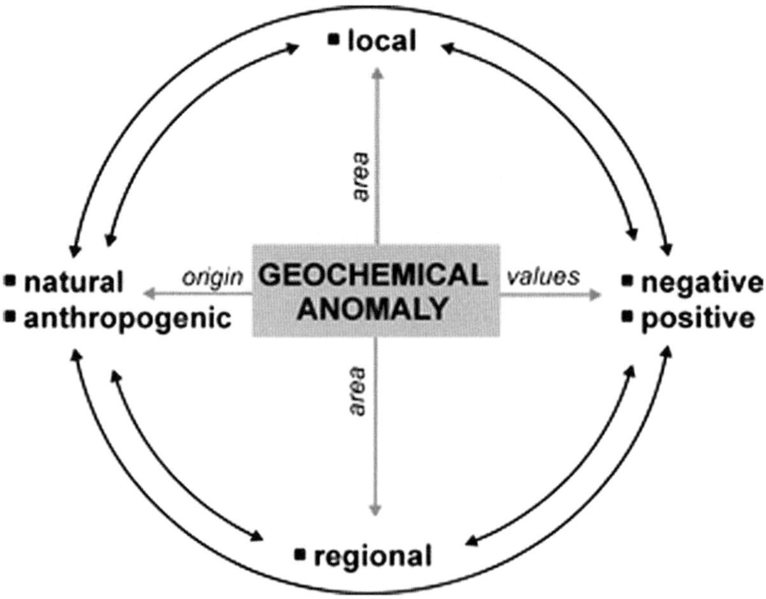

Fig. 2. Systematics of geochemical anomalies.

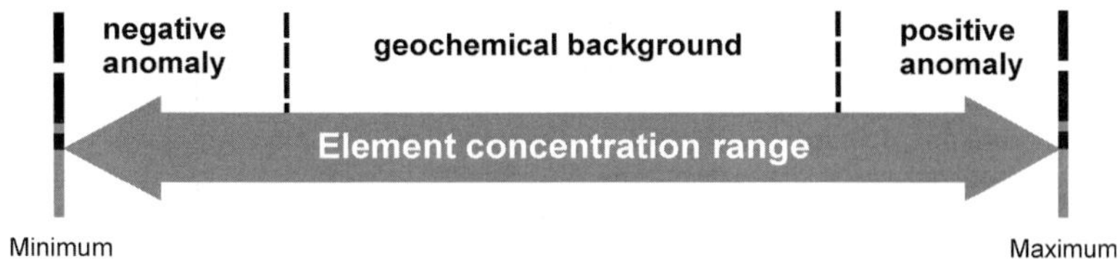

Fig. 3. Geochemical background and anomalies v. element concentrations.

researchers have also used traditional geochemical measures such as Clarke values (Rudnick & Gao 2003) or the average shale value of elements (Turekian & Wedepohl 1961) as an approximation of the upper limit of the geochemical background range. The Clarke values are average concentrations of elements occurring in the Earth's crust. These values were first published by Clarke (1924) and have been updated several times (e.g. Wedepohl 1995; Rudnick & Gao 2003). They are often applied in the calculation of geochemical parameters, as discussed below.

The geochemical background may be evaluated directly through chemical analyses of samples collected in areas not heavily influenced by anthropogenic activity or in samples representing the pre-industrial period (dated sediment and ice cores, archival materials). Another widely used approach for geochemical background evaluation involves univariate and multivariate statistical methods. Here, geochemical anomalies are considered as the outliers (extreme values) in the distribution of geochemical datasets. In the earliest methodologies (e.g. Hawkes & Webb 1962) it was recommended that all the values outside two standard deviations from the mean should be considered a geochemical anomaly.

In recent years, the geochemical background and its evaluation in the context of anthropogenic pollution has been widely discussed (Matschullat *et al.* 2000; Reimann & de Caritat 2005; Reimann & Garret 2005; Reimann *et al.* 2005; Gałuszka 2007; Gałuszka & Migaszewski 2011). It should be noted that knowledge of the geochemical background range allows detection of geochemical anomalies, but is not sufficient to establish the origin of the anomaly. The Anthropocene may be interpreted from the presence of anthropogenic anomalies recorded in recent sediments and in other environmental compartments. Distinguishing between anthropogenic and natural anomalies requires knowledge of natural pollution sources. Natural anomalies are most often of geological origin and their presence is usually limited to the areas of mineralized zones, geothermal fields and so on. Mining activity increases the area of the geochemical anomaly and represents combined natural and anthropogenic influence on the environment (Runkel *et al.* 2007; Fernández-Caliani *et al.* 2009; Davis *et al.* 2010). Current anthropogenic fluxes of several elements, including helium, chromium, antimony, rhenium, platinum group metals and gold, predominate over their natural counterparts (Sen & Peucker-Ehrenbrink 2012). Except for rhenium, the biogeochemical cycle of which is influenced mostly by petroleum burning, mining plays the major role in modification of these element cycles at the Earth's surface.

In contrast to most natural anomalies, anthropogenic ones are recorded in surficial environments (soils, recent sediments) that have been subjected to pollution from industrial sources. However, the use of top/bottom ratios, that is, the ratio of element concentration in surface and deep soil horizons, as a measure of anthropogenic pollution, has been criticized (Reimann & de Caritat 2005; Sucharová *et al.* 2012) on the basis that the natural properties of organic soil horizons favour accumulation of excessive amounts of pollutants.

Geochemical parameters

Several parameters have been proposed for evaluation of the anthropogenic influence on different environmental compartments. The highest number of contamination indices has been developed for sediments. Concentrations of potentially toxic trace metals (e.g. As, Cd, Cr, Cu, Hg, Zn) and persistent organic pollutants (e.g. polycyclic aromatic hydrocarbons and polychlorinated biphenyls) determined in a large number of sediment samples are used in assessment. Different approaches in the methodology of index evaluation have been the basis for distinguishing three types of indices based on comparison (Caeiro *et al.* 2005) of (1) the lowest and highest contaminant concentrations in the samples examined (contamination indices); (2) contaminant concentrations with their geochemical background value (background enrichment indices); and (3) the results obtained using environmental quality guidelines and toxicity data (ecological risk indices).

Published values of geochemical parameters as indicators of anthropogenic pollution calculated for different environmental samples are of potential application to defining the Anthropocene. Most data are published in terms of the following parameters: enrichment factor (EF), contamination factor (CF), pollution load index, or index of geoaccumulation. The stratigraphic use of these parameters is enhanced by their having been applied to different environmental samples (soils, sediments, peats, particulate matter and so on).

Enrichment factor

One of the most commonly used geochemical parameters is the enrichment factor, which is a typical background enrichment index. The formula applied for its calculation is:

$$\mathrm{EF} = \frac{A_e \times B_c}{A_c \times B_e} \quad (1)$$

where A_e is element concentration in an environmental sample, B_e is reference element concentration in an environmental sample, A_c is the Clarke value or average shale value of the element, and B_c is the Clarke value or average shale value of the reference element.

The reference elements should be characterized by high resistance to weathering (they should be conservative), have an exclusively crustal source, and their concentrations in the environment should not be variable. The following elements are most often used as reference: Si, Al, Fe, Li, Sc, Zr, Cs and Ti.

Interpretation of EF values is not uniform, and different cutoff values have been used as indicators of enrichment in a certain element. Generally, it is assumed that EF values close to 1 are indicative of a crustal origin, whereas those greater than 5 (Halstead *et al.* 2000) are considered to be of a non-crustal origin. Sutherland (2000) proposed a five-category scale for evaluation of the intensity of element enrichment on the basis of EF values (EF < 2, deficiency to minimal enrichment; $2 \leq$ EF < 5, moderate enrichment; $5 \leq$ EF < 20, significant enrichment; $20 \leq$ EF < 40, very high enrichment; EF ≥ 40, extremely high enrichment). Interpretation of high EF values as indicative of anthropogenic pollution may be difficult and usually requires additional data or expert knowledge (Reimann & de Caritat 2000). Table 1 summarizes the results of EF calculations in different environmental samples and of various locations.

Contamination factor

The contamination factor (CF) was introduced by Hakanson (1980) for the classification of the quality of sediments and its use was progressively extended to other samples (e.g. soils and road dust) (Table 2). For calculation of the CF, a mean concentration of a given element in at least five samples is divided by the pre-industrial concentration or Clarke value of the element:

$$\mathrm{CF} = \frac{C_i}{C_n} \quad (2)$$

where C_i is the mean content of the element in samples taken from at least five sampling sites, and C_n is the pre-industrial (background) concentration of the element.

A scale is used for an assessment of contamination by CFs. It is proposed that CF < 1 indicates low contamination, $1 \leq$ CF < 3 moderate contamination, $3 \leq$ CF < 6 considerable contamination and CF ≥ 6 very high contamination. In environmental studies, more than one element is usually determined. Evaluation of the degree of contamination with CFs, calculated for more than one element in the samples, is possible by summing CF values for eight pollutant species (As, Cd, Cu, Cr, Hg, Pb, Zn and polychlorinated biphenyls) to obtain a parameter known as the degree of contamination (C_{deg}) (Hakanson 1980). The value of C_{deg} is categorized into one of four classes: $C_{\mathrm{deg}} < 8$, low degree of contamination; $8 \leq C_{\mathrm{deg}} < 16$, moderate degree of contamination; $16 \leq C_{\mathrm{deg}} < 32$, considerable degree of contamination; $C_{\mathrm{deg}} \geq 32$, very high degree of contamination.

Pollution load index

In the calculation of pollution load index (PLI) (Tomlinson *et al.* 1980), the concentration factors of different elements in the same sample are needed (ConcF) for each determined element with respect to its background value:

$$\mathrm{PLI} = \sqrt[n]{\mathrm{Conc}F_1 \times \mathrm{Conc}F_2 \times \cdots \times \mathrm{Conc}F_{\mathrm{n}}} \quad (3)$$

where Conc$F$ is a quotient of the concentration of the element in the sample and its background concentration (usually understood as the world average shale value or Clarke value).

Samples with PLI values greater than 1 are considered to be polluted. Although PLI provides information about the extent of pollution with potentially toxic elements, this parameter has commonly been used for evaluation of the effectiveness of clean-up operations (Galán *et al.* 2002). Examples of PLI values for different elements, sample types and locations are shown in Table 3.

Index of geoaccumulation

The geoaccumulation index (I_{geo}), introduced by Müller (1969), can be considered a quantitative

Table 1. *Examples of enrichment factor values in different environmental samples*

Sampled material	Location	Reference element	Element	EF† value	References
Sediments	Puck Bay, Poland	Al	Zn	1.7–14.6	Szefer *et al.* (1998)
			Ag	4.1–300	
			Cd	3.8–28	
			Pb	2.4–24	
	Tamaki Estuary, New Zealand	Fe	Cd	1.4–10	Abrahim & Parker (2008)
			Cu	0.5–14.7	
			Pb	0.7–6.8	
			Zn	2.2–4.0	
	Kent, UK	Li	Cu	0.9–2.0	Spencer *et al.* (2003)
			Pb	0.9–10	
			Zn	0.8–2.6	
	Wadi Al-Arab Dam, Jordan	Fe	Cd	30*	Ghrefat & Yusuf (2006)
			Cu	1.33*	
			Zn	3.6*	
Soils	Zlatibor, Serbia	Fe	Cd	106*	Dragović *et al.* (2008)
			Cr	5.62*	
			Cu	2.31*	
			Mn	9.18*	
			Ni	65.1*	
			Pb	11.5*	
			Zn	1.72*	
	Northern France	Al	Cd	7.91*	Bourennane *et al.* (2010)
			Hg	3.26*	
			Pb	3.65*	
			Se	2.04*	
			Zn	2.15*	
	Southern Poland	Ca	As	8.07*	Loska *et al.* (2004)
			Cd	11.1*	
	Seville, Spain	Fe	Ba	5.9*	Oliva & Fernándes Espinosa (2007)
			Cu	3.9*	
			Ni	3.0*	
			Pb	12.7*	
Peat profiles	Switzerland	Sc or Al	Pb	10–207	Weiss *et al.* (1999)
Particulate matter	La Plata, Argentina	Fe	Pb	8–860	Bilos *et al.* (2001)
			Zn	4–1700	
			Cd	10–456	
			Cu	5–429	

*Mean value.
†EF, enrichment factor.

measure of the extent of pollution with potentially toxic elements:

$$I_{\text{geo}} = \log_2 \frac{C_e}{1.5 \times \text{GB}} \quad (4)$$

where C_e is the concentration of the examined element in the sample, GB is the geochemical background concentration, and 1.5 is the correction factor for possible variations in the geochemical background.

According to I_{geo} values (from 0 to 10), there are seven classes of sample pollution, varying from 0 (unpolluted) to 6 (extremely polluted). A score of 6 in the I_{geo} scale refers to a 100-fold enrichment above the background value. In its earliest applications, I_{geo} was calculated only for sediment samples, and the C_e value was understood to be the element concentration in fine-grained sediment. After modification this parameter may be applied to other environmental samples. Although the majority of published data on I_{geo} values refer to sediments, there are also examples of the use of this parameter for evaluation of soil and road dust contamination (Table 4).

Comparison of the values of different geochemical parameters may be useful for their potential application in assessment of the Anthropocene.

Table 2. *Contamination factor values from different studies*

Sampled material	Location	Element	CF† value	References
Sediments	Bosphorus, Turkey	Pb	0.9–3.4	Balkıs *et al.* (2012)
		Cd	1.3–1.5	
		Hg	0.1–0.5	
	India	Cu	0.3–1.4	Ray *et al.* (2006)
		Pb	0.1–4.7	
		Cd	1.7–57	
	Tamaki Estuary, New Zealand	Cd	1.3–7.8	Abrahim & Parker (2008)
		Cu	1.1–15	
		Pb	2.2–5.7	
		Ni	1.5–6.5	
	Norway and Russia	Ni	0.3–120	Dauvalter & Rognerud (2001)
		Cu	0.1–76	
		Hg	0.1–4.3	
	Portugal	Cd	0.1–262	Soares *et al.* (1999)
		Cr	0.5–48	
		Cu	0.5–609	
		Ni	0.7–18	
		Pb	0.6–40	
Soils	Southern Poland	As	3.1–11	Loska *et al.* (2004)
		Cd	2.0–21	
		Pb	1.0–12	
	Southern India	Cr	4.3–12	Krishna & Govil (2008)
		Cu	0.9–15	
		Ni	0.6–3.9	
		Pb	0.1–7	
	Beijing, China	Cd	2.6*	Liu *et al.* (2005)
		Cr	1.5*	
		Cu	2.0*	
		Zn	1.7*	
		Pb	1.6*	
	Southern Poland	Cd	6.9–520	Diatta *et al.* (2008)
		Cu	0.1–4.0	
		Pb	5.7–176	
		Zn	2.7–72	
Road dust	Southern Korea	Cd	1.1–12	Duong & Lee (2011)
		Cu	1.7–8.1	
		Ni	1.5–6.6	
		Pb	1.0–4.4	
		Zn	1.2–4.2	

*Mean value.
†CF, contamination factor.

Elements such as Pb, Cd, Cu and Zn, which are usually attributed to anthropogenic pollution sources, exhibit high values of the parameters discussed in different environmental compartments. Thus, elevated levels of these elements are likely to be evident in future geological records of present-day sediments.

Of the other trace metals, Hg may be recognized as a potential indicator of the Anthropocene. Human influence is clear in its far higher abundance ratio in recent times compared to ancient deposits (Aston *et al.* 1973). However, due to its high mobility and volatility (biomethylation), this element must be considered with caution. When assessing the usefulness of Hg and other trace metals, their geochemical background in a study area needs consideration. It is interesting to note that concentrations of Hg in recent sediments of Lake Wigry, northeastern Poland (11–95 $\mu g\ kg^{-1}$) (Migaszewski *et al.* 2003), are distinctly lower than the natural geochemical background concentrations of this element in ancient sediments of Windemere Lake, Great Britain (110 $\mu g\ kg^{-1}$) (Aston *et al.* 1973) and Helgoland Bay, the North Sea (200–800 $\mu g\ kg^{-1}$) (Förstner 1990).

Table 3. *Pollution load index values reported by different authors*

Sampled material	Location	Elements	PLI value	References
Sediments	India	Fe, Mn, Cu, Co, Ni, Pb, Cd, Cr	0.34–1.03	Ray *et al.* (2006)
	Hunan Province, China	Be, V, Cr, Co, Ni, Cu, Zn, Rb, Mo, Cd, Cs, Ba, Nd, Pb, Fe	2.22–3.46	Zhang *et al.* (2011)
	Nigeria	Cu, Cr, Fe, Ni, Pb, V, Zn	1.26–1.76	Essien *et al.* (2009)
	Yokohama, Japan	As, Br, Cd, Cr, Cu, I, Ni, Pb, Sr, V, Zn	1.24–7.65	Mohiuddin *et al.* (2010)
	Baja California, Mexico	Cd, Co, Cu, Ni, Pb, Zn	1.01–4.29	Huerta-Diaz *et al.* (2008)
Soils	Beijing, China	Cd, Cr, Cu, Pb, Zn	2.39–3.43	Liu *et al.* (2005)
	Northern Bangladesh	K, Ca, Ti, Mn, Fe, Zn, As, Rb, Sr, Pb, Zr, Nb	1.67–4.02	Bhuiyan *et al.* (2010)
	Southeastern Egypt	Ag, As, Au, Cd, Co, Cu, Cr, Hg, Mn, Mo, Ni, Pb, Zn	0.64–2.92	Rashed (2010)
	Southwestern Spain	As, Cu, Pb, Zn	1.71–4.19	Fernández-Caliani *et al.* (2009)
Road dust	Ontario, Canada	Cd, Cr, Cu, Fe, K, Mg, Mn, Ni, Pb, Zn	2.0–7.9	Nazzal *et al.* (2012)

Geochemical signals of the Anthropocene

Geochemistry of nutrient elements

Industrial activity influences the global biogeochemical cycles of major nutrient elements including C, N and P. There are many anthropogenic sources of these elements, among which fossil-fuel combustion, agriculture and fertilizer production are predominant. The significance of anthropogenic pollution as regards changes in the global carbon cycle is well known and has been discussed widely, mostly in the context of climate change and ocean acidification (Falkowski *et al.* 2000; Orr *et al.* 2005), but in the last decade the importance of anthropogenic N sources in the biogeochemistry of this element has been highlighted (Elser *et al.* 2009; Krishnamurthy *et al.* 2009; Schlesinger 2009; Holtgrieve *et al.* 2011; Tyrrell 2011).

An important issue in the context of the geochemistry of major nutrient elements is the complexity and interrelationships between their biogeochemical cycles. According to Hutchins *et al.* (2009), acidification of the oceans caused by anthropogenic CO_2 will have a serious impact on the biogeochemistry of major nutrients in marine environments. For example, in the nitrogen cycle, nitrification will be suppressed whereas nitrogen fixation and denitrification will increase. These changes can cause structural and functional fluctuations in marine ecosystems and may have serious consequences for primary production. The decrease in ocean pH will cause under-saturation of $CaCO_3$, resulting in changes of sediment composition from predominantly calcareous to siliciclastic (Tyrrell 2011), possibly leading to the destruction of coral reefs. There are also significant interactions of anthropogenically changed trace element cycles with major nutrient cycles in the oceans. Chemical reactions of desert dust with acid aerosols of anthropogenic origin cause elevated concentrations of soluble Fe in the ocean. In regions of Fe deficiency, increased deposition of Fe-rich desert dust, observed since 1870, has intensified the growth of phytoplankton and consequently enhanced nitrogen fixation in the ocean (Mahowald 2011).

It has been suggested that the most profound impact on the nitrogen cycle over the past 2.5 billion years has been caused by humans (Canfield *et al.* 2010). Three anthropogenic sources of N are responsible for more than doubling the amount of reactive N at the Earth's surface over the past century: (1) combustion of fossil fuels and production of NO_x; (2) the Haber–Bosch process, applied mostly for the production of nitrogenous fertilizers; and (3) biological nitrogen fixation by several species of cultivated plants (Galloway *et al.* 2003). Although anthropogenic N is mostly introduced to the atmosphere and soils, much is subsequently transported to the ocean by rivers, which receive N loads either from atmospheric deposition or as a consequence of wash-out from fertilized soils (Elser *et al.* 2009; Tyrrell 2011).

Table 4. *Index of geoaccumulation values for different elements, samples and locations*

Sampled material	Location	Element	I_{geo} value	References
Sediments	Tamaki Estuary, New Zealand	Cd	0.08–2.37	Abrahim & Parker (2008)
		Cu	0.42–2.06	
		Pb	0.52–1.94	
		Zn	0.01–2.12	
	Japan	As	1.2–3.4	Ahmed *et al.* (2011)
		Cu	0.4*	
		Cr	1.1*	
		Ni	0.4*	
	Southern Poland	N	2.81*	Loska & Wiechuła (2003)
		P	4.60*	
		Cd	5.72*	
		Cu	2.39*	
		Zn	3.40*	
	Germany	Cd	2–6	Förstner *et al.* (1993)
		Cu	0–2	
		Hg	0–6	
		Pb	1–2	
		Zn	1–4	
	Southwestern Spain	Cu	0.4–2.2	Santos Bermejo *et al.* (2003)
		Mn	1.1–2.2	
		Pb	2.7–4.0	
		Zn	3.5–4.0	
	Toulon Bay, France	Cu	2.7†	Tessier *et al.* (2011)
		Hg	5.4†	
		Pb	2.6†	
		Zn	2.5†	
Soils	Northern Bangladesh	Mn	1.24*	Bhuiyan *et al.* (2010)
		Zn	1.49*	
		Pb	1.63*	
	Central Jordan	Cd	1.01*	Banat *et al.* (2005)
		Hg	0.48*	
		Pb	0.32*	
		Zn	0.01*	
	20 cities in China	Cd	2.39*	Wei & Yang (2010)
		Cu	1.22*	
		Pb	3.54*	
		Zn	0.57*	
Road dust	Ontario, Canada	Cd	1.31–2.80	Nazzal *et al.* (2012)
		Cu	5.12–5.38	
		Pb	4.05–4.36	
		Zn	5.14–5.66	

*Mean value.
†Median value.

Sustained influx of excess reactive N to the sea by rivers causes increasing eutrophication of coastal environments. This process has substantially intensified over the last three decades and seems set to continue (Seitzinger *et al.* 2010). The eutrophication of surface waters increases algal blooms, leading to a deficit of oxygen from increased biomass decomposition and changing the redox potential. Increased levels of major nutrients in rivers are accompanied by a decrease in dissolved silica input to the sea. This phenomenon is an effect of increased fluvial diatom populations, whereby the silica, instead of being transported to the sea, is buried in river sediments as diatom tests (Seitzinger *et al.* 2010).

Phosphorus is another major nutrient that contributes to eutrophication of surface waters. Anthropogenic sources of this element constitute 4.8% of global yearly input of atmospheric P and 14% of soluble P (Mahowald *et al.* 2008). Anthropogenic impact on the nitrogen cycle is much higher than on the phosphorus cycle, and this difference may lead to high disproportion in availability of these nutrients to marine biota, creating P limitation in the future. The large natural marine N and P budgets may not be significantly affected by anthropogenic activity (Mahowald 2011). However, significant changes have been observed in lake sediments, even in remote areas (Holtgrieve *et al.* 2011), in relation to pre-industrial times. It has been suggested that these changes are sufficiently well marked to provide a reference level, that is, a Global Boundary Stratotype Section and Point (GSSP = 'golden spike') for the beginning (or base) of the Anthropocene (Wolfe *et al.* 2013).

Trace elements

Some issues of trace element geochemistry with respect to anthropogenic impact have been discussed above. In this section we focus on the use of trace element determinations in the study of pollution chronology. Interpretation of historical pollution trends with consistent patterns over large areas may help provide geochemical indicators of the Anthropocene.

Dated sediment and peat core samples allow the study of the history of environmental pollution. Most studies have focused on trace elements, but less has been done on organic pollutants and stable isotopes. Sediment samples of different origins (lake, stream, river, marine) have been analysed, but some authors suggest that marine sediments are better indicators of pollution chronology because they are less influenced by variations in pH, pE and other factors at the water–sediment interface that affect element mobility (Valette-Silver 1993; Buckley *et al.* 1995). The results of different studies carried out in Europe and North America have shown similar trends in contaminant concentrations, with a constant growth from the late 1800s and the highest values in the period 1940–1980. It has been shown that Pb and Zn concentrations in sediments are strongly correlated with population density and traffic intensity (Callender & Rice 2000).

Antimony can be as good an indicator of anthropogenic pollution as Pb or Hg (Shotyk *et al.* 2005*a*). This trace element shows a similar spatial and temporal distribution in dated peat bogs as Pb, thus suggesting their common origin. For example, the sources of Pb and Sb are ore mining and smelting, and fossil-fuel combustion, but these elements also have similar geochemical behaviour, namely relative immobility (Shotyk *et al.* 2005*a*, *b*; Cloy *et al.* 2005). Owing to the increased industrial use of Sb compounds for manufacturing plastics (as a flame retardant), alloys, semiconductors, pigments and so on, a growth in its concentration in the environment is expected. Cloy *et al.* (2005) showed a substantial (up to 100-fold) enrichment in both Sb and Pb in the atmosphere in the twentieth century, making these elements potential indicators of the Anthropocene.

There is a new trend in pollution relating to the study of the origin and fate of substances referred to as 'emerging pollutants' (e.g. La Farré *et al.* 2008). These substances are not currently covered by existing regulations, although they may cause potential threats to ecosystems and human health, and some may be important indicators of the Anthropocene. Rare earth elements (REE), including lanthanides (La through to Lu) and sometimes also Y and/or Sc, are significant in this respect. Increasing use of these elements in different high-technology industries and medicine has been accompanied by their enrichment in the environment. Gadolinium was the first anthropogenic REE anomaly recorded in river waters. Elevated levels of this element were found in the Havel River (Bau & Dulski 1996), and a similar positive Gd anomaly has been reported in river waters throughout the world (Kulaksız & Bau 2011). The anomaly mainly reflects wide use of different Gd complexes as contrast agents in magnetic resonance imaging. These chemical compounds are very persistent, not biodegradable, and cannot be removed in traditional wastewater treatment plants. Gadolinium entering rivers can hence be carried across large distances. More REE anomalies will probably be found in the environment as a consequence of the growing industrial applications of these elements. Indeed, a recent study of the Rhine River, Germany, showed an anthropogenic La anomaly (Kulaksız & Bau 2011). The natural sources of soluble REE are very scarce, so their anthropogenic origin may be clearly recognizable in the environment, therefore classifying these elements as potential indicators of the Anthropocene.

Isotopes

Stable isotopes. Stable isotopes of several elements are widely used in the study of industrial pollution (Oulhote *et al.* 2011; Négrel *et al.* 2012). Isotope signatures of S, C, O and N can fingerprint emission sources and environmental receptors to evaluate relative source strengths. In addition, the specific isotope signature of a substance derived from a given source allows identification of its natural and/or anthropogenic origin and of transport pathways (de Caritat *et al.* 1997; Wadleigh 2003; Górka *et al.* 2008, 2012). These studies are most successful when the isotopic ratio of the source emissions is substantially different from those ratios occurring naturally in the receptors. In studies involving complex ecosystems, fingerprinting anthropogenic pollution sources using stable isotopes is not always straightforward. Moreover, due to the common occurrence of these isotopes in various chemical species usually derived from many overlapping sources, both natural and anthropogenic, interpretation of the results can be difficult. To overcome this problem, a multi-isotope approach may be used (Hosono *et al.* 2011). Another possible solution is to combine isotopic and elemental signatures. For example, determinations of trace element (e.g. rare earth elements) contents and stable isotope ratios of airborne particulates are often used to pinpoint anthropogenic sources (Grobéty *et al.* 2010). Holtgrieve *et al.* (2011) observed that the N stable isotope fingerprint highlighted by depletion in the $\delta^{15}N$ values in remote lake sediments of the northern hemisphere correlated with increased anthropogenic CO_2 emissions in the twentieth century. They also suggested that this similarity could reflect a common anthropogenic source of CO_2 and reactive N, that is, biomass and fossil-fuel combustion.

Stable isotopes can also be useful for tracing sources of persistent organic compounds and their degradation rate in the environment (Philp 2007; Vetter *et al.* 2008). In environmental pollution studies, stable isotopes of Pb ($^{206}Pb/^{204}Pb$, $^{207}Pb/^{206}Pb$, $^{207}Pb/^{204}Pb$ and $^{208}Pb/^{204}Pb$) have found the widest application (Komárek *et al.* 2008). Global pollution of the environment with Pb took place in the period 1940–1980, when alky-lead additives were used in gasoline as antiknock agents. Since the ban on leaded gasoline use was imposed, Pb concentrations have decreased in the atmosphere. However, Pb enrichment of the Earth's surface systems and its specific isotope ratios have still been noted in sediments, peats and soils (Teutsch *et al.* 2001; Renberg *et al.* 2002; Shotyk *et al.* 2002). The spatial and temporal distribution of Pb in the environment combined with well-defined Pb isotope ratios from different anthropogenic sources may be used in assessing the Anthropocene. In a study of the Clyde estuary, UK (Vane *et al.* 2011), sediment core samples were analysed for $^{207}Pb/^{206}Pb$ and other geochemical indicators of anthropogenic pollution (polycyclic aromatic hydrocarbons, polychlorinated biphenyls and total petroleum hydrocarbons). This multi-tracer approach allowed reconstruction of a pollution history comprising signals recognizable from the coal, petroleum and electrical ages.

Recent attention has focused on transition-metal stable isotopes with a short residence time (Baskaran 2011). Isotope ratios of Cr, Cu, Zn and Cd, that is, metals with wide industrial applications, are useful as non-conservative tracers for assessing metal pollution sources (Bullen & Walczyk 2009). A worldwide database of metal stable isotope characteristics of different environmental samples and anthropogenic sources has not yet been created. However, specific substances have been analysed on a local scale, for instance B, an element used in production of detergents and fertilizers. Boron isotopes have shown the impact of anthropogenic sources on groundwater and have a possible application in tracing agricultural return flows (Vengosh 1998).

Radioisotopes. Global-scale enrichment of sediments in artificial radioisotopes such as ^{90}Sr, ^{137}Cs, ^{238}Pu, ^{239}Pu, ^{240}Pu and ^{241}Am could help indicate the beginning of the Anthropocene (Zalasiewicz *et al.* 2010; Hancock *et al.* this volume, in review). Large amounts of these radionuclides were introduced to the environment in atmospheric nuclear weapon testing, the highest activity being in 1952–1958 and 1961–1962. In 1945–1980, a total of 543 tests were carried out at different locations, including Novaya Zemlya, the Marshall Islands, Christmas Island, French Polynesia, Lop Nop, Semipalatinsk, Johnson Atoll and the Nevada test site (Livingston & Povinec 2000; Lindahl *et al.* 2010). Nuclear weapons testing left its imprint in the sediment record globally, because the radioactive waste reached the stratosphere and was then deposited widely. Nuclear reactor accidents comprise another large-scale impact on Earth's surface radioactivity, among which Chernobyl in 1986 and Fukushima Daiichi in 2011 were the most serious, scattering a large amount of radionuclides over large areas. Some rivers are still transporting radioactive particles several years after accidents have occurred (Gulin *et al.* 2002).

Of the radioisotopes emitted from anthropogenic sources, ^{137}Cs is believed to be the most representative of anthropogenic marine radioactivity as it has a high fission yield and a relatively long half-life, 30.17 ± 0.03 years (Livingston & Povinec 2000). This isotope also found its application in radiochronological dating of sediments as a good time marker. In terms of marking the beginning of the Anthropocene, long-lived isotopes such as ^{239}Pu (half-life of 24 100 years) or other transuranic element isotopes may be considered because currently over 70% of ^{137}Cs originating from nuclear weapons testing has already decayed (Erten 2011).

Organic pollutants

Organic pollutants are a large group of chemical compounds with different structures (aliphatic, aromatic, polymers), which consist mostly of biogenic elements (C, H and O). Organic compounds move relatively easily from phase to phase and can partition between air, water and organic species (Wania & Mackay 1999). Some organic pollutants are synthetic (e.g. organochlorine pesticides, polybrominated diphenyl ethers) and others have both natural and anthropogenic origins (e.g. polycyclic aromatic hydrocarbons). Of the different organic compounds, persistent organic pollutants (POP) have aroused the highest interest in environmental and health sciences due to their long residence time in different environmental compartments, as well as their accumulative and toxic properties. These compounds are resistant to degradation (biodegradation, photolysis), oxidation and hydrolysis, and thus may persist in the environment for decades. The most often determined organic pollutants are polycyclic aromatic hydrocarbons and polychlorinated biphenyls. For example, abundance ratios of polycyclic aromatic hydrocarbons in recent v. ancient sediments of Wigry Lake, north-eastern Poland, varied from 2 (calcareous ooze) to 15 (gyttja) (Gałuszka & Migaszewski 2004).

During the twentieth century, a substantial increase in concentrations of organic pollutants was reported as part of a planetary inventory (Steffen *et al.* 2011). We consider, though, that dynamic temporal changes in concentrations of organic pollutants, brought about by phase partitioning and biodegradation, means their use as indicators of the Anthropocene should be treated cautiously.

Modification of sediments by chemical pollution

The influence of industrial pollution on the chemistry of modern sediments has been widely documented (Buckley *et al.* 1995; Gulin *et al.* 2002; Huerta-Diaz *et al.* 2008; Li *et al.* 2012). Maximum concentrations of elements originating from industrial emissions are recorded in marine, lake and river sediments. This enrichment is so evident that some trace element patterns (e.g. Cu, Cd and Hg) have been suggested as chronostratigraphic markers (Lepland *et al.* 2010).

Recent trends in pollution on different scales (from local to global) are clearly recognized worldwide through environmental monitoring, with studies conducted both for governmental agencies and for scientific purposes. The beginnings of human influence on the environment, including historical atmospheric industrial emissions, may be seen, for instance, as enrichments of pollutants in well-dated sediment cores. Accumulation of these substances in recent sediments is generally highest in industrialized areas or in close vicinity to pollution sources. However, long-range atmospheric transport along prevailing air-mass trajectories leads to the presence of pollutants in the remotest regions of the world. The most pervasive pollutants that undergo long-range transport in the atmosphere are potentially toxic trace metals, such as As, Cd, Cu, Hg, Ni and Pb (Marx & McGowan 2010).

Interpretation of metal enrichment in sediments reveals a spatial variation caused not only by a human-induced influence on the region, but also by some important natural factors, such as climate, geology, land-cover changes and so on. A good example is the varied sensitivity of surface waters for acidification, depending on whether carbonate rocks are present, to increase the natural buffering capacity of the water (Meybeck 2003). These spatial differences are important for tracking the Anthropocene with geochemical methods. In most developed countries in Europe and North America, industrial emissions have declined in recent years, whereas in developing countries a rapid increase has been recorded (Schuster *et al.* 2002; Li *et al.* 2012). Thus, temporal trends in accumulation of pollutants in sediments will differ from one region to another. Nevertheless, the sediment records of historical pollution reported from many studies in both industrialized and remote areas (Fig. 4) show clear evidence of anthropogenic influence and can be used in stratigraphic assessment of the Anthropocene.

Results of geochemical studies with large datasets are often used for preparation of geochemical maps on different scales – from local (Lis & Pasieczna 1995) to continental (Salminen 2005; de Vos & Tarvainen 2006). The spatial data allow identification of geochemical anomalies and are invaluable for assessment of levels of pollution, especially regarding their sources, spatial distribution and so on, but they do not provide the evidence of historical trends necessary to help constrain the Anthropocene.

Secondary effects of anthropogenic pollution. Enrichment of the Earth's surface systems with pollutants is usually induced by direct anthropogenic activity. There are also important effects arising from influx of pollutants into different environmental compartments. Changes in geochemical parameters, such as pH, redox potential, chemical composition, temperature or pressure, resulting from anthropogenic pollution, may greatly influence the geoavailability, mobility, translocation and deposition of elements.

For instance, acidification of the Earth's surface comes from many natural and anthropogenic sources. A decrease in pH is induced by deposition of acidifying pollutants or microbial oxidation of sulphides and ammonia to form sulphuric and nitric acids (Rice & Herman 2012). The lower pH affects the mobility and bioavailability of different elements, leading to element enrichment or depletion in some environmental compartments. As a result of acid rain, trace metals are remobilized from the sediment and their concentrations

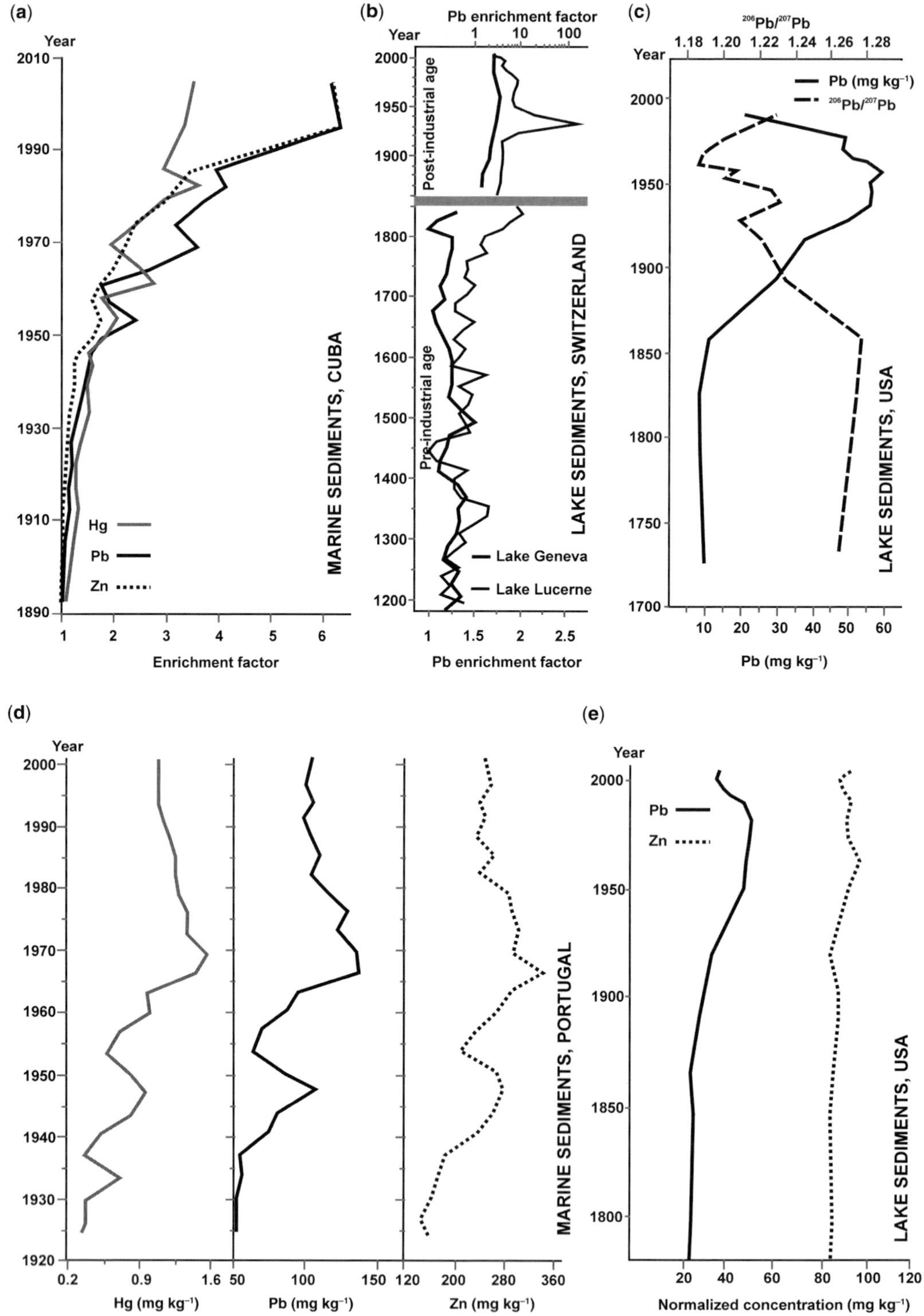

Fig. 4. Historical trends in accumulation of Pb, Hg and Zn in sediments (**a**, after Díaz-Asencio *et al.* 2011; **b**, after Thevenon *et al.* 2011; **c**, after Balogh *et al.* 2009; **d,** after Mil-Homens *et al.* 2009; **e**, after Mast *et al.* 2010).

increase in groundwater with the drop in pH values (Kjøller *et al.* 2004).

Oxidation and reduction reactions are also important for element mobility. The changes in the pH and redox potential of sediments may cause remobilization of substances from buried layers with subsequent precipitation at the sediment–water interface. Importantly, increases in pH and redox potential commonly lead to oxidation of sulphides and other species of reduced inorganic S and precipitation of Fe and Mn oxides and hydroxides (Buckley *et al.* 1995). These compounds have high sorptive properties and may cause a natural enrichment of surface sediment layers in many trace elements such as Cu, Ni and Pb. In contrast, very strong reducing conditions, usually combined with the presence of H_2S, may act as a geochemical barrier and bring about precipitation of many highly mobile trace metals such as Cd and Zn as sulphides. This process may also cause diagenetic remobilization and translocation of elements from subsurface sediment layers to the water column.

Conclusions

Human influence on the environment exceeds the changes caused by any other species in Earth's history. Mining, mineral processing and many other industrial activities have historically had a detrimental impact on the environment and have also left characteristic signatures in the sediment record. To formally recognize the Anthropocene as an epoch of Earth's history, quantitative markers of its beginning are needed. A review of different geochemical approaches to the study of environmental pollution provides some guidance to the selection of the most reliable geochemical signals of the Anthropocene.

The human-induced mobilization of trace elements on a global scale is much more evident than the anthropogenic mobilization of major nutrients (Rauch 2012). Of the trace elements discussed, Pb seems to be the best-known indicator of the Anthropocene due to (1) anthropogenic enrichment being well documented on a global scale, the history of pollution since pre-industrial time being known from many studies of dated sediment and peat cores; (2) possession of four stable isotopes with Pb isotope ratios being characteristic for given pollution sources; and (3) analysis of different samples by various instrumental methods being well-established. As an example, unleaded gasoline was introduced in the 1980s, and since then, Pb concentrations have substantially decreased while $^{206}Pb/^{207}Pb$ ratios have significantly increased.

A review of different case studies indicates that there are at least two reasons why sediments show high enrichments in trace metals especially. First, most trace elements tend to be immobilized in deep organic-rich sediments by chelation with organic matter, adsorption on clay minerals and precipitation of insoluble oxides, hydroxides, phosphates and carbonates. Second, a small rate of sediment deposition favours higher concentrations of many elements.

There has been an assumption that diagenetic processes do not change concentrations of elements and organic compounds in Holocene waterlain sediments (Förstner 1990). However, changes in physicochemical parameters, for example in the pH and redox potential of sediment and water column, may lead not only to different rates of deposition, but also to remobilization and interactions of chemical species.

Human activity can trigger a chain of physical, chemical and biological feedback processes that influence the spatial and temporal distribution of elements and their isotopes in sediments. Elemental concentrations and stable isotope ratios vary from site to site, under the influence of bedrock and parent material lithology, as well as physicochemical and biological factors. The indicators described in this report produce distinct human-induced signals in recent sediments, but cannot be taken as benchmarks for determining the base of the Anthropocene, most being diachronous. Given the complex array of natural and anthropogenic factors, it is best to use multiple indicators (geochemical, palaeontological and sedimentological) in assessing the Anthropocene.

The more holistic the approach to analysis, the more complex the methods and techniques used and media studied. In the future, determinations of a wider range of trace elements, organics and stable isotopes will encompass host rocks, colloids, waters and airborne emissions in addition to sediments. Background concentrations of elements always need to be taken into account.

We thank Guest Editor Dr C. Waters for the invitation to prepare this article.

References

Abrahim, G. M. S. & Parker, R. J. 2008. Assessment of heavy metal enrichment factors and the degree of contamination in marine sediments from Tamaki Estuary, Auckland, New Zealand. *Environmental Monitoring and Assessment*, **136**, 227–238, http://dx.doi.org/10.1007/s10661-007-9678-2

Ahmed, F., Bibi, M. H., Fukushima, T., Seto, K. & Ishiga, H. 2011. Recent sedimentary environment of coastal lagoon in southwestern Japan: evidence from major and trace elements. *Environmental Monitoring and Assessment*, **173**, 167–180, http://dx.doi.org/10.1007/s10661-010-1379-6

ALLABY, A. & ALLABY, M. 1999. *A Dictionary of Earth Sciences*. Oxford University Press, Oxford, New York.

ASTON, S. R., BRUTY, D., CHESTER, R. & PADGHAM, R. C. 1973. Mercury in lake sediments: a possible indicator of technological growth. *Nature*, **246**, 450–451.

BALKIS, N., AKTAN, Y. & BALKIS, N. 2012. Toxic metal (Pb, Cd and Hg) levels in the nearshore surface sediments from the European and Anotolian Shores of Bosphorus, Turkey. *Marine Pollution Bulletin*, **64**, 1938–1939, http://dx.doi.org/10.1016/j.marpolbul.2012.04.012

BALOGH, S. J., ENGSTROM, D. R. *ET AL.* 2009. A sediment record of trace metal loadings in the Upper Mississippi River. *Journal of Paleolimnology*, **41**, 623–639, http://dx.doi.org/10.1007/s10933-008-9295-2

BANAT, K. M., HOWARI, F. M. & AL-HAMAD, A. A. 2005. Heavy metals in urban soils of central Jordan: should we worry about their environmental risks? *Environmental Research*, **97**, 258–273, http://dx.doi.org/10.1016/j.envres.2004.07.002

BASKARAN, M. 2011. Environmental isotope geochemistry: past, present and future. *In*: BASKARAN, M. (ed.) *Handbook of Environmental Isotope Geochemistry*. Springer, Berlin, Heidelberg, 3–10, http://dx.doi.org/10.1007/978-3-642-10637-8_1

BAU, M. & DULSKI, P. 1996. Anthropogenic origin of positive gadolinium anomalies in river waters. *Earth and Planetary Science Letters*, **143**, 245–255.

BHUIYAN, M. A. H., PARVEZ, L., ISLAM, M. A., DAMPARE, S. B. & SUZUKI, S. 2010. Heavy metal pollution of coal mine-affected agricultural soils in the northern part of Bangladesh. *Journal of Hazardous Materials*, **173**, 384–392, http://dx.doi.org/10.1016/j.jhazmat.2009.08.085

BILOS, C., COLOMBO, J. C., SKORUPKA, C. N. & RODRIGUEZ PRESA, M. J. 2001. Sources, distribution and variability of airborne trace metals in La Plata City area, Argentina. *Environmental Pollution*, **111**, 149–158.

BOURENNANE, H., DOUAY, F., STERCKEMAN, T., VILLANNEAU, E., CIESIELSKI, H., KING, D. & BAIZE, D. 2010. Mapping of anthropogenic trace elements inputs in agricultural topsoil from Northern France using enrichment factors. *Geoderma*, **157**, 165–174, http://dx.doi.org/10.1016/j.geoderma.2010.04.009

BUCKLEY, D. E., SMITH, J. N. & WINTERS, G. V. 1995. Accumulation of contaminant metals in marine sediments of Halifax Harbour, Nova Scotia: environmental factors and historical trends. *Applied Geochemistry*, **10**, 175–195.

BULLEN, T. D. & WALCZYK, T. 2009. Environmental and biomedical applications of natural metal stable isotope variations. *Elements*, **5**, 381–385, http://dx.doi.org/10.2113/gselements.5.6.381

CAEIRO, S., COSTA, M. H. *ET AL.* 2005. Assessing heavy metal contamination in Sado Estuary sediment: an index analysis approach. *Ecological Indicators*, **5**, 151–169, http://dx.doi.org/10.1016/j.ecolind.2005.02.001

CALLENDER, E. & RICE, K. C. 2000. The urban environmental gradient: anthropogenic influences on the spatial and temporal distributions of lead and zinc in sediments. *Environmental Science and Technology*, **34**, 232–238, http://dx.doi.org/10.1021/es990380s

CANFIELD, D. E., GLAZER, A. N. & FALKOWSKI, P. G. 2010. The evolution and future of Earth's nitrogen cycle. *Science*, **330**, 192–196, http://dx.doi.org/10.1126/science.1186120

CHOI, S. D. & WANIA, F. 2011. On the reversibility of environmental contamination with persistent organic pollutants. *Environmental Science and Technology*, **45**, 8834–8841, http://dx.doi.org/10.1021/es2017544

CLARKE, F. W. 1924. *The Data of Geochemistry. US Geological Survey Bulletin 770*. 5th edn. US Government Printing Office, Washington.

CLOY, J. M., FARMER, J. G., GRAHAM, M. C., MACKENZIE, A. B. & COOK, G. T. 2005. A comparison of antimony and lead profiles over the past 2500 years in Flanders Moss ombrotrophic peat bog, Scotland. *Journal of Environmental Monitoring*, **7**, 1137–1147, http://dx.doi.org/10.1039/b510987f

DAUVALTER, V. & ROGNERUD, S. 2001. Heavy metal pollution in sediments of the Pasvik river drainage. *Chemosphere*, **42**, 9–18.

DAVIS, A., HEATWOLE, K., GREER, B., DITMARS, R. & CLARKE, R. 2010. Discriminating between background and mine-impacted groundwater at the Phoenix mine, Nevada USA. *Applied Geochemistry*, **25**, 400–417, http://dx.doi.org/10.1016/j.apgeochem.2009.12.007

DE CARITAT, P., KROUSE, H. R. & HUTCHEON, I. 1997. Sulphur isotope composition of streamwater, moss and humus from eight Arctic catchments in the Kola Peninsula region (NW Russia, N Finland, NE Norway). *Water, Air, and Soil Pollution*, **94**, 191–208.

DE VOS, W. & TARVAINEN, T. (eds) 2006. *Geochemical Atlas of Europe, Part 2: Interpretation of Geochemical Maps, Additional Tables, Figures, Maps, and Related Publications*. Geological Survey of Finland, Espoo.

DIATTA, J. B., CHUDZINSKA, E. & WIRTH, S. 2008. Assessment of heavy metal contamination of soils impacted by a zinc smelter activity. *Journal of Elementology*, **13**, 5–16.

DÍAZ-ASENCIO, M., CORCHO ALVARADO, J. A. *ET AL.* 2011. Reconstruction of metal pollution and recent sedimentation processes in Havana Bay (Cuba): a tool for coastal ecosystem management. *Journal of Hazardous Materials*, **196**, 402–411, http://dx.doi.org/10.1016/j.jhazmat.2011.09.037

DRAGOVIĆ, S., MIHAILOVIĆ, N. & GAJIC, B. 2008. Heavy metals in soils: distribution, relationship with soil characteristics and radionuclides and multivariate assessment of contamination sources. *Chemosphere*, **72**, 491–495, http://dx.doi.org/10.1016/j.chemosphere.2008.02.063

DUONG, T. T. T. & LEE, B.-K. 2011. Determining contamination level of heavy metals in road dust from busy traffic areas with different characteristics. *Journal of Environmental Management*, **92**, 554–562, http://dx.doi.org/10.1016/j.jenvman.2010.09.010

ELSER, J. J., ANDERSEN, T., BARON, J. S., BERGSTROM, A.-K., JANSSON, M., MELACK, J. & DOWNING, J. A. 2009. Shifts in lake N:P stoichiometry and nutrient limitation driven by atmospheric nitrogen deposition. *Science*, **326**, 835–837, http://dx.doi.org/10.1126/science.1176199

ERTEN, H. N. 2011. Radiochronological methods as tools to study environmental pollution. *In*: GÖKÇEKUS, H., TÜRKER, U. & LAMOREAUX, J. W. (eds) *Survival and Sustainability*. Springer, Berlin, Heidelberg, 1015–1022, http://dx.doi.org/10.1007/978-3-540-95991-5_96

ESSIEN, J. P., ANTAI, S. P. & OLAJIRE, A. A. 2009. Distribution, seasonal variations and ecotoxicological significance of heavy metals in sediments of Cross River estuary mangrove swamp. *Water, Air, and Soil Pollution*, **197**, 91–105, http://dx.doi.org/10.1007/s11270-008-9793-x

FALKOWSKI, P., SCHOLES, R. J. ET AL. 2000. The global carbon cycle: a test of our knowledge of Earth as a system. *Science*, **290**, 291–296, http://dx.doi.org/10.1126/science.290.5490.291

FERNÁNDEZ-CALIANI, J. C., BARBA-BRIOSO, C., GONZÁLEZ, I. & GALÁN, E. 2009. Heavy metal pollution in soils around the abandoned mine sites of the Iberian Pyrite Belt (Southwest Spain). *Water, Air, and Soil Pollution*, **200**, 211–226, http://dx.doi.org/10.1007/s11270-008-9905-7

FÖRSTNER, U. 1990. Inorganic sediment chemistry and elemental speciation. *In*: BAUDO, R., GIESY, J. P. & MUNTAU, H. (eds) *Sediments: Chemistry and Toxicity of In-Place Pollutants*. Lewis Publishers, Boston, 61–105.

FÖRSTNER, U., AHLF, W. & CALMANO, W. 1993. Sediment quality objectives and criteria development in Germany. *Water Science and Technology*, **28**, 307–316.

GALÁN, E., GONZÁLEZ, I. & FERNÁNDEZ-CALIANI, J. C. 2002. Residual pollution load of soils impacted by the Aznalcóllar (Spain) mining spill after clean-up operations. *The Science of the Total Environment*, **286**, 167–179, http://dx.doi.org/10.1016/S0048-9697(01)00974-3.

GALLOWAY, J. N., ABER, J. D., ERISMAN, J. W., SEITZINGER, S. P., HOWARTH, R. W., COWLING, E. B. & COSBY, B. J. 2003. The nitrogen cascade. *BioScience*, **53**, 341–356.

GAŁUSZKA, A. 2007. A review of geochemical background concepts and an example using data from Poland. *Environmental Geology*, **52**, 861–870, http://dx.doi.org/10.1007/s00254-006-0528-2

GAŁUSZKA, A. & MIGASZEWSKI, Z. M. 2004. Concentrations of polynuclear aromatic hydrocarbons in sediments of Lake Wigry and selected rivers of the Suwałki area (NE Poland). *Prace Komisji Paleogeografii Czwartorzędu, Polska Akademia Umiejętności*, **II**, 49–54 (in Polish).

GAŁUSZKA, A. & MIGASZEWSKI, Z. M. 2011. Geochemical background – an environmental perspective. *Mineralogia*, **42**, 7–17, http://dx.doi.org/10.2478/v10002-011-0002-y

GHREFAT, H. & YUSUF, N. 2006. Assessing Mn, Fe, Cu, Zn, and Cd pollution in bottom sediments of Wadi Al-Arab Dam, Jordan. *Chemosphere*, **65**, 2114–2121, http://dx.doi.org/10.1016/j.chemosphere.2006.06.043

GÓRKA, M., JĘDRYSEK, M. O. & STRĄPOĆ, D. 2008. Isotopic composition of sulphates from meteoric precipitation as an indicator of pollutant origin in Wrocław (SW Poland). *Isotopes in Environmental and Health Studies*, **44**, 177–188, http://dx.doi.org/10.1080/10256010802066307

GÓRKA, M., ZWOLIŃSKA, E., MALKIEWICZ, M., LEWICKA-SZCZEBAK, D. & JĘDRYSEK, M. O. 2012. Carbon and nitrogen isotopic analysis coupled with palynological data of PM10 in Wrocław city (SW Poland) – assessment of anthropogenic impact. *Isotopes in Environmental and Health Studies*, **48**, 327–344, http://dx.doi.org/10.1080/10256016.2012.639449

GROBÉTY, B., GIERÉ, R., DIETZE, V. & STILLE, P. 2010. Airborne particles in the urban environment. *Elements*, **6**, 229–234. http://dx.doi.org/10.2113/gselements.6.4.229

GULIN, S. B., POLIKARPOV, G. G., EGOROV, V. N., MARTIN, J. M., KOROTKOV, A. A. & STOKOZOV, N. A. 2002. Radioactive contamination of the northwestern Black Sea sediments. *Estuarine, Coastal and Shelf Science*, **54**, 541–549, http://dx.doi.org/10.1006/ecss.2000.0663

HAKANSON, L. 1980. An ecological risk index for aquatic pollution control: a sedimentological approach. *Water Research*, **14**, 975–1001.

HALSTEAD, M. J. R., CUNNINGHAME, R. G. & HUNTER, K. A. 2000. Wet deposition of trace metals to a remote site in Fjordland, New Zealand. *Atmospheric Environment*, **34**, 665–676, http://dx.doi.org/10.1016/S1352-2310(99)00185-5

HANCOCK, G. J., TIMS, S. G., FIFIELD, L. K. & WEBSTER, I. K. In review. The release and persistence of radioactive anthropogenic nuclides. *In*: WATERS, C. N., ZALASIEWICZ, J. A., WILLIAMS, M., ELLIS, M. A. & SNELLING, A. M. (eds) *A Stratigraphical Basis for the Anthropocene*. Geological Society, London, Special Publications, **395**.

HAWKES, H. E. & WEBB, J. E. 1962. *Geochemistry in Mineral Exploration*. Harper, New York.

HOLTGRIEVE, G. W., SCHINDLER, D. E. ET AL. 2011. A coherent signature of anthropogenic nitrogen deposition to remote watersheds of the Northern Hemisphere. *Science*, **334**, 1545–1548, http://dx.doi.org/10.1126/science.1212267

HOSONO, T., WANG, C.-H. ET AL. 2011. Multiple isotope (H, O, N, S and Sr) approach elucidates complex pollution causes in the shallow groundwaters of the Taipei urban area. *Journal of Hydrology*, **397**, 23–36, http://dx.doi.org/10.1016/j.jhydrol.2010.11.025

HUERTA-DIAZ, M. A., DELGADILLO-HINOJOSA, F. ET AL. 2008. Diagnosis of trace metal contamination in sediments: the example of Ensenada and El Sauzal, two harbors in Baja California, Mexico. *Marine Environmental Research*, **66**, 345–368, http://dx.doi.org/10.1016/j.marenvres.2008.05.008

HUTCHINS, D. A., MULHOLLAND, M. R. & FU, F. 2009. Nutrient cycles and marine microbes in a CO_2-enriched ocean. *Oceanography*, **22**, 128–145.

KJØLLER, C., POSTMA, D. & LARSEN, F. 2004. Groundwater acidification and the mobilization of trace metals in a sandy aquifer. *Environmental Science and Technology*, **38**, 2829–2835, http://dx.doi.org/10.1021/es030133v

KOMÁREK, M., ETTLER, V., CHRASTNÝ, V. & MIHALJEVIČ, M. 2008. Lead isotopes in environmental sciences: A review. *Environment International*, **34**, 562–577, http://dx.doi.org/10.1016/j.envint.2007.10.005

Krishna, A. K. & Govil, P. K. 2008. Assessment of heavy metal contamination in soils around Manali industrial area, Chennai, Southern India. *Environmental Geology*, **54**, 1465–1472, http://dx.doi.org/10.1007/s00254-007-0927-z

Krishnamurthy, A. J. K., Moore, N., Mahowald, C., Luo, S. C., Doney, K. L. & Zender, C. S. 2009. Impacts of increasing anthropogenic soluble iron and nitrogen deposition on ocean biogeochemistry. *Global Biogeochemical Cycles*, **23**, GB3016, http://dx.doi.org/10.1029/2008GB003440

Kulaksiz, S. & Bau, M. 2011. Rare earth elements in the Rhine River, Germany: first case of anthropogenic lanthanum as a dissolved microcontaminant in the hydrosphere. *Environment International*, **37**, 973–979, http://dx.doi.org/10.1016/j.envint.2011.02.018

La Farré, M., Pérez, S., Kantiani, L. & Barceló, D. 2008. Fate and toxicity of emerging pollutants, their metabolites and transformation products in the aquatic environment. *Trends in Analytical Chemistry*, **27**, 991–1007, http://dx.doi.org/10.1016/j.trac.2008.09.010

Lepland, A., Andersen, T. J., Lepland, A., Arp, H. P. H., Alve, E., Breedveld, G. D. & Rindby, A. 2010. Sedimentation and chronology of heavy metal pollution in Oslo harbor, Norway. *Marine Pollution Bulletin*, **60**, 1512–1522, http://dx.doi.org/10.1016/j.marpolbul.2010.04.017

Li, X., Liu, L. et al. 2012. Integrated assessment of heavy metal contamination in sediments from a coastal industrial basin, NE China. *PLoS ONE*, **7**, e39690, http://dx.doi.org/10.1371/journal.pone.0039690

Lindahl, P., Lee, S.-H., Worsfold, P. & Keith-Roach, M. 2010. Plutonium isotopes as tracers for ocean processes: a review. *Marine Environmental Research*, **69**, 73–84, http://dx.doi.org/10.1016/j.marenvres.2009.08.002

Lis, J. & Pasieczna, A. 1995. *Geochemical Atlas of Upper Silesia 1:200 000*. Polish Geological Institute, Warszawa.

Liu, W., Zhao, J., Ouyang, Z., Söderlund, L. & Liu, G. 2005. Impacts of sewage irrigation on heavy metal distribution and contamination in Beijing, China. *Environment International*, **31**, 805–812, http://dx.doi.org/10.1016/j.envint.2005.05.042

Livingston, H. D. & Povinec, P. P. 2000. Anthropogenic marine radioactivity. *Ocean & Coastal Management*, **43**, 689–712.

Loska, K. & Wiechuła, D. 2003. Application of principal component analysis for the estimation of source of heavy metal contamination in surface sediments from the Rybnik Reservoir. *Chemosphere*, **51**, 723–733, http://dx.doi.org/10.1016/S0045-6535(03)00187-5

Loska, K., Wiechuła, D. & Korus, I. 2004. Metal contamination of farming soils affected by industry. *Environment International*, **30**, 159–165, http://dx.doi.org/10.1016/S0160-4120(03)00157-0

Mahowald, N. 2011. Aerosol indirect effect on biogeochemical cycles and climate. *Science*, **334**, 794–796, http://dx.doi.org/10.1126/science.1207374

Mahowald, N., Jickells, T. D. et al. 2008. Global distribution of atmospheric phosphorus sources, concentrations and deposition rates, and anthropogenic impacts. *Global Biogeochemical Cycles*, **22**, GB4026, http://dx.doi.org/10.1029/2008GB003240

Marx, S. K. & McGowan, H. A. 2010. Long-distance transport of urban and industrial metals and their incorporation into the environment: sources, transport pathways and historical trends. *In*: Zereini, F. & Wiseman, C. L. S. (eds) *Urban Airborne Particulate Matter, Environmental Science and Engineering*. Springer-Verlag, Berlin, Heidelberg, 103–124, http://dx.doi.org/10.1007/978-3-642-12278-1_6

Mast, M. A., Manthorne, D. J. & Roth, D. A. 2010. Historical deposition of mercury and selected trace elements to high-elevation National Parks in the Western U.S. inferred from lake-sediment cores. *Atmospheric Environment*, **44**, 2577–2586, http://dx.doi.org/10.1016/j.atmosenv.2010.04.024

Matschullat, J., Ottenstein, R. & Reimann, C. 2000. Geochemical background – can we calculate it? *Environmental Geology*, **39**, 990–1000, http://dx.doi.org/0.1007/s002549900084

McQueen, K. 2009. Regolith geochemistry. *In*: Scott, K. & Pain, C. (eds) *Regolith Science*. CSIRO Publishing, Melbourne, 73–104.

Meybeck, M. 2003. Global analysis of river systems: from Earth system controls to Anthropocene syndromes. *Philosophical Transactions of the Royal Society B*, **358**, 1935–1955, http://dx.doi.org/10.1098/rstb.2003.1379

Miesch, A. T. 1967. *Methods of Computation for Estimating Geochemical Abundance*, US Geological Survey Professional Papers 574-B. US Government Printing Office, Washington.

Migaszewski, Z. M., Gałuszka, A. & Pasławski, P. 2003. Baseline v. background concentrations of trace elements in sediments of Lake Wigry, NE Poland. *Limnological Review*, **3**, 165–171.

Mil-Homens, M., Branco, V., Lopes, C., Vale, C., Abrantes, F., Boer, W. & Vicente, M. 2009. Using factor analysis to characterise historical trends of trace metal contamination in a sediment core from the Tagus Prodelta, Portugal. *Water, Air, and Soil Pollution*, **197**, 277–287, http://dx.doi.org/10.1007/s11270-008-9810-0

Mohiuddin, K. M., Zakir, H. M., Otomo, K., Sharmin, S. & Shikazono, N. 2010. Geochemical distribution of trace metal pollutants in water and sediments of downstream of an urban river. *International Journal of Environmental Science and Technology*, **7**, 17–28.

Müller, G. 1969. Index of geoaccumulation in sediments of the Rhine River. *Geojournal*, **2**, 108–118.

Nazzal, Y., Rosen, M. A. & Al-Rawabdeh, A. M. 2012. Assessment of metal pollution in urban road dusts from selected highways of the Greater Toronto Area in Canada. *Environmental Monitoring and Assessment*, **185**, 1847–1858, http://dx.doi.org/10.1007/s10661-012-2672-3

Négrel, P., Blessing, M., Millot, R., Petelet-Giraud, E. & Innocent, C. 2012. Isotopic methods give clues about the origins of trace metals and organic pollutants in the environment. *Trends in Analytical Chemistry*, **38**, 143–153, http://dx.doi.org/10.1016/j.trac.2012.03.017

OLIVA, S. R. & FERNÁNDES ESPINOSA, A. J. F. 2007. Monitoring of heavy metals in topsoils, atmospheric particles and plant leaves to identify possible contamination sources. *Microchemical Journal*, **86**, 131–139, http://dx.doi.org/10.1016/j.microc.2007.01.003

ORR, J. C., FABRY, V. J. ET AL. 2005. Anthropogenic ocean acidification over the twenty-first century and its impact on calcifying organisms. *Nature*, **437**, 681–686, http://dx.doi.org/10.1038/nature04095

OULHOTE, Y., LE BOT, B., DEGUEN, S. & GLORENNEC, P. 2011. Using and interpreting isotope data for source identification. *Trends in Analytical Chemistry*, **30**, 302–312, http://dx.doi.org/10.1016/j.trac.2010.10.015

PHILP, R. P. 2007. The emergence of stable isotopes in environmental and forensic geochemistry studies: a review. *Environmental Chemistry Letters*, **5**, 57–66, http://dx.doi.org/10.1007/s10311-006-0081-y

PRICE, S. J., FORD, J. R., COOPER, A. H. & NEAL, C. 2011. Humans as major geological and geomorphological agents in the Anthropocene. The significance of artificial ground in Great Britain. *Philosophical Transactions of the Royal Society A*, **369**, 1056–1084, http://dx.doi.org/10.1098/rsta.2010.0296

RASHED, M. N. 2010. Monitoring of contaminated toxic and heavy metals, from mine tailings through age accumulation, in soil and some wild plants at Southeast Egypt. *Journal of Hazardous Materials*, **178**, 739– 746, http://dx.doi.org/10.1016/j.jhazmat.2010.01.147

RAUCH, J. N. 2012. The present understanding of Earth's global anthrobiogeochemical metal cycles. *Mineral Economics*, **25**, 7–15, http://dx.doi.org/10.1007/s13563-011-0011-8

RAY, A. K., TRIPATHY, S. C., PATRA, S. & SARMA, V. V. 2006. Assessment of Godavari estuarine mangrove ecosystem through trace metal studies. *Environment International*, **32**, 219–223, http://dx.doi.org/10.1016/j.envint.2005.08.014

REIMANN, C. & DE CARITAT, P. 2000. Intrinsic flaws of element enrichment factors (EFs) in environmental geochemistry. *Environmental Science and Technology*, **34**, 5084–5091.

REIMANN, C. & DE CARITAT, P. 2005. Distinguishing between natural and anthropogenic sources of element in the environment: regional geochemical surveys v. enrichment factors. *Science of the Total Environment*, **337**, 91–107, http://dx.doi.org/10.1016/j.scitotenv.2004.06.011

REIMANN, C. & GARRET, R. G. 2005. Geochemical background – concept and reality. *Science of the Total Environment*, **350**, 12–27, http://dx.doi.org/10.1016/j.scitotenv.2005.01.047

REIMANN, C., FILZMOSER, P. & GARRET, R. G. 2005. Background and threshold: critical comparison of methods of determination. *Science of the Total Environment*, **346**, 1–16, http://dx.doi.org/10.1016/j.scitotenv.2004.11.023

RENBERG, I., BRÄNVALL, M. L., BINDLER, R. & EMTERYD, O. 2002. Stable lead isotopes and lake sediments – a useful combination for the study of atmospheric lead pollution history. *Science of the Total Environment*, **292**, 45–54, http://dx.doi.org/10.1016/S0048-9697(02)00032-3

RICE, K. C. & HERMAN, J. S. 2012. Acidification of Earth: an assessment across mechanisms and scales. *Applied Geochemistry*, **27**, 1–14, http://dx.doi.org/10.1016/j.apgeochem.2011.09.001

RUDNICK, R. L. & GAO, S. 2003. Composition of the continental crust. *In*: HOLLAND, H. D. & TUREKIAN, K. K. (eds) *Treatise on Geochemistry*. Elsevier–Pergamon, Oxford, **3**, 1–64.

RUNKEL, R. L., KIMBALL, B. A., WALTON-DAY, K. & VERPLANCK, P. L. 2007. A simulation-based approach for estimating premining water quality: Red Mountain Creek, Colorado. *Applied Geochemistry*, **22**, 1899–1918, http://dx.doi.org/10.1016/j.apgeochem.2007.03.054

SALMINEN, R. (ed.) 2005. *FOREGS Geochemical Atlas of Europe, Part 1: Background Information, Methodology and Maps*. Geological Survey of Finland, Espoo.

SANTOS BERMEJO, J. C., BELTRÁN, R. & GÓMEZ ARIZA, J. L. 2003. Spatial variations of heavy metals contamination in sediments from Odiel River (southwest Spain). *Environment International*, **29**, 69–77.

SCHLESINGER, W. H. 2009. On the fate of anthropogenic nitrogen. *Proceedings of the National Academy of Sciences of the USA*, **106**, 203–208, http://dx.doi.org/10.1073/pnas.0810193105

SCHRAG, D. P. 2012. Geobiology of the Anthropocene. *In*: KNOLL, A. H., CANFIELD, D. E. & KONHAUSER, K. O. (eds) *Fundamentals of Geobiology*. Wiley, Chichester, 425–436.

SCHUSTER, P. F., KRABBENHOFT, D. P. ET AL. 2002. Atmospheric mercury deposition during the last 270 years: a glacial ice core record of natural and anthropogenic sources. *Environmental Science and Technology*, **36**, 2303–2310.

SEITZINGER, S. P., MAYORGA, E. ET AL. 2010. Global river nutrient export: a scenario analysis of past and future trends. *Global Biogeochemical Cycles*, **24**, GB0A08, http://dx.doi.org/10.1029/2009GB003587

SELINUS, O. S. & ESBENSEN, K. 1995. Separating anthropogenic from natural anomalies in environmental geochemistry. *Journal of Geochemical Exploration*, **55**, 55–66, http://dx.doi.org/10.1016/0375-6742(95)00034-8

SEN, I. S. & PEUCKER-EHRENBRINK, B. 2012. Anthropogenic disturbance of element cycles at the Earth's surface. *Environmental Science and Technology*, **46**, 8601–8609, http://dx.doi.org/10.1021/es301261x

SHOTYK, W., WEISS, D., HEISTERKAMP, M., CHEBURKIN, A. K. & ADAMS, F. C. 2002. A new peat bog record of atmospheric lead pollution in Switzerland: Pb concentrations, enrichment factors, isotopic composition, and organolead species. *Environmental Science and Technology*, **36**, 3893–3900.

SHOTYK, W., KRACHLER, M. & CHEN, B. 2005*a*. Antimony: global environmental contaminant. *Journal of Environmental Monitoring*, **7**, 1135–1136, http://dx.doi.org/10.1039/B515468P

SHOTYK, W., CHEN, B. & KRACHLER, M. 2005*b*. Lithogenic, oceanic and anthropogenic sources of atmospheric Sb to a maritime blanket bog, Myrarnar, Faroe Islands. *Journal of Environmental Monitoring*, **7**, 1148–1154, http://dx.doi.org/10.1039/B509928P

SOARES, H. M. V., BONAVENTURA, R. A. R., MACHADO, A. A. S. C. & ESTEVES DA SILVA, J. C. G. 1999.

Sediments as monitors of heavy metal contamination in the Ave river basin (Portugal): multivariate analysis of data. *Environmental Pollution*, **105**, 311–323.

SPENCER, K. L., CUNDY, A. B. & CROUDACE, I. W. 2003. Heavy metal distribution and early-diagenesis in salt marsh sediments from the Medway Estuary, Kent, UK. *Estuarine, Coastal and Shelf Science*, **57**, 43–54, http://dx.doi.org/10.1016/S0272-7714(02) 00324-4

STEFFEN, W., GRINEVALD, J., CRUTZEN, P. & MCNEILL, J. 2011. The Anthropocene: conceptual and historical perspectives. *Philosophical Transactions of the Royal Society A*, **369**, 842–867, http://dx.doi.org/10.1098/rsta.2010.0327

SUCHAROVÁ, J., SUCHARA, I. *ET AL.* 2012. Top-/bottom-soil ratios and enrichment factors: what do they really show? *Applied Geochemistry*, **27**, 138–145, http://dx.doi.org/10.1016/j.apgeochem.2011.09.025

SUTHERLAND, R. A. 2000. Bed sediment-associated trace metals in an urban stream, Oahu, Hawaii. *Environmental Geology*, **39**, 611–627, http://dx.doi.org/10.1007/s002540050473

SZEFER, P., KUSAK, A., SZEFER, K., GLASBY, G. P., JANKOWSKA, H., WOŁOWICZ, M. & ALI, A. A. 1998. Evaluation of the anthropogenic influx of metallic pollutants into Puck Bay (southern Baltic). *Applied Goechemistry*, **13**, 293–304.

TESSIER, E., GARNIER, C. *ET AL.* 2011. Study of the spatial and historical distribution of sediment inorganic contamination in the Toulon bay (France). *Marine Pollution Bulletin*, **62**, 2075–2086, http://dx.doi.org/10.1016/j.marpolbul.2011.07.022

TEUTSCH, N., EREL, Y., HALICZ, L. & BANIN, A. 2001. Distribution of natural and anthropogenic lead in Mediterranean soils. *Geochimica et Cosmochimica Acta*, **65**, 2853–2864, http://dx.doi.org/10.1016/S0016-7037(01)00607-X

THEVENON, F., GRAHAM, N. D., CHIARADIA, M., ARPAGAUS, P., WILDI, W. & POTÉ, J. 2011. Local to regional scale industrial heavy metal pollution recorded in sediments of large freshwater lakes in central Europe (lakes Geneva and Lucerne) over the last centuries. *Science of the Total Environment*, **412–413**, 239–247, http://dx.doi.org/10.1016/j.scitotenv.2011.09.025

TOMLINSON, D. L., WILSON, J. G., HARRIS, C. R. & JEFFREY, D. W. 1980. Problems in the assessment of heavy-metal levels in estuaries and the formation of a pollution index. *Helgoland Marine Research*, **33**, 566–575.

TUREKIAN, K. K. & WEDEPOHL, K. H. 1961. Distribution of the elements in some major units of the Earth's crust. *Geological Society of America Bulletin*, **72**, 175–192.

TYRRELL, T. 2011. Anthropogenic modification of the oceans. *Philosophical Transactions of the Royal Society A*, **369**, 887–908, http://dx.doi.org/10.1098/rsta.2010.0339

VALETTE-SILVER, N. J. 1993. The use of sediment cores to reconstruct historical trends in contamination of estuarine and coastal sediments. *Estuaries and Coasts*, **16**, 577–588.

VANE, C. H., CHENERY, S. R., HARRISON, I., KIM, A. W., MOSS-HAYES, V. & JONES, D. G. 2011. Chemical signatures of the Anthropocene in the Clyde estuary, UK: sediment-hosted Pb, $^{207/206}$Pb, total petroleum hydrocarbon, polyaromatic hydrocarbon and polychlorinated biphenyl pollution records. *Philosophical Transactions of the Royal Society A*, **369**, 1085–1111, http://dx.doi.org/10.1098/rsta.2010.0298

VENGOSH, A. 1998. The isotopic composition of anthropogenic boron and its potential impact on the environment. *Biological Trace Element Research*, **66**, 145–151, http://dx.doi.org/10.1007/BF02783134

VETTER, W., GAUL, S. & ARMBRUSTER, W. 2008. Stable carbon isotope ratios of POPs – a tracer that can lead to the origins of pollution. *Environment International*, **34**, 357–362, http://dx.doi.org/10.1016/j.envint.2007.03.010

WADLEIGH, M. A. 2003. Lichens and atmospheric sulphur: what stable isotopes reveal. *Environmental Pollution*, **126**, 345–351, http://dx.doi.org/10.1016/S0269-7491(03)00247-1

WANIA, F. & MACKAY, D. 1999. The evolution of mass balance models of persistent organic pollutant fate in the environment. *Environmental Pollution*, **100**, 223–240.

WEDEPOHL, H. 1995. The composition of the continental crust. *Geochimica et Cosmochimica Acta*, **59**, 1217–1239.

WEI, B. & YANG, L. 2010. A review of heavy metal contaminations in urban soils, urban road dusts and agricultural soils from China. *Microchemical Journal*, **94**, 99–107.

WEISS, D., SHOTYK, W., APPLEBY, P. G., KRAMERS, J. D. & CHEBURKIN, A. K. 1999. Atmospheric Pb deposition since the industrial revolution recorded by five Swiss peat profiles: enrichment factors, fluxes, isotopic composition, and sources. *Environmental Science and Technology*, **33**, 1340–1352.

WOLFE, A. P., HOBBS, W. O. *ET AL.* 2013. Stratigraphic expressions of the Holocene–Anthropocene transition revealed in sediments from remote lakes. *Earth-Science Reviews*, **116**, 17–34.

ZALASIEWICZ, J., WILLIAMS, M., STEFFEN, W. & CRUTZEN, P. 2010. The new world of the Anthropocene. *Environmental Science & Technology*, **44**, 2228–2231, http://dx.doi.org/10.1021/es903118j

ZALASIEWICZ, J., WILLIAMS, M. *ET AL.* 2011*a*. Stratigraphy of the Anthropocene. *Philosophical Transactions of the Royal Society A*, **369**, 1036–1055, http://dx.doi.org/10.1098/rsta.2010.0315

ZALASIEWICZ, J., WILLIAMS, M., HAYWOOD, A. & ELLIS, M. 2011*b*. The Anthropocene: a new epoch of geological time? *Philosophical Transactions of the Royal Society A*, **369**, 835–841, http://dx.doi.org/10.1098/rsta.2010.0339

ZHANG, C., QUIAO, Q., PIPER, J. D. A. & HUANG, B. 2011. Assessment of heavy metal pollution from a Fe-smelting plant in urban river sediments using environmental magnetic and geochemical methods. *Environmental Pollution*, **159**, 3057–3070, http://dx.doi.org/10.1016/j.envpol.2011.04.006

Definition of the Anthropocene: a view from the underworld

IAN J. FAIRCHILD[1]* & SILVIA FRISIA[2]

[1]*School of Geography, Earth and Environmental Science, University of Birmingham B15 2TT, UK*

[2]*School of Environmental and Life Sciences, University of Newcastle, Callaghan 2308 NSW, Australia*

**Corresponding author (e-mail: i.j.fairchild@bham.ac.uk)*

Abstract: Annually laminated stalagmites from natural caves and limestone mines capture a number of significant environmental and climatic signals during the anthropogenically disturbed era. The effects of forest clearance, or development of agricultural or industrial practices, can be marked by changes in soil or hydrological responses leading to shifts in both chemical (e.g. carbon and oxygen isotope ratios or trace elements) and physical (e.g. fabric, thickness of laminae) signals. However, these signals are diachronous because of the spatial heterogeneity of human societies. Twentieth-century changes in atmospheric composition are known from speleothems at several sites and demonstrate pollution disturbance of the sulphur cycle and the signal provided by the 1950s rise in radiocarbon caused by atmospheric nuclear tests. This latter is a global signal and hence a strong candidate to define the start of the Anthropocene, although other considerations, including comparison with instrumental archives, would favour an earlier timing. An attractive option is the climate amelioration marking the end of the Little Ice Age in the mid-nineteenth century, which is marked in Alpine and other Northern Hemisphere areas. Examples are illustrated from the Grotta di Ernesto cave to illustrate the appearance of a putative mid-nineteenth-century boundary.

This article considers how one could, on the basis of speleothems (precipitates forming in caves), usefully construct a geological boundary to mark the onset of a profoundly anthropogenically disturbed period, the Anthropocene (Crutzen & Stoermer 2000; Zalasiewicz *et al.* 2011). Records formed of calcium carbonate are among those depicted to be particularly useful in subdividing our present interglacial, the Holocene (Walker *et al.* 2012), but defining the Anthropocene is recognized as an enterprise of a different character. Consideration of this boundary is not just a narrow semantic issue, but, in a broader sense, there is no greater challenge facing humanity than to reflect on the profound and partially irreversible change to which we have subjected the Earth. This leads to actions to mitigate/restore where possible and otherwise adapt to the new conditions. The contrast between the Anthropocene and the pre-Anthropocene is therefore of fundamental importance. As Steffen *et al.* (2011*a*) have cautioned, 'The Anthropocene is a reminder that the Holocene, during which complex human societies have developed, has been a stable, accommodating environment and is the only state of the Earth System that we know for sure can support contemporary society'.

Dark, silent and apparently unchanging underground environments have represented a transcendental otherworld since antiquity and their labyrinthine networks represent the only real frontier for human geographical exploration on Earth (Orrock 1999). Cave environments were the habitat of the famous experiments on human subject Michel Siffre to test for the existence of circadian rhythms independent of the stimulation of daylight (e.g. Chouvet *et al.* 1974). If long-term human observers deep in a cave interior could remain sane, they would hear seasonal or spasmodic changes in the influx of dripping or flowing water. In appropriate locations, they could also feel changes in air circulation on a seasonal or even diurnal basis. Nowadays, an observer would deploy sensitive, possibly automated instruments and record hourly multi-annual changes in the levels of gases such as radon and CO_2, in the discharge as well as chemical properties of dripping waters. Speleothems respond to such *physiological* changes in the cave environment (Fairchild & Baker 2012) and act as natural recorders of underground environmental conditions, which represent modified signals of the changing pedosphere–biosphere–hydrosphere–atmosphere above the cave (Fairchild *et al.* 2006). Much of the focus of speleothem research is on palaeoclimatic reconstruction, but the palaeoclimate proxies are, in reality, often more nuanced indicators of environmental change rather than

From: Waters, C. N., Zalasiewicz, J. A., Williams, M., Ellis, M. A. & Snelling, A. M. (eds) 2014. *A Stratigraphical Basis for the Anthropocene*. Geological Society, London, Special Publications, **395**, 239–254.
First published online October 24, 2013, http://dx.doi.org/10.1144/SP395.7

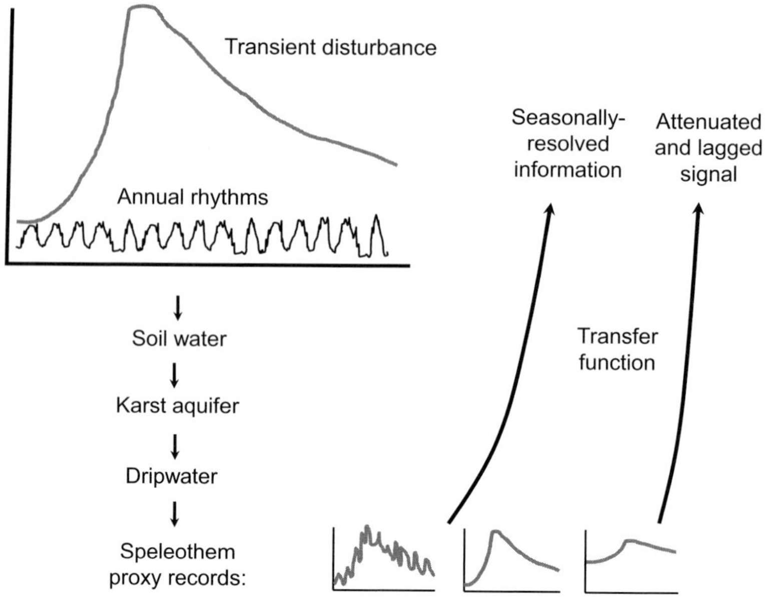

Fig. 1. Speleothems as recorders of environmental (including climatic) signals. Here, a multiyear transient environmental perturbation (e.g. anthropogenic) is depicted. The vegetation–soil–karst system transforms the signals into various proxies whose sensitivity depends on the proxy and the properties of the karst system. At one extreme, an intricate record in which annual noise is superimposed on the disturbance could be preserved, or a smoother record or, more typically, an attenuated and lagged signal. The least sensitive parameter just gives information on long-term mean properties.

of surface temperature and rainfall variability (Fig. 1). Speleothems offer a huge variety of physicochemical properties for study and so are capable of recording many facets of environmental change, which is helpful in view of the wide range of types of disturbance to natural conditions that are characteristic of the Anthropocene.

In this article, where we make generalized statements in the text about speleothems without specific references, the justification is found in Fairchild & Baker (2012). We focus on calcareous ($CaCO_3$) speleothems, rather than other cave mineral deposits. This is because they can be readily dated by combinations of radiometric and lamina-counting techniques and they also display many properties that are proxies for past environment and climate conditions. For the Anthropocene definition, we are particularly interested in samples exhibiting visible stacking of annual laminae, analogous to sediment varves, which attest to their steady growth and permit the exact year of deposition to be determined. For the last 200 years of environmental archives, radiometric dating using the $^{210}Pb-^{226}Ra$ method has much potential (Baskaran 2011). Over a wider time range, the primary incorporation of ^{234}U, and its subsequent decay to ^{230}Th, is the usual means of obtaining absolute ages, and steady growth over periods of up to 10^5-10^6 years has been demonstrated.

There have also been a range of direct anthropogenic disturbances to caves throughout prehistory, albeit ultimately limited by the skill and ingenuity of hominins to provide light, manufacture aids such as ropes to traverse cave interiors, and imagination to decorate them with paintings. Breakage and removal of speleothems started in the Palaeolithic and is only now being curbed by environmental considerations. It might be thought that the creation of showcaves would represent a profound anthropogenic disturbance, yet monitoring studies demonstrate that speleothem properties are not necessarily fundamentally disturbed by these changes, if the proliferation of plants and bacteria is controlled.

Climate and environmental proxies in speleothems

The driver for speleothem growth is the generation and supply of *either* dripping waters supersaturated with $CaCO_3$ from a cave ceiling to generate downward-growing stalactites and upward-growing stalagmites *or* broader thin sheets of water moving over surfaces to generate flowstones. The supersaturation arises by a two-stage process. Initially, downwards percolating calcareous fluids are generated by carbonate mineral dissolution enhanced by the presence of high pCO_2 in a soil zone, or more rarely pCO_2 raised by acid generated from pyrite oxidation. When the water emerges into an underground space with lower pCO_2 than that which the water has previously equilibrated, CO_2-degassing occurs and the solution tends to become supersaturated for $CaCO_3$ deposition. Such degassing is usually a much more important process than evaporation. Stalagmites are generally preferred to stalactites for study because of their simpler geometry and preferred to flowstones because of their faster growth rate. Although long-term upward accretion rates of <0.01 mm yr^{-1} are known, for study of the Anthropocene in particular, a relatively fast growth rate (typically 0.05–1 mm yr^{-1}) is desirable, and this is typical of speleothems in temperate environments.

Table 1 lists a wide range of parameters that can be determined for stalagmite samples and indicates the important processes influencing their values and the natural and anthropogenic drivers for change. In this section, we focus on the natural drivers that provide Holocene (pre-Anthropocene) background signals. Samples displaying episodic growth reflect either long-term, orbitally controlled environmental fluctuations, or tectonic or other controls on water flow through karstified carbonate bedrock. The changing growth rate of annually laminated samples provides a high-resolution climatic proxy. Examples of climatic controls that have been inferred from growth rate at suitably sensitive sites include winter temperature changes correlated to solar variability (Frisia *et al.* 2003) and rainfall modulated by North Atlantic Oscillation variability (Proctor *et al.* 2000). The effects of changing controlling parameters on upward growth rate and overall morphology (e.g. width of stalagmite column) have been the subject of numerical modelling studies (Dreybrodt 1988; Kaufmann 2003). The internal architecture of speleothems includes the generation of distinctive crystal fabrics whose characteristics are determined by dripwater composition and the variability of time elapsed between two drops through the year (Frisia *et al.* 2000; Frisia & Borsato 2010; Boch *et al.* 2011).

To date, stable isotope studies of speleothems have provided the main lines of evidence for palaeoclimate interpretation, yet the controls on $\delta^{18}O$ and $\delta^{13}C$ compositions, including cave-specific parameters, are complex, and the dominant factors vary from case to case (McDermott 2004; Lachniet 2009). In the case of $\delta^{18}O$, the paradigm in earlier work was the role of temperature as a master variable in controlling speleothem $\delta^{18}O$ by a combination of atmospheric and in-cave processes, which was used to interpret temperature changes over time in speleothems (Lauritzen & Lundberg 1999). Now there is rather more emphasis on factors on changing atmospheric circulation patterns, changes in seasonality of atmospheric precipitation, and changing spatial gradients of $\delta^{18}O$ in atmospheric moisture (Fairchild *et al.* 2006; Lachniet 2009; McDermott *et al.* 2011). For $\delta^{13}C$, earlier work emphasized effects of significant changes in vegetation abundance (Hellstrom & McCulloch 2000) and type (Dorale *et al.* 1992). Monitoring studies have, however, clarified the role of kinetic in-cave processes (Spötl *et al.* 2005; Frisia *et al.* 2011), which cause a major shift in values from soil to speleothem and can impart seasonal noise, although under favourable circumstances this may not overwrite significant environmental shifts.

Radioactive carbon (^{14}C) is generated in the stratosphere and is rapidly mixed through Earth's atmosphere; it has a half-life of 5730 years. Modern vegetation obtains carbon by photosynthesis and so generates organic carbon with an activity equivalent to that in the atmosphere. The dissolved inorganic carbon in soil waters is derived from a combination of sources: carbonate bedrock (free of ^{14}C; i.e. all its carbon is dead carbon) and soil organic carbon (containing ^{14}C activities corresponding to modern values where the carbon is fresh, to rather lower values in cases where organic matter is up to a few thousand years old). Where CO_2 production by respiration and organic decay is limited, the inorganic and organic sources of carbon are similar in abundance and the dead carbon percentage (dcp) can range up to 50% or potentially more. Such conditions cover a range from the classic *closed system* to what Fairchild & Baker (2012) termed the *undersupplied open system*. Where soil CO_2 production is higher, a significant proportion is lost to the atmosphere by diffusion and the ^{14}C activity approaches that of the organic end-member – this is the *oversupplied* open system and dcp tends towards zero, although in practice it does not fall much below 5% (Genty *et al.* 1998). Values of dcp tend to be constant over short time periods, but are expected to vary with major changes in soil processes that could affect rates of CO_2 production, and recent work has demonstrated an overall hydrological control with higher rainfall leading to a relative reduction in the importance of carbon sourced from organic matter compared with bedrock and hence higher dcp (Griffiths *et al.* 2012). Recently, studies combining radiocarbon and $\delta^{13}C$ systematics have identified the climate-controlled role of recalcitrant v. labile organic matter available for microbial degradation as a factor affecting $\delta^{13}C$ (Fohlmeister

Table 1. *Parameters observed in speleothems, and the natural and anthropogenic driving factors for change*

Parameter	Process	Natural drivers for change	Specific anthropogenic drivers for change
Growth rate	Growth rate limited by (1) availability of dripwater (2) supersaturation of dripwater (3) presence of biofilm	(1) Drying or wetting climate, changing utilization of moisture by vegetation, clogging or opening of hydrological pathways (2) Cooling or warming climate, change in vegetation abundance or extent of prior calcite precipitation; seasonal changes in cave ventilation (3) Tectonic (seismic) disruption of the feeding system	(1) Introduction of water-hungry vegetation (e.g. crops, eucalypts), conversely deforestation; soil disturbance affecting hydrological pathways (2) Change in vegetation; soil disturbance; artificial changes to cave ventilation, for example, by opening or closing of entrance; anthropogenic warming
Macro-morphology	Reduction in width of speleothem because of lowered water flux	As (1) above	As (1) above
Crystal fabric	Change in: (1) impurity content of dripwater and/or (2) supersaturation of dripwater	(1) See (1) above: increased impurity content associated with higher water flux (2) See (2) above	Potentially, agricultural practices in general; anthropogenic warming; introduction of new microbial species into caves
$\delta^{13}C$	Higher $\delta^{13}C$ related to (1) reduction in soil pCO_2 (2) more extensive degassing or prior calcite precipitation in cave related to for example, slower drip rate (3) change from C3 to C4 (photosynthetic pathway) vegetation types	Range of factors from (1) and (2) above. Onset of C4 vegetation is specifically an aridity indicator	(1) Reduction in biomass by forest clearance (2) Planting of C4 crops (3) Other factors promoting degassing
$\delta^{18}O$	Lower $\delta^{18}O$ related to (1) warmer conditions (2) lower $\delta^{18}O$ in dripwater (3) reduction in kinetic factors	(1) Warmer external climate (2) Lower mean annual $\delta^{18}O$ in rainfall or greater relative influence of winter precipitation (3) Reduced evaporation of soil water or in-cave degassing	(1) Anthropogenic warming; deforestation (2) Additional change in atmospheric circulation patterns (3) Agricultural or cave ventilation disturbance

^{14}C	^{14}C activity increases (1) in younger samples that have experienced less radioactive decay (2) with decreased dead (rock-derived) carbon	Increased soil CO_2 production (root-bearing, warm and moist, but not waterlogged soil) will decrease dead carbon percentage and increase ^{14}C activity	Post-1950 atmospheric nuclear tests, peaking in 1962
$^{87}Sr/^{86}Sr$	$^{87}Sr/^{86}Sr$ reflects balance of sources of Sr (e.g. silicates, carbonates, marine aerosol)	Change in dust supply (e.g. glacial–interglacial cycles); change in weathering regime (hydrological effects on carbonate v. silicate weathering)	(1) Soil disturbance affecting availability of minerals for leaching (2) Fertilizer addition
Alkaline earth elements	(1) Change in ratio of trace element (Tr) to Ca in water (2) Significant change in temperature (Mg) (3) Change in crystal growth kinetics (Sr, Ba)	(1) Tr/Ca raised by reduction in driprate and increased proportion of water from slowflow parts of aquifer or by increased prior calcite precipitation (2) Seasonal change (near cave entrance) (3) Seasonal change in supersaturation or input of competitor ions	Hydrological or cave entrance factors mentioned above
Colloidally transported elements	Seasonally high fluxes of colloids in dripwater leads to higher element concentrations in speleothem	Seasonal water infiltrations; availability of organic colloids in soils; changes in soil drainage	Deforestation liming of soil fertilizers (e.g. P, N)
Fluorescence and organic biomarkers	Incorporation of colloidal or dissolved organics in speleothem reflecting abundance in soil and transport through dripwater	As for colloids plus change in vegetation and soil organisms	Agricultural practices
Sulphate	Depends on (1) abundance of sulphate sourced from atmosphere/geosphere, mediated by storage in soil-ecosystem (2) dripwater pH (low pH raises SO_4/CO_3 ratio and sulphate content of calcite)	(1) Volcanic eruptions; input of marine aerosol; degree of vegetation storage (2) Seasonal changes in cave ventilation	Atmospheric sulphate pollution
$\delta^{34}S$ of sulphate	Distinctively low $\delta^{34}S$ in volcanic, pollution and pyrite sources compared with carbonate, evaporite and aerosol sources	Potentially volcanic eruptions; long-term changes in weathering or aerosol sources	Atmospheric sulphate pollution

et al. 2011; Rudzka *et al.* 2011; Griffiths *et al.* 2012; Hua *et al.* 2012).

The $^{87}Sr/^{86}Sr$ ratio characterizes the different minerals that are leached to supply the Sr that is ultimately incorporated into speleothems. The ratio tends to be higher in silicate phases compared with carbonate or marine aerosol phases because of production of ^{87}Sr by radioactivity decay of ^{87}Rb, which is found substituted for potassium in silicate minerals. Changes in the ratio over time imply either that the relative abundance of the phases has changed (e.g. variable atmospheric dust supply, Goede *et al.* 1998), or that one phase has become more or less readily leached, which could reflect changes in hydrology (Verheyden *et al.* 2000).

Trace element (Tr) concentrations of alkaline earth trace elements in calcite (and also Sr in aragonite) bear a relationship to the Tr/Ca ratio of the formation water, with the Tr/Ca ratio in calcite being much lower than in solution. A variety of factors tend to cause trace elements to be more abundant in absolute terms or more abundant in relation to Ca in drier climate conditions, particularly in sites with a significant soil moisture deficit during part of the year (Fairchild & Treble 2009). These factors include derivation of a high proportion of dripwater from slow-flow portions of the aquifer where there is enhanced dolomite or clay dissolution, and enhanced *prior calcite precipitation*, that is, the formation of calcite en route before reaching the studied speleothem (Fairchild *et al.* 2000). A seasonal change in mineralogy can even be triggered (Railsback *et al.* 1994).

Many other trace elements are now known to be mobilized and transported into cave dripwaters bound to colloids (0.001–1 μm entities) or even particulates (>1 μm). Such colloids typically consist of organic molecules, such as soil-derived humic substances, often in combination with iron or manganese oxides, or silicates in larger colloids. There is a non-linear relationship between drip rate and element mobilization and transport through to the cave, so the colloidally bound elements are particularly abundant during times of high water infiltration (Hartland *et al.* 2012). The elements then enter the $CaCO_3$ of speleothems either still bound to the colloid, or by competition as free ions where they are particularly attracted to defect sites in the crystal lattice. Fluorescence studies demonstrate that humic substances likewise can be focused within the time of year of maximum water infiltration. Organic biomarkers can only be analysed on large samples, so their abundance variation has only been determined at decadal or longer timescales.

A final constituent of environmental interest is the sulphate ion, which substitutes for carbonate in $CaCO_3$ and whose abundance is thought to be proportional to the SO_4/CO_3 ratio in solution (Fairchild *et al.* 2010). Accordingly, sulphate concentration in speleothems varies seasonally in caves where there is a strong variation in cave ventilation and hence pCO_2 of cave air and dripwater and thus pH and HCO_3/CO_3 of dripwater (Frisia *et al.* 2005). Atmospheric archives of sulphate recorded in ice cores are important archives of volcanic eruption activity (as well as pollution, see below). Volcanic events have also been recognized in speleothems (Frisia *et al.* 2005, 2008), but the sulphur records in speleothems can be complicated, as issues arise because of assimilation of sulphate by vegetation, which leads to a smearing of the atmospheric signal when transmitted underground (Wynn *et al.* 2013). A particularly distinctive attribute of sulphate is its $\delta^{34}S$ composition (Wynn *et al.* 2008, 2010), which reflects the sulphate source and will assume importance when we consider atmospheric inputs later.

Human disturbance of soil and ecosystems

Ellis (2011) constructed a framework for understanding human systems as agents of change to terrestrial biomes, distinguishing the original wildlands from semi-natural landscapes, rangelands, croplands and densely settled landscapes. The latter four distinctively show development of forestry, pasture, crops, and finally ornamental plants and built structures. The modified ecosystem can characteristically be identified by plant remains, of which there is a pollen record in speleothems (McGarry & Caseldine 2004), but concentrations are low. Alternatively, there are indicators of changed ecosystem process, and Ellis (2011) lists them as follows: net primary production, combustion, organic carbon accumulation, reactive nitrogen, soil phosphorus, geomorphology, hydrology, biodiversity, biogeochemistry and residual effects.

Net primary production is reflective of the plant cover and nutrients. Setting aside the croplands, the key issue is deforestation, which in less managed landscapes has been induced mainly by fire. In lacustrine and peat archives, charcoal evidence can be used as a proxy for fire frequency (Marlon *et al.* 2009). Distinctive biogeochemical changes occur in soils after combustion (Certini 2005), but products leached from the soil have not yet been specifically sought in speleothems. High rates of decomposition of organic matter and root respiration in soils lead to open system dissolution conditions, which should result in lower $\delta^{13}C$ values in percolating waters compared with a closed system that may characterize sparsely vegetated environments. A link between vegetation abundance and $\delta^{13}C$ values of calcite has been made in speleothem literature. For example, the coincidence of rapidly falling $\delta^{13}C$ with independent pollen evidence of revegetation in New Zealand was recorded in late glacial records of Hellstrom *et al.* (1998).

A study by Domínguez-Villar *et al.* (2013) at Eagle Cave in central Spain demonstrates the importance of the quality of vegetation cover for the temperature of the surface environment, a thermal signal that is transmitted to the cave by conduction. Revegetation at this site has led to a reduction in surface (and hence cave) temperature since the 1970s, despite approximately steady atmospheric temperatures. The predicted impact on the $\delta^{18}O$ signal of speleothems, via the temperature-dependent $\delta^{18}O$ fractionation during calcite precipitation, is significant in relation to other short- to medium-term factors affecting $\delta^{18}O$. Another case study of revegetation, this time at a former mine site at Brown's Folly, Avon, England, provided some clear trends during the twentieth century (Baldini *et al.* 2005). A reduction in $\delta^{13}C$ values in stalagmites from around −3‰ to −3.5‰ in 1915–1930 to −4.5‰ in 1990 is observed, and attributed to reestablishment of vegetation on a previously barren site. Smaller and less reproducible changes in $\delta^{18}O$ were also observed, which were tentatively related to a change in the proportion of winter to summer recharge to the site associated with a vegetation-induced stronger summer soil moisture deficit.

The nutrient elements may in principle also be reflective of ecosystem changes, but a problem is the independent effects of rainfall variations and hydrological changes. Phosphorus is detectable in speleothems, and in European examples typically displays a seasonal variation (Fairchild *et al.* 2001), but this is strongly modulated by autumnal die-back of vegetation coupled with increased infiltration. A multidecadal (twentieth century) trend to lower phosphorus in an Australian example was shown to correspond to decreasing rainfall (Treble *et al.* 2003), but opposite trends can occur in condensed sections (Frisia *et al.* 2012). Phosphorus is mono-isotopic, but nitrogen offers the possibility of determining both abundance and characteristic isotope fractionations, for example from fertilizer application (Mayer *et al.* 2002). Even remote watersheds are known to show characteristic nitrogen isotope changes in lacustrine archives from around 1900, reflective of atmospheric delivery of nitrogen with isotope contribution modified by fossil-fuel burning and the industrial production and use of nitrogen fertilizers (Holtgrieve *et al.* 2011). However, nitrogen remains uninvestigated in speleothem archives.

Geomorphological effects of human activity comprise structural changes in the land surface, for example, levelling of land, improving drainage or various local projects (Ellis 2011). Each case will have unique characteristics (e.g. addition of specific pollutants), but improved drainage is likely to be the most common change. This may result in the drip regime becoming more variable if a waterlogged soil reservoir is no longer present, and trace elements in solution resulting from prolonged water–mineral reaction may diminish. Colloidal and particulate impurities may increase during hydrological events (i.e. 'flushing'), but, in overall dry periods, hiatuses due to clogging of flow pathways may become more common. There are only a few worked examples from speleothems where the cause-and-effect relationships of ground disturbance are discussed, and even then the relationships are inferential. A possible example of a chemical signal of disturbance has been provided at Crag Cave, Ireland, by a strontium isotope study by Tooth (2000), which is summarized in Fairchild & Baker (2012). The site is distinctive in the presence of *c.* 1 m of carbonate-free till overlying the karstic bedrock, which historically limited drainage. Modern dripwater signatures, reflecting a mixture of carbonate bedrock and silicate from the till, differ from those in early Holocene speleothem samples, which imply a higher silicate contribution. This difference was attributed to a bypassing of water flow through the till by the construction of drainage ditches. There is much potential for using this powerful chemical tracer to document such soil disturbances. A contrasting example is provided in a speleothem from Grotta di Carburangeli, Sicily (Frisia *et al.* 2006). Growth of a distinctive dendritic calcite ceased at 6 ka and restarted around 3.5 ka with calcite rich in impurities derived as insoluble residue from the overlying epikarst, and suggesting stripping of the original soil at the site. Another hiatus followed from 2.8 to 1.4 ka, and growth resumed with a different, impurity-free columnar fabric implying changed hydrological conditions probably associated with the development of an agricultural soil as found today at the site.

A mature ecosystem in temperate and tropical environments is associated with cycling of humic substances and trace elements, but, if disturbed, there can be significant leaching of these components to underlying karst (Hartland *et al.* 2012). Hence, hydrological changes can be a characteristic consequence of deforestation or afforestation. A good example of the specific effect of such disturbance is provided at Grotta di Ernesto cave, at *c.*1200 m elevation in northern Italy, where a strong annual signal of fluorescent laminae associated with a range of colloid-transported elements occurs. The trace element concentrations are significantly enhanced during a time interval of known deforestation in the early twentieth century (Borsato *et al.* 2007). Another example is at the Browns Folly Mine Site mentioned above, where a major decline in the mean annual concentration of colloid-associated elements such as lead, zinc and bromine was observed over a growth period of *c.* 20 years from *c.* 1915 AD (Fairchild & Hartland 2010), a period when vegetation was starting to be reestablished at the site following a period of forest clearance associated with limestone working. The less efficient delivery of these elements underground was associated with a progressive increase in the ability of the soil system to retain organic colloids.

In speleothem records, organic molecules may be considered in bulk, using fluorescence of humic substances as a surrogate for the total organic load, including the non-fluorescent polysaccharides. Alternatively biomarker molecules can be sought (Blyth *et al.* 2008). Blyth *et al.* (2007) investigated changes in lipid composition over the last century from a speleothem from the Mechara karst of Ethiopia. A reduction in the ratio of C_{27}/C_{31} *n*-alkanes, and increase in the proportion of high-molecular-weight *n*-fatty acids and *n*-alkanols around the 1930s coincided with the known change from scrubland to cropland usage.

The final process indicator of an anthropogenically disturbed ecosystem, according to Ellis (2011), is that of residual effects following disturbance, specifically the presence of artificial substances or, more generally, changed concentrations of element or isotopes. Such effects could be highly localized, and a direct linkage between surface and sub-surface could be hard to achieve given that water flowlines may have a lateral as well as a vertical component. An example of a possible effect is the reduction in lead and zinc over the past 150 years in speleothems from Obir cave, SE Austria (Fairchild *et al.* 2010), a location where mining of lead and zinc ores in the local bedrock occurred in the nineteenth century, and where increased fluxes of these elements to the underlying cave system in the past could have been accentuated by soil disturbance.

Atmospheric pollution: nature and timings

Ice-core records from ice sheets and Alpine glaciers have provided the definitive records to date of atmospheric change, complementing networks of the direct observations of recent decades. Naturally, attention focuses on CO_2 levels, which have risen from the pre-industrial level of 280 ppm, with a strong acceleration over the past century to the current levels of nearly 400 ppm, which are associated with a reduction in $\delta^{13}C$ of atmospheric pCO_2 from −6.4‰ to −8.2‰ related to the emission of CO_2 generated by the burning of isotopically light (^{13}C-depleted) fossil fuels (Keeling *et al.* 2010; Schmitt *et al.* 2012). Speleothems, however, are insensitive to these changes because of the passage of dripwater through soils with pCO_2 at much higher levels than those encountered in the atmosphere followed by extensive in-cave processes which involve modifications to $\delta^{13}C$. There remains a theoretical possibility that the change in atmospheric $\delta^{13}C$ could be recognized in cases where dripwaters strongly equilibrate with a low-pCO_2 cave atmosphere (Baskaran & Krishnamurthy 1993; Frisia *et al.* 2011).

A more distinctive attribute of young atmospheric carbon is its content of radiocarbon, as this was greatly raised during the period of atmospheric nuclear testing (Hua 2009). Atmospheric measurements display an exponential decline in abundance of ^{14}C since it peaked at the time of the test-ban treaty in 1962 at *c.* 190% of 1950 values (Fig. 2). This pollution pulse was first recognized in a Bermudan coral archive by Nozaki *et al.* (1978) and is now widely known in sediment records such as those of salt marshes (Marshall *et al.* 2007). Genty *et al.* (1998) constructed the first time series to demonstrate that this atmospheric signal was preserved in modified form in speleothems (Fig. 2). The timing of the onset of the ^{14}C rise in the speleothem is within 1–2 years of the atmospheric rise in 1955 (Genty & Massault 1999), allowing this date to be specifically identified in speleothem records (Hua 2009; Hua *et al.* 2012). The peak of the signal is lagged and attenuated compared with the atmosphere properties, which have been attributed to storage of soil organic carbon and which can be modelled using a combination of soil carbon reservoirs of different lability. Under conditions where the carbon is derived primarily from cave air rather than dripwater, it could be anticipated that attenuation and lag might be diminished and this appears to apply in some cases (Mattey *et al.* 2008; Smith *et al.* 2009) where dynamic cave ventilation occurs, or potentially where deeply rooted plants are present that may facilitate the direct introduction of modern CO_2 in the gas phase to the cave environment (Breecker *et al.* 2012).

A distinctive attribute of twentieth-century atmosphere has been the period of burning of sulphur-bearing coal in power stations without effective treatment of the resulting sulphur dioxide emissions. This led to the acid rain phenomenon whereby carbonate-poor catchments became strongly acidified through production of sulphuric acid and ecosystems were badly damaged. In the worst-affected regions, limestone building stones were badly damaged. Although sulphur emissions are still increasing in some emergent economies, in much of the world effective treatments have been made to prevent significant SO_2 emissions.

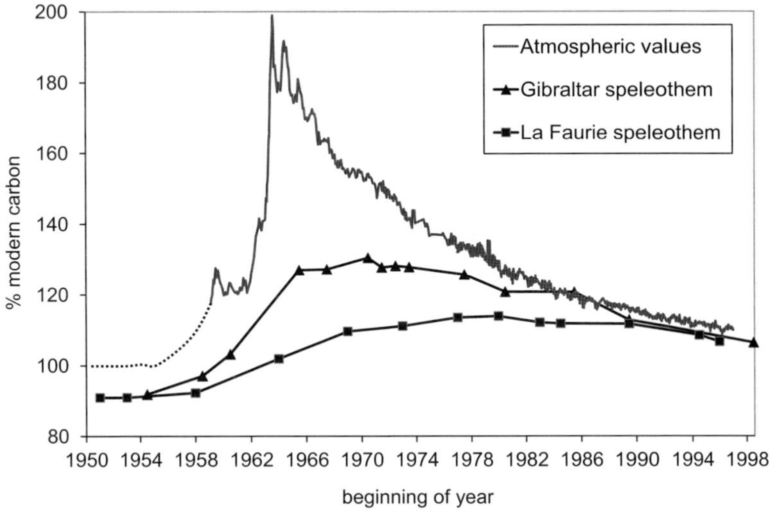

Fig. 2. Comparison of northern hemisphere atmospheric $\delta^{14}C$ values (German–Austrian composite; see Smith *et al.* 2009) expressed as percent modern carbon (defined as 1950 activities) with speleothems from La Faurie Cave, SW France (Genty & Massault 1999) and Lower St Michael's Cave, Gibraltar (Mattey *et al.* 2008). The speleothem records show differing degrees of attenuation of the atmospheric signal and the La Faurie record has a lagged peak. However, the initial rise is seen by 1958 in both records and this can be regarded as a reliable marker for correlation (Hua *et al.* 2012).

Karstic areas were not significantly affected by the pollution because of its rapid neutralization, yet the sulphur flux left an imprint on the soil ecosystem and underlying speleothem deposits. Figure 3 illustrates schematically both natural and anthropogenic sulphur sources with their characteristic isotope compositions.

Speleothem records (Frisia *et al.* 2005; Wynn *et al.* 2008, 2010) commonly display a pollution imprint, but the sulphur records are not identical to the atmospheric changes. In the Crag Cave (Ireland) catchment, strong redox effects in the waterlogged aquifer led to changes in concentration and isotopic composition of sulphate, nullifying the anthropogenic signal. In three other cases, however (Browns Folly Mine, Bath, UK; Grotta di Ernesto Cave, NE Italy; Obir Cave, east Austria), a modified atmospheric signal was found to be present both in terms of sulphate concentration and in terms of the $\delta^{34}S$ isotopic signal of the sulphate (Fig. 4, Wynn *et al.* 2010), reflective of input of isotopically light sulphur input from a pollution source. A comparative study of sulphur accumulation in conifers above Ernesto Cave revealed that the sulphur peak occurred sooner and more distinctly in wood and was present in a range of redox states (Fairchild *et al.* 2009). Data on $\delta^{18}O$ and $\delta^{34}S$ of sulphate at the site demonstrate that sulphate is reduced to be assimilated into organic matter and is later re-released with a novel $\delta^{18}O$ composition reflective of the local environment (Wynn *et al.* 2013). The effect has been an attenuation and delay of the pollution signal with a peak in dripwaters more than 20 years after the atmospheric peak.

Nineteenth-century warming

As reported below, speleothems at diverse sites show some distinctive changes in the mid-nineteenth century, in several cases consistent with warming, but the wider evidence base concerning late-nineteenth-century climate conditions has inconsistencies. Specifically, there are some puzzling disagreements between instrumental and climate-proxy data for the late-nineteenth-century climate, as well as uncertainties regarding anthropogenic effects. Vincent *et al.* (2005) argued that glacier retreat in the Alps from 1830 onwards was primarily related to a reduction in winter snowfall. However, such a pattern would be expected to be a regional, not global phenomenon, and yet glaciers in all regions globally show marked retreat in the second half of the nineteenth century (Leclercq & Oerlemans 2012), which could be argued to be more consistent with a global rise in temperature. However, hemispheric instrumental temperature records imply no northern hemisphere mean temperature rise until the twentieth century and yet tree-ring temperature reconstructions do show a warming trend (IPCC 2007, fig. 6.10).

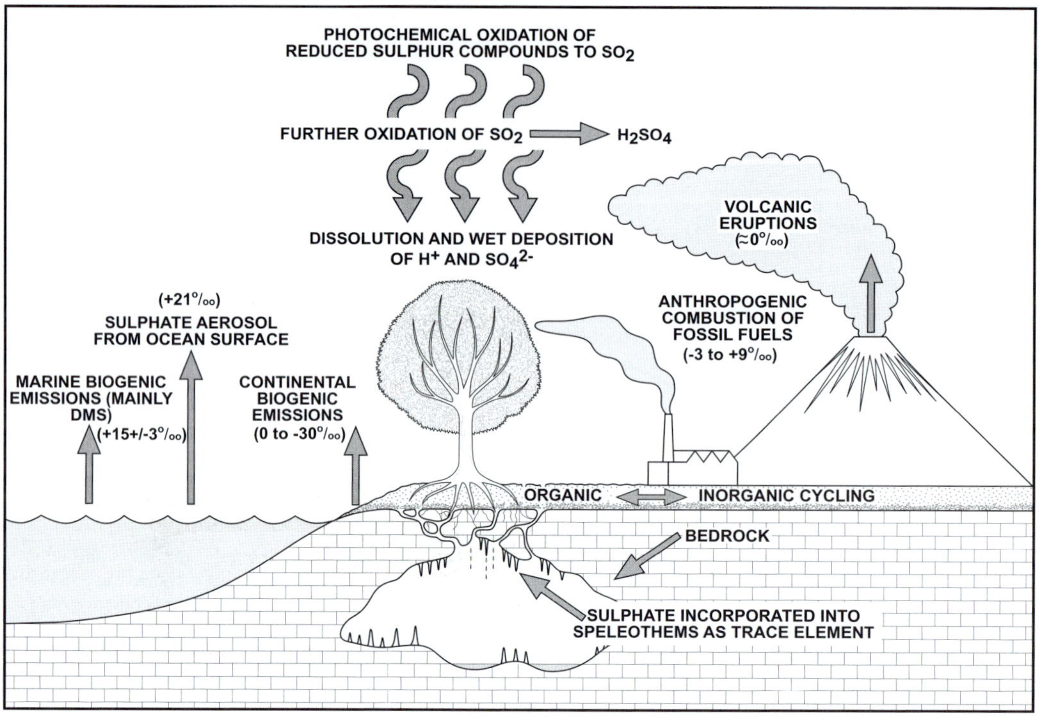

Fig. 3. Sources and cycling of sulphur contributing to bulk stalagmite sulphate composition with characteristic $\delta^{34}S$ compositions (from Wynn *et al.* 2008).

Radiative forcing by increased CO_2 concentrations does become higher by the end of the nineteenth century than in pre-industrial times, but this forcing is weak, being only *c.* 0.2 W m^{-2} by 1900 compared with 1.8 W m^{-2} by 2000 (IPCC 2007, fig. 2.23). A new hypothesis has emerged by Painter *et al.* (2013), who propose that enhanced snowmelt caused by deposition of industrial black

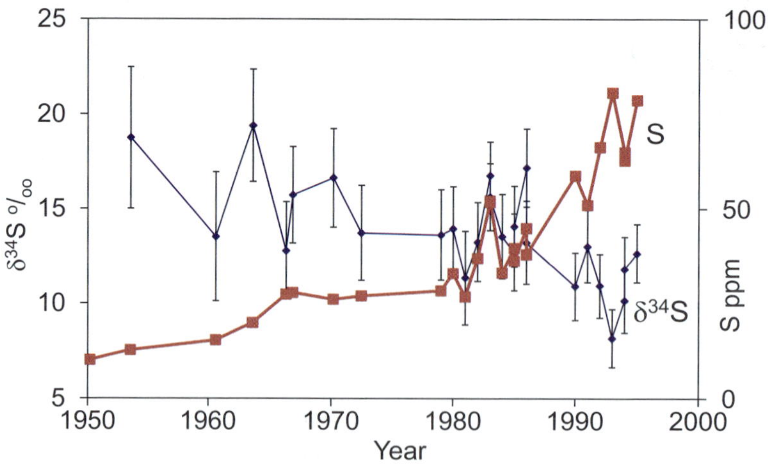

Fig. 4. Sulphur concentration and $\delta^{34}S$ record from ion microprobe spot analyses of stalagmite ER78 from Grotta di Ernesto, illustrating an increasing pollution signal through the late twentieth century (replotted from Wynn *et al.* 2010). Monitoring shows that the pollution signal reduced after 2000 AD (Wynn *et al.* 2013).

carbon was important. Clearly this issue needs more careful study to resolve (1) when widespread warming started and (2) when a discernable anthropogenic signal is first recognizable.

Triplicated Grotta di Ernesto records shows a pronounced change in stalagmite fabric since *c.*1840 AD from more open and impurity-rich crystals to translucent, relatively clean calcite with progressively thicker laminae (representing higher growth rate). The change in fabric and lamina thickness (Figs 5 & 6) observed in the three Ernesto stalagmites over the last *c.* 170 years has a good positive correlation with mean annual temperature, and even better with winter temperature (Frisia *et al.* 2003). Independent evidence is provided by banded soda straw stalactites from two different chambers in the same cave (Fig. 7). Hence, the mid-nineteenth-century transition is interpreted to be temperature-driven, marking a change in the efficiency of soil microbial degradation, which is known to be strongly temperature-dependent (Frisia *et al.* 2003). Soil CO_2 production controls the maximum rate of speleothem growth and this production is primarily related to the seasonal maximum temperature achieved. In contrast, changes in winter snowfall would not be expected to influence speleothem growth because melt is rapid in the spring.

The mean annual temperature at the Milan (Brera) weather station, used as a regional control by Frisia *et al.* (2003), rose by a degree between 1850 and 1900, including a jump in the 1850s

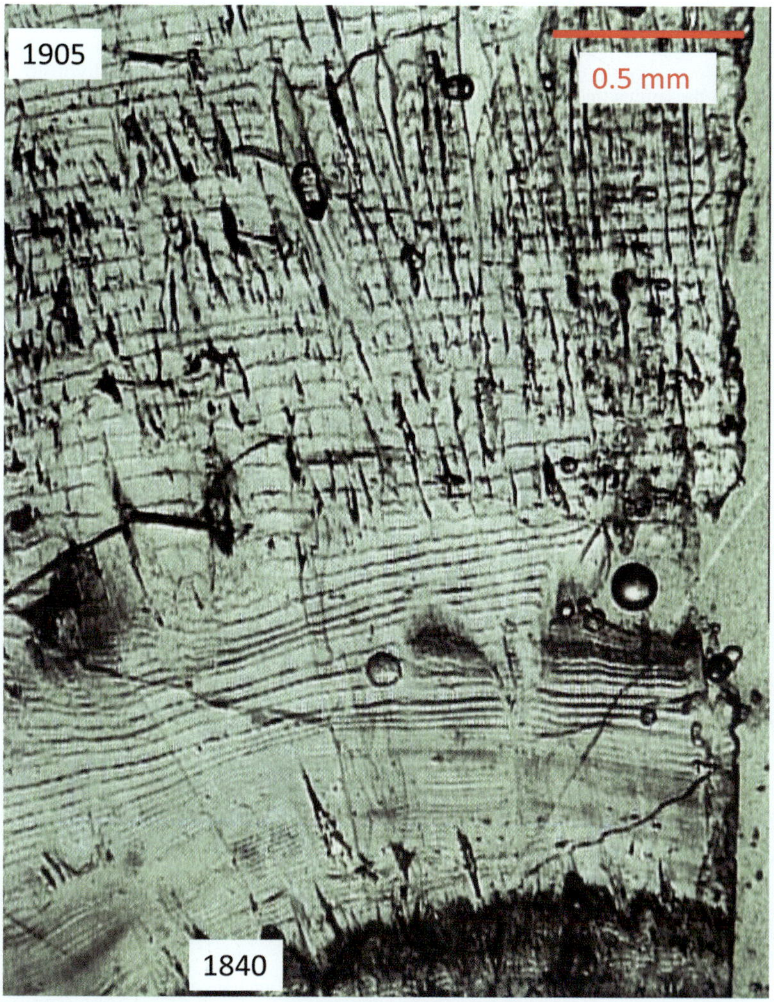

Fig. 5. Thin section of sample ER77 illustrating annual laminae. Crystal textures change from microcrystalline columnar calcite (inclusion-rich at base) to columnar calcite, using the classification of Frisia & Borsato (2010).

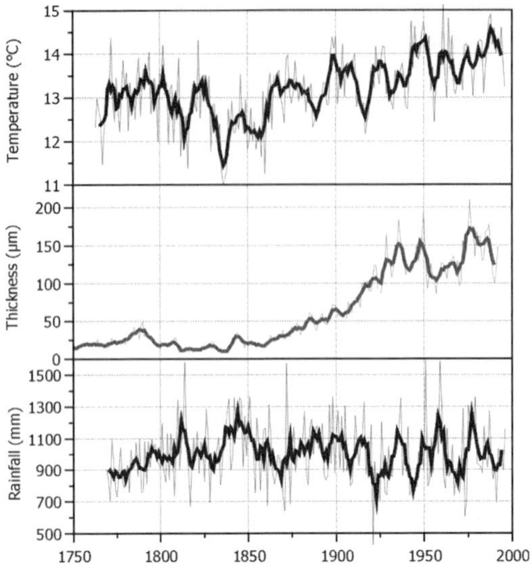

Fig. 6. Comparison of the lamina thickness series from ER77 (centre), the instrumental mean annual temperature (top) and the instrumental mean annual rainfall (bottom) series from Milan (Brera). From Frisia *et al.* (2003).

that may have permitted a threshold to growth to be crossed in terms of soil CO_2 production. It is noteworthy that change in fabric (mediated via changes in soil CO_2 production and dripwater supersaturation for $CaCO_3$) coincides with the initial increase in sulphate concentration (Frisia *et al.* 2005) that is attributed in the following to atmospheric pollution. Similarly, the growth rate increases in speleothems from Scotland (Proctor *et al.* 2000) and China (Tan *et al.* 2003). These two examples, together with those from Grotta di Ernesto, have been used to build the first northern-hemisphere speleothem-based temperature record, which matches several features of the instrumental and tree-ring-based temperature reconstruction and models (Smith *et al.* 2006). Other caves in Europe, such as Spannagel (Austria), record in their oxygen isotope composition a mid-eighteenth-century temperature rise (Vollweiler *et al.* 2006). In other regions of the Northern Hemisphere, a distinct increase in hydrological variability related to temperature increase has been documented in a speleothem from Sequoia National Park in California (McCabe-Glynn *et al.* 2011) since the mid-nineteenth century.

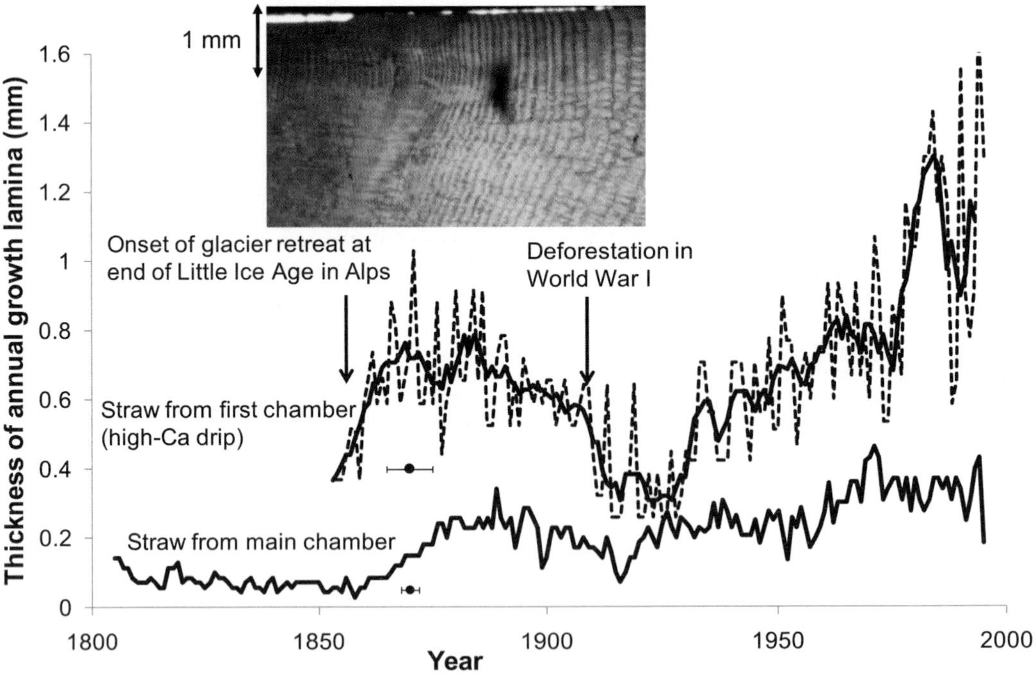

Fig. 7. Annual extension rate variation in two soda straw stalactites with annual growth ridges from Grotta di Ernesto; an image of one is shown. As with stalagmites from the same cave, growth rates increase after 1850; there is also a decrease around the time of a known period of deforestation, before a further warming-related increase up to the late twentieth century. From Fairchild & Baker (2012).

Discussion

Speleothem evidence for anthropogenic disturbance of the ground surface is typically indirect, but dates at least to the Bronze Age in Europe. However, given the diversity of human societies and feeding strategies, changes in speleothem fabrics or chemical changes are bound to be site-specific and recorded in periods that are likely to be non-correlatable worldwide. Thus we feel, in agreement with Gale & Hoare (2012), that the use of criteria related to soil disturbance does not meet the essential criterion of synchroneity to establish a chronological boundary for a new epoch.

If correlation of a synchronous horizon marking profound human impact were the most important criterion then a late-twentieth-century date might seem appropriate. A wide range of parameters show an exponential rise in the post-World War II era, a period referred to as the Great Acceleration (Steffen *et al.* 2011*b*). Of 24 representatives plotted by these authors, five (fertilizer consumption, atmospheric CO_2 concentration, hemispheric mean surface temperatures, loss of forest, and amount of domesticated land) refer to phenomena that could influence speleothem properties. We have illustrated two examples – radiocarbon and sulphate pollution – where speleothems can display regional to global records with great clarity, albeit with some attenuation and lag of peak effects. Nevertheless, the initial rise in ^{14}C can be assumed to be at *c.* 1955 AD, even in the absence of annual phenomena preserved in the stalagmite to count back years from the present (Hua *et al.* 2012). Similarly, Zalasiewicz *et al.* (2011) recognized that the spread of radioactive isotopes from atmospheric atomic bomb tests provide a practical global time marker, and Wolfe *et al.* (2013) developed such an argument in detail in their synthesis of records from remote lakes.

However, there are drawbacks to recognizing a boundary that in itself does not mark a long-term rise in the rate of pollution, or a mass extinction, or change in climate. It is merely a spike, and a return to background levels is continuing apace. Giving preeminence to radiocarbon would create an emphasis on a time period that post-dates many industrial activities that could be regarded as arguably distinctively Anthropocene in nature. In Greenland ice cores and terrestrial archives generally, an earlier such attribute is the spread of aerosol-derived atmospheric lead (Bindler 2011). This phenomenon will necessarily be of regional significance, with local hot-spots near to industrial centres such as the Scottish Clyde estuary (Vane *et al.* 2011) where anthropogenic lead is recognized from *c.* 1750–1835 AD (along with polyaromatic hydrocarbons from coal combustion) and increases distinctively from 1915 AD. Much earlier atmospheric lead adjacent to mine workings can be anticipated and McFarlane *et al.* (2013), in a novel reconnaissance study, interpret distinctive pre-Roman, Roman and mediaeval patterns of variation in lead from a speleothem in a known mining district (GB cave, Charterhouse, Somerset, UK), as indicative of local atmospheric lead input from mining activity. It is notable that lacustrine and peat deposits in Europe record the Roman mining peak in terms of lead concentrations and isotopes, even in remote areas (Renberg *et al.* 2001).

Changes in physical and chemical characteristics associated with the past *c.* 150 years of warming are more convincing, because these have been recorded in other geological and biological archives and coincide with the instrumental record. Although, to date, there are still few high-resolution temperature records for the last 500 years from speleothems, the body of evidence points to a marked change since the mid-nineteenth century, that is, at the end of the Little Ice Age (LIA) and the beginning of industrialization in Europe and North America. The end of the LIA, thus, may constitute a recognizable geological boundary that marks the Anthropocene. In this article we have provided a clear illustration of evidence from a particular site, the Grotta di Ernesto, illustrating what a cave stratotype could look like in terms of changes in fabric, growth rate and chemical (sulphate) tracers.

Conclusion

Whenever it can be deemed to have started, it is clear that we are experiencing a once-and-for-all depletion of Earth's resources and a dramatic decline in biodiversity and species abundance in our very own Anthropocene. In considering how the base of the Anthropocene might appear from a cave point of view, we recognize that the journey is more important than the destination. Thus Gale & Hoare (2012) argue, in discussing an attempt to define the Anthropocene base in soil sequences, that 'a stratigraphic definition of this epoch may be unnecessary, constraining and arbitrary'. Zalasiewicz *et al.* (2011) note that a stratigraphic tie-point, a GSSP (Global Boundary Stratotype Section and Point), may not be necessary, a numerical age being sufficient. We have illustrated that a wide variety of human disturbances to the land and the atmosphere can be recorded in cave precipitates, although the sensitivity to record information is highly variable between speleothems (Fairchild & Baker 2012). Nevertheless, we have shown, through a variety of studies, including particularly extensive work at the Grotto di Ernesto site, that the same story can be told by different samples under favourable circumstances. We also point out the utility of using the excellent capabilities of speleothems to produce well-dated archives to understand more of the climatic response to anthropogenic activity over the past 200 years, particularly in the earlier period when instrumental measurements were sparse. We can also reflect that our modern cave monitoring work characterizes the Anthropocene rather than the Holocene.

References

BALDINI, J. U. L., MCDERMOTT, F., BAKER, A., BALDINI, L. M., MATTEY, D. P. & RAILSBACK, L. B. 2005. Biomass effects on stalagmite growth and isotope ratios: a 20th century analogue from Wiltshire, England. *Earth and Planetary Science Letters*, **240**, 486–494.

BASKARAN, M. 2011. Dating of biogenic and inorganic carbonates using $^{210}Pb-^{226}Ra$ disequilibrium method: a review. *In*: BASKARAN, M. (ed.) *Handbook of Environmental Isotope Geochemistry*. Springer-Verlag, New York, Chapter 37, 789–809.

BASKARAN, M. & KRISHNAMURTHY, R. V. 1993. Speleothems as proxy for the carbon isotope composition of atmospheric CO_2. *Geophysical Research Letters*, **20**, 2905–2908.

BINDLER, R. 2011. Contaminated lead environments of man: reviewing the lead isotopic evidence in sediments, peat, and soils for the temporal and spatial patterns of atmospheric lead pollution in Sweden. *Environmental Geochemistry & Health*, **33**, 311–329.

BLYTH, A. J., ASRAT, A., BAKER, A., GULLIVER, P., LENG, M. J. & GENTY, D. 2007. A new approach to detecting vegetation and land-use change using high-resolution lipid biomarker records in stalagmites. *Quaternary Research*, **68**, 314–324.

BLYTH, A. J., BAKER, A. ET AL. 2008. Molecular organic matter in speleothems and its potential as an environmental proxy. *Quaternary Science Reviews*, **27**, 905–921.

BOCH, R., SPÖTL, C. & FRISIA, S. 2011. Origin and palaeoenvironmental significance of lamination in stalagmites from Katerloch Cave, Austria. *Sedimentology*, **58**, 508–531.

BORSATO, A., FRISIA, S., FAIRCHILD, I. J., SOMOGYI, A. & SUSINI, J. 2007. Trace element distribution in annual stalagmite laminae mapped by micrometer-resolution X-ray fluorescence: implications for incorporation of environmentally significant species. *Geochimica et Cosmochimica Acta*, **71**, 1494–1512.

BREECKER, D. O., PAYNE, A. E., QUADE, J., BANNER, J. L., BALL, C. E., MEYER, K. W. & COWAN, B. D. 2012. The sources and sinks of CO_2 in caves under mixed woodland and grassland vegetation. *Geochimica et Cosmochimica Acta*, **96**, 230–246.

CERTINI, G. 2005. Effects of fire on properties of forest soils: a review. *Oecologia*, **143**, 1–10.

CHOUVET, G., MOURET, J., COINDET, J., SIFFRE, M. & JOUVET, M. 1974. Periodité bicircadienne du cycle veille-sommeil dans les conditions hors du temps. Étude polygraphique. *Electroencephaloraphy and Clinical Neurophysiology*, **37**, 367–380.

CRUTZEN, P. J. & STOERMER, E. F. 2000. The Anthropocene. *Global Change Newsletter*, **41**, 17–18.

DOMÍNGUEZ-VILLAR, D., FAIRCHILD, I. J., BAKER, A., CARRASCO, R. M. & PEDRAZA, J. 2013. Reconstruction of cave air temperature based on surface atmosphere temperature and vegetation changes: implications for speleothem palaeoclimate records. *Earth and Planetary Science Letters*, **369**, 158–168.

DORALE, J. A., GONZALEZ, L. A. ET AL. 1992. A high-resolution record of Holocene climate change in speleothem calcite from cold water cave, Northeast Iowa. *Science*, **258**, 1626–1630.

DREYBRODT, W. 1988. *Processes in Karst Systems*. Springer, Berlin.

ELLIS, E. C. 2011. Anthropogenic transformation of the terrestrial biosphere. *Philosophical Transactions of the Royal Society A*, **369**, 1010–1035.

FAIRCHILD, I. J. & BAKER, A. 2012. *Speleothem Science. From Process to Past Environment*. Wiley-Blackwell, Chichester.

FAIRCHILD, I. J. & HARTLAND, A. 2010. Trace element variations in stalagmites: controls by climate and by karst system processes. *In*: STOLL, H. & PRIETO, M. (eds) *Ion Partitioning in Ambient Temperature Aqueous Systems: From Fundamentals to Applications in Climate Proxies and Environmental Geochemistry*. EMU Notes in Mineralogy, London, **10**, 259–287.

FAIRCHILD, I. J. & TREBLE, P. C. 2009. Trace elements in speleothems as recorders of environmental change. *Quaternary Science Reviews*, **28**, 449–468.

FAIRCHILD, I. J., BORSATO, A. ET AL. 2000. Controls on trace element (Sr–Mg) compositions of carbonate cave waters: implications for speleothem climatic records. *Chemical Geology*, **166**, 255–269.

FAIRCHILD, I. J., BAKER, A., BORSATO, A., FRISIA, S., HINTON, R. W., MCDERMOTT, F. & TOOTH, A. F. 2001. High-resolution, multiple-trace-element variation in speleothems. *Journal of the Geological Society, London*, **158**, 831–841.

FAIRCHILD, I. J., SMITH, C. L. ET AL. 2006. Modification and preservation of environmental signals in speleothems. *Earth-Science Reviews*, **75**, 105–153.

FAIRCHILD, I. J., LOADER, N. J. ET AL. 2009. Sulfur fixation in wood mapped by synchrotron X-ray studies: implications for environmental archives. *Environmental Science and Technology*, **43**, 1310–1315.

FAIRCHILD, I. J., SPÖTL, C. ET AL. 2010. Petrology and geochemistry of annually laminated stalagmites from an Alpine cave (Obir, Austria): seasonal cave physiology. *In*: PEDLEY, H. M. & ROGERSON, M. (eds) *Tufas and Speleothems: Unravelling the Microbial and Physical Controls*. Geological Society, London, Special Publications, **336**, 295–321.

FOHLMEISTER, J., KROMER, B. & MANGINI, A. 2011. The influence of soil organic matter age spectrum on the reconstruction of atmospheric ^{14}C levels via stalagmites. *Radiocarbon*, **53**, 99–115.

FRISIA, S. & BORSATO, A. 2010. Karst. *In*: ALONSO-ZARZA, A. M. & TANNER, L. H. (eds) *Carbonates in Continental Settings*. Elsevier, Amsterdam, 269–318.

FRISIA, S., BORSATO, A., FAIRCHILD, I. J. & MCDERMOTT, F. 2000. Calcite fabrics, growth mechanisms and environments of formation in speleothems from the Italian Alps and southwest Ireland. *Journal of Sedimentary Research*, **70**, 1183–1196.

Frisia, S., Borsato, A., Preto, N. & McDermott, F. 2003. Late Holocene annual growth in three Alpine stalagmites records the influence of solar activity and the North Atlantic Oscillation on winter climate. *Earth and Planetary Science Letters*, **216**, 411–424.

Frisia, S., Borsato, A., Fairchild, I. J. & Susini, J. 2005. Variations in atmospheric sulphate recorded in stalagmites by synchrotron micro-XRF and XANES analyses. *Earth and Planetary Science Letters*, **235**, 729–740.

Frisia, S., Borsato, A., Mangini, A., Spötl, C., Madonia, G. & Sauro, U. 2006. Holocene climate variability in Sicily from a discontinuous stalagmite record and the Mesolithic to Neolithic transition. *Quaternary Research*, **66**, 388–400.

Frisia, S., Borsato, A. & Susini, J. 2008. Synchrotron radiation applications to past volcanism archived in speleothems: an overview. *Journal of Volcanology and Geothermal Research*, **177**, 96–100.

Frisia, S., Fairchild, I. J., Fohlmeister, J., Miorandi, R., Spötl, C. & Borsato, A. 2011. Carbon mass-balance modelling and carbon isotope exchange processes in dynamic caves. *Geochimica et Cosmochimica Acta*, **75**, 380–400.

Frisia, S., Borsato, A., Drysdale, R. N., Paul, B., Greig, A. & Cotte, M. 2012. A re-evaluation of the palaeoclimatic significance of phosphorus variability in speleothems revealed by high-resolution synchrotron micro XRF mapping. *Climates of the Past*, **8**, 2039–2051.

Gale, S. J. & Hoare, P. G. 2012. The stratigraphic status of the Anthropocene. *The Holocene*, **22**, 1491–1494.

Genty, D. & Massault, M. 1999. Carbon transfer dynamics from bomb-^{14}C and $\delta^{13}C$ time series of a laminated stalagmite from SW France: modelling and comparison with other stalagmite records. *Geochimica et Cosmochimica Acta*, **63**, 1537–1548.

Genty, D., Vokal, B., Obelic, B. & Massault, M. 1998. Bomb ^{14}C time history recorded in two modern stalagmites – importance for soil organic matter dynamics and bomb ^{14}C distribution over continents. *Earth and Planetary Science Letters*, **160**, 795–809.

Goede, A., McCulloch, M., McDermott, F. & Hawkesworth, C. 1998. Aeolian contribution to strontium and strontium isotope variations in a Tasmanian speleothem. *Chemical Geology*, **149**, 37–50.

Griffiths, M. L., Fohlmeister, J. *et al.* 2012. Hydrological control of the dead carbon fraction in a Holocene tropical speleothem. *Quaternary Geochronology*, **14**, 81–93.

Hartland, A., Fairchild, I. J., Lead, J. R., Borsato, A., Baker, A., Frisia, S. & Baalousha, M. 2012. From soil to cave: transport of trace metals by natural organic matter in karst dripwaters. *Chemical Geology*, **304–305**, 68–82.

Hellstrom, J. C. & McCulloch, M. T. 2000. Multiproxy constraints on the climatic significance of trace element records from a New Zealand speleothem. *Earth and Planetary Science Letters*, **179**, 287–297.

Hellstrom, J. C., McCulloch, M. T. & Stone, J. 1998. A 31 000 year high-resolution record of southern hemisphere maritime climate changes, from the stable isotope geochemistry of New Zealand speleothems. *Quaternary Research*, **50**, 167–178.

Holtgrieve, G. W., Schindler, D. E. *et al.* 2011. A coherent signature of anthropogenic nitrogen deposition to remote watersheds of the northern hemisphere. *Science*, **334**, 1545–1548.

Hua, Q. 2009. Radiocarbon: a chronological tool for the recent past. *Quaternary Geochronology*, **4**, 378–390.

Hua, Q., McDonald, J. *et al.* 2012 Robust chronological reconstruction for young speleothems using radiocarbon. *Quaternary Geochronology*, **14**, 67–80.

IPCC (Intergovernmental Panel on Climate Change) 2007. *Climate Change 2007: The Physical Science Basis*. Cambridge University Press, Cambridge.

Kaufmann, G. 2003. Stalagmite growth and paleoclimate: the numerical perspective. *Earth and Planetary Science Letters*, **214**, 251–266.

Keeling, R. F., Piper, S. C., Bollenbacher, A. F. & Walker, S. J. 2010. Monthly atmospheric $^{13}C/^{12}C$ isotopic ratios for 11 SIO stations. *In*: *Trends: A Compendium of Data on Global Change*. Carbon Dioxide Information Analysis Center, Oak Ridge National Laboratory, US Department of Energy, Oak Ridge. http://cdiac.ornl.gov/trends/co2/iso-sio/iso-sio.html

Lachniet, M. S. 2009. Climatic and environmental controls on speleothem oxygen-isotope values. *Quaternary Science Reviews*, **28**, 412–432.

Lauritzen, S.-E. & Lundberg, J. 1999. Calibration of the speleothem delta function: an absolute temperature record for the Holocene in northern Norway. *The Holocene*, **9**, 659–669.

Leclercq, P. W. & Oerlemans, J. 2012. Global and hemispheric temperature reconstruction from glacier length fluctuations. *Climate Dynamics*, **38**, 1065–1079.

Marlon, J. R., Bartlein, P. J. *et al.* 2009. Climate and human influences on global biomass burning over the past two millennia. *Nature Geoscience*, **1**, 697–702.

Marshall, W. A., Gehrels, W. R. *et al.* 2007. The use of 'bomb pike' calibration and high-precision AMS ^{14}C analyses to date salt marsh sediments deposited during the last three centuries. *Quaternary Research*, **68**, 325–337.

Mattey, D. M., Lowry, D. *et al.* 2008. A 53 year seasonally resolved oxygen and carbon isotope record from a modern Gibraltar speleothem: reconstructed drip water and relationship to local precipitation. *Earth and Planetary Science Letters*, **269**, 80–95.

Mayer, B., Boyer, E. W. *et al.* 2002. Sources of nitrate in rivers draining sixteen watersheds in the northeastern U.S.: isotopic constraints. *Biogeochemistry*, **57/58**, 171–197.

McCabe-Glynn, S. E., Johnson, K. R., Berkelhammer, M. B., Sinha, A., Cheng, H. & Edwards, R. 2011. *A multi-proxy reconstruction of hydrologic variability over the last millennium from a Sierra Nevada mountain stalagmite*. American Geophysical Union, Fall Meeting, San Francisco, abstract #PP51D-1908.

McDermott, F. 2004. Palaeo-climate reconstruction from stable isotope variations in speleothems: a review. *Quaternary Science Reviews*, **23**, 901–918.

McDermott, F., Atkinson, T. C., Fairchild, I. J., Baldini, L. M. & Mattey, D. P. 2011. A first evaluation of the spatial gradients in $\delta^{18}O$ recorded by European Holocene speleothems. *Global and Planetary Change*, **79**, 275–287.

McFarlane, D. A., Lundberg, J. & Neff, H. 2013. A speleothem record of early British and Roman mining at Charterhouse, Mendip, England. *Archaeometry*. http://dx.doi.org/10.1111/arcm.12025

McGarry, S. F. & Caseldine, C. J. 2004. Speleopalynology: a neglected tool in British Quaternary studies. *Quaternary Science Reviews*, **23**, 2389–2404.

Nozaki, Y., Rye, D. M., Turekian, K. K. & Dodge, R. E. 1978. A 200 year record of carbon-13 and carbon-14 variations in a Bermuda coral. *Geophysical Research Letters*, **5**, 825–828.

Orrock, C. 1999. The thirst to be first. *Caves and Caving*, **83**, 5.

Painter, T. H., Flanner, M. G. *et al.* 2013. End of the Little Ice Age in the Alps forced by industrial black carbon. *Proceedings of the National Academy of Sciences*, **110**, 15 216–15 221.

Proctor, C. J., Baker, A., Barnes, W. L. & Gilmour, M. A. 2000. A thousand year speleothem proxy record of North Atlantic climate from Scotland. *Climate Dynamics*, **16**, 815–820.

Railsback, L. B., Brook, G. A. *et al.* 1994. Environmental controls on the petrology of a late Holocene speleothem from Botswana with annual layers of aragonite and calcite. *Journal of Sedimentary Research*, **A64**, 147–155.

Renberg, I., Bindler, R. & Brannvall, M. L. 2001. Using the historical atmospheric lead-deposition records as a chronological marker in sediment deposits in Europe. *The Holocene*, **11**, 511–516.

Rudzka, D., McDermott, F., Baldini, L. M., Fleitmann, D., Moreno, A. & Stoll, H. 2011. The coupled $\delta^{13}C$-radiocarbon systematics of three Late Glacial/early Holocene speleothems; insights into soil and cave processes at climatic transitions. *Geochimica et Cosmochimica Acta*, **75**, 4321–4339.

Schmitt, J., Schneider, R. *et al.* 2012. Carbon isotope constraints on the deglacial CO_2 rise from ice cores. *Science*, **336**, 711–713.

Smith, C. L., Baker, A. *et al.* 2006. Reconstructing hemispheric scale climates from multiple stalagmite records. *International Journal of Climatology*, **26**, 1417–1424.

Smith, C. L., Fairchild, I. J. *et al.* 2009. Chronology-building using objective identification of annual signals in trace element profiles of stalagmites. *Quaternary Geochronology*, **4**, 11–21.

Spötl, C., Fairchild, I. J. & Tooth, A. F. 2005. Cave air control on dripwater geochemistry, Obir Caves (Austria): Implications for speleothem deposition in dynamically ventilated caves. *Geochimica et Cosmochimica Acta*, **69**, 2451–2468.

Steffen, W., Persson, Å. *et al.* 2011*a*. The Anthropocene: from global change to planetary stewardship. *Ambio*, **40**, 739–761.

Steffen, W., Grinevald, J., Crutzen, P. & McNeil, J. 2011*b*. The Anthropocene: conceptual and historical perspectives. *Philosophical Transactions of the Royal Society A*, **369**, 842–867.

Tan, M., Liu, T. *et al.* 2003. Cyclic rapid warming on centennial scale revealed by a 2650-year stalagmite record of warm season temperature. *Geophysical Research Letters*, **30**, 1617. http://dx.doi.org/10.1029/2003GL017352

Tooth, A. F. 2000. *Controls on the geochemistry of speleothem-forming karstic drip waters*. PhD thesis, Keele University.

Treble, P., Shelley, J. M. G. & Chappell, J. 2003. Comparison of high resolution subannual records of trace elements in a modern (1911–1992) speleothem with instrumental climate data from southwest Australia. *Earth and Planetary Science Letters*, **216**, 141–153.

Vane, C. H., Chenery, S. R., Harrison, I., Kim, A. W., Moss-Hayes, V. & Jones, D. G. 2011. Chemical signatures of the Anthropocene in the Clyde estuary, UK: sediment-hosted Pb, $^{207/206}Pb$, total petroleum hydrocarbon, polyaromatic hydrocarbon and polychlorinated biphenyl pollution records. *Philosophical Transactions of the Royal Society A*, **369**, 1085–1111.

Verheyden, S., Keppens, E. *et al.* 2000. Mg, Sr and Sr isotope geochemistry of a Belgian Holocene speleothem: implications for paleoclimate reconstructions. *Chemical Geology*, **169**, 131–144.

Vincent, C., Le Meur, E. & Six, D. 2005. Solving the paradox of the end of the Little Ice Age in the Alps. *Geophysics Research Letters*, **32**, L09706. http://dx.doi.org/10.1029/2005GL022552

Vollweiler, N., Scholz, D. *et al.* 2006. A precisely dated climate record for the last 9 kyr from three high alpine stalagmites, Spannagel cave, Austria. *Geophysical Research Letters*, **33**, L20703. http://dx.doi.org/10.1029/2006GL027662

Walker, M. J. C., Berkelhammer, M. *et al.* 2012. Formal subdivision of the Holocene Series/Epoch: a discussion paper by a working group of INTIMATE (Integration of Ice-core, Marine and Terrestrial Records) and the Subcommission on Quaternary Stratigraphy (International Commission on Stratigraphy). *Journal of Quaternary Science*, **27**, 649–659.

Wolfe, A. P., Hobbs, W. O. *et al.* 2013. Stratigraphic expressions of the Holocene–Anthropocene transition revealed in sediments from remote lakes. *Earth-Science Reviews*, **116**, 17–34.

Wynn, P. M., Fairchild, I. J. *et al.* 2008. Isotopic archives of sulphur in speleothems. *Geochimica et Cosmochimica Acta*, **72**, 2465–2477.

Wynn, P. M., Fairchild, I. J. *et al.* 2010. High-resolution sulphur isotope analysis of speleothem carbonate by secondary ionisation mass spectrometry. *Chemical Geology*, **271**, 101–107.

Wynn, P. M., Borsato, A., Baker, A., Frisia, S., Miorandi, R. & Fairchild, I. J. 2013 Biogeochemical cycling of sulphur in karst and transfer into speleothem archives at Grotta di Ernesto, Italy. *Biogeochemistry*, **114**, 255–267.

Zalasiewicz, J., Williams, M. *et al.* 2011. Stratigraphy of the Anthropocene. *Philosophical Transactions of the Royal Society A*, **369**, 1036–1055.

Ice Sheets and the Anthropocene

ERIC W. WOLFF[1,2]

[1]*British Antarctic Survey, High Cross, Madingley Road, Cambridge CB3 0ET, UK*

[2]*Department of Earth Sciences, University of Cambridge, Downing Street, Cambridge CB2 3EQ, UK (e-mail: ew428@cam.ac.uk)*

Abstract: Ice could play a role in identifying and defining the Anthropocene. The recurrence of northern hemisphere glaciation and the stability of the Greenland Ice Sheet are both potentially vulnerable to human impact on the environment. However, only a very long hiatus in either would be unusual in the context of the Quaternary Period, requiring the definition of a geological boundary. Human influence can clearly be discerned in several ice-core measurements. These include a sharp boundary in radioactivity due to atmospheric nuclear testing; increases, unprecedented at least in the Holocene, in Greenland concentrations of sulphate, nitrate and metals such as lead; the appearance in ice-core air bubbles of previously undetectable compounds such as SF_6; and the rise, unprecedented in the last 800 ka, in concentrations of carbon dioxide and methane. Some combination of these changes could be used by future generations to clearly identify the onset of a new epoch defined at a particular calendar date. However, it is not yet clear what the character of the fully developed Anthropocene will be, and it might be wise to let future generations decide, with hindsight, when the Anthropocene started, acknowledging only that we are in the transition towards it.

Lifeless planets evolve in response to changing energy levels from their Sun, bombardment by material from space, and internal geological forces. Many aspects of the Earth, however, have long been influenced by the appearance of new forms of life. For example, the emergence of cyanobacteria led to the formation of an oxygen-rich atmosphere over 2 billion years ago, while the emergence of land plants significantly altered the Earth's albedo. Individual species can dominate local habitats, and significantly change their appearance. However, in the recent past, a single species (*Homo sapiens*) has begun to alter major aspects of the Earth system at a global scale. Perhaps the most obvious manifestation of this occurs in the carbon cycle, where increased concentrations of carbon dioxide (CO_2), well beyond the natural range, are due to anthropogenic emissions. These increases are expected to lead to global-scale climate change and acidification of the ocean, with effects lasting for thousands of years.

In this and a number of other aspects of the Earth system, human activity is now a force that is stronger than geological activity, and one that overwhelms the buffering abilities of natural systems. It has therefore been proposed that we are now in a new period of time, coined the 'Anthropocene', in which humans have become a controlling influence of the Earth system (Crutzen 2002; Steffen *et al.* 2011). This is an interesting concept for describing the extent of human influence on the environment, and therefore for emphasizing the extent of human responsibility for managing that environment. However, it has now been taken further, with some authors proposing that the Anthropocene should be formally recognized as a geological epoch (Zalasiewicz *et al.* 2008, 2011), on a level with the Holocene and Pleistocene that precede it.

Such an idea has to be considered from the viewpoint of a geologist viewing sequences thousands or millions of years in the future. Would they be able to recognize that defining characteristics of the preceding epochs had been definitively replaced by new ones in which anthropogenic forces dominated? What would they characterize as the features of the new epoch? What date would they assign as the start of the new epoch, and what feature would they want to use to delineate that start?

One difficulty in answering these questions is that, in contrast to all other geological time periods, we do not yet know what the Anthropocene will look like. CO_2 concentrations have so far risen from about 280 ppmv in pre-industrial times to just under 400 ppmv in 2013. Under current trends, they could easily rise to between 500 and 1000 ppmv. Climate change is under way, and although global mean temperature remains within the bounds already seen during the Holocene (Marcott *et al.* 2013), this is unlikely to be the case later this century. The Greenland and Antarctic Ice Sheets have not yet created large changes in landscape or sea level, but some projections suggest they might in the next few centuries.

From: WATERS, C. N., ZALASIEWICZ, J. A., WILLIAMS, M., ELLIS, M. A. & SNELLING, A. M. (eds) 2014. *A Stratigraphical Basis for the Anthropocene*. Geological Society, London, Special Publications, **395**, 255–263.
First published online November 25, 2013, http://dx.doi.org/10.1144/SP395.10

Alternatively, political and technological mitigation could kick in quickly, such that the present level of perturbation of the carbon cycle represents the major effect of the Anthropocene. One could argue that only when all these issues play out will scientists see in retrospect what the environment controlled by humans looked like, and be able to make an objective assessment of what the Anthropocene was and when it began.

In this paper I consider the different roles that ice plays in this debate. First, how, if at all, should the existence and evolution of ice sheets play into the definition of the new epoch? Second, what do ice-core records of atmospheric composition tell us about the unusual nature of the current period in comparison to the preceding epochs. Third, are there signals in ice cores that can be used to define the start of the new epoch, in the same way that ice-core signals have been used to define the start of the Holocene epoch (Walker *et al.* 2009)?

Ice sheets

Ice sheets are a major facet of the Earth system. Their existence (or not) on Earth is one of the signs of global climate state that is visible from outer space. Ice sheets, mainly because of their impact on albedo, form a very important climate feedback. They control the landscape in areas that they periodically cover. They also influence the nature of sediments in areas around them, and in the case of ice-rafted debris carried by icebergs, across large areas of the ocean. Finally, because polar ice sheets always consist of water molecules that are depleted in heavy isotopes (i.e. their water has a negative oxygen isotope ratio, $\delta^{18}O$), they form a principal control over the oxygen isotope content of the ocean, which is reflected (but only as one factor) in the $\delta^{18}O$ preserved in shells within marine sediments globally. The appearance or disappearance of different ice sheets is therefore an attribute, widely detectable, that might be used to define large-scale changes sufficient to warrant that a change in epoch had occurred.

Before discussing whether ice sheets can be of any help in defining the existence of the Anthropocene, we need to consider their role in shaping the preceding epochs. The Quaternary Period, consisting so far of the Pleistocene and Holocene epochs, was recently redefined as starting at a Global Stratigraphic Section and Point (GSSP) 2.588 Ma ago (Gibbard *et al.* 2010). This point is formally defined as the base of a layer in a rock succession in Sicily. However, from the discussions that surrounded the adoption of this date, it is clear that one of the concepts behind the new definition of the Quaternary Period (and hence the Pleistocene Epoch) is that of the onset of major northern hemisphere glaciations. If this is the case, then we might wonder if the end of the sequence of northern glaciations would signal the end of the Quaternary. Would the loss of the Antarctic Ice Sheet, which has apparently been present in some form for at least 35 Ma (Zachos *et al.* 2008), signal another geological boundary?

The role of climate and ice in defining a more limited boundary, the end of the Holocene, should have been clear. This is because, in reality, the natural Holocene (i.e. that which would have occurred without human intervention) would have been just another unremarkable interglacial in a long sequence of events of a broadly similar nature. These other interglacials have regional stage names and marine isotope stage (MIS) numbers, but the Holocene is unique in being an epoch, named only because the first geologists lived during it. It would have been ended conceptually by the next glacial inception, and, by analogy, as with its start (Walker *et al.* 2009), would probably have been defined by an abrupt climate cooling seen in a Greenland ice core. Because we are in a time of low eccentricity, the precessional forcing that controls summer northern hemisphere insolation, and hence the inception of ice sheets, is unusually weak. As a result, some modelling studies suggest that the Holocene would have been an unusually long interglacial, delayed by one or two precessional cycles (Berger & Loutre 2002); that is, for 50 ka.

We can consider three ice-sheet issues that are so significant that they might require some kind of geological boundary to be declared: the occurrence of glacial cycles affecting northern-hemisphere ice sheets; the stability of the Greenland Ice Sheet; and the stability of the Antarctic Ice Sheet.

In relation to the first issue, already discussed above, if the glacial/interglacial cycles that have characterized the Quaternary ceased because the Laurentide and Fennoscandian ice sheets did not return (or alternatively became permanent), this would appear to be a conceptual change that could, according to the reasoning that decided the start of the Quaternary, signal its end. Modelling studies show, as expected, that the insolation reduction required for glacial inception, as well as the ice volume that can be achieved, has a strong dependence on CO_2 concentration (Berger & Loutre 2002; Ganopolski & Calov 2011), although the exact concentration required to deny inception remains uncertain. However, we can observe that, in the last 800 ka, glacial inception (by one definition) has never occurred at a CO_2 concentration above 260 ppmv (Tzedakis *et al.* 2012). One way to describe what happens in 'normal' interglacials is that during and at the end of an interglacial, natural processes bring the CO_2 concentration down from an interglacial peak to a level at which changes in northern hemisphere insolation can induce inception to occur. It is therefore reasonable to assume that these processes will be insufficient as long as the CO_2 concentration remains more than a few tens of ppmv above normal interglacial levels. In a recent synthesis (Archer *et al.* 2009) of modelling studies in which a pulse of CO_2 was treated with a range of models with an interactive carbon cycle, a pulse took the atmospheric concentration from a baseline of 280 ppmv to around 700 ppmv. It remained between 330 and 400 ppmv after 10 ka in the studies examined, and the silicate weathering feedback, with a lifetime on the order of 100 ka, was needed to remove the remaining CO_2. This discussion suggests that human activity on the scale envisaged in the next century might delay the next glacial inception by as much as 100 ka. However, unless other changes are also induced, then it is likely that glacial cycles would resume after that.

The Greenland Ice Sheet contains enough ice to raise the sea level by *c.* 7 m if it were all to melt. Modelling studies (Gregory *et al.* 2004; Robinson *et al.* 2012) suggest that it would be vulnerable to sustained temperature increases of the order of 1–5°, a range that is clearly within the realms of possible outcomes for the next few centuries. However, a recent study of a new Greenland ice core (NEEM Community Members 2013) suggested that the altitude-corrected temperature above at least part of Greenland could have been as much as 8° higher than the present value during the last interglacial (*c.* 125 ka ago), yet much of the Greenland Ice Sheet clearly survived. The evidence for the very high temperature of this period needs further confirmation before the true significance of this result can be assessed. Very high concentrations of pollen during MIS 11 (*c.* 400 ka ago) in a marine core from just south of Greenland (de Vernal & Hillaire-Marcel 2008) indicate the establishment of forests over southern Greenland at that time, suggesting that a significant portion of the ice sheet was lost in response to temperatures that were probably not much higher than those of today, but sustained through a particularly long interglacial. The loss of a significant part of the Greenland Ice Sheet during MIS11 is consistent with the most recent estimates of the MIS11 sea-level being 6–13 m above present (Raymo & Mitrovica 2012).

The evidence is thus somewhat contradictory in terms of the influence of small changes in temperature on the stability of the Greenland Ice Sheet, but it seems probable, based on the evidence above, that it suffered significant loss in MIS 11. It has been asserted that the initial growth of the Greenland Ice Sheet in the late Pliocene (*c.* 3 Ma ago) was the result of a reduction in atmospheric CO_2 concentration to below 400 ppmv. This, along with the earlier estimates of the lifetime of CO_2 enhancements, suggests that any loss of the ice sheet could be reversed after a few thousand years (once CO_2 returned to below 400 ppmv). This means that it would not be dissimilar to what we believe occurred during MIS11, and could not therefore be considered unusual enough to warrant declaration of a geological boundary.

Finally, the West Antarctic Ice Sheet (containing 4.5 m sea-level equivalent, including the Antarctic Peninsula; Fretwell *et al.* 2013) is considered vulnerable to future climate change, and indeed may have contributed to sea-level rise during both the last interglacial (NEEM Community Members 2013) and MIS11 (Raymo & Mitrovica 2012). However, the threshold for loss of the much larger East Antarctic Ice Sheet (53 m sea-level equivalent; Fretwell *et al.* 2013) is considered much further away. Furthermore, the timescale for dynamic wastage of the ice sheet is very long. Loss of this ice sheet would apparently be unprecedented in 40 Ma, would have widespread implications (easily detected in a range of measures), and would probably warrant the declaration of a geological boundary. However, at present, the most likely warming scenarios are not expected to lead to significant loss of this ice sheet.

In summary, the recurrence time of northern-hemisphere glaciation and the stability of the Greenland Ice Sheet are both potentially vulnerable to anthropogenic climate change. However, neither event appears as if it would be completely irreversible (on timescales of 10–100 ka), and neither is so qualitatively different from previous events in the Pleistocene that they would necessarily warrant the erection of a geological boundary.

Signals of anthropogenic change in ice-core records

Ice cores record signals directly from the atmosphere, and are well suited for recording changes in atmospheric composition. They record atmospheric composition in two ways. First, aerosol material is collected by falling snow and dry deposition at the surface of the ice sheet. Because aerosol has a short lifetime in the atmosphere, this material is often deposited heterogeneously, and may only record local sources. Only if multiple sources lead to pervasive change at a global scale, or if material enters the stratosphere, will signals be recorded in ice cores worldwide. An advantage of the material trapped in the snow is that the time resolution is excellent; at some sites, with tens of centimetres of snowfall each year, it is capable of resolving events shorter than a year, with the age of the material being the same as the snow, which can often be well dated.

The second recording medium in ice takes the form of air bubbles trapped below the porous firn layer in the ice sheet. These contain a record of the stable gases in the atmosphere (N_2, O_2 and Ar, but also trace gases such as CO_2 and CH_4). Although latitudinal gradients exist for some of these gases, most are rather well mixed, and an ice core at either pole will reflect global-scale temporal variations. The disadvantage of records from air bubbles is that the gas enclosure process imposes a natural smoothing (ranging from a decade to over a millennium, depending on the site), and the gas record has an age offset compared to the climate and other records in the snow phase.

Numerous aerosol chemical species have been affected, directly or indirectly, by human activity. Here I summarize only a few of the more direct ones. One very clear anthropogenic signal that can be detected in ice is radioactive material from the atmospheric nuclear bomb tests that occurred from the 1950s until (formally) 1986, when the last country carrying out atmospheric tests abandoned their use. This shows up in ice-core records worldwide (summarized, for example, in Wolff & Peel 1985) as a huge jump in beta radioactivity in 1954, a further jump in 1964 in response to very large tests in the early 1960s, and then a slow decay towards natural background levels as the material was removed from the atmosphere and its radioactivity decayed (Fig. 1). At their peak in 1966, the levels of beta radioactivity and tritium in snow were almost a factor of 100 above background levels. The signal can also be observed in radioisotopes with longer half-lives; for example, a clear signal in ^{239}Pu has been observed in an Alpine ice core (Gabrieli *et al.* 2011). This can certainly be seen as a signal of one aspect of humanity's potential to overwhelm its environment, in this case through the destructive ability of nuclear weapons and the associated ability to pollute the environment with radioactive material.

Industrial activity has taken over the sulphur cycle in many parts of the world, with the result that ice-core records from ice sheets and glaciers across the northern hemisphere at least show sustained increases that have not been observed before in the Holocene. In Greenland ice, in 1980, sulphate concentration rose to about a factor of four above its pre-industrial peak (Fig. 1), with the main increases in the periods 1900–1920 and 1940–1980 (Fischer *et al.* 1998). The peak concentration remains below the brief excursions in sulphate concentration that occur after the largest volcanic eruptions, and below the elevated concentrations that occurred during the last glacial maximum (Legrand *et al.* 1997). No significant recent increase in sulphate concentration has been observed in Antarctica, which still remains too remote to see the effects of a species with a rather short atmospheric lifetime.

Human activity has similarly taken over the nitrogen cycle in many parts of the world (Wolff 2013). Elevated ammonium concentrations in mid-latitude glaciers are ascribed to agricultural emissions, while increased nitrate concentrations across at least the northern hemisphere are due to NO_x emissions from combustion. Nitrate in Greenland snow and ice rose by a factor of two (Fig. 1) during the twentieth century (Fischer *et al.* 1998), but in this case the main period of increase was 1950–1980, and levels in the 1980s were probably somewhat higher than at any point in the past 100 ka. Again, no significant increased signal is observed in Antarctic ice cores (Wolff 2013).

The cycles of many metals have also been taken over by human activity. A prime example is lead. Smelting and mining activities as long ago as the Greek and Roman periods elevated lead concentrations in Greenland snow above their Holocene background concentration (Hong *et al.* 1994), suggesting that, at least for this metal, anthropogenic control across much of the northern hemisphere was already in place 2000 years ago. However much more significant increases took place in the eighteenth century and later; these were hugely enhanced from the 1950s due to emissions from leaded petrol. At the high point, in the 1960s (Boutron *et al.* 1991), the concentration of lead in Greenland snow was elevated above its Holocene background level by a factor of 200, and just exceeded the maximum concentrations seen in the last glacial maximum, when concentrations of many metals were elevated due to increased dust transport (Hong *et al.* 1996). Removal of lead from petrol has led to a very significant subsequent decrease. The twentieth-century elevation of lead in the atmosphere was global, with an increase by a factor of five over the pre-industrial Holocene level observed in Antarctic snow (Wolff & Suttie 1994).

From atmospheric gases recorded as air bubbles in polar ice sheets, ice cores provide critical records of increasing concentrations of long-lived greenhouse gases in the recent past (Fig. 1). CO_2 (Lüthi *et al.* 2008) and CH_4 (Loulergue *et al.* 2008) (Fig. 2), and also N_2O (Schilt *et al.* 2010), are all now at concentrations higher than observed in any ice cores for the last 800 ka. The ice cores contain no data for earlier periods, but pCO_2 inferred from measurements of the isotope ^{11}B in marine sediments appears to extend this observation to at least 2 Ma (Hönisch *et al.* 2009). The ice-core data (MacFarling Meure *et al.* 2006), supplemented by measurements in the atmosphere over the last

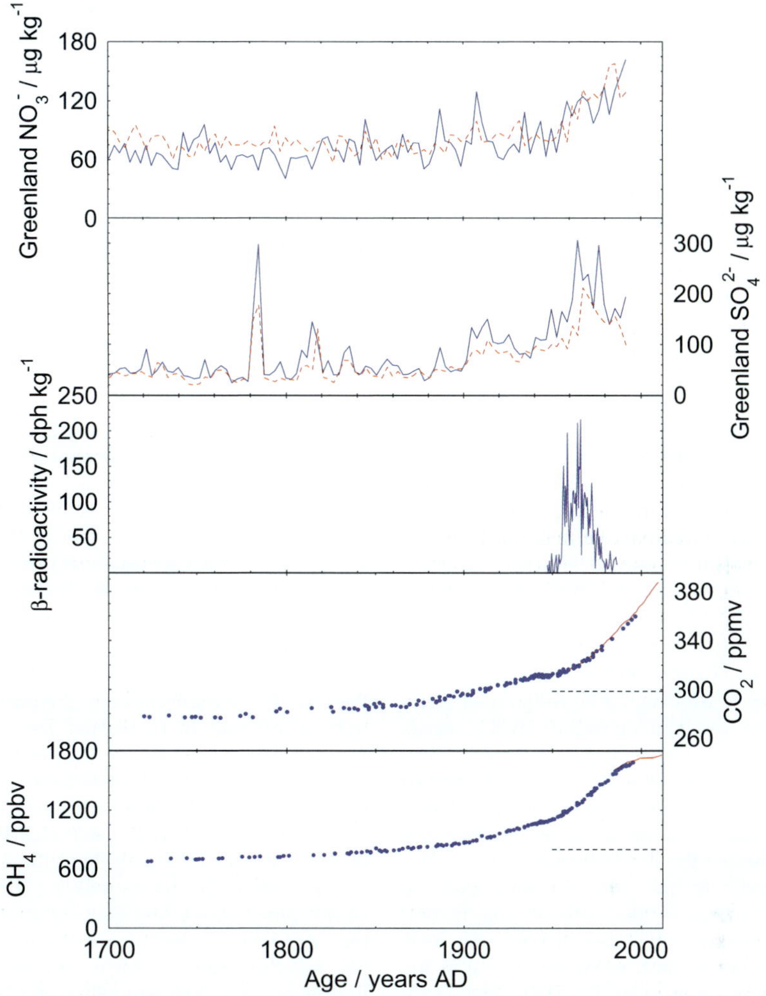

Fig. 1. A range of indicators of human influence since 1700 AD. For CH_4 and CO_2, the blue dots are ice-core data from Law Dome, Antarctica (MacFarling Meure *et al.* 2006), the red lines are recent atmospheric data from Mauna Loa (CO_2, courtesy of NOAA ESRL) and Cape Grim Observatory (CH_4, courtesy of CSIRO Marine and Atmospheric Research and Australian Bureau of Meteorology). The horizontal dashed lines are the highest values observed in ice cores of the last 800 000 years prior to the period shown. Beta radioactivity is shown from a snow pit in Coats Land, Antarctica (Wolff *et al.* 1999): values are assumed to remain low prior to the measurement period. Sulphate and nitrate are shown for two cores from Greenland: B16 (dashed red, 73.94°N, 37.63°W) and B21 (solid blue, 80.00°N, 41.14°W) (Fischer *et al.* 1998). The eighteenth- and nineteenth-century spikes in sulphate are signals of volcanic eruptions.

c. 50 years, show that CO_2 concentrations rose above the previous Holocene range in the early nineteenth century and have risen 40% above any previous Holocene level and 30% above the highest value of the last 800 ka. CH_4 concentrations similarly rose above their Holocene range in the early nineteenth century, and are now double any value observed in the last 800 ka. The isotopic content of CO_2 in the atmosphere today is also highly unusual, with a $\delta^{13}CO_2$ value below $-8‰$ (the range in the previous 25 ka was always between $-6.3‰$ and $-6.8‰$; Schmitt *et al.* 2012). There is some discussion in the literature as to whether rising levels of CO_2 during the last 8 ka, and of CH_4 during the last 6 ka, could also be the result of human activity (land clearance and the widespread adoption of agriculture; Ruddiman 2003). However, this idea of an early to mid-Holocene start to anthropogenic control of the carbon and methane cycle is controversial (Claussen *et al.*

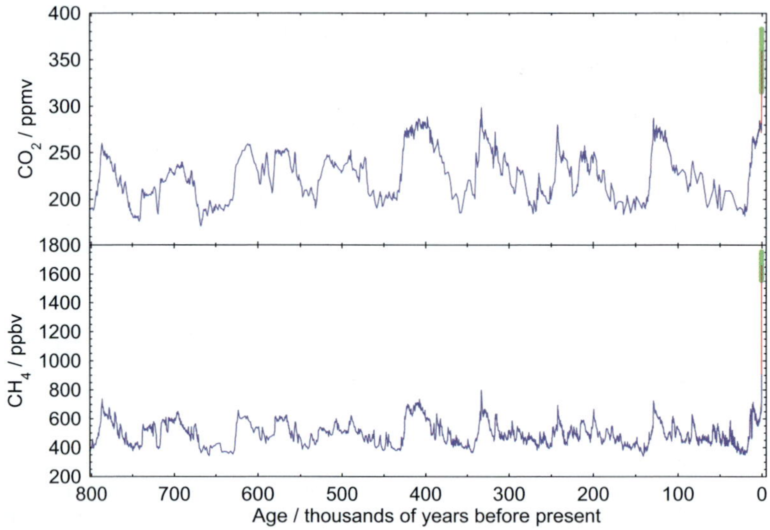

Fig. 2. CO_2 and CH_4 over the last 800 ka from Antarctic ice cores. The blue lines are data from Dome C (CH_4) (Loulergue *et al.* 2008) and from Dome C and Vostok (CO_2) (Lüthi *et al.* 2008). The red lines are high-resolution data from Law Dome (MacFarling Meure *et al.* 2006 and papers therein). The green data are recent atmospheric measurements.

2005; Broecker & Stocker 2006; Singarayer *et al.* 2011). Because the concentration levels, trends and rates of change of CO_2 and CH_4 between 8 ka and 200 years ago are not unique in the 800 ka ice-core record, there is no clear basis for asserting that humans had taken these cycles outside any natural range before the last 200 years.

A final candidate indicating an anthropogenic influence is the range of chemicals with no natural sources that are now detectable in ice cores. This includes persistent organic pollutants found in the snow phase (e.g. Gabrieli *et al.* 2010), but also a group of long-lived gases that have become measurable in firn air, and with large enough samples could be measured in ice cores. Of particular interest are the halogenated gases, such as chlorofluorocarbons (CFCs), hydrofluorocarbons (HFCs) and SF_6 (Butler *et al.* 1999), many of which are detectable in air dated to the 1960s and later. When viewed from the perspective of the far future, many of these gases will show a rise from undetectable levels, and in the case of those with very long (millennial and beyond) atmospheric lifetimes their continuing presence will be a long-lasting reminder of humanity's abilities to fill the atmosphere with new molecules.

In summary, a wide range of chemicals measured in ice cores have left an imprint in the ice that shows humankind's influence on the atmosphere and on important biogeochemical cycles. Of those I have discussed, only lead can be said to show a clear and unequivocally anthropogenic imprint before the last 200 years. Even then, the pre-1800 concentrations of lead in Greenland ice do not exceed those found naturally in the last glacial period. Some of the records discussed above are shown in Figure 1. CO_2 profiles in ice cores are those that future generations will most clearly be able to identify as showing anthropogenic control of the Earth system. Concentrations are clearly elevated significantly above any concentration observed in the last 800 ka, and this conclusion can probably be extended to 2 Ma and beyond through marine records. The isotopic change is also highly unusual, indicating that the release of fossil carbon is involved. Concentrations are approaching levels at which significant impacts on the climate and the carbonate chemistry of the ocean are inevitable, and the effective lifetime of CO_2 in the atmosphere (Archer *et al.* 2009) means that the concentration profile will appear as a step increase, followed by a slow rampdown towards typical interglacial levels, which will probably take thousands of years.

Should ice cores be used to define the start of the Anthropocene?

Finally, I will discuss whether it would make sense to use an ice-core signal as a 'golden spike' for the start of the Anthropocene, as has been done for the start of the Holocene (Walker *et al.* 2009). Several of the signals described above give a clear signal of the dominance of human activity, and any one or combination of them could be used to define the start of a new Anthropocene epoch. However, some have significant drawbacks.

The sharp increase in radioactivity seen in ice cores at 1954 is a very sharp boundary, and has the advantage that it is correlatable across numerous glaciers and ice sheets, is precisely dated, and is also recorded in other media. However, the main measured species (beta radioactivity based mainly on isotopes with half-lives of 30 years and tritium with a half-life of 12.3 years) will decay beyond detection limit within a few half-lives. Although there are longer-lived isotopes in the ice (e.g. ^{36}Cl from bomb tests carried out at oceanic sites; Elmore *et al.* 1982), these have not been extensively characterized and are not yet in a state to be used as a boundary marker.

Twentieth-century increases in nitrate, sulphate or lead all clearly indicate the impact of human activity on important biogeochemical cycles. However, because of the short lifetime of aerosol in the atmosphere, these are all influenced by regional rather than global emissions and will show different patterns (at decadal resolution) in different glaciers. One could therefore only adopt the increase above some threshold in a particular location (probably Greenland). Thinking from the perspective of a geologist many millennia in the future, care must be taken, as the sustained late-twentieth-century concentrations of sulphate and lead, although unique in the context of the Holocene, are not unusual in the context of the late Pleistocene.

The appearance of new chemicals (such as SF_6) in air bubbles in ice cores forms a sharp indicator of industrial activity. Owing to their long lifetime (at least several centuries), their onset would, with analytical improvements, form an easily recognized boundary that could be found in both Greenland and Antarctic ice, offering the potential for secondary correlations with other records in both hemispheres. However, because air bubbles are only closed off below the permeable firn layer (typically 60–100 m depth), the boundary containing a sharp change in gas concentrations is offset from the ice layer containing climatic and other information of the same age. In addition, the boundary is smoothed over decades to centuries depending on the snow accumulation at the ice-core sites.

Finally, the increase in CO_2 above a threshold (say 300 ppmv) offers a unique (within probably the last 2 Ma) marker that is most relevant to the environmental changes that anthropogenic activity may induce. However, it suffers from the same issues that have been described in the previous paragraph, meaning that a sharp and well-dated boundary could be used only in a core with very high snow accumulation rate.

Ice cores present several possible candidates to identify the start of the time at which humans controlled major features of the environment. We can, in principle, choose a date on which we say that the Anthropocene started. A golden spike could then be placed in a Greenland or Antarctic ice core sequence at that date, and we could describe it as having a particular age relationship with any of the sharp changes above (e.g. sharp jump in a long-lived isotope, SF_6 onset, CO_2 exceedance of a threshold), so that the same age can easily be found in other ice cores (using increasingly well-understood relationships between the age of air and the age of ice to remove the gas age–ice age offset). This layer could be correlated to other (non-ice) sequences through a range of measures (e.g. known volcanic eruption spikes in ice cores correlated to their geological manifestation; ^{10}Be patterns in ice cores correlated to ^{14}C patterns in tree rings; Steinhilber *et al.* 2012). It is not clear to me whether such a definition, essentially using a calendar date and a series of identifiers, meets the official criteria for a GSSP (Remane *et al.* 1996), but that definition was not designed for a period where we have observed changes in real time and can place calendar dates on many of our records.

Some philosophical issues

Despite the possibility of using ice cores to help future generations identify a defined-date start of the Anthropocene, there are some philosophical issues to consider. The first concerns the potential impermanence of the ice-sheet record. Ice layers are preserved in monotonous sequences for typically 100 ka in central Greenland and 1 million years in central Antarctica, but then become folded or melted in older ice near the ice-sheet bed. This means that the GSSP can be preserved in the field only for this length of time – not long in geological terms. Clearly, the wastage of the Greenland or Antarctic ice sheets in response to climate change, were it to occur, would be another threat to the sequence in the field. This is not a fatal drawback if the boundary has been transferred to secondary and more permanent sequences.

A second issue concerns what we should consider the Anthropocene to be. All other epochs except the Holocene were defined only after they were complete and their dominant characteristics were clear. We do not yet know if the period of human dominance of the environment will be one in which CO_2 concentrations peak at just above 400 ppmv, or will rise to as much as 1000 ppmv. If the former applies, then we are at the start of a long plateau; if the latter is relevant, we are at the bottom of a long rise towards a later plateau. For the base of the Holocene, a point has been chosen (Walker *et al.* 2009) that, for most characteristics, is the very end of a several-thousand-year transition. If we choose a time that has already occurred for the start of the Anthropocene, then we are making a very different choice, in which the start of the transition is defined as the base of the new epoch. While acknowledging that humans have taken control of important parts of the Earth system, I wonder if it may be more sensible to describe ourselves as being firmly in the transition into the Anthropocene, but to leave future generations to define, with the benefit of hindsight, when the epoch actually starts or started.

This work was funded by the Natural Environment Research Council.

References

Archer, D., Eby, M. *et al.* 2009. Atmospheric lifetime of fossil fuel carbon dioxide. *Annual Review of Earth and Planetary Sciences*, **37**, 117–134, http://dx.doi.org/10.1146/annurev.earth.031208.100206

Berger, A. & Loutre, M. F. 2002. An exceptionally long interglacial ahead? *Science*, **297**, 1287–1288.

Boutron, C. F., Görlach, U., Candelone, J.-P., Bolshov, M. A. & Delmas, R. J. 1991. Decrease in anthropogenic lead, cadmium and zinc in Greenland snows since the late 1960s. *Nature*, **353**, 153–156.

Broecker, W. S. & Stocker, T. F. 2006. The Holocene CO_2 rise: anthropogenic or natural? *EOS Transactions*, **87**, 27.

Butler, J. H., Battle, M. *et al.* 1999. A record of atmospheric halocarbons during the twentieth century from polar firn air. *Nature*, **399**, 749–755.

Claussen, M., Brovkin, V., Calov, R., Ganopolski, A. & Kubatzki, C. 2005. Did humankind prevent a Holocene glaciation? *Climatic Change*, **69**, 409–417.

Crutzen, P. J. 2002. Geology of mankind. *Nature*, **415**, 23–23, http://dx.doi.org/10.1038/415023a

de Vernal, A. & Hillaire-Marcel, C. 2008. Natural variability of Greenland climate, vegetation, and ice volume during the past million years. *Science*, **320**, 1622–1625, http://dx.doi.org/10.1126/science.1153929

Elmore, D., Tubbs, L. E. *et al.* 1982. ^{36}Cl bomb pulse measured in a shallow ice core from Dye 3, Greenland. *Nature*, **300**, 735–737.

Fischer, H., Wagenbach, D. & Kipfstuhl, J. 1998. Sulfate and nitrate firn concentrations on the Greenland ice sheet. 2. Temporal anthropogenic deposition changes. *Journal of Geophysical Research*, **103**, 21 935–21 942.

Fretwell, P., Pritchard, H. D. *et al.* 2013. Bedmap2: improved ice bed, surface and thickness datasets for Antarctica. *The Cryosphere*, **7**, 375–393, http://dx.doi.org/10.5194/tc-7-375-2013

Gabrieli, J., Vallelonga, P. *et al.* 2010. Post 17th-century changes of European PAH emissions recorded in high-altitude alpine snow and ice. *Environmental Science & Technology*, **44**, 3260–3266, http://dx.doi.org/10.1021/es903365s

Gabrieli, J., Cozzi, G. *et al.* 2011. Contamination of Alpine snow and ice at Colle Gnifetti, Swiss/Italian Alps, from nuclear weapons tests. *Atmospheric Environment*, **45**, 587–593, http://dx.doi.org/10.1016/j.atmosenv.2010.10.039

Ganopolski, A. & Calov, R. 2011. The role of orbital forcing, carbon dioxide and regolith in 100 kyr glacial cycles. *Climate of the Past*, **7**, 1415–1425, http://dx.doi.org/10.5194/cp-7-1415-2011

Gibbard, P. L., Head, M. J., Walkers, M. J. C. & QUATERNARY SUBCOMMISSION 2010. Formal ratification of the Quaternary System/Period and the Pleistocene Series/Epoch with a base at 2.58 Ma. *Journal of Quaternary Science*, **25**, 96–102, http://dx.doi.org/10.1002/jqs.1338

Gregory, J. M., Huybrechts, P. & Raper, S. C. B. 2004. Climatology – threatened loss of the Greenland ice-sheet. *Nature*, **428**, 616.

Hong, S., Candelone, J.-P., Patterson, C. C. & Boutron, C. F. 1994. Greenland ice evidence of hemispheric lead pollution two millenia ago by Greek and Roman civilizations. *Science*, **265**, 1841–1843.

Hong, S., Candelone, J.-P., Turetta, C. & Boutron, C. F. 1996. Changes in natural lead, copper, zinc and cadmium concentrations in central Greenland ice from 8250 to 149,100 years ago: their association with climatic changes and resultant variations of dominant source contributions. *Earth and Planetary Science Letters*, **143**, 233–244.

Hönisch, B., Hemming, N. G., Archer, D., Siddall, M. & McManus, J. F. 2009. Atmospheric carbon dioxide concentration across the Mid-Pleistocene transition. *Science*, **324**, 1551–1554, http://dx.doi.org/10.1126/science.1171477

Legrand, M., Hammer, C., DeAngelis, M., Savarino, J., Delmas, R., Clausen, H. & Johnsen, S. J. 1997. Sulfur-containing species (methanesulfonate and SO_4) over the last climatic cycle in the Greenland Ice Core Project (central Greenland) ice core. *Journal of Geophysical Research*, **102**, 26 663–26 679.

Loulergue, L., Schilt, A. *et al.* 2008. Orbital and millennial-scale features of atmospheric CH_4 over the last 800,000 years. *Nature*, **453**, 383–386.

Lüthi, D., Le Floch, M. *et al.* 2008. High-resolution carbon dioxide concentration record 650,000–800,000 years before present. *Nature*, **453**, 379–382.

MacFarling Meure, C., Etheridge, D. *et al.* 2006. Law Dome CO_2, CH_4 and N_2O ice core records extended to 2000 years BP. *Geophysical Research Letters*, **33**, L14810, doi: 10.1029/2006GL026152.

Marcott, S. A., Shakun, J. D., Clark, P. U. & Mix, A. C. 2013. A reconstruction of regional and global temperature for the past 11,300 years. *Science*, **339**, 1198–1201, http://dx.doi.org/10.1126/science.1228026

NEEM COMMUNITY MEMBERS 2013. Eemian interglacial reconstructed from a Greenland folded ice core. *Nature*, **493**, 489–494, http://dx.doi.org/10.1038/nature11789

Raymo, M. E. & Mitrovica, J. X. 2012. Collapse of polar ice sheets during the stage 11 interglacial. *Nature*, **483**, 453–456, http://dx.doi.org/10.1038/nature10891

REMANE, J., BASSETT, M. G., COWIE, J. W., GOHRBANDT, K. H., LANE, H. R., MICHELSEN, O. & NAIWEN, W. 1996. Revised guidelines for the establishment of global chronostratigraphic standards by the International Commission on Stratigraphy (ICS). *Episodes*, **19**, 77–81.

ROBINSON, A., CALOV, R. & GANOPOLSKI, A. 2012. Multistability and critical thresholds of the Greenland ice sheet. *Nature Climate Change*, **2**, 429–432, http://dx.doi.org/10.1038/nclimate1449

RUDDIMAN, W. F. 2003. The anthropogenic greenhouse era began thousands of years ago. *Climatic Change*, **61**, 261–293.

SCHILT, A., BAUMGARTNER, M., BLUNIER, T., SCHWANDER, J., SPAHNI, R., FISCHER, H. & STOCKER, T. F. 2010. Glacial–interglacial and millennial-scale variations in the atmospheric nitrous oxide concentration during the last 800,000 years. *Quaternary Science Reviews*, **29**, 182–192, http://dx.doi.org/10.1016/j.quascirev.2009.03.011

SCHMITT, J., SCHNEIDER, R. *ET AL.* 2012. Carbon isotope constraints on the deglacial CO_2 rise from ice cores. *Science*, **336**, 711–714, http://dx.doi.org/10.1126/science.1217161

SINGARAYER, J. S., VALDES, P. J., FRIEDLINGSTEIN, P., NELSON, S. & BEERLING, D. J. 2011. Late-Holocene methane rise caused by orbitally-controllled increase in tropical sources. *Nature*, **470**, 82–85.

STEFFEN, W., GRINEVALD, J., CRUTZEN, P. & MCNEILL, J. 2011. The Anthropocene: conceptual and historical perspectives. *Philosophical Transactions of the Royal Society A – Mathematical Physical and Engineering Sciences*, **369**, 842–867, http://dx.doi.org/10.1098/rsta.2010.0327

STEINHILBER, F., ABREU, J. A. *ET AL.* 2012. 9,400 years of cosmic radiation and solar activity from ice cores and tree rings. *Proceedings of the National Academy of Sciences of the United States of America*, **109**, 5967–5971, http://dx.doi.org/10.1073/pnas.1118965109

TZEDAKIS, P. C., WOLFF, E. W., SKINNER, L. C., BROVKIN, V., HODELL, D. A., MCMANUS, J. F. & RAYNAUD, D. 2012. Can we predict the duration of an interglacial? *Climate of the Past*, **8**, 1473–1485.

WALKER, M., JOHNSEN, S. *ET AL.* 2009. Formal definition and dating of the GSSP (Global Stratotype Section and Point) for the base of the Holocene using the Greenland NGRIP ice core, and selected auxiliary records. *Journal of Quaternary Science*, **24**, 3–17, http://dx.doi.org/10.1002/jqs.1227

WOLFF, E. W. 2013. Ice sheets and nitrogen. *Philosophical Transactions of the Royal Society B*, **368**, 20130127, http://dx.doi.org/10.1098/rstb.2013.0127

WOLFF, E. W. & PEEL, D. A. 1985. The record of global pollution in polar snow and ice. *Nature*, **313**, 535–540.

WOLFF, E. W. & SUTTIE, E. D. 1994. Antarctic snow record of southern hemisphere lead pollution. *Geophysical Research Letters*, **21**, 781–784.

WOLFF, E. W., SUTTIE, E. D. & PEEL, D. A. 1999. Antarctic snow record of cadmium, copper, and zinc content during the twentieth century. *Atmospheric Environment*, **33**, 1535–1541.

ZACHOS, J. C., DICKENS, G. R. & ZEEBE, R. E. 2008. An early Cenozoic perspective on greenhouse warming and carbon-cycle dynamics. *Nature*, **451**, 279–283, http://dx.doi.org/10.1038/nature06588

ZALASIEWICZ, J., WILLIAMS, M. *ET AL.* 2008. Are we now living in the Anthropocene? *GSA Today*, **18**, 4–8, http://dx.doi.org/10.1130/GSAT01802A.1

ZALASIEWICZ, J., WILLIAMS, M., HAYWOOD, A. & ELLIS, M. 2011. The Anthropocene: a new epoch of geological time? Introduction. *Philosophical Transactions of the Royal Society A – Mathematical Physical and Engineering Sciences*, **369**, 835–841, http://dx.doi.org/10.1098/rsta.2010.0339

The release and persistence of radioactive anthropogenic nuclides

GARY J. HANCOCK[1]*, STEPHEN G. TIMS[2], L. KEITH FIFIELD[2] & IAN T. WEBSTER[3]

[1]*CSIRO Land and Water, Canberra, ACT 2601, Australia*

[2]*Department of Nuclear Physics, Research School of Physics and Engineering, The Australian National University, Canberra, ACT 2601, Australia*

[3]*1 Scaddan Place, Curtin, ACT 2605, Australia*

**Corresponding author (e-mail: gary.hancock@csiro.au)*

Abstract: Atmospheric testing of nuclear weapons during the period 1945–1980 ushered in the 'atomic age' and released large quantities of anthropogenic radiogenic nuclides into the atmosphere. These radionuclides were subsequently deposited as fallout to the entire surface of the planet. While many have decayed to negligible levels, long-lived radionuclides persist and will do so for thousands of years. Isotopes of plutonium, ^{239}Pu (half-life 24 100 years) and ^{240}Pu (half-life 6563 years), provide the best chronological markers for the onset of this anthropogenic event both now and into the future due to their long half-lives, particle-reactivity, and the fact that they were present in negligible quantities prior to anthropogenic production and release. Chronostratigraphic markers established by distinct Pu concentration profiles and Pu isotope changes in sediment sequences and ice and coral cores can provide high-resolution dating over the last 60 years. However, even though fallout has ceased, it is found that the Pu inventory currently held in surface soil layers and the oceans will continue to supply Pu to sediment deposition zones for millennia and centuries, respectively. The delivery of this Pu will depend on soil erosion and bioturbation rates, and the rate of removal of dissolved Pu from the ocean.

The discovery and practical implementation of nuclear fission at Los Alamos in the 1940s lead to the onset of the 'atomic age' and the associated staged release of radiogenic nuclides into the Earth's biosphere. The release processes were initially mostly unrestrained, via testing of nuclear weapons in the atmosphere, but later included releases from other sources such as underground nuclear testing, nuclear weapons fabrication, nuclear fuel reprocessing and radioisotope production, and included both controlled and uncontrolled (accidental) releases. These episodes released anthropogenic radioactivity into the environment through a variety of pathways, and with different release histories.

Some of the radionuclides released were already present in the environment prior to the onset of anthropogenic production, but many were not or were present in only trace quantities. For those nuclides which have persisted in the environment over the last few decades, i.e. those with radioactive half-lives of a few years or greater, clear evidence of their presence can be seen in measurements made on a variety of environmental samples. For many of these nuclides, especially those with an affinity for soil and sediment particles, a clear pattern corresponding with the history of release can be seen in the sedimentary record where sedimentation processes are favourable. It is the aim of this paper to examine the ability of anthropogenic radionuclides to provide an unequivocal marker of this human-induced event in sediment records. We also assess their persistence such that they are able to provide a unique and recognizable signal for the Anthropocene from the mid-1900s onwards. In this context we look at Pu isotopes in detail, particularly the most abundant and longest lived isotope, ^{239}Pu. We consider soil and sediment concentrations of Pu both before and after the onset of the atomic age, and predict likely concentration trends in the sediment record into the distant future.

Anthropogenic fallout radionuclides

Radioactive atoms (radionuclides) are inherently unstable, releasing energy as they decay to more stable states. Many radionuclides present in the environment occur naturally; for example, isotopes of uranium (^{238}U, ^{235}U) and thorium (^{232}Th) and many of their decay products. Even plutonium occurs naturally (see below). However, radionuclides are also produced by anthropogenic activities, with the vast majority of anthropogenic radionuclides in existence in the environment today originating from atmospheric weapons testing during the period 1945–1980 (UNSCEAR 2000). Around 500 tests were conducted at various

From: Waters, C. N., Zalasiewicz, J. A., Williams, M., Ellis, M. A. & Snelling, A. M. (eds) 2014. *A Stratigraphical Basis for the Anthropocene*. Geological Society, London, Special Publications, **395**, 265–281.
First published online February 28, 2014, http://dx.doi.org/10.1144/SP395.15

locations on and above the Earth's surface. These tests released radioactive debris into the local, regional and global environment. The larger thermonuclear tests commencing after 1950 injected most of their debris into the stratosphere where it was dispersed and deposited as fallout over the entire surface of the Earth.

Other anthropogenic sources include controlled liquid discharges from the nuclear reprocessing industry and accidental releases, including the Chernobyl power reactor in 1986 and the recent nuclear power plant disaster at Fukushima. The main controlled release of radioactivity has occurred by liquid discharge from the Sellafield reprocessing plant into the Irish Sea from 1952 until the mid-1980s. While these accidental and controlled releases have contributed significant anthropogenic radioactivity on a regional scale, and the extent of Fukushima contamination is still being assessed, the current radioactive inventory associated with these releases is small compared to the combined atmospheric input from thermonuclear explosions.

Although many of the radionuclides produced during weapons testing had short half-lives and decayed to safe levels within weeks to months of release, others persist in the environment to this day and will do so for centuries or millenia into the future. Fallout radioactivity estimates are mostly based on direct measurements of ^{90}Sr, and to a lesser extent, ^{137}Cs, with estimates of other radionuclides based on assessments of their activity ratios to ^{90}Sr in the bomb debris (UNSCEAR 2000). The half-lives of ^{90}Sr and ^{137}Cs are similar (29 and 30 years), and, while their residual fallout radioactivity is still easily measured today, their total fallout supply (inventory) existing on the Earth's surface has decayed by around 70% over the 50–60 years since most of the atmospheric fallout occurred. In some regions where global fallout was low, the measurement of ^{137}Cs, a convenient tracer of soil erosion (Zapata 2003), is already approaching its limit of detection. Clearly, for environmental applications such as this, decay of fallout ^{137}Cs and ^{90}Sr will become a problem of increasing magnitude into the twenty-first and twenty-second centuries.

Long-lived fallout radionuclides include ^{129}I (half-life 1.6×10^7 years), ^{14}C (5730 years), and isotopes of Pu, ^{239}Pu (half-life 24 110 years) and ^{240}Pu (half-life 6563 years). ^{129}I is soluble and is readily dispersed in the environment, and ^{14}C has been incorporated into terrestrial and marine carbon where its fate is subject to carbon uptake and release cycles of biota. Pu, on the other hand, is highly particle reactive, binding strongly to the surfaces of soil and sediment particles. Its high distribution coefficient, a measure of the relative concentrations of a species in the solid and liquid phases (Christensen *et al.* 1998), and low solubility has led to most Pu on land and along the ocean margins being associated with soils and sediment. While the total Pu radioactivity released was much less than ^{137}Cs and ^{90}Sr (Table 1), its fallout mass was much greater and its sensitivity of measurement by mass spectrometric techniques more than makes up for this limitation. Thus, ^{239}Pu and ^{240}Pu are currently comparable with ^{137}Cs in their utility as sediment chronometers pertaining to the last 60 years (e.g. Jaakkola *et al.* 1983; Hancock *et al.* 2012), and we propose that their long half-lives will make them far superior as future markers of the Anthropocene. Fallout activity of a third Pu isotope, ^{241}Pu, was significantly higher than ^{239}Pu and ^{240}Pu (UNSCEAR 2000), but it has a much shorter half-life (14.3 years) and much of its activity has already decayed. A fourth isotope, ^{238}Pu (half-life 88 years), was released in low activities during weapons testing, although a significant additional atmospheric input occurred during the disintegration of the navigational satellite SNAP-9A in 1964. This input has provided important dating information for the 1960s (Koide *et al.* 1977, 1979; Smith *et al.* 1987; Hancock *et al.* 2012) but, once again, the usefulness of ^{238}Pu into the future will be limited by its half-life. By contrast, the activities of long-lived ^{239}Pu and ^{240}Pu have barely changed since fallout began in 1946, and will decrease only slightly over the next millennia.

Table 1. *Fallout of long-lived radionuclides resulting from atmospheric weapons testing (UNSCEAR 2000)*

Radionuclide	Half-life (years)	Global fallout (PBq)	Global fallout (g)
^{90}Sr	28.8	622	0.12×10^6
^{137}Cs	30.1	948	0.30×10^6
^{14}C	5730	213	1.29×10^6
^{241}Pu	14.4	142	0.04×10^6
^{238}Pu*	88	0.3	0.4×10^3
^{240}Pu†	6563	4.4	0.52×10^6
^{239}Pu†	24 110	6.5	2.84×10^6

*Global ^{238}Pu fallout does not include fallout from SNAP-9A, estimated at 0.6 PBq.
†Global ^{239}Pu and ^{240}Pu fallout does not include close-in fallout at the Pacific Proving Grounds estimated at 5.1 PBq (Hamilton 2005).

History and spatial variation of atmospheric fallout

The ^{137}Cs fallout history for the northern and southern hemisphere is shown in Figure 1; the plot is generated from data reported by UNSCEAR

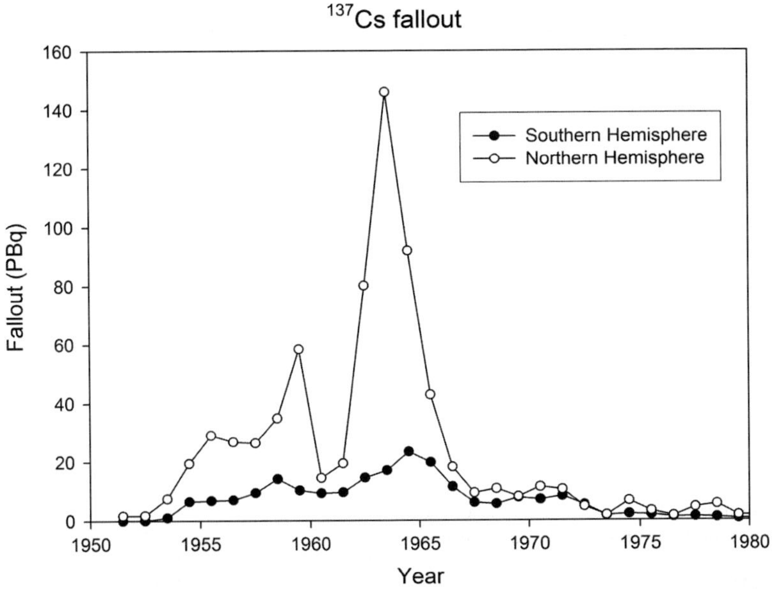

Fig. 1. History of ^{137}Cs fallout. Data are based on a combination of ^{90}Sr and ^{137}Cs measurements (UNSCEAR 2000).

(2000), which includes direct measurements of annual ^{137}Cs fallout together with estimates generated from ^{90}Sr measurements. Fallout commenced in 1945 during World War II, but it was not until around 1952 and the onset of high-yield thermonuclear tests that fallout accelerated to the point where retrospective studies could positively identify weapons fallout on a global scale. A lull in testing during the 1958–1960 moratorium reduced the rate of fallout, but it accelerated again in the 1960s, reaching a maximum in 1963–1964. The fallout rate rapidly declined during the late 1960s and effectively ceased in 1980. Total fallout is approximately three times higher in the northern hemisphere, where most of testing occurred, compared to the south, although the rate is seven times higher during the peak fallout period of the early 1960s. The latitudinal variation of global fallout is explored further in Figure 2, where ^{90}Sr fallout is seen to be concentrated in the mid-latitude bands of each hemisphere.

Although the total activity of fallout Pu is two orders of magnitude lower than ^{137}Cs or ^{90}Sr, the mass of Pu deposited is one order of magnitude higher (Table 1). Much of this plutonium originated from the capture of fast neutrons by ^{238}U (the dominant isotope of uranium) in the uranium tamper of fusion devices. The role of this tamper was to confine the fusion reaction for long enough to ensure that a significant fraction of the heavy isotopes of hydrogen (deuterium ^{2}H and tritium ^{3}H) were able to fuse. Uranium metal was optimal for this role due to its high density. The fusion reactions $^{2}H + {}^{2}H \rightarrow {}^{3}He + n$ and $^{2}H + {}^{3}H \rightarrow {}^{4}He + n$, where n denotes a neutron, release large amounts of energy. Hence, fusion devices can have much larger yields than pure fission devices such as the Hiroshima or Nagasaki bombs. Much of the energy released in the fusion reactions is contained in the kinetic energy of the neutrons produced, and these neutrons may be absorbed by the ^{238}U in the tamper to produce ^{239}U and ^{240}U, which ultimately decay after the explosion to ^{239}Pu and ^{240}Pu. In addition to production of plutonium, these neutrons also have enough energy to initiate fission of the ^{238}U and, paradoxically in some devices, more energy was derived from this fission of the ^{238}U of the tamper than from the fusion reactions themselves. However, because the fission products are spread over many possible nuclei, the mass of individual products such as ^{90}Sr and ^{137}Cs was substantially less than the mass of plutonium. It is only because of their relatively rapid rates of decay (relatively short half-lives) that the fallout activities of ^{90}Sr and ^{137}Cs are initially higher than Pu.

Few direct measurements of Pu fallout were made during the 1950s and 1960s, but the historical fallout pattern of $^{239+240}$Pu is believed to be similar to that of ^{137}Cs, especially after 1960. The Pu fallout record shown in Figure 3 is based on the assumption that the trend in Pu fallout follows that of ^{137}Cs and ^{90}Sr, but having a $^{137}Cs/^{239+240}Pu$ activity ratio of 50 for the pre-moratorium period (pre-1958) and an activity ratio

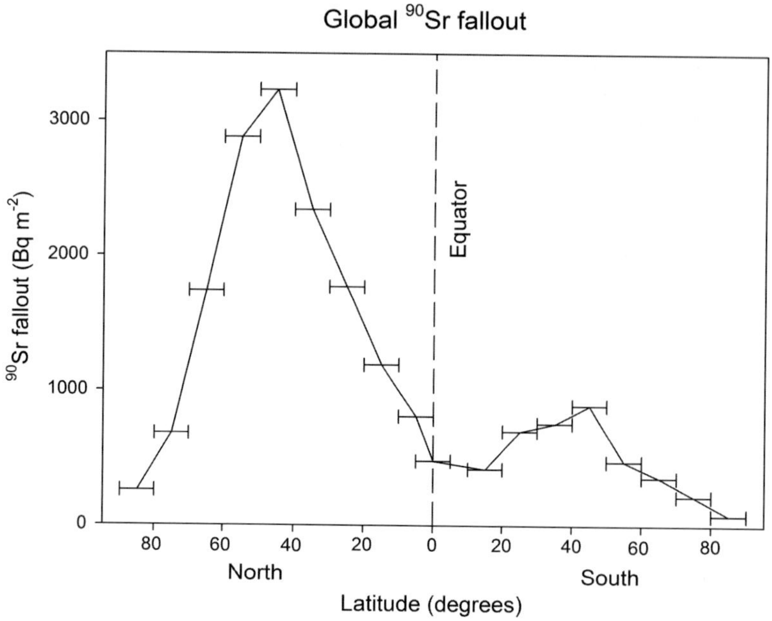

Fig. 2. Latitudinal variation of total ^{90}Sr fallout from 1945 to 1980.

of 83 for post-1959. This assumption is broadly consistent with Pu activity ratios measured in ice and sediment core profiles reported in the literature (Koide *et al.* 1979, 1982, 1985; Krey *et al.* 1990). The low pre-moratorium activity ratio is due to the relatively high Pu activity of fallout from the US-dominated tests in the period 1952–1958 (Koide *et al.* 1979, 1985).

Also shown in Figure 3 is a plot of ^{137}Cs fallout corrected for radioactive decay to the year 2012. The measurable activity of ^{137}Cs in soils and sediment has decreased by a factor of approximately three compared to its original activity. Nevertheless, due to the low cost and ease of measurement of ^{137}Cs by gamma spectrometry, its fallout record has been extensively used to establish known-age

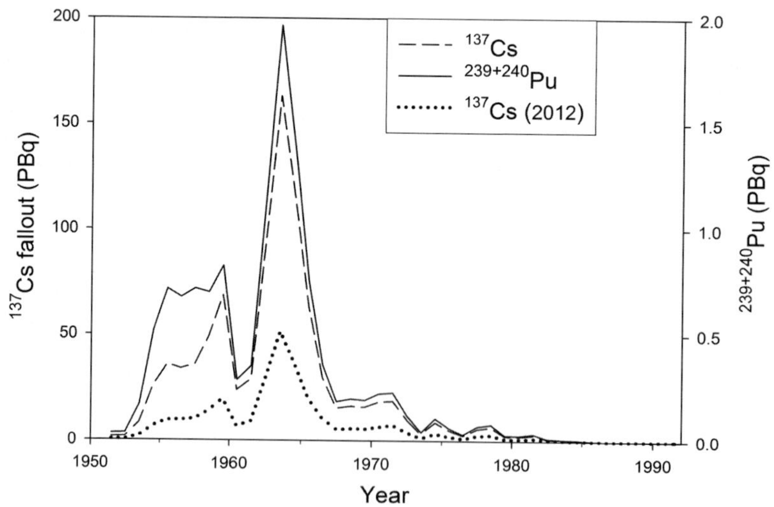

Fig. 3. The global ^{137}Cs and $^{239+240}$Pu fallout record. The Pu record is estimated from ^{90}Sr and ^{137}Cs fallout data and using their ratios to Pu shown in ice cores. The dotted line shows ^{137}Cs activity in 2012.

stratigraphic horizons for sediment geochronology studies in all parts of the globe. In particular, the fallout peak corresponding to 1963–1964 and the onset of detectable fallout in the early 1950s are important markers. The 1963–1964 ^{137}Cs fallout peak is usually clearly seen in undisturbed sediments in the northern hemisphere, but is much less obvious in the south. This is attributed to the lower fallout there, and confounding processes such as high sediment fluxes associated with subsurface soil erosion within the water catchment during periods of agricultural development and clearing of native forest (Hancock *et al.* 2002). A few studies have used Pu (e.g. Jaakkola *et al.* 1983) and even ^{241}Am, a long-lived decay product of ^{241}Pu (Appleby *et al.* 1991), for similar chronological purposes to that of ^{137}Cs, with Hancock *et al.* (2012) showing the advantages of Pu over ^{137}Cs in Australian and New Zealand sediments (see the following section). While the ease and relatively low cost of ^{137}Cs measurement by gamma spectrometry often makes it the method of choice, the advantages of Pu include easier identification of the 1964 fallout peak in southern hemisphere sediment, and the use of historical shifts in the Pu isotope ratios of fallout to establish dated stratigraphic horizons (e.g. Koide *et al.* 1985; Smith *et al.* 1987; Tims *et al.* 2010*b*; Hancock *et al.* 2012). Importantly, the first appearance of Pu fallout in the sediment sequence is potentially more definitive than ^{137}Cs owing to the greater sensitivity of measurement for Pu (Leslie & Hancock 2008).

Plutonium measuring techniques and detection limits

A variety of techniques can be used for measuring plutonium isotopes. Where plutonium concentrations are relatively high, α-spectroscopy has been the traditional method employed. Here the α-particle activity of the combined activity of ^{239}Pu + ^{240}Pu is measured after radiochemical purification of the sample; a tracer (usually ^{242}Pu) is used to monitor chemical yield. This method is suitable for higher activity soil and sediment samples, and for tracing discharges from nuclear fuel reprocessing plants. The very low background activity of the α-particle spectrum also provides sufficient sensitivity for many environmental samples even in the southern hemisphere; although, for soils exposed to typical levels of global fallout, large sample masses and long count times are often required to obtain sufficient measurement sensitivity. At these lower plutonium concentrations, mass spectrometric (MS) techniques, which literally count Pu atoms, become the methods of choice. Thermal-ionization mass spectrometry (TIMS), resonance-ionization mass spectrometry (RIMS) and inductively-coupled plasma mass spectrometry (ICP-MS) can readily measure the long-lived isotopes at the concentrations present in typical modern surface soils. However, these techniques are sensitive to interference from polyatomic ions that arise in the ion plasma: for example, $^{238}UH^+$ and $^{238}UH_2^+$ can masquerade in the ion counter as $^{239}Pu^+$ and $^{240}Pu^+$, respectively. Modern high-resolution sector-field ICP-MS (Wyse *et al.* 2001) and in-source RIMS (Raeder *et al.* 2012) instruments can, at least partially, overcome this issue. Accelerator mass spectrometry (AMS) has the advantage that molecular species are dissociated as part of the acceleration process, and hence it removes interferences that may be missed with the other MS methods (Fifield 2008).

As the above discussion suggests, each technique has its own advantages and disadvantages. For example, the infrastructure costs increase markedly with the complexity of the technique, making AMS the most expensive method. The simplicity and much lower cost of α-spectroscopy relative to those of MS is a clear advantage, as is the ability to measure ^{238}Pu, which is not generally accessible with the MS instruments owing to interference from ^{238}U. However, while MS can measure ^{239}Pu and ^{240}Pu individually, separation of ^{239}Pu from ^{240}Pu is at best imprecise with α-spectroscopy because the α-particle energies of the decays of the two isotopes overlap and are poorly resolved. Consequently, data on the ^{240}Pu/^{239}Pu ratio, a useful diagnostic for determining the source of the plutonium, is unavailable by α-spectroscopy. Furthermore, the α-spectroscopy detection limit is approximately 0.05 mBq of $^{239+240}$Pu, equivalent to around 20 fg (2×10^{-14} g) of ^{239}Pu, the isotope measured with the greatest sensitivity by MS. This is a factor of at least 5 greater than that for the MS methods (Ketterer & Szechenyi 2008; Raeder *et al.* 2012), and up to a factor of 100 greater than that achievable with the best sector-field ICP-MS, in-source RIMS or AMS instruments.

Pre-anthropogenic Pu

^{239}Pu occurs naturally in the Earth's crust as a result of the fission of ^{238}U, and has been measured in samples of uranium ore. The pre-anthropogenic concentration of ^{239}Pu in typical soils is estimated at about 0.05 mBq kg^{-1} (Taylor 2001). In principle, this activity should be measureable given the current detection limits of the best techniques (*c.* 100 nBq ^{239}Pu) and the fact that the sample mass (and hence the mass of Pu) taken for analysis can be increased in accordance with measurement sensitivity requirements. Nevertheless, instrumental and practical limitations exist due to polyatomic ion interference, mentioned above, and the chemical separation procedures required prior to MS analysis. Certainly, the requisite sensitivity to measure natural ^{239}Pu at concentrations as low as 0.02 mBq kg^{-1} (*c.* 9 fg kg^{-1}) in a 20 g soil sample is currently available, although we know of no measurements reported to date.

Sediment profiles of Pu

Crater lake

The record of Pu deposition is best preserved at sites where atmospheric fallout Pu delivered directly to that site makes up most of the Pu in the sediment profile. At such sites the fallout record is not complicated by variable additional inputs of Pu from sediment transported from a distance by wind and water. The best fallout records are seen in the ice sheets of polar regions (Koide *et al.* 1977, 1979) and lakes where the area of water catchment draining to the lake is equal to or only slightly larger than the area of the lake. Such lakes include crater lakes of extinct volcanoes. Figure 4 shows a $^{239+240}$Pu sediment core profile from Blue Lake, a crater lake at Mount Gambier in SE Australia (Vanderzalm *et al.* 2013). The core was collected in 2008 and is mainly composed of authigenic calcareous material precipitated from the water column. The $^{239+240}$Pu profile was measured by α-spectrometry and shows a recognizable pattern, with Pu concentrations initially increasing from undetectable concentrations below 78 cm to a peak in the 60–70 cm depth section, before decreasing to almost undetectable levels in the upper 30 cm. The depth horizon of 78 cm is given the Pu first-detection date of 1954, as calculated by Leslie & Hancock (2008). Verification of the 1964 peak at 60–70 cm and the high resolution of the chronology are demonstrated by the shift in the ^{238}Pu/$^{239+240}$Pu activity ratio from low values prior to the SNAP-9A disintegration to a peak in the 55–60 cm depth interval. The depth of 30 cm corresponds to the mid- to late 1970s as determined from the core chronology, and very low concentrations seen in the upper 30 cm are therefore assumed to correspond to the end of fallout from weapons tests. The fact that Pu concentrations do

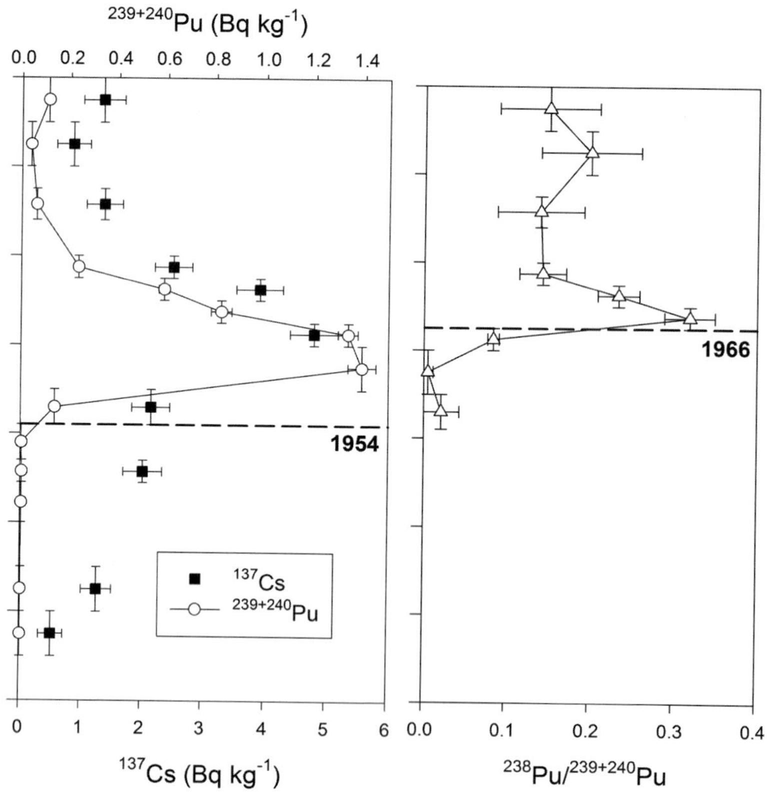

Fig. 4. Core profile of sediment collected from Blue Lake, a crater lake in SE Australia (Vanderzalm *et al.* 2013).

not decline back to wholly undetectable levels is an indication that low amounts of eroded surface soil from the slopes of the crater and wind-blown dust are being deposited.

It is noted that in this case the ^{137}Cs profile (closed symbols) only poorly corresponds to the Pu profile, with ^{137}Cs being seen below the horizon attributed to onset of nuclear tests (78 cm). This is ascribed to post-depositional mobility of sediment-bound ^{137}Cs resulting from weak bonding between ^{137}Cs and the calcareous sediment. Desorption of ^{137}Cs from sediment surfaces enables it to diffuse vertically through sediment pore water, a phenomena also seen in studies of marine (saline) sediment (Beasley *et al.* 1982; Sholkovitz & Mann 1984). These studies also conclude that Pu profiles show no evidence for alteration and, together with Blue Lake, attest to the ability of Pu to act as a reliable chronometer on a timescale of decades, and probably much longer.

Estuarine lake

An example of a core profile collected in 2004 from a lake draining a large catchment is shown in Figure 5. Lake Victoria is one of the lakes at the outlet of the East Gippsland catchment (area 10 000 km^2) in SE Australia. The lake water is usually brackish, with ocean water exchanging with lake water through a narrow entrance. Once again, a high-resolution chronology is determined using $^{239+240}Pu$ concentrations and the $^{238}Pu/^{239+240}Pu$ activity ratio (Hancock *et al.* 2012). The 1964 fallout peak is clearly visible. In a similar fashion to Blue Lake, the Pu concentration declines towards the upper surface of the core, although in this case it does not approach its detection limit but follows ^{137}Cs, reaching relatively constant concentrations in the upper 25–30 cm. The concentrations of ^{137}Cs and Pu are easily measurable and are well in excess of their pre-1950s concentrations. It is argued that the upper 25–30 cm of Lake Victoria sediment strata represents the most recent two or three decades since weapons testing ceased, as determined by ^{210}Pb geochronology, and that the ^{137}Cs and Pu in the upper sediment strata reflect ongoing input of ^{137}Cs and Pu from broad-scale catchment sources. The reason that recent input of Pu is much higher for Lake Victoria than Blue Lake, with the current sediment Pu concentrations being about 20% of the 1964 peak concentration compared to 3% for Blue

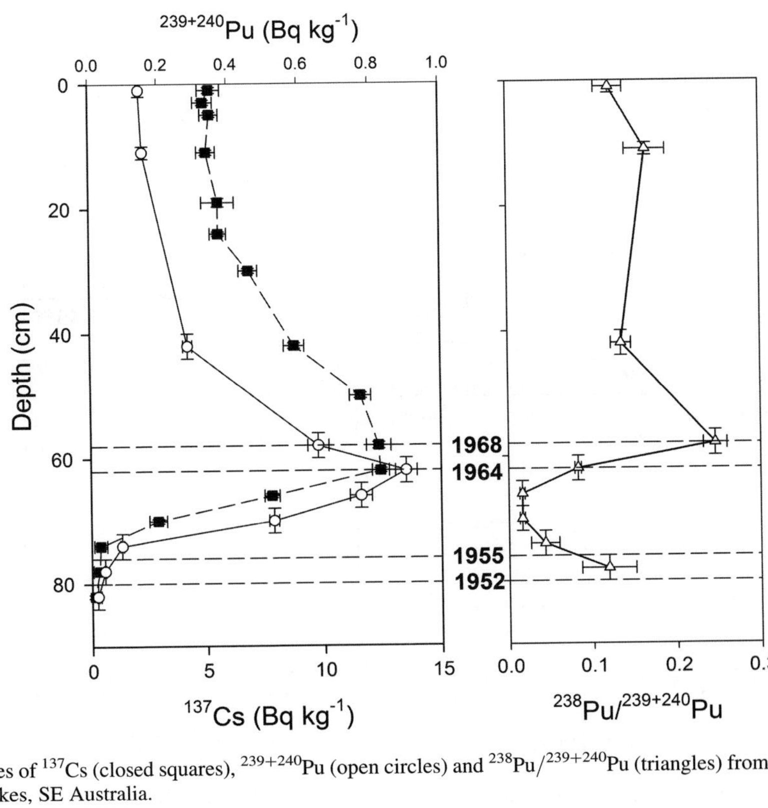

Fig. 5. Profiles of ^{137}Cs (closed squares), $^{239+240}Pu$ (open circles) and $^{238}Pu/^{239+240}Pu$ (triangles) from Lake Victoria, Gippsland Lakes, SE Australia.

Lake, is the much larger East Gippsland catchment area and low authigenic sediment production.

The ^{137}Cs peak in Lake Victoria is much broader than Pu and is due to a higher proportion of ^{137}Cs being delivered from the catchment compared to Pu (Hancock *et al.* 2012). It is proposed that delivery of direct-fallout Pu to bottom sediment is facilitated by the more efficient scavenging of Pu in the saline water of Lake Victoria compared to ^{137}Cs, with a large proportion of direct fallout ^{137}Cs remaining soluble in the lake water.

The process responsible for the continued delivery of Pu to Lake Victoria – that is, ongoing erosion of surface soil from hillslopes and delivery to sediment deposition zones – will obviously continue into the future. Accordingly, prediction of the shape of future Pu sediment profiles will need to consider catchment sources. Assuming the proportions of soil eroded from surface and subsurface sources in the catchment remain relatively constant, changes in the concentration of Pu in catchment surface soils will be the primary factor controlling the future Pu content of sediments derived from large catchments.

Future profiles; the Pu content of surface soils

The examples above show that fallout radionuclides accumulating at sites of sediment deposition can have two sources: direct fallout of the nuclide from the atmosphere into the zone of deposition (e.g. a lake) where it is scavenged from the water column by suspended sediment particles and settles by gravity to the lake bottom; or, alternatively, the nuclide may have deposited elsewhere within the water catchment where it became attached to soil or sediment particles before being mobilized and transported with these particles to the deposition site. The latter process usually occurs as the result of water erosion of surface soil and subsequent transport of this soil through the river network, although in some cases wind erosion may be significant.

In large catchments the dominant source of fallout Pu and ^{137}Cs to downstream water bodies is the surface layer of hillslope soils (Ritchie & McHenry 1990; Wallbrink *et al.* 1998; Tims *et al.* 2010*a*). Downslope runoff of water after rainfall can erode soil and deliver it to stream lines. The source of the soil could be surface or subsurface erosion, with the former mainly associated with sheet erosion of hillslope surfaces, and the latter sourced from deep rilling and drainage-line incision. Although Pu and ^{137}Cs attached to soil particles has migrated downwards through the soil profile in the 60 years since deposition began, most of the Pu and ^{137}Cs still remains in the upper 0–20 cm. This downward migration will continue into the future. As shown by Walling & He (1992), the shape of ^{137}Cs profiles in sediment deposits receiving inputs of eroded soil from river catchments depends strongly on the ^{137}Cs concentration of surface soils in the catchment. Similarly, the prediction of the Pu distribution in future sediment deposits in terrestrial and some near-shore marine systems will depend largely on the depth distribution in the erodible zone of surface soils.

Downward migration of anthropogenic fallout radionuclides through the soil profile has been widely observed (e.g. Walling *et al.* 1995; He & Walling 1997; Wallbrink *et al.* 1999), and has been attributed to infiltration of rainfall and bioturbation. While bioturbation by vertebrates can cause localized disturbance of surface soils, invertebrates such as burrowing worms, ants and termites are the most common causes of surficial soil displacement. Burrowing and feeding activities deliver soil from depth to the surface, diluting the surface soil concentrations of ^{137}Cs and Pu, and leading to downwardly migrating subsurface concentration peaks of ^{137}Cs and Pu. Below this peak the concentrations generally decline exponentially with depth. The depth of bioturbation varies greatly, ranging from about 10 to 100 cm, but most soils are not bioturbated significantly below 30 cm (Rodriguez 2006). It is therefore expected that, for typical surface soils, fallout Pu will be retained within the 0–30 cm depth interval with the concentration profile becoming flatter over time. Bioturbation depths of between 10 and 30 cm are similar to tillage depths reached during soil cultivation. As shown in Figure 6, cultivated soils already show approximately constant concentrations of Pu and ^{137}Cs in the upper 20 or 30 cm (He & Walling 1997; Tims *et al.* 2010*a*). This is due to rapid mixing by tilling, with complete mixing having occurred within a few decades of fallout.

Another factor affecting the shape and concentration of Pu in soil profiles is the rate of surface soil erosion. Loss of Pu associated with topsoil erosion will reduce the Pu soil inventory at a given site. In natural systems, soil erosion can be estimated by assuming soil erosion is in steady state with soil production; that is, the rates of soil loss and soil production from source rocks are approximately equal. In a summary of Australian studies, Bui *et al.* (2011) reported soil production rates in the range 0.004–0.60 mm a^{-1}, with a mean of 0.015 mm a^{-1}, equivalent to 20 g m^{-2} a^{-1}. Clearly, where farming practices have increased erosion, soil (and Pu) loss will be greater.

The evolving distribution of anthropogenic radionuclides in soil profiles can be modelled by assuming that radionuclides are entirely bound to

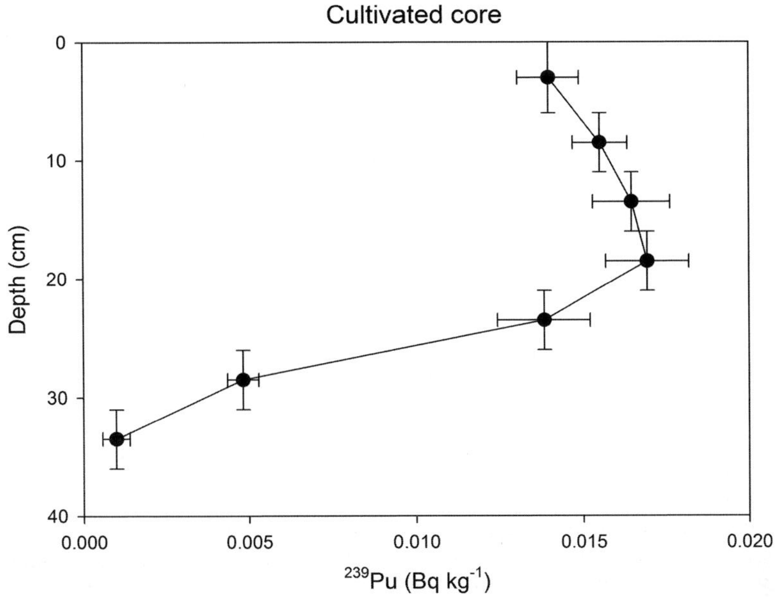

Fig. 6. A ^{239}Pu profile of a cultivated soil core from northern Australia collected in 2008 (Tims *et al.* 2010*b*). Within measurement uncertainties, the concentration is approximately constant in the upper 25 cm.

soil particles, and that transport of the particles through the soil profile occurs by processes akin to diffusion and advection. Boudreau (1986) described a one-dimensional (1D) model that predicts the evolution of tracer concentrations in a sediment column by bioturbation. The following equation, based on Boudreau's general formulation, describes the time evolution of the vertical Pu profile within a soil column due to: (a) localized soil mixing by bioturbation between adjacent sediment layers; (b) 'conveyor belt' exchange between sediment at depth and the surface layer arising from feeding, and tunnel excavation and infilling; (c) the vertical advection of particles in the soil column due to sedimentation/erosion at the surface of the soil column; and (d) radioactive decay:

$$\frac{\partial A}{\partial t} = \frac{\partial}{\partial z}\left(D\frac{\partial A}{\partial z}\right) - eA - \frac{\partial wA}{\partial z} - \lambda A. \qquad (1)$$

In this equation, $A(z, t)$ is the activity of the radionuclide in the bulk sediment (Bq m^{-3}), t is time, z is the distance below the sediment surface, D (m^2 a^{-1}) is the bioturbation diffusivity describing local mixing, e (a^{-1}) is the proportional excavation rate for sediment at depth, w (m a^{-1}) is the local vertical movement of the sediment column due to sedimentation/erosion and local excavation, and λ is the radioactive decay rate of Pu. A proportional excavation rate of $e = 0.01$ a^{-1}, for example, means that 1% of soil particles are excavated to the surface layer every year. Key assumptions in the derivation of equation (1) are that the porosity of the soil is constant and uniform, that erosion and deposition in the soil column can be described as vertical translation of the coordinate system, and that horizontal divergence of Pu transport is negligible. These are common assumptions for modelling sediment profiles (e.g. Christensen 1982).

For solution, equation (1) was discretized into N sediment layers of uniform thickness $\delta = 0.01$ m in an Excel spreadsheet. The first (diffusion) term on the right-hand side was represented using a central difference formulation, and the third (advection) term was represented using an upwind formulation (Roache 1982). The surface layer was treated differently to all the deeper cells as it received all of the sediment excavated from below. Sedimentation was specified as S in m a^{-1} (negative for erosion), so that at $z = 0$, $w = S$. Thus, if $E = \sum_{i=2}^{N} e_i$ is the total proportional rate of sediment column excavation where e_i is the excavation rate from sediment layer i, then at the bottom of the surface layer $w_1 = \delta E + S$. Since sediment is being excavated from deeper layers, the vertical velocity across the lower boundary of layer k gradually reduces with depth as $w_k = \delta(E - \sum_{i=2}^{k} e_i) + S$.

The scheme specifies input of Pu into the surface layer at a rate $F = SA_0$, where A_0 is the concentration of Pu in the depositing or eroding sediment. For the period 1950–1980, Pu fallout

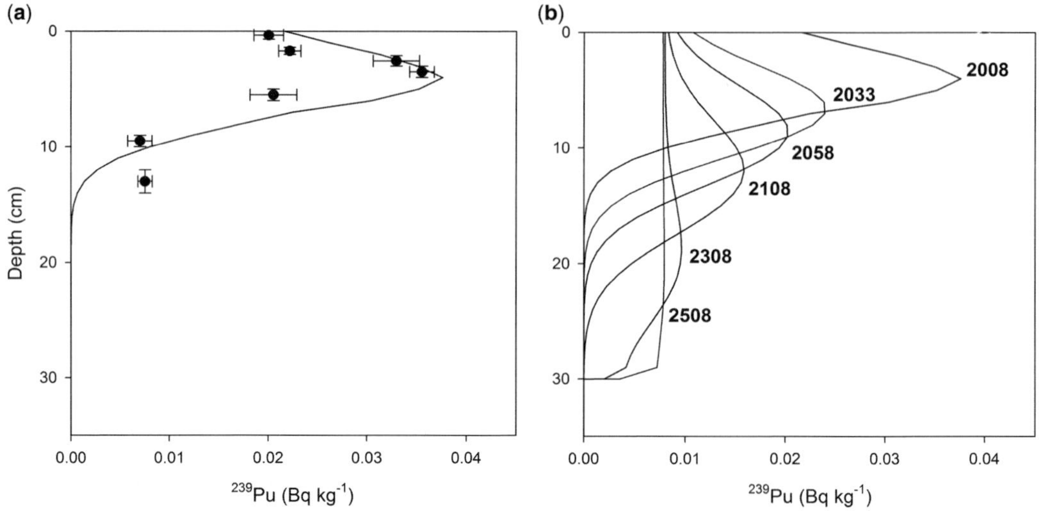

Fig. 7. (**a**) ^{239}Pu profile for a northern Australian pasture soil collected in the year 2008 (Tims *et al.* 2010*b*), data shown as closed circles. The depth of soil bioturbation is assumed to be 30 cm. The line is the best model fit. (**b**) Future profiles predicted using parameters determined from the 2008 best fit. The approximately vertical line predicted for the year 2508 indicates complete mixing of ^{239}Pu through the bioturbated layer.

activity is given as a time function and is assumed to mix instantaneously through the surface layer. Bioturbation appears in equation (1) both through the diffusion parameter, *D*, and through the excavation parameter, *e*. Within the upper bioturbated part of the soil column of thickness *d*, we set *D* and *e* to be uniform and constant, but below this layer, we set *D* and $e = 0$.

Figure 7a shows the ^{239}Pu soil profile of uncultivated native pasture collected from tropical eastern Australia. Erosion is accommodated by setting *S* to -0.015 mm a^{-1}; *d* is set to 0.30 m. A subsurface peak is seen at 4 cm; below this depth the concentration declines to zero. The solid line shows the model best fit to the data determined using equation (1) and varying the values of *D* and *e* such that $D = 1.0 \times 10^{-5}$ m^2 a^{-1} and $e = 0.004$ a^{-1}. The average soil turnover rate within the bioturbated zone due to excavation is estimated by $W = \delta E/2$ and is found to be 0.06 mm a^{-1}. This rate is at the lower end of values measured for soils (Paton *et al.* 1995; Rodriguez 2006), although it is typical of values for Australian pasture soils (Eldridge & Pickard 1994; Debruyn & Conacher 1994; Paton *et al.* 1995 and references therein). Note that the mixing due to local diffusion results in a timescale for complete mixing within the bioturbated zone of approximately $d^2D^{-1} = 9000$ years, compared to the turnover timescale of around $dW^{-1} = 250$ years. Thus, advective transport dominates soil turnover at this location.

Adopting the best-fit parameter values and limiting *d* to 30 cm allows equation (1) to predict the future shape of this soil profile. Figure 7b shows the subsurface ^{239}Pu peak moving downwards and broadening over upcoming decades with ^{239}Pu becoming detectable at the base of *d* by the year 2108. Continuing a further 400 years into the future (AD 2508) sees ^{239}Pu eventually being mixed throughout the bioturbated layer such that its concentration is approximately constant in that layer. An equilibrium (mixed) soil ^{239}Pu concentration of about 0.008 Bq kg^{-1} (3.5 pg kg^{-1}) is predicted for this north Australian pasture soil (Fig. 7b).

While the equilibrium ^{239}Pu concentration within the soil bioturbation zone and the time frame for achieving this concentration will vary significantly with soil type and geographical location, this model shows that the likely outcome for soils in a non-accreting location is the attainment of a constant concentration of ^{239}Pu within the bioturbated (A-horizon) layer of the soil. Such a situation will occur when the turnover timescale (e^{-1}) is much less than the timescale of erosion ($d/|S|$). The equilibrium ^{239}Pu concentration determined in this example is easily detectable by current analytical methods. Since this illustration is sourced from a latitudinal zone of relatively low Pu fallout (southern hemisphere tropics), it is expected that future concentrations of ^{239}Pu in soils from most other regions of the world would be even easier to detect.

The soil turnover rate in this example is relatively slow, with rates being 1–2 orders of magnitude faster in many other regions (Paton *et al.*

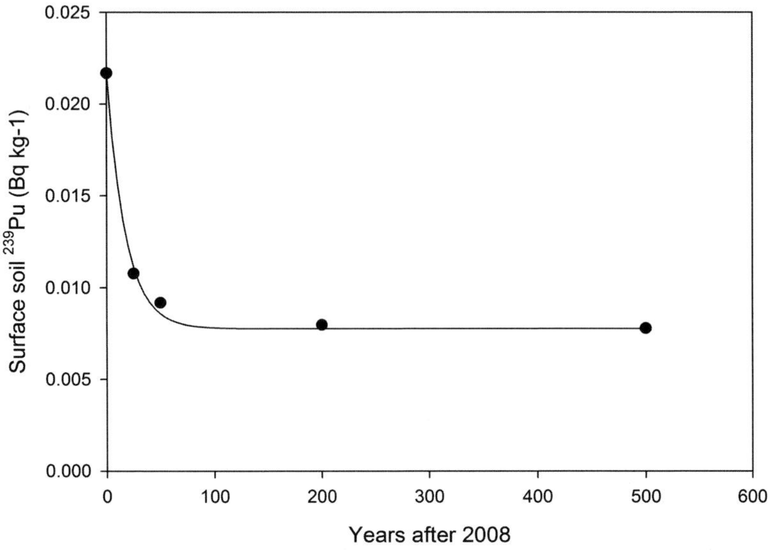

Fig. 8. Predicted ^{239}Pu surface soil concentrations for the upcoming 500 years. The best-fit line to the data is shown.

1995; Rodriguez 2006). Consequently, for many soils the time taken for complete mixing of fallout Pu may be significantly shorter than estimated here, possibly less than 100 years in highly bioturbated soils. The example presented here indicates that approximately constant concentrations of ^{239}Pu would be attained in most surface soils within 500 years. Figure 8 shows the predicted change in the surface soil concentration of ^{239}Pu for the upcoming 500 years. An approximately exponential decline is seen towards the mixed layer concentration. A best-fit line is shown, of the form:

$$C_t = \left((C_0 - C_s)(e^{-bt})\right) + C_s e^{\propto t} \qquad (2)$$

where

$$\propto = \frac{S}{d} - \lambda$$

and C_t is the ^{239}Pu concentration at time t, C_0 is the surface soil concentration in 2008, C_s is the surface soil concentration in the year 2508 (complete mixing), λ is the ^{239}Pu decay constant and b is a constant. For the best-fit line shown in Figure 8, $b = 0.057\ a^{-1}$. Hence, the exponential on the left in equation (2) gives the short-term concentration decline due to bioturbation, and the exponential on the right gives the longer-term decline due to erosion ($S < 0$) and radioactive decay. Equation (2) shows that from the time of complete mixing onwards the ^{239}Pu concentration will be controlled by just radioactive decay and erosion, with both processes diminishing the Pu soil inventory and the Pu soil concentration within the mixed layer.

Figure 9 shows the Lake Victoria profile using equation (2) to predict future ^{239}Pu concentrations of sediment delivered from the East Gippsland catchment using the value of b determined above. Erosion sources are assumed to remain constant. Concentrations of ^{239}Pu are given here rather than the $^{239+240}$Pu data shown earlier due to the differing half-lives of the two isotopes and the greater sensitivity of measurement of ^{239}Pu by MS techniques. C_0 is determined from ^{239}Pu concentrations at the core surface and C_s is scaled according to the C_0/C_s ratio seen in the pasture core. A log scale is used for the vertical axis (years into the future). The predicted profile shows that ^{239}Pu in sediment delivered to the lake is still readily detectable 1000 years into the future, with surficial lake sediment declining to a concentration of 0.032 Bq kg^{-1}, well above the current detection limit by MS techniques (*c.* 0.02 mBq kg^{-1}). In reality, the timescale for further prediction is not relevant to Lake Victoria since the lake will have reached its sediment storage capacity within 1000 years at the current rate of sediment delivery. However, it is instructive to consider the ^{239}Pu concentration tens of thousands of years into the future. As shown in equation (2), the erosion term in the exponent $\propto$ reduces C_t (and the Pu inventory in the soil column) by S/d each year; that is, using the example above, erosion losses akin to the natural soil-forming rate (0.015 mm a^{-1}) will reduce the total amount of ^{239}Pu within the active bioturbated region by 0.015/300, or 0.005% each year.

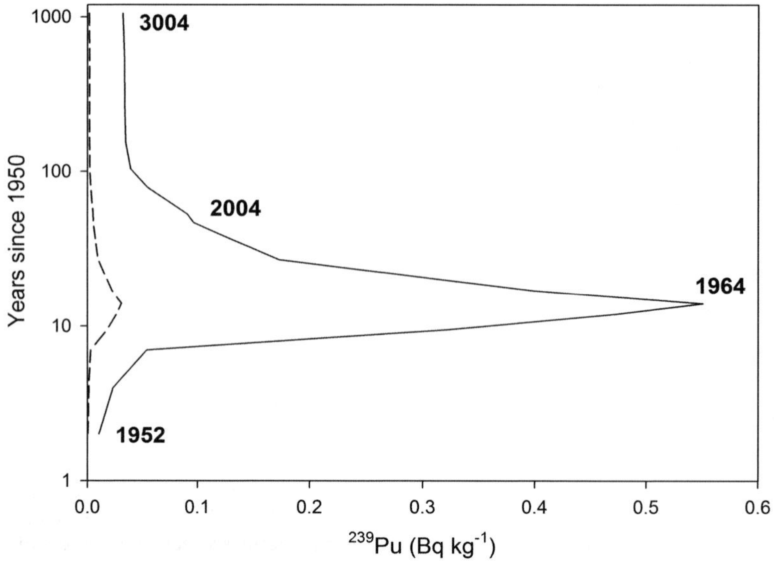

Fig. 9. Modelled future concentrations of ^{239}Pu in Lake Victoria sediment assuming current sedimentation rates (solid line). Note the log scale for the vertical axis. The lower curve (dashed line) shows the same profile 100 000 years into the future, assuming stratigraphic preservation. The ^{239}Pu detection limit is approximately 0.02 mBq kg^{-1}.

Radioactive decay will reduce it by 0.003% each year, with the combined annual reduction being 0.008%, equivalent to a soil half-life for ^{239}Pu of 8800 years. Thus, the equilibrium (year 2508) soil concentration of ^{239}Pu (0.032 Bq kg^{-1}) will remain detectable for about 10 ^{239}Pu soil half-lives, or about 90 000 years. The consequence of this analysis is that some river catchments may continue to deliver measurable Pu to deposited sediment in water bodies for many tens of thousands of years. Clearly these time estimates are location-specific, and subject to the same caveats on geographical location and surface soil erosion rates described above, with soil erosion being the factor most likely to limit the persistence of Pu in soils. For example, a high erosion rate typical of some agricultural practices (*c.* 0.3 mm a^{-1} or 4.0 t ha a^{-1}: Bui *et al.* 2011) would see ^{239}Pu stripped from the surface soil layer such that it would be undetectable within 7000 years.

Owing to the very high fallout occurring during the 1960s, ^{239}Pu in preserved sediment strata will remain detectable for even longer periods. The low concentration curve (dashed line) in Figure 9 shows what the 1000 year sediment profile predicted for Lake Victoria will look like in 100 000 years (about four ^{239}Pu half-lives) assuming that the original sediment strata are preserved and Pu remains immobile. While the onset of testing in the early 1950s will be more difficult to discern using current instruments, Pu input in the late 1950s is still easily detected, and the distinctive 1964 peak remains.

Ocean sources of Pu to sediment

Much of the Pu released by nuclear weapons testing and coastal nuclear facilities resides in the ocean where it is subject to a range of physical and biogeochemical processes. Owing to its unique origins and variety of oxidation states, the oceanic behaviour of Pu has been intensively studied. For example, the particle-reactive nature of Pu(IV) allows the determination of rates of scavenging of dissolved species by suspended particles (Hirose *et al.* 1992, 2001; Lee *et al.* 2003). The known time of delivery of Pu fallout can be used with water-column inventories to provide estimates of the rate of suspended particle sinking through the oceanic water column, and deposition onto the seafloor. Ocean circulation patterns of relatively soluble Pu(V) can be elucidated from known fallout patterns, and the location of isolated Pu inputs and isotopic abundances (e.g. Yamada *et al.* 2007; Lindahl *et al.* 2010; Gastaud *et al.* 2011).

Scavenging of Pu from surface ocean water along the turbid coastal fringes has contributed to the accumulation of Pu in coastal sediment. The NW Pacific region is particularly well studied, with many publications describing the delivery of Pu from central Pacific Ocean water via the Northern

Equatorial Current and the Kuroshio Current to the East China Sea and the Sea of Japan (e.g. Wang & Yamada 2005; Zheng & Yamada 2005, 2006; Yamada *et al.* 2007). High sediment inventories of Pu, well in excess of that expected from direct atmospheric fallout, attest to the role that advective ocean input has played in the supply of Pu to the sediments of these regions. High Pu inventories have also been reported in sediment in other coastal locations well removed from testing sites, including the South Pacific (Hancock *et al.* 2012).

While Pu remains in the surface ocean this water it will continue to be a supply of Pu to coastal marine sediments. Despite losses of Pu to coastal sediments and the floor of the deep ocean (Yamada *et al.* 2007; Yamada & Zheng 2012), much of the Pu delivered to the ocean as global fallout still resides there in both soluble and particulate forms. Recent estimates of the water-column inventories at various locations in the Pacific range from 50 to 90% of the total Pu inventory at that location (Livingstone *et al.* 2001). Estimates of Pu residence time in the ocean are uncertain; Livingstone & Anderson (1983) reported values of approximately 450 years for equatorial oceans, although sediment inventories of Pu did not corroborate this value, being 2–3 times lower than expected (Aarkrog 1998). Based on the reduction of the water inventory of Pu over time, Livingstone *et al.* (2001) reported a half-removal time of 50 years for Pu in the central NW Pacific, although the authors note that this value is probably too low since it is probably affected by mixing in the upper water column as low Pu ocean water is delivered advectively into the region.

Removal of soluble Pu from the ocean water column to sediment occurs via two main pathways: scavenging by relatively turbid near-shore ocean water followed by deposition onto the floor of the coastal margin; and scavenging by biogenically produced particles in the deep ocean. Pu is seen throughout the ocean water column, but a pronounced subsurface concentration maximum is widely reported (e.g. Livingstone *et al.* 2001; Lee *et al.* 2003). This concentration peak has broadened and moved deeper over the decades, and currently lies at a depth of 600–800 m in the central North Pacific and at a similar depth range in the South Pacific (Hirose *et al.* 2011). Hirose *et al.* (1992, 2001) have shown that the concentrations of Pu in Pacific Ocean surface water have decreased exponentially over the last few decades. Some of this Pu removal is associated with adsorption to sinking particles, but dilution by vertical advection from waters with low Pu is also proposed as a factor (Hirose *et al.* 2001; Livingstone *et al.* 2001). Recent data have shown that North Pacific Ocean water is also a likely source of Pu to Southern Hemisphere Oceans via the Indonesian islands (Gastaud *et al.* 2011), raising the likelihood that significant cross-hemisphere export of Pu has occurred.

At present, the existing very low concentrations seen in Pacific Ocean surface water appear to represent a balance between Pu removal processes and supply from upwelling of deeper Pu-rich water and from a discrete North Pacific source, close-in fallout from the Pacific Proving Grounds (PPG) in the Marshall Islands. While most of the atmospheric weapons testing delivered radioactive debris into the stratosphere, an extensive series of ground-level and low-altitude tests were undertaken at the PPG, mainly in the early 1950s. These tests delivered a significant local injection of Pu, much of it as close-in fallout (Aarkrog 1998), and much of it with an isotopic signature quite different to global fallout (Muramatsu *et al.* 2001; Lindahl *et al.* 2011). Thus, Pu fallout and transport in the North Pacific has been the subject of many scientific publications, with two isotopically distinct Pu sources being identified, global fallout and the PPG, the latter continuing to release Pu into the North Equatorial Current. Studies of sediment deposits and coral bands using the isotope ratio $^{240}Pu/^{239}Pu$ show that, since atmospheric weapons testing ceased, between 20 and 60% of Pu deposited in sediment along the continental margins of the western Pacific originated as close-in fallout in the PPG. More recently, dating of coral bands from the NW Pacific Ocean region have indicated that the proportion of PPG Pu supplied to that region could be as high as 70% (Lindahl *et al.* 2012).

Given a constant sediment supply to the coastal zone, it is reasonable to assume that a decline in the dissolved Pu concentration of ocean surface water will lead to a reduction in the Pu concentration seen in coastal sediment deposits. Figure 10 shows the Pu profile of a sediment core collected in 1996 off the east Australian coast. The characteristic 1964 concentration peak and Pu isotope shifts described earlier are clearly seen providing a high-resolution chronology (Hancock *et al.* 2012). Based on the relative concentrations of ^{137}Cs and $^{239+240}Pu$, it is calculated that more than 80% of the Pu delivered to this sediment is ocean-derived. The accumulation at this sedimentation 'hotspot' is around 4 cm a^{-1} and it can be seen that the $^{239+240}Pu$ concentration has decreased by a factor of 2.5 since the cessation of testing, from a concentration of 0.28 Bq kg^{-1} at 90 cm (dated to be the mid-1970s) to 0.11 Bq kg^{-1} at the sediment surface (1996).

Perhaps the best illustration of the temporal variation of Pu in surface ocean water is shown by coral cores. Lindahl *et al.* (2011, 2012) concluded that the Pu content of coral bands showed good

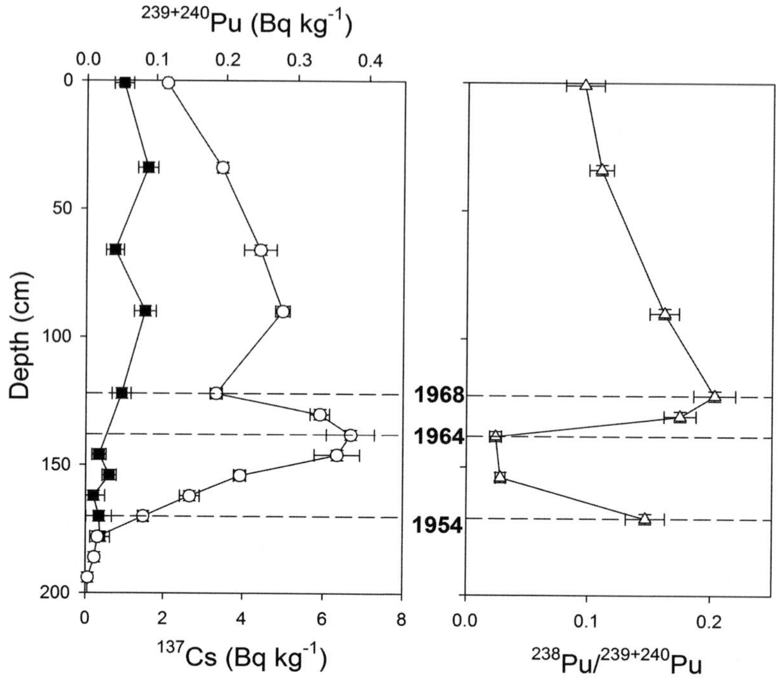

Fig. 10. Dated horizons for a coastal sediment core collected in NE Australia determined from profiles of ^{137}Cs (filled squares), $^{239+240}$Pu (unfilled circles) and (right) the ^{238}Pu/$^{(239+240)}$Pu activity ratio.

correlation with ocean water Pu activity, with coral Pu concentrations of the NW Pacific decreasing by two orders of magnitude since the 1960s. Over the last decade (up to 2007) the coral band concentrations have decreased to the lowest values seen since the 1950s, although they still appear to be up to an order of magnitude higher than prior to anthropogenic input (pre-1945). Given the continued supply of Pu from the PPG, it is hypothesized that the Pu concentration of surface Pacific Ocean water will decline to low but finite values over the coming decades and, perhaps, centuries. For example, it is estimated that about 90 TBq of $^{239+240}$Pu remains in lagoon sediment of the PPG test sites at Enewetak and Bikini Atoll, and that about 0.2 TBq is released to the ocean each year (Noshkin *et al.* 1987; Hamilton *et al.* 1996; Hamilton 2005). Based on these data, it is calculated that the PPG could provide a source of Pu to the Pacific Ocean well into this millennium.

Sediment in the eastern Irish Sea is another point source that is likely to continue to supply Pu for decades or longer. Much of the Pu discharged from Sellafield is now concentrated in a subtidal mud-patch of the eastern Irish Sea, and this mud reservoir acts as a continual supply of Pu to regional coastal sediments and the North Atlantic (Kershaw *et al.* 1999*b*). It is estimated that 435 TBq of $^{239+240}$Pu is held in the subtidal sediments (Kershaw *et al.* 1999*a*), and that dissolution of this Pu is resulting in the annual loss of about 0.6 TBq through the North Channel to the Arctic and North Atlantic. Again, barring significant changes to the direction of in currents in the Irish Sea, these data indicate that supply of Pu from this point source to the ocean is likely to continue for centuries.

Pu has been measured in sediment of the deep Pacific Ocean (Livingstone *et al.* 2001). High variability is seen in the sediment inventories reflecting PPG input and ocean current movement. Enhanced Pu removal in the higher productivity waters of the Equator is another likely cause of variability. Obviously, as long as dissolved Pu exists in the ocean, the potential exists for Pu to be scavenged and deposited in regions of the deep ocean where sediment accumulates. Owing to the low flux of particles, Pu concentrations may be higher in these sediments compared to the continental margins, and detectable concentrations may therefore persist longer.

Conclusions

Release of radioactivity from the detonation of atmospheric nuclear weapons during the period

1945–1980 has supplied a source of anthropogenic radionuclides to the surface of the entire globe. Of these radionuclides, Pu isotopes (^{239}Pu and ^{240}Pu) will provide the best long-term (>100 years) signal of the onset of the atomic age in natural sedimentary, ice and coral archives. This is due to their high sensitivity of measurement, extremely low pre-1945 concentrations and long half-lives. Owing to the longer-lived nature of ^{239}Pu compared to ^{240}Pu (*c.* 4 times) and its greater abundance, ^{239}Pu measured by mass spectrometric techniques is predicted to be the chronometer of choice into the distant future.

The future shape of sediment profiles of Pu will depend on the future Pu concentrations of surface soil being eroded and delivered to the sites of sediment deposition. Concentrations of the ^{239}Pu in surface soils experiencing low (natural) rates of erosion will remain detectable by mass spectrometric techniques for millenia and water catchments will therefore continue to deliver Pu to zones of sediment deposition for this period of time. The time frame for the continued detection of Pu in soils where erosion rates are high will be much lower, perhaps less than a few thousand years.

It is likely that the continued supply of ocean-derived Pu to near-shore and deep ocean sediment will occur over a shorter time frame than that for terrestrial Pu sources. Owing to the complexities controlling the transport and fate of Pu in the ocean, we have not attempted to model future Pu concentrations of ocean water. However, as Pacific and Atlantic Ocean water is mixed globally, as global fallout Pu is removed to marine sediments, and as the supply of Pu from the PPG and the eastern Irish Sea diminish, the long-term ^{239}Pu concentration of surface ocean water can be expected to continue to decline. While the length of time before Pu becomes undetectable in accreting coral bands and marine sediments is not known, it is likely to be centuries into the future.

The persistence of ^{239}Pu in sediments will depend on factors such as soil Pu concentration (the global position and extent of fallout), and the size and definition of the 1960s fallout peak. Provided that sediment strata are preserved, it is likely that in many regions of the globe the signal of anthropogenic ^{239}Pu will be detectable for 100 000 years.

References

AARKROG, A. 1998. A retrospect of anthropogenic radioactivity in the global marine environment. *Radiation Protection Dosimetry*, **75**, 23–31.

APPLEBY, P. G., RICHARDSON, N. & NOLAN, P. J. 1991. ^{241}Am dating of sediments. *Hydrobiologia*, **214**, 35–42.

BEASLEY, T. M., CARPENTER, R. & JENNINGS, C. D. 1982. Plutonium, ^{241}Am and ^{137}Cs ratios, inventories and vertical profiles in Washington and Oregon continental shelf sediments. *Geochimica et Cosmochimica Acta*, **46**, 1931–1946.

BOUDREAU, B. P. 1986 Mathematics of tracer mixing in sediments II. Nonlocal mixing and biological conveyor-belt phenomena. *American Journal of Science* **286**, 199–238.

BUI, E. N., HANCOCK, G. J. & WILKINSON, S. N. 2011. 'Tolerable' hillslope soil erosion rates in Australia: linking science and policy. *Agriculture, Ecosystems and Environment*, **144**, 136–149.

CHRISTENSEN, E. R. 1982. A model for radionuclides in sediments influenced by mixing and compaction. *Journal of Geophysical Research*, **87**, 566–572.

CHRISTENSEN, G. C., BERGAN, T. D. S., BERGE, D. & BAEKKEN, T. 1998. A comparison of sea water and fresh water in a study of sediment-water exchange of radionuclides. *Radiation Protection Dosimetry*, **75**, 107–109.

DEBRUYN, L. A. L. & CONACHER, A. J. 1994. The bioturbation activity of ants in agricultural and naturally vegetated habitats in semi-arid environments. *Australian Journal of Soil Research*, **32**, 555–570.

ELDRIDGE, D. J. & PICKARD, J. 1994. Effects of ants on sandy soils in semi-arid eastern Australia: II. Relocation of nest entrances and consequences for bioturbation. *Australian Journal Soil Research*, **32**, 323–333.

FIFIELD, L. K. 2008. Accelerator mass spectrometry of the actinides. *Quaternary Geochronology*, **3**, 276–290.

GASTAUD, J., POVINEC, P. P. ET AL. 2011. Transport and scavenging of Pu in surface waters of the Southern Hemisphere Oceans. *In*: AOYAMA, M., POVINEC, P. P. & SANCHEZ-CABEZA, J.-A. (eds) *The Southern Hemisphere Ocean Tracer Studies. Progress in Oceanography*, **89**, 92–100.

HAMILTON, T. F. 2005. Linking legacies of the cold war to arrival of anthropogenic radionuclides in the oceans through the 20th century. *In*: LIVINGSTONE, H. D. (ed.) *Marine Radioactivity, Radioactivity in the Environment*. Elsevier, Amsterdam, 23–28.

HAMILTON, T. F., MILLIES-LACROIX, J. C. & HONG, G. H. 1996. ^{137}Cs (^{90}Sr) and Pu isotopes in the Pacific Ocean: sources and trends. *In*: GUEUENIAT, P., GERMAINE, P. & METIVIER, H. (eds) *Radionuclides in the Oceans Inputs and Inventories*. Institut de Protection et de Sureté Nucléaire (IPSN), Les Ulis, France, 29–58.

HANCOCK, G., EDGINGTON, D. N., ROBBINS, J. A., SMITH, J. N., BRUNSKILL, G. & PFITZNER, J. 2002. Workshop on radiological techniques in sedimentation studies: methods and applications. *In*: FERNANDEZ, J. M. & FICHEZ, R. (eds) *Environmental Changes and Radioactive Tracers*. IRD editions, Paris, 233–251.

HANCOCK, G. J., LESLIE, C., EVERETT, S. E., TIMS, S. G., BRUNSKILL, G. J. & HAESE, R. 2012. Plutonium as a chronomarker in Australian and New Zealand sediments: a comparison with ^{137}Cs. *Journal of Environmental Radioactivity*, **102**, 919–929, http://dx.doi.org/10.1016/j.jenvrad.2009.09.008

HE, Q. & WALLING, D. E. 1997. The distribution of fallout ^{137}C and ^{210}Pb in undisturbed and cultivated soils. *Applied Radiation and Isotopes*, **48**, 677–690.

Hirose, K., Sugumura, Y. & Aoyama, M. 1992. Plutonium and ^{137}Cs in the Western North Pacific: estimation of residence time of plutonium in surface waters. *Applied Radiation and Isotopes*, **43**, 349–359.

Hirose, K., Aoyama, M., Miyao, T. & Igarashi, Y. 2001. Plutonium in seawaters of the western North Pacific 2001. *Journal of Radioanalytical and Nuclear Chemistry*, **248**, 771–776.

Hirose, K., Kim, C. S. et al. 2011. Vertical profiles of plutonium in the central South Pacific. *Progress in Oceanography*, **89**, 101–107.

Jaakkola, T., Tolonen, K., Huttunen, P. & Leskinen, S. 1983. The use of fallout ^{137}Cs and $^{239,240}Pu$ for dating of lake sediments. *Hydrobiologia*, **103**, 15–19.

Kershaw, P. J., Denoon, D. C. & Woodhead, D. S. 1999*a*. Observations on the redistribution of plutonium and americium in the Irish Seas sediments, 1978 to1996: concentrations and inventories. *Journal of Environmental Radioactivity*, **44**, 191–221.

Kershaw, P. J., McCubbin, D. & Leonard, K. S. 1999*b*. Continuing contamination of north Atlantic and Arctic waters by Sellafield radionuclides. *Science of the Total Environment*, **237/238**, 119–132.

Koide, M., Goldberg, E. D., Herron, M. M. & Langway, C. C., Jr. 1977. Transuranic depositional history in South Greenland firn layers. *Nature*, **269**, 137–139.

Koide, M., Michel, R. & Goldberg, E. D. 1979. Depositional history of artificial radionuclides in the Ross Ice Shelf, Antarctica. *Earth and Planetary Science Letters*, **44**, 205–223.

Koide, M., Michel, R., Goldberg, E. D., Herron, M. M. & Langway, C. C., Jr. 1982. Characterisation of radioactive fallout from pre- and post-moratorium tests to polar ice caps. *Nature*, **296**, 544–547.

Koide, M., Bertine, K. K., Chow, T. J. & Goldberg, E. D. 1985. The $^{240}Pu/^{239}Pu$ ratio, a potential chronometer. *Earth and Planetary Science Letters*, **72**, 1–8.

Krey, P. W., Heit, M., Liningston, H. D. & Miller, K. M. 1990. History of Plutonium composition of fallout in the northeastern U.S. from contemporary measumrents. *Journal of Radioanalytical and Nuclear Chemistry Articles*, **138**, 385–406.

Ketterer, M. E. & Szechenyi, S. C. 2008. Determination of plutonium and other transuranic elements by inductively-coupled mass spectrometry: a historical perspective and new frontiers in the environmental sciences. *Spectrochimica Acta Part B – Atomic Spectroscopy*, **63**, 719–737.

Lee, S.-H., Gastaud, J. et al. 2003. Distribution of plutonium and americium in the marginal seas of the northwest Pacific Ocean. *Deep-Sea Research II*, **50**, 2727–2750.

Leslie, C. & Hancock, G. J. 2008. Estimating the date corresponding to the horizon of the first detection of ^{137}Cs and $^{239+240}Pu$ in sediment cores. *Journal of Environmental Radioactivity*, **99**, 483–490.

Lindahl, P., Lee, S-H., Worsfold, P. & Keith-Roach, M. 2010. Plutonium isotopes as tracers of ocean processes: a review. *Marine Environmental Research*, **69**, 73–84.

Lindahl, P., Worsfold, P. et al. 2011. Temporal record of Pu isotopes in the inter-tidal sediments from the northeastern Irish Sea. *Science of the Total Environment*, **409**, 5020–5025.

Lindahl, P., Andersen, M. B., Keith-Roach, M., Worsfold, P., Hyeong, K., Choi, M.-S. & Lee, S.-H. 2012. Spatial and temporal distribution of Pu in the northwest Pacific Ocean using modern coral archives. *Environment International*, **40**, 196–201.

Livingstone, H. D. & Anderson, R. F. 1983. Large particle transport of plutonium and other fallout radionuclides to the deep ocean. *Nature*, **303**, 228–231.

Livingstone, H. D., Povinec, P. P., Ito, T. & Togawa, O. 2001. The behaviour of plutonium in the Pacific Ocean. *In*: Kudo, A. (ed.) *Plutonium in the Environment*. Elsevier, Amsterdam, 267–292.

Muramatsu, Y., Hamilton, T., Uchida, S., Tagami, K., Yoshida, S. & Robison, W. 2001. Measurement of $^{240}Pu/^{239}Pu$ isotopic ratios in soils from the Marshall Islands using ICP-MS. *Science of the Total Environment*, **278**, 151–159.

Noshkin, V. E., Wong, K. M., Jokela, T. A., Brunk, J. L. & Eagle, R. J. 1987. Plutonium and americium behaviour in coral atoll environments. *In*: Conner, T. P., Burt, C. W. & Duedall, L. A. (eds) *Oceanic Processes in Marine Pollution, Volume 2. Physicochemical Processes and Wastes in the Ocean*. Krieger, Malabar, FL, 159–174.

Paton, T. R., Humphreys, G. S. & Mitchell, P. B. 1995. *Soils. A New Global View*. UCL Press, London.

Raeder, S., Hakimi, A., Stöbener, N., Trautmann, N. & Wendt, K. 2012. Detection of plutonium isotopes at lowest quantities using in-source resonance ionization mass spectrometry. *Analytical Bioanalytical Chemistry*, **404**, 2163–2172.

Ritchie, J. C. & McHenry, J. R. 1990. Application of radioactive fallout Cesium-137 for measuring soil erosion and sediment accumulation rates and patterns: a review. *Journal of Environmental Quality*, **19**, 215–233.

Roache, P. J. 1982. *Computational Fluid Dynamics*. Hermosa, Albuquerque, NM.

Rodriguez, M. D. 2006. *The Bioturbation and transport of chemicals in surface soils*. MSc thesis, Louisiana State University.

Sholkovitz, E. R. & Mann, D. R. 1984. The pore water chemistry of $^{239,240}Pu$ and ^{137}Cs in sediments of Buzzards Bay, Massachusetts. *Geochimica et Cosmochimica Acta*, **48**, 1107–1114.

Smith, J. N., Ellis, K. M. & Nelson, D. N. 1987. Time-dependent modeling of fallout radionuclide transport in a drainage basin: significance of 'slow' erosional and 'fast' hydrological components. *Chemical Geology*, **63**, 157–180.

Taylor, D. M. 2001. Environmental plutonium – creation of the universe to twenty-first century mankind. *In*: Kudo, A. (ed.) *Plutonium in the Environment*. Elsevier, Amsterdam, 1–14.

Tims, S. G., Everett, S. E., Fifield, L. K., Hancock, G. J. & Bartley, R. 2010*a*. Plutonium as a tracer of soil and sediment movement in the Herbert River, Australia. *Nuclear Instruments and Methods in Physics Research B*, **268**, 1150–1554.

Tims, S. G., Pan, S. M., Zhang, Z., Fifield, L. K., Wang, Y. P. & Gao, J. H. 2010*b*. Plutonium AMS measurements in Yangtze River estuary sediment. *Nuclear*

Instruments and Methods in Physics Research B, **268**, 1155–1558.

UNSCEAR 2000. *Sources and Effects of Ionizing Radiation, United Nations Scientific Committee on the Effects Atomic Radiation Exposure from Man-made Sources of Radiation, Volume 1*. Report to the General Assembly with Scientific annexes, Annex C. http://www.unscear.org/unscear/en/publications/2000_1.html

Vanderzalm, J. L., Dillon, P. J., Hancock, G. J., Leslie, C., Dighton, J., Smith, J. C. & Pearce, G. 2013. Using elemental profiles in the sediment of a lake used to supply drinking water to understand the impacts of urban stormwater recharge. *Marine and Freshwater Research*, **64**, 493–506, http://dx.doi.org/10.1071/MF12215

Wallbrink, P. J., Murray, A. S. & Olley, J. M. 1998. Determining sources and transit times of suspended sediment in the Murrumbidgee River, New South Wales, Australia, using fallout ^{137}Cs and ^{210}Pb. *Water Resources Research*, **34**, 879–887.

Wallbrink, P. J., Murray, A. S. & Olley, J. M. 1999. Relating suspended sediment to its original soil depth using fallout radionuclides. *Soil Science Society of America Journal*, **63**, 369–378.

Walling, D. E. & He, Q. 1992. Interpretation of caesium-137 profiles in lacustrine and other sediments: the role of catchment-derived inputs. *Hydrobiologia*, **235/236**, 219–230.

Walling, D. E., He, Q. & Quine, T. A. 1995. *Use of Caesium-137 and Lead-210 as Tracers of Soil Erosion Investigations*. IAHS Publication, **229**. IAHS Press, Wallingford.

Wang, Z.-L. & Yamada, M. 2005. Plutonium activities and $^{240}Pu/^{239}Pu$ atom ratios in sediment cores from the East China Sea and Okinawa Trough: sources and inventories. *Earth and Planetary Science Letters*, **233**, 441–453.

Wyse, E. J., Lee, S. H., Rosa, J. L., Povinec, P. & Mora, S. J. D. 2001. ICP sector field mass spectrometry analysis of plutonium isotopes: recognizing and resolving potential interferences. *Journal of Analytical Atomic Spectrometry*, **16**, 1107–1111.

Yamada, M. & Zheng, J. 2012. ^{239}Pu and ^{240}Pu inventories and $^{240}Pu/^{239}Pu$ atom ratios in the equatorial Pacific Ocean water column. *Science of the Total Environment*, **430**, 20–27.

Yamada, M., Zheng, J. & Wang, Z.-L. 2007. $^{240}Pu/^{239}Pu$ atom ratios in seawater from Sagami Bay, western Northwest Pacific Ocean: sources and scavenging. *Journal of Environmental Radioactivity*, **98**, 274–294.

Zapata, F. 2003. The use of environmental radionuclides as tracers in soil erosion and sedimentation investigations: recent advances and future developments. *Soil and Tillage Research*, **69**, 3–13.

Zheng, J. & Yamada, M. 2005. Vertical distributions of $^{239+240}Pu$ activities and $^{240}Pu/^{239}Pu$ atom ratios in sediment cores: implications for the sources of Pu in the Japan Sea. *Science of the Total Environment*, **340**, 199–211.

Zheng, J. & Yamada, M. 2006. Plutonium isotopes in settling particles: transport and scavenging of Pu in the Western Northwest Pacific. *Environmental Science and Technology*, **40**, 4103–4108.

Volcanic markers for dating the onset of the Anthropocene

VICTORIA C. SMITH

Research Laboratory for Archaeology and the History of Art, University of Oxford, Dyson Perrins Building, South Parks Road, Oxford OX1 3QY, UK
(e-mail: victoria.smith@rlaha.ox.ac.uk)

Abstract: Volcanic ash is dispersed over thousands of kilometres during large-scale eruptions, forming sedimentary layers. These ash (tephra) deposits are increasingly being used as unique marker layers in a variety of sedimentary archives including ice cores, and terrestrial and marine records. Tephra dispersed during large explosive eruptions that coincide with the defined beginning of the Anthropocene could therefore be used to help identify this event in various archives, and assess the relative spatial differences in marked anthropogenic change. The 1815 eruption of Tambora, Indonesia, was the largest in historical time and occurred in the middle of Europe's Industrial Revolution. Volatile emissions injected into the atmosphere during this eruption caused widespread effects including the 'year without a summer' during which there were anomalously cooler temperatures recorded across much of North America and Europe. Sulphate aerosols associated with the eruption were dispersed by stratospheric and tropospheric winds across the entire globe. Deposits of these are clearly recorded in the Earth's key palaeoclimatic records: polar ice cores. Significantly, the Tambora eruption occurred immediately prior to substantial increases in greenhouse gases, a defining feature of the Anthropocene.

The onset of an Epoch cannot be defined by palaeoclimatic change alone, as not only do the different components of the system (ocean and atmosphere) respond at different rates, but the drivers for climate change are not spatially synchronous. Therefore, it is important that the Global Stratotype Section and Point (GSSP) for the Anthropocene be accurately dated. The study of tephra may provide a robust chronology and marker layer. The deposits of very large explosive eruptions are found across the globe as layers of ash (tephra) or sulphate aerosol, and occasionally both. The volcanic glass shards found within the tephra layers are often found to be compositionally unique (e.g. Shane 2000). This allows for robust correlation over large areas and hence the synchronization of environmental archives, leading to the identification of spatial leads and lags in palaeoclimate. Furthermore, volcanic deposits can be dated using radiometric methods and have been used to constrain the ages of GSSPs, such as the base of the Triassic at Meishan (Bowring *et al.* 1998) and the base of the Jurassic (Hettangian) at Kunga Island (Pálfy *et al.* 2000). Accordingly, tephra may comprise ideal marker layers for defining the onset of the Anthropocene, in particular because they are found in a variety of records, from terrestrial and marine sequences to ice cores from high altitudes and the poles.

Explosive volcanic eruptions

The volcanic explosivity index (VEI) is typically used to express the magnitude of an eruption (Newhall & Self 1982). The scale for this index is logarithmic, with the largest eruptions on Earth having a VEI of 8. Small eruptions, those with a VEI less than 4, erupt $<0.1\ km^3$ of dense rock equivalent (DRE) magma during strombolian and vulcanian activity, and typically only generate small eruption columns (<15 km high). Eruptions with VEI of 5 or greater have sustained Plinian columns that extend above 25 km, ejecting more than $1\ km^3$ of DRE magma and generating fine material that is dispersed hundreds of kilometres from the vent (Newhall & Self 1982). For example, the 1980 VEI 5 eruption of Mount St Helens in the USA deposited more than 1 cm of ash 500 km downwind of the volcano (Sarna-Wojcicki *et al.* 1981). The most powerful VEI 8 events erupt volumes of more than $300\ km^3$ DRE magma and disperse ash thousands of kilometres from the vent. For example, the 75 ka Toba eruption in Sumatra, Indonesia (Mark *et al.* 2013), erupted more than $2500\ km^3$ DRE magma, dispersing 5 cm of ash over an area extending more than 2500 km from the volcano (Fig. 1; Matthews *et al.* 2011).

Identification and characterization

Tephra forms both visible and non-visible (cryptotephra) layers in sedimentary sequences. The thickness of the layers is greatest close to the source and they become exponentially thinner towards distal regions (Thorarinsson 1954). The deposits closest (proximal) to the volcano (up to several metres thick) comprise fine ash and clasts ranging

From: Waters, C. N., Zalasiewicz, J. A., Williams, M., Ellis, M. A. & Snelling, A. M. (eds) 2014. *A Stratigraphical Basis for the Anthropocene*. Geological Society, London, Special Publications, **395**, 283–299.
First published online November 21, 2013, http://dx.doi.org/10.1144/SP395.11

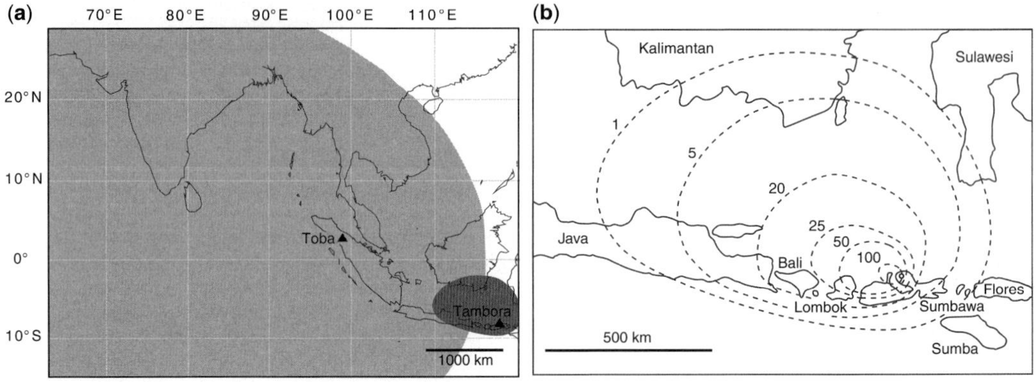

Fig. 1. (**a**) Area covered by at least 1 cm of volcanic ash during the 75 ka Toba (VEI 8) supereruption (light grey; modelled distribution from Matthews *et al.* 2011) compared with that of the 1815 Tambora (VEI 7) eruption (dark grey; from Self *et al.* 1984). (**b**) Dispersal and thickness (from Self *et al.* 1984) of the Tambora tephra deposit in centimetres (figure modified from Gertisser *et al.* 2012).

from a few millimetres to centimetres in diameter, which also tend to be vesicular (e.g. pumice) (Fig. 2). These deposits comprise volcanic glass (usually >75% of the volume) and crystals that include feldspars, pyroxenes, amphiboles, Fe–Ti oxides and other phases. The ratio of glass to crystals increases with distance from the vent, as the crystal phases are denser and are therefore deposited closer to the source. In distal sediments the tephra layers are often invisible to the naked eye, and are only identified by continuous sampling and extraction by density separation methods (Merkt *et al.* 1993; Turney 1998; Blockley *et al.* 2005). The glass shards (typically *c.* 10–80 μm in diameter) are identified and counted using a microscope, and peak concentrations in the sediments are then referred to as a cryptotephra (commonly also called microtephra). The development of cryptotephra identification methods has therefore increased the geographical scope for tephrochronology. Cryptotephras are increasingly being used to correlate palaeoenvironmental and archaeological archives, even on a continental scale, providing both absolute and relative chronology (e.g. Lane *et al.* 2011; Lowe *et al.* 2012).

Tephra layers are correlated to each other and to specific eruptions using the composition of the volcanic glass shards, which is typically unique for a given eruption, providing an effective chemical 'fingerprint' (Shane 2000; Lowe 2011). However, successive eruption deposits from the same volcano can be compositionally similar over thousands of years. For example, similarities in major element glass compositions have been observed in the consecutive units from Campi Flegrei caldera, Italy (see Smith *et al.* 2011*a*), the large eruptions from Toba caldera, Indonesia (Smith *et al.* 2011*b*), and many widespread tephra from Iceland (Lane *et al.* 2012). However, some of the consecutive units have distinct trace element compositions (e.g. Tomlinson *et al.* 2012*a*) or distinct crystal phase compositions (Shane 1998; Smith *et al.* 2011*b*), or both, reconciling the unique identification of these tephra layers.

The composition of individual glass shards and minerals may be characterized using an electron microprobe (EMP) and laser ablation inductively coupled mass spectrometry (LA-ICPMS). Bulk analytical techniques, such as X-ray fluorescence, cannot be used for tephrochronology as they only provide an average composition of a volume (micrograms) of sample. Because the ratio of glass and crystals changes with distance from the vent, distal and proximal tephra will have different bulk compositions. Furthermore, using the same bulk analyses on glass separates can still be misleading, as the composition of glass shards within a tephra deposit may not be completely homogeneous (e.g. Shane *et al.* 2008; Tomlinson *et al.* 2012*b*). Thus, analyses of individual glass shards are required for reliable correlation. In addition, one must identify whether or not the tephra layer has undergone any post-depositional contamination, as this will also affect the bulk chemical composition.

Major and minor oxide compositions such as SiO_2, CaO, FeO, K_2O, Na_2O, Al_2O_3, MgO, MnO, TiO_2, P_2O_5 and Cl are routinely determined for individual glass shards using an EMP. However, the glass is damaged by the electron beam and it loses mobile elements like sodium (Miotello & Mazzoldi 1982), so instrument setup and operating conditions must be carefully selected to obtain accurate data. The best analytical protocols are outlined in detail in Kuehn *et al.* (2011).

Fig. 2. (**a**) Thick proximal pyroclastic fall and flow (pyroclastic density current) deposits of the Minoan eruption on Santorini, Greece (people for scale). (**b**) A proximal tephra fallout deposit comprising pumice on Ulleungdo, South Korea (scale bar in centimetres). (**c**) Photograph and (**d**) scanning electron microscope image of distal tephra layers in sedimentary cores. These thin ash layers predominantly comprise glass shards and the occasional mineral phase. (**e**) A tephra layer of one of the largest eruptions in the Holocene, the *c.* 7.2 ka K–Ah eruption (VEI 7) from Kyushu, Japan (Machida 1999; Smith *et al.* 2013), in an outcrop a few hundred kilometres from the source.

Advances in mass spectrometry methods now allow for small single shards (<40 μm) to be routinely analysed on conventional LA-ICPMS instruments (Tomlinson *et al.* 2010; Pearce *et al.* 2011), giving a full suite of trace element (including rare earth elements) analyses. These analyses of individual glass shards replace bulk trace element analysis by instrumental neutron activation, which has been used to correlate tephra layers in the past (e.g. Sarna-Wojcicki *et al.* 1997). The incompatible trace elements are particularly useful for discriminating eruption units from the same volcano, especially the high field strength elements (HFSE) such as U, Th, Nb, Ta, Zr and Y (e.g. Tomlinson *et al.* 2012*a*).

Using volcanic markers for chronology

Tephra layers offer relative and absolute chronology provided they are distinct and can be unequivocally correlated to the volcanic deposits near the source volcano (proximal) or others in medial to distal settings.

Absolute chronology

The most commonly used methods for dating volcanic eruptions are radiocarbon and $^{40}Ar/^{39}Ar$. However, it is rare that there is sufficient appropriate material to date in the distal units, so they typically need to be correlated to the coarser eruption deposits found closer to the source volcano. The potassium-rich crystals typically found in proximal deposits, such as biotite and sanidine, can be extracted and dated using $^{40}Ar/^{39}Ar$ methods (e.g. Smith *et al.* 2011*c*; Mark *et al.* 2013). ^{40}K decays to ^{40}Ca and ^{40}Ar with a long half-life of 1.248×10^9 years, allowing $^{40}Ar/^{39}Ar$ methods to be used to date eruptions from *c.* 10 ka (provided that there are K-rich phases) through to some of the oldest volcanic deposits preserved on Earth. Eruption deposits containing charred material can be dated using radiocarbon methods. Carbonized material is commonly found in proximal eruption deposits (e.g. Di Vito *et al.* 1999) and especially those of pyroclastic density currents (PDC) as they are hot when they flow over the landscape and burn vegetation. However, the short half-life of ^{14}C (5730 years) means that radiocarbon methods can only be used to date eruptions within the last 50 kyr.

The age of an eruption can also be constrained if the tephra is preserved in a well-dated sedimentary sequence or ice core. For example, the SG06 lake core from Lake Suigetsu, Japan, has a very detailed chronology, with over 600 radiocarbon dates from terrestrial macrofossils. These dates have been modelled with layer counts of the varved sediment sequence to provide a robust age–depth model that extends back to the radiocarbon limit (Bronk-Ramsey *et al.* 2012). This high-resolution chronology provides very accurate and precise dates for all the post-50 ka tephra units within this SG06 archive, many of them being widespread tephra markers from large volcanic eruptions in Japan (Smith *et al.* 2013).

Relative chronology

Correlating the tephra layers in different archives provides a relative chronology, effectively synchronizing these records. Tephra from moderate to large eruptions may be used for correlation on the regional scale, where correlation on the continental scale can only be achieved by identifying cryptotephra from some of the largest eruptions on Earth, for example the *c.* 75 ka Toba eruption (Matthews *et al.* 2011; Lane *et al.* 2013). Tephra layers also allow other sedimentary sequences, such as archaeological sites, to be tied to environmental archives, which is important for assessing the relationship between cultural changes and palaeoclimate (e.g. Lowe *et al.* 2012). Synchronizing palaeoenvironmental records using marker layers allows spatial variations in palaeoclimate to be assessed, providing information on leads and lags (e.g. Lane *et al.* 2011). Setting the onset of the Anthropocene to coincide with a large-scale volcanic eruption would conveniently allow the timing of anthropogenic change around the globe to be assessed.

Assessing or defining the onset of the Anthropocene using tephra layers

There have been numerous large eruptions in the last 2 kyr with a VEI of 6 or higher (Newhall & Self 1982) – appropriate candidates for marking the onset of the Anthropocene. Many had Plinian phases with >25 km-high eruption columns, erupted more than *c.* 10 km^3 of tephra, and produced regional tephra marker layers (listed in Table 1). The dates of recent eruptions are accurately recorded in historic texts, but those of older events are typically poorly constrained. For example, the radiocarbon dates of eruptions in the first Millennium have often been reported as uncalibrated dates, giving inaccurate ages and artificially high levels of precision. Table 1 presents some newly calibrated and modelled radiocarbon dates, based on published data, for some of these prehistoric eruptions. The IntCal09 calibration curve (Reimer *et al.* 2009) was used, and modelling was performed using Bayesian statistics implemented within OxCal 4.2 (Bronk-Ramsey 2009) in order to produce the accurate age ranges (95.4% probability, *c.* 2σ, is reported throughout). The largest eruptions in the last 2 kyr, all listed in Table 1, include: the AD 232 ± 5 Taupo eruption from Taupo caldera, New Zealand (Wilson & Walker 1985; Hogg *et al.* 2012), the Millennium eruption of Baitoushan (also known as Tianchi, Baekdusan or Changbaishan) on the China and North Korea border (AD 921–941; Machida 1999; Yin *et al.* 2012), the AD 1366–1484 eruption of Kuwae, Vanuatu (Monzier *et al.* 1994), the AD 1600 eruption of Huaynaputina, Peru (de Silva & Zielinski 1998), the 1815 eruption of Tambora, Indonesia (Self *et al.* 1984), the 1883 eruption of Krakatau, Indonesia (Self & Rampino 1981), the 1912 Katmai Novarupta (Valley of Ten Thousand Smokes) eruption, Alaska, USA (Hildreth 1983), and the 1991 eruption of Mt Pinatubo, Philippines (McCormick *et al.* 1995). Ideal markers are ones that occur in key archives that record anthropogenic changes, such as the polar ice cores. Unfortunately, all the large eruptions in the last 2 kyr would have only dispersed glass shards to a maximum of a couple of thousand kilometres from the vent and into regional palaeoenvironmental archives. However, sulphate peaks are found in the ice cores that may be related to specific volcanic eruptions (Zielinski 1995; Sigl *et al.* 2013).

Sulphate peaks in ice cores

Volatile species such as H_2O, CO_2, Cl, F and S (as H_2S and SO_2) dissolved in magma are released during ascent, making the melt buoyant and increasing the overpressure, eventually resulting in eruption. During large explosive eruptions these gases are injected into both the troposphere (10–13 km above the Earth's surface) and the stratosphere (between *c.* 13 and 50 km), and are transported around the globe by the respective winds. Sulphur dioxide reacts with hydroxide (OH^-) in the atmosphere (generated by the breakdown of water vapour in sunlight) to form sulphuric acid (H_2SO_4), referred to as sulphate aerosol. Droplets of the aerosol scatter all radiation with a wavelength less than a couple of micrometres, which includes visible light. The amount of solar radiation reaching the lower atmosphere is therefore reduced. The longer wavelengths are scattered at a higher angle, resulting in the unusually red sunsets that have often been observed after large eruptions and recorded in historical texts and paintings. The lifetime of H_2SO_4 in the troposphere is on the order of days as it is readily removed by precipitation. In contrast, sulphate aerosols in the stratosphere remain suspended, gradually settling into the troposphere over a period of years, with *c.* 35% lost every *c.* 8–10 months (Grainger & Highwood 2003). The presence of the sulphate aerosol results in heating of the stratosphere, associated with the absorption of long-wave radiation, and a cooling of the troposphere as the scattering reduces the amount of radiation making it to the surface of the Earth. For example, the sulphate aerosol generated from the volatiles injected into the atmosphere during the *c.* 9 km^3 DRE, VEI 6, 1991 eruption of Pinatubo in the Philippines resulted in a 1 °C increase in the stratospheric temperature that lasted *c.* 2 years (Bekki *et al.* 1996; Robock 2000). The temperature of the lower troposphere, and global surface temperature, decreased by 0.5 °C and lasted for a year after the eruption (McCormick *et al.* 1995).

The sulphate aerosol settles out of the atmosphere and is deposited on the surface, recorded as peaks in the sulphate signature of ice cores. The peaks rise above a background largely attributed to sea salt, starting just after the large volcanic event and lasting 1 to 2 years. Volcanic aerosol deposits in ice cores may also be identified by electrical conductivity measurements (ECM), which are collected continuously along the cores and reflect the acidity of the ice. However, like direct sulphate measurements, ECM values are affected by a large background associated with the sulphuric acid generated from dimethyl sulphide produced by marine algae and phytoplankton. Therefore, only large volcanic eruptions generate enough aerosols to be clearly identified in ice-core sulphate and ECM records. These aerosol signatures do not have a distinct geochemical fingerprint that can link them to a particular eruption or volcanic centre (Lemieux-Dudon *et al.* 2010), but they can be ascribed to particular eruptions in one of two indirect ways: (1) by identifying and analysing volcanic glass shards in the same position (Davies *et al.* 2010) and using their composition to correlate to a known eruption deposit, as described above, and (2) by using the ice-core chronology based on annual layer counting to obtain an age for comparison with the ages of large eruptions recorded in historic accounts or determined using radiometric methods (see Absolute chronology section). The latter is speculative for eruptions further back in time, as the counting errors associated with the ice-core chronology become larger, and the precision of the radiometric ages is reduced. Furthermore, detailed work on the Greenland ice cores by Davies *et al.* (2010) has shown that not all tephra units are associated with peaks in sulphate concentration, and vice versa. Some magmas are sulphur-poor and do not generate enough aerosol to produce a significant peak in the cores, even though volcanic glass shards are present (Davies *et al.* 2010). Also, the absence of glass shards at a sulphate peak may simply reflect the distance from the source, as it is unlikely that tephra is transported thousands of kilometres from the vent, except in the case of VEI 8 supereruptions.

Numerous sulphate peaks recorded in both the Greenland and Antarctic records (Zielinski 1995; Kurbatov *et al.* 2006) (Table 1 and Fig. 3), and other ice cores (e.g. Nevado Illimani, Bolivia, Ramirez *et al.* 2003; Everest and Svalbard, Moore *et al.* 2012) have been attributed to large volcanic eruptions. Many of these peaks are common to multiple ice-core records (Delmas 1992; Moore *et al.* 2012) and are robustly correlated as the errors for ice-core chronologies in the last *c.* 1 kyr are very small (typically ± 1 year). Some of these sulphate peaks have also been correlated to particular eruptions using chronology (Severi *et al.* 2007; Sigl *et al.* 2013), as the dates of the large eruptions in the last few hundred years are recorded in historic texts. However, correlating the peaks to eruptions becomes more ambiguous with time due to poorer historic accounts and increasing uncertainty in radiometric age determinations (see Table 1).

Eruption signatures present in the ice records from both poles are likely to represent large significant eruptions that are potentially recorded in other palaeoenvironmental archives around the globe, and are therefore more suitable for use as markers for defining the onset of the Anthropocene. It should be noted, however, that many of the sulphate peaks observed in either Greenland or Antarctica could just be associated with small to moderate eruptions from proximal volcanoes. For example, many Icelandic eruptions have been recorded in Greenland cores (Davies *et al.* 2010), and numerous eruptions from Mt Erebus, Mt Melbourne and The Pleiades volcanic centre (Dunbar & Kurbatov 2011) are observed in the Antarctic ice cores. In particular, a very large sulphate peak is observed

Table 1. *The largest eruptions in the last two millennia*

Volcano	Location	Date of eruption (calibrated)	Latitude (°)	Longitude (°)	VEI	Recorded in ice cores (sulphate or ECM)			Date of sulphate peak§	References
						Antarctica*	Greenland†	Other‡		
Pinatubo	Philippines	June 1991	15.1 N	120.4 E	6	Yes	Yes	SV & EV	1991–1992	McCormick *et al.* (1995), Luhr & Melson (1996)
Katmai Novarupta	Alaska, USA	June 1912	58.3 N	155.2 W	6	No	Yes	NA & SV	1912–1914	Hildreth (1983), Westrich *et al.* (1991)
Santa Maria	Guatemala	October 1902	14.8 N	91.6 W	6	Yes		NA	1902–1904	Rose (1972)
Krakatau	Indonesia	August 1883	6.1 S	105.4 E	6	Yes	Yes	NA, SV & EV	1884–1886	Devine *et al.* (1984), Camus *et al.* (1987)
Tambora	Indonesia	April 1815	8.3 S	118.0 E	7	Yes	Yes	NA & SV	1815–1818	Self & Rampino (1981), Gertisser *et al.* (2012)
Huaynaputina	Peru	February 1600	16.6 S	70.9 W	6	Yes	Yes	SV	1600–1603	de Silva & Zielinski (1998), Dietterich & de Silva (2010)
Long Island	Papua New Guinea	AD 1520–1654¶	5.4 S	147.1 E	6	Possibly	Possibly			Pain *et al.* (1981)
Billy Mitchell	Papua New Guinea	AD 1450–1630¶	6.1 S	155.2 E	6	Possibly	Possibly			McKee *et al.* (1990)
Bárdarbunga (Veidivötn)	Iceland	February 1477	64.6 N	17.5 W	6	No	Yes		1476–1478	Mørk (1984)
Kuwae	Vanuatu	AD 1366–1484¶	16.8 S	168.5 E	7	Probably	Probably	SV & EV	AD 1453–1454	Monzier *et al.* (1994), Gao *et al.* (2006), Witter & Self (2006)

Quilotoa	Ecuador	AD 1194–1317[¶]	0.9 S	78.9 W	6	Possibly	Possibly			Mothes & Hall (2008)
Ceboruco (Jala)	Mexico	AD 688–1263[¶]	21.1 N	104.5 W	6	Possibly	Possibly			Nelson (1980)
Baitoushan/ Baekdusan	China/North Korea	AD 921–941	42.0 N	128.1 E	7	Possibly	Probably		AD 922–926?	Machida (1999), Horn & Schmincke (2000), Yin *et al.* (2012)
Churchill (White River)	Alaska, USA	AD 694–936	61.4 N	141.8 W	6	Possibly	Probably			Clague *et al.* (1995), Robinson (2001)
Ilopango	El Salvador	AD 122–324[¶]	13.7 N	89.1 W	6	Probably	Probably			Rose *et al.* (1999), Mehringer *et al.* (2005)
Taupo	New Zealand	AD 232 ± 5	38.8 S	176.0 E	7	Probably	Possibly		AD 231–235?	Wilson & Walker (1985), Smith *et al.* (2005), Hogg *et al.* (2012)
Ksudach	Kamchatka, Russia	AD 184–330[¶]	51.8 N	157.5 E	6	Possibly	Probably			Braitseva *et al.* (1996)

*Age ranges (*c.* 2σ) for these eruptions were obtained by calibrating (using IntCal09; Reimer *et al.* 2009) and modelling the published radiocarbon dates in OxCal 4.2 (Bronk-Ramsey 2009).
[†]Antarctic ice-core information is from the Siple, EDML, EDL, Law Dome and WDC06A ice cores, published by Palmer *et al.* (2001), Kurbatov *et al.* (2006), Severi *et al.* (2007) and Sigl *et al.* (2013).
[‡]Greenland ice-core information is from the GISP2 and NEEM S1 cores, published by Zielinski (1995) and Sigl *et al.* (2013).
[§]NA, central North America (Fremont Glacier; Schuster *et al.* 2000); EV, Everest; SV, Svalbard (Moore *et al.* 2012).
[¶]Taken from Sigl *et al.* (2013).

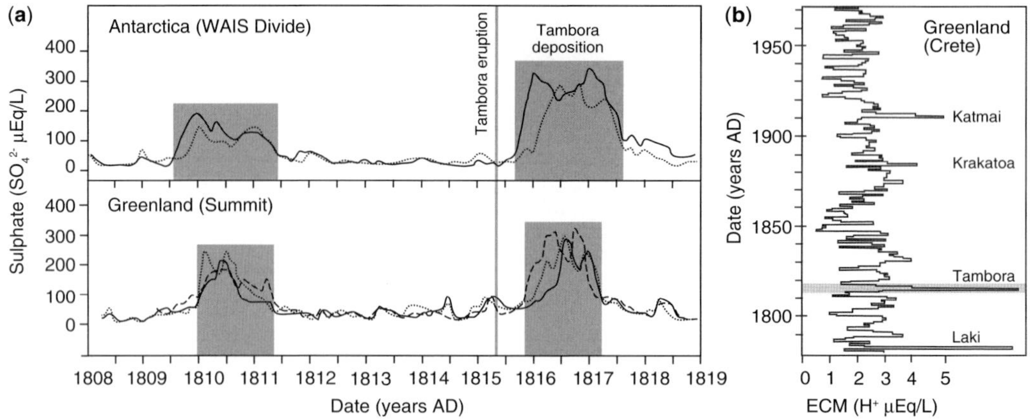

Fig. 3. (**a**) Sulphate measurements of Antarctic (WAIS Divide) and Greenland (Summit) ice cores (figure modified from Cole-Dai *et al.* 2009). The lines (solid and dashed) are sulphate measurements from parallel cores at each pole. The elevated concentrations (grey boxes) in 1816–1817 are associated with the 1815 Tambora eruption. The high concentrations also observed in 1810–1811 are thought to be associated with another volcanic eruption; however, there are no known eruptions that could be responsible for this increase (Cole-Dai *et al.* 2009). (**b**) Variations in ECM in the Greenland Crête ice core (figure modified from Delmas 1992). The elevated ECM concentrations can be linked to large historic eruptions using the ice-core chronology.

in the Greenland ice cores that is associated with the predominantly effusive VEI 4, AD 1783–1784 Laki (Skaftár Fires) eruption from Iceland (Thordarson & Self 1993; Thordarson *et al.* 1996). Remarkably, the Laki sulphate concentration in the Greenland core was found to be five times larger than that associated with bigger and more explosive eruptions closer to the tropics (e.g. the VEI 6, AD 1815 Tambora eruption), but, critically, it has not been recorded in Antarctica (Sigl *et al.* 2013). To ensure that the sulphate aerosol markers transported to the poles in both hemispheres are from a single large eruption, we should choose an event from a volcano located close to the equator.

A tephra marker close to the onset of anthropogenic change

Anthropogenic changes, such as the dramatic increase in greenhouse gases at the start of the nineteenth century (Friedli *et al.* 1986; MacFarling *et al.* 2006) (Fig. 4), are clearly observed in the Earth's key climate records – the polar ice cores. Any global marker used to define the onset of the Anthropocene should therefore also be evident in the ice-core records, located as close as possible to the defining changes in atmospheric composition. Deposits of many of the large recent historic eruptions could be used as a marker, but the 1815 Tambora eruption is probably the best candidate. It is one of the largest eruptions in the last two millennia, and it occurred in the later part of the industrial revolution. It is also clearly recorded by sulphate peaks in ice cores from both poles (Delmas 1992), and glaciers in both the northern and southern hemispheres (Schuster *et al.* 2000; Ramirez *et al.* 2003). Furthermore, the impact of the eruption is recorded in historic records and other archives, such as tree rings (Briffa *et al.* 1998).

The 1815 Tambora eruption

The VEI 7, 1815 eruption of Tambora occurred on Sumbawa island, Indonesia, and erupted *c.* 33 km^3 DRE of magma (Self *et al.* 1984; Self 2004). The first significant phase of the eruption happened on 5th April, and it moved into its climactic phase five days later. The climactic phase generated a 43-km-high Plinian column (Sigurdsson & Carey 1989) and PDC that persisted for a few days. Large co-ignimbrite ('phoenix') clouds rose off the top of the PDC, which dispersed fine ash over wide areas. In addition, light monsoonal winds blew from the SE, dispersing ash towards the NW (Fig. 1b). More than 1 cm of ash was deposited on western Java and southeastern Kalimantan, more than 650 km away, and the ash was also reported as far as 1800 km west of Tambora (Crawfurd 1856).

The major component of pyroclastic deposits is vesicular brown to black pumice with crystal contents less than 15 vol% (Gertisser *et al.* 2012). The mineral assemblage includes plagioclase, clinopyroxene, magnetite, apatite and the occasional biotite and olivine crystal (Gertisser *et al.* 2012). The matrix glass compositions range from 57.8 to

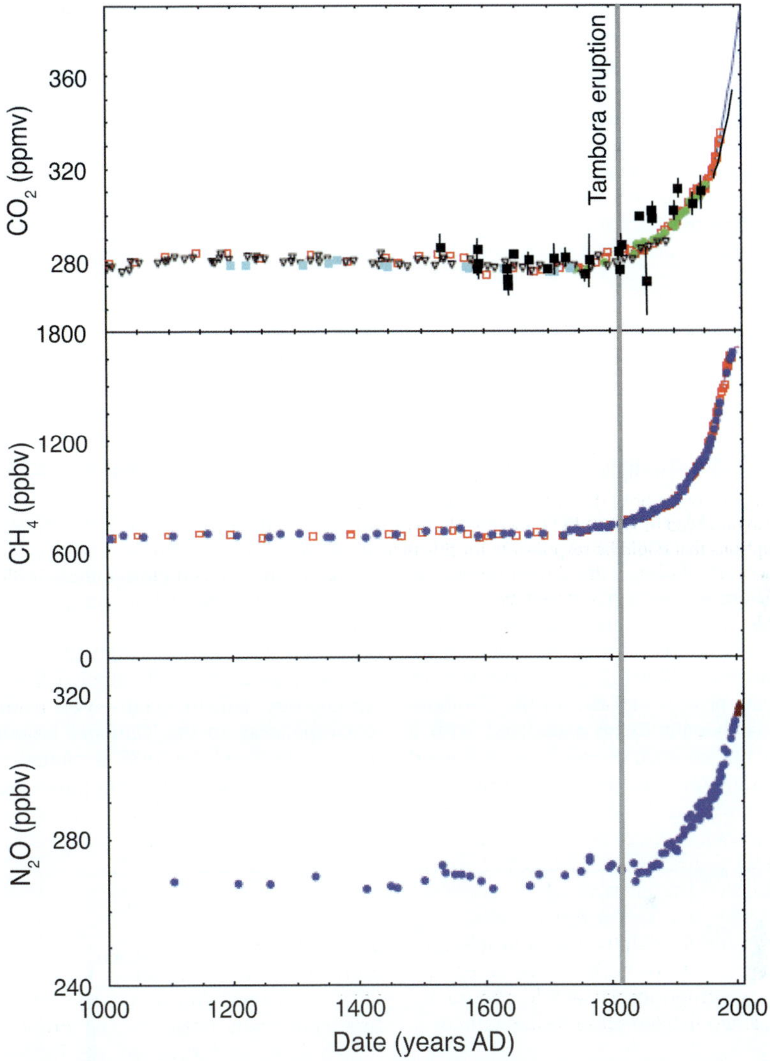

Fig. 4. Changes in greenhouse gases recorded in ice cores in the last 1000 years (figure modified from Wolff 2011). Coloured symbols reflect data from Antarctic ice cores (red open squares are from Law Dome, green circles are from Siple, purple triangles are from EPICA DML, and light blue boxes are from the South Pole), and the black boxes are data from GISP2, Greenland (Wahlen *et al.* 1991). Atmospheric CO_2 measurements from Mauna Loa are also plotted (solid blue line) for the last *c.* 50 years. The vertical grey line is the date of the Tambora eruption, which is recorded as a sulphate/ECM peak in the ice cores.

58.8 wt% SiO_2, 2.91 to 3.57 wt% CaO, and total alkali contents ($Na_2O + K_2O$) range between 11.1 and 12.3 wt% (Gertisser *et al.* 2012). The glass compositions straddle the tephriphonoite, phonolite, trachyandesite and trachytic compositional fields.

The eruption released *c.* 53–58 Tg of SO_2, which formed 93–118 Tg of sulphate aerosols (Self 2004). These aerosols were dispersed around the globe, although, because the source is south of the equator, more was distributed in the southern hemisphere than in the northern hemisphere. These aerosols reduced radiative forcing and resulted in global cooling (Self 2004).

In the aftermath, more than 70 000 people were killed on the Indonesian islands of Sumbawa and Lombok (Oppenheimer 2003). The year that followed is known as 'the year without a summer' and anomalies in the climate were recorded at many meteorological observatories. The summer temperatures recorded across much of Europe were 1–2 °C cooler than the average for the decade (1810–1819), and rainfall was significantly higher in the Mediterranean during the 1816 summer. Furthermore, cool weather was also observed across North America, with numerous exceptional frosts and snow falls in the spring and summer, which greatly impacted the length of the growing season and therefore crop yields (Baron 1992; Briffa & Jones 1992; discussed in detail in Oppenheimer 2003).

Tambora in archives that record anthropogenic change

The position of the Tambora eruption can be deduced in many records, including ice cores (as already discussed). ECM and sulphate peaks are observed just after the eruption in ice cores from Antarctica (Severi *et al.* 2007) and Greenland (Zielinski 1995). There is also a prominent peak in the ice-core records from both poles in 1809, a few years before the Tambora eruption (Dai *et al.* 1991; Cole-Dai *et al.* 2009) (Fig. 3). The size of the 1809 sulphate peak is half that of the Tambora eruption, and is thought to be associated with a large eruption, but no such events were recorded (VEI > 4) in 1809 or the preceding years (Dai *et al.* 1991; Cole-Dai *et al.* 2009). Hence, the peak in both Greenland and Antarctica is either associated with a large eruption from a source where historical records are lacking (e.g. the Pacific) or is related to multiple small eruptions from sources close to each pole.

The easily identifiable double sulphate peak associated with the 1809 (unknown) and 1815 (Tambora) eruptions (Fig. 3) occurs just before major changes in atmospheric composition that may denote the onset of the Anthropocene. Remarkably, the changes associated with such a large eruption are insignificant in comparison to the steady increase in greenhouse gases (CO_2, CH_4 and N_2O) observed after the eruption that still continues today (Wahlen *et al.* 1991; Sowers *et al.* 2002; Wolff 2011 and references therein). These increases are associated with anthropogenic activities such as the burning of fossil fuels, the increase in agriculture, and chemical emissions associated with manufacturing.

The Tambora eruption can also be identified in other ice records, such as the glacier ice in North America and in the Bolivian Andes. The ECM from an ice record obtained from the Upper Fremont Glacier in Wyoming, USA, has a known chronology refined against two prominent peaks corresponding to the Tambora eruption (Schuster *et al.* 2000) and the 1883 Krakatau eruption. The glacier core also provides a palaeoclimatic record

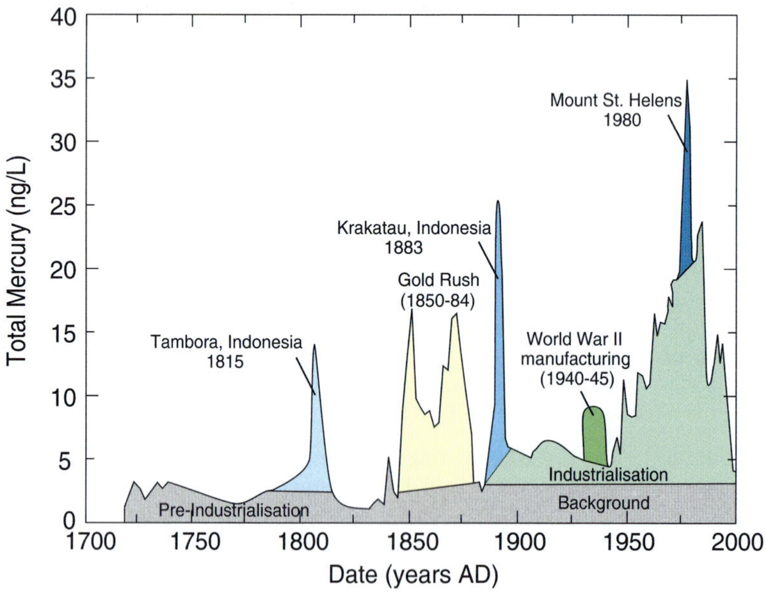

Fig. 5. Changes in total mercury deposited in the Upper Freemont Glacier, Wyoming, USA (data from Schuster *et al.* 2002, and plot modified from DeWayne *et al.* 2010). Large eruptions are clearly marked by an increase in mercury deposition. Other more sustained increases in total mercury are linked to anthropogenic activities.

Table 2. *Moderate to large eruptions (VEI $\geq$ 4) between 1790 and 1840 (from Newhall & Self 1982)*

Volcano	Location	Date	Latitude (°)	Longitude (°)	Elevation (m)	VEI	References
Cosiguina	Northern Nicaragua	June 1835	13.0 N	87.6 W	859	5	Hradecký & Rapprich (2008)
Isanotski	Aleutians, USA	1825	54.8 N	163.7 W	2480	4	
Galunggung	Java, Indonesia	October 1822	7.3 S	108.1 E	2168	5	Gourgaud *et al.* (2000)
Usu	Northern Japan	March 1822	42.5 N	140.8 E	725	4	
Colima	Western Mexico	February 1818	19.4 N	103.7 E	3960	4	González *et al.* (2002)
Beerenberg	Jan Mayen, East of Greenland	1818	71.1 N	8.2 W	2277	4	Wastegård & Davies (2009)
Tambora	Lesser Sunda, Indonesia	April 1815	8.3 S	118.0 E	2851	7	
Mayon	Philippines	February 1814	13.3 N	123.7 E	2462	4	
Soufriere (St Vincent)	Caribbean	April 1812	13.3 N	16.2 W	1178	4	
Unknown	Unknown	1809					Cole-Dai *et al.* (2009)
St Helens	Washington state, USA	1802	46.2 N	122.2 W	2549	4	
Pogromni	Aleutians, USA	1795	54.6 N	164.7 W	2286	4	
San Martin	Eastern Mexico	March 1793	18.6 N	95.2 W	1550	4	Espíndola *et al.* (2010)
Alaid	Kuril islands, Russia	1793	50.8 N	155.5 E	2339	4	

and a high-resolution record of total atmospheric mercury deposition between 1720 and 1993 for North America (Schuster *et al.* 2002) (Fig. 5). There is a clear change in the climate registered by a number of standard proxies (e.g. $\delta^{18}O$), and the significant changes in Hg in the mid-1800s, directly linked to anthropogenic activity.

The Nevado Illimani core from a glacier in the Eastern Bolivian Andes shows changes in SO_4 and stable sulphur isotopes at a position in the age–depth model that is consistent with the Tambora eruption (Ramirez *et al.* 2003). In 1815, there was an increase in calcium sulphate, fluorine and chlorine deposits, and in the following two years (1816–1817) there were elevated sulphur and halogen acids. These changes are interpreted to reflect the tropospheric and stratospheric aerosol deposits, respectively, associated with the eruption (De Angelis *et al.* 2003). It is likely that evidence (sulphate aerosols) for the Tambora eruption is also preserved in many of the non-polar ice-core records, such as those from China, Peru and the Himalayas (see Thompson 2000 and references therein).

The effects of the Tambora eruption are observed in other records that have a detailed chronology. For example, integrated tree ring data from the northern hemisphere, which mainly reflects changes in the summer temperature (Briffa *et al.* 1998), shows the colder temperatures related to the Tambora eruption. This is consistent with what was observed.

Volcanic ash from the 1815 Tambora eruption is preserved in sedimentary sequences such as land and marine sediments, hundreds of kilometres from the volcano (Kandlbauer *et al.* 2013). Given the magnitude of the eruption, it is also likely to be preserved in a cryptotephra horizon up to thousands of kilometres from the source. The Tambora tephra therefore provides a key marker layer for correlating regional records.

Other eruptions between 1790 and 1840

Other moderately sized eruptions ($4 < VEI < 6$) from different volcanic sources within 25 years of the Tambora eruption (the typical error associated with a radiocarbon date) can be used as local marker layers within hundreds of kilometres from vent. Identifying tephra from these eruptions in the respective regional palaeoenvironmental records would provide a precise chronology, enabling them to be compared to others in different locations. Correlation of these archives will help assess the timing of human-induced environmental change around the globe. For example, the tephra associated with the 1814 Mayon eruption can be used as a marker in the Philippines (Newhall 1979), the 1793 San Martin and the 1818 Colima eruption deposits can be used in Mexico, the 1795 Pogromni and the 1825 Isanotski deposits in the Aleutians, 1822 Usu eruption in Northern Japan, and the 1835 Cosigüina Volcano from western Nicaragua (Hradecký & Rapprich 2008) (see Table 2). However, many of these deposits are not adequately characterized, with very little information published. Further research would need to be done to achieve a reliable correlation.

Conclusions

Large explosive eruptions are recorded in palaeoclimatic archives across the globe as layers of compositionally distinct glass in sediments or sulphate peaks in ice cores. They can therefore be used as key marker layers, provided they can be robustly identified. The 1815 Tambora eruption is an ideal marker for defining the onset of the Anthropocene as it occurs just before major anthropogenic changes observed in key climate records. However, if the consensus of the geological community is to place the beginning of the Anthropocene further back in time, other volcanic layers may be used but with greater uncertainty in the absolute chronology.

I would like to thank C. Waters for inviting me to write this chapter, and A. Snelling, S. Wastegård, R. Johnson, P. Albert and an anonymous reviewer for constructive comments on the manuscript.

References

Baron, W. R. 1992. 1816 in perspective: the view from the northeastern USA. *In*: Harington, C. R. (ed.) *The Year Without a Summer? World Climate in 1816*. Canadian Museum of Nature, Ottawa, 124–144.

Bekki, S., Pyle, J. A., Zhong, W., Toumi, R., Haigh, J. D. & Pyle, D. M. 1996. The role of microphysical and chemical processes in prolonging the climate forcing of the Toba eruption. *Geophysical Research Letters*, **23**, 2669–2672, http://dx.doi.org/10.1029/96GL02088

Blockley, S. P. E., Pyne-O'Donnell, S. *et al.* 2005. A new and less destructive laboratory procedure for the physical separation of distal glass tephra shards from sediments. *Quaternary Science Reviews*, **24**, 1952–1960, http://dx.doi.org/10.1016/j.quascirev.2004.12.008

Braitseva, O. A., Melekestsev, I. V., Ponomareva, V. V. & Kirianov, V. Y. 1996. The caldera-forming eruption of Ksudach volcano about cal. AD 240: the greatest explosive event of our era in Kamchatka, Russia. *Journal of Volcanology and Geothermal Research*, **70**, 49–65, http://dx.doi.org/10.1016/0377-0273(95)00047-X

Bowring, S. A., Erwin, D. H., Jin, Y. G., Martin, M. W., Davidek, K. & Wang, W. 1998. U/Pb zircon geochronology and tempo of the end-Permian mass extinction. *Science*, **280**, 1039–1045, http://dx.doi.org/10.1126/science.280.5366.1039

BRIFFA, K. R. & JONES, P. D. 1992. The climate of Europe during the 1810s with special reference to 1816. *In*: HARINGTON, C. R. (ed.) *The Year Without a Summer? World climate in 1816*. Canadian Museum of Nature, Ottawa, 372–391.

BRIFFA, K. R., JONES, P. D., SCHWEINGRUBER, F. H. & OSBORN, T. J. 1998. Influence of volcanic eruptions on Northern Hemisphere summer temperature over the past 600 years. *Nature*, **393**, 450–455, http://dx.doi.org/10.1038/30943

BRONK-RAMSEY, C. 2009. Bayesian analysis of radiocarbon dates. *Radiocarbon*, **51**, 337–360.

BRONK-RAMSEY, C., STAFF, R. A. *ET AL*. 2012. A complete terrestrial radiocarbon record for 11.2 to 52.8 kyr B.P. *Science*, **338**, 370–374, http://dx.doi.org/10.1126/science.1226660

CAMUS, G., GOURGAUD, A. & VINCENT, P. M. 1987. Petrologic evolution of Krakatau (Indonesia): implications for a future activity. *Journal of Volcanology and Geothermal Research*, **33**, 299–316, http://dx.doi.org/10.1016/0377-0273(87)90020-5

CLAGUE, J. J., EVANS, S. G., RAMPTON, V. N. & WOODSWORTH, G. J. 1995. Improved age estimates for the White River and Bridge River tephras, western Canada. *Canadian Journal of Earth Sciences*, **32**, 1172–1179, http://dx.doi.org/10.1139/e95-096

COLE-DAI, J., FERRIS, D., LANCIKI, A., SAVARINO, J., BARONI, M. & THIEMENS, M. H. 2009. Cold decade (AD 1810–1819) caused by Tambora (1815) and another (1809) stratospheric volcanic eruption. *Geophysical Research Letters*, **36**, L22703, http://dx.doi.org/10.1029/2009GL040882

CRAWFURD, J. 1856. *A Descriptive Dictionary of the Indian Islands and Adjacent Countries*. Bradbury and Evans, London.

DAI, J., MOSLEY THOMPSON, E. & THOMPSON, L. G. 1991. Ice core evidence for an explosive tropical volcanic eruption 6 years preceding Tambora. *Journal of Geophysical Research*, **96**, 17 361–17 366, http://dx.doi.org/10.1029/91JD01634

DAVIES, S. M., WASTEGÅRD, S. *ET AL*. 2010. Tracing volcanic events in the NGRIP ice-core and synchronising North Atlantic marine records during the last glacial period. *Earth and Planetary Science Letters*, **294**, 69–79, http://dx.doi.org/10.1016/j.epsl.2010.03.004

DE ANGELIS, M., SIMOES, J., BONNAVEIRA, H., TAUPIN, J. D. & DELMAS, R. J. 2003. Volcanic eruptions recorded in the Illimani ice core (Bolivia): 1918–1998 and Tambora periods. *Atmospheric Chemistry and Physics*, **3**, 1725–1741, http://dx.doi.org/10.5194/acp-3-1725-2003

DE SILVA, S. L. & ZIELINSKI, G. A. 1998. Global influence of the AD1600 eruption of Huaynaputina, Peru. *Nature*, **393**, 455–458, http://dx.doi.org/10.1038/30948

DEVINE, J. D., SIGURDSSON, H., DAVIS, A. N. & SELF, S. 1984. Estimates of sulphur and chlorine yield to the atmosphere from volcanic eruptions and potential climatic effects. *Journal of Geophysical Research*, **89**, 6309–6325, http://dx.doi.org/10.1029/JB089ib07p06309

DEWAYNE, C. L., DAVID, L., NAFTZ, D. L., SCHUSTER, P. F., SUSONG, D. D. & GREEN, J. R. 2010. The palaeoenvironmental record preserved in middle latitude, high-mountain glaciers – an overview of the U.S. Geological Survey experience in the Central Asia and the United States. *In*: WILLIAMS, R. S. Jr & FERRIGNO, J. G. (eds) *Glaciers of Asia*. US Geological Survey Professional Paper **1386–F**, United States Government Printing Office, Washington, DC.

DELMAS, R. J. 1992. Environmental information from ice cores. *Reviews in Geophysics*, **30**, 1–21, http://dx.doi.org/10.1029/91RG02725

DI VITO, M. A., ISAIA, R. *ET AL*. 1999. Volcanism and deformation since 12,000 years at the Campi Flegrei caldera (Italy). *Journal of Volcanology and Geothermal Research*, **91**, 221–246, http://dx.doi.org/10.1016/S0377-0273(99)00037-2

DIETTERICH, H. & DE SILVA, S. 2010. Sulphur yield of the 1600 eruption of Huaynaputina, Peru: contributions from magmatic, fluid-phase, and hydrothermal sulphur. *Journal of Volcanology and Geothermal Research*, **197**, 303–312, http://dx.doi.org/10.1016/j.jvolgeores.2010.01.003

DUNBAR, N. W. & KURBATOV, A. V. 2011. Tephrochronology of the Siple Dome ice core, West Antarctica: correlations and sources. *Quaternary Science Reviews*, **30**, 1602–1614, http://dx.doi.org/10.1016/j.quascirev.2011.03.015

ESPÍNDOLA, J. M., ZAMORA-CAMACHO, A., GODINEZ, M. L., SCHAAF, P. & RODRÍGUEZ, S. R. 2010. The 1793 eruption of San Martín Tuxtla volcano, Veracruz, Mexico. *Journal of Volcanology and Geothermal Research*, **197**, 188–208, http://dx.doi.org/10.1016/j.jvolgeores.2009.08.005

FRIEDLI, H., LÖTSCHER, H., OESCHGER, H., SIEGENTHALER, U. & STAUFFER, B. 1986. Ice core record of the $^{13}C/^{12}C$ ratio of atmospheric CO_2 in the past two centuries. *Nature*, **324**, 237–238, http://dx.doi.org/10.1038/324237a0

GAO, C., ROBOCK, A. *ET AL*. 2006. The 1452 or 1453 A.D. Kuwae eruption signal derived from multiple ice core records: greatest volcanic sulfate event of the past 700 years. *Journal of Geophysical Research*, **111**, D12107, http://dx.doi.org/10.1029/2005JD00 6710

GERTISSER, R., SELF, S., THOMAS, L. E., HANDLEY, H. K., VAN CALSTEREN, P. & WOLFF, J. A. 2012. Processes and timescales of magma genesis and differentiation leading to the great Tambora Eruption in 1815. *Journal of Petrology*, **53**, 271–297, http://dx.doi.org/10.1093/petrology/egr062

GONZÁLEZ, M. B., RAMÍREZ, J. J. & NAVARRO, C. 2002. Summary of the historical eruptive activity of Volcán de Colima, Mexico 1519–2000. *Journal of Volcanology and Geothermal Research*, **117**, 21–46, http://dx.doi.org/10.1016/S0377-0273(02)00233-0

GOURGAUD, A., THOURET, J. C. & BOURDIER, J. L. 2000. Stratigraphy and textural characteristics of the 1982–83 tephra of Galunggung volcano (Indonesia): implications for volcanic hazards. *Journal of Volcanology and Geothermal Research*, **104**, 169–186, http://dx.doi.org/10.1016/S0377-0273(00)00205-5

GRAINGER, R. G. & HIGHWOOD, E. J. 2003. Changes in stratospheric composition, chemistry, radiation and climate caused by volcanic eruptions. *In*: Oppenheimer, C., Pyle, D. & Barclay, J. (eds) *Volcanic Degassing*. Geological Society, London, Special Publications, **213**, 329–347, http://dx.doi.org/10.1144/GSL.SP.2003.213.01.20

Hildreth, W. 1983. The compositionally zoned eruption of 1912 in the Valley of Ten Thousand Smokes, Katmai National Park, Alaska. *Journal of Volcanology and Geothermal Research*, **18**, 1–56, http://dx.doi.org/10.1016/0377-0273(83)90003-3

Hogg, A., Lowe, D. J., Palmer, J., Boswijk, G. & Ramsey, C. B. 2012. Revised calendar date for the Taupo eruption derived by ^{14}C wiggle-matching using a New Zealand kauri ^{14}C calibration data set. *The Holocene*, **22**, 439–449, http://dx.doi.org/10.1177/0959683611425551

Horn, S. & Schmincke, H. U. 2000. Volatile emission during the eruption of Baitoushan Volcano (China/North Korea) ca. 969 AD. *Bulletin of Volcanology*, **61**, 537–555, http://dx.doi.org/10.1007/s004450050004

Hradecký, P. & Rapprich, V. 2008. Historical tephrastratigraphy of the Cosigüina Volcano (Western Nicaragua). *Revista Geológica de América Central*, **38**, 65–79.

Kandlbauer, J., Carey, S. N. & Sparks, R. S. J. 2013. The 1815 Tambora ash fall: implications for transport and deposition of distal ash on land and in the deep sea. *Bulletin of Volcanology*, **75**, 708, http://dx.doi.org/10.1007/s00445-013-0708-3

Kuehn, S., Froese, D. G. & Shane, P. A. R. 2011. The INTAV intercomparison of electron-beam microanalysis of glass by tephrochronology laboratories: results and recommendations. *Quaternary International* **246**, 19–47, http://dx.doi.org/10.1016/j.quaint.2011.08.022

Kurbatov, A. V., Zielinski, G. A., Dunbar, N. W., Mayewski, P. A., Meyerson, E. A., Sneed, S. B. & Taylor, K. C. 2006. A 12,000 year record of explosive volcanism in the Siple Dome Ice Core, West Antarctica. *Journal of Geophysical Research*, **111**, D12307, http://dx.doi.org/10.1029/2005JD006072

Lane, C. S., Blockley, S. P. E., Ramsey, C. B. & Lotter, A. F. 2011. Tephrochronology and absolute centennial scale synchronisation of European and Greenland records for the last glacial to interglacial transition: a case study of Soppensee and NGRIP. *Quaternary International*, **246**, 145–156, http://dx.doi.org/10.1016/j.quaint.2010.11.028

Lane, C. S., Blockley, S. P. E. *et al.* 2012. Was the 12.1 ka Icelandic Vedde Ash one of a kind? *Quaternary Science Reviews*, **33**, 87–99, http://dx.doi.org/10.1016/j.quascirev.2011.11.011

Lane, C. S., Chorn, B. T. & Johnson, T. C. 2013. Ash from the Toba supereruption in Lake Malawi shows no volcanic winter in East Africa at 75 ka. *Proceedings of the National Academy of Sciences of the USA*, **110**, 8025–8029, http://dx.doi.org/10.1073/pnas.1301474110

Lemieux-Dudon, B., Blayo, E. *et al.* 2010. Consistent dating for Antarctic and Greenland ice cores. *Quaternary Science Reviews*, **29**, 8–20, http://dx.doi.org/10.1016/j.quascirev.2009.11.010

Lowe, D. J. 2011. Tephrochronology and its application: a review. *Quaternary Geochronology* **6**, 107–153, http://dx.doi.org/10.1016/j.quageo.2010.08.003

Lowe, J., Barton, N. *et al.* 2012. Volcanic ash layers illuminate the resilience of Neanderthals and early modern humans to natural hazards. *Proceedings of the National Academy of Sciences of the USA*, **109**, 13 532–13 537, http://dx.doi.org/10.1073/pnas.1204579109

Luhr, J. F. & Melson, W. G. 1996. Mineral and glass compositions in June 15, 1991, pumices: evidence for dynamic disequilibrium in the dacite of Mount Pinatubo. *In*: Newhall, C. G. & Punongbayan, R. S. (eds) *Fire and Mud: Eruptions and Lahars of Mount Pinatubo, Philippines*. University of Washington Press, Seattle, 733–750.

MacFarling, C., Etheridge, D. *et al.* 2006. Law Dome CO_2, CH_4 and N_2O ice core records extended to 2000 years BP. *Geophysical Research Letters*, **33**, L14810, http://dx.doi.org/10.1029/2006GL026152

Machida, H. 1999. The stratigraphy, chronology and distribution of distal marker-tephras in and around Japan. *Global and Planetary Change*, **21**, 71–94, http://dx.doi.org/10.1016/S0921-8181(99)00008-9

Mark, D. F., Petraglia, M. *et al.* 2013. A high-precision $^{40}Ar/^{39}Ar$ age for the Young Toba Tuff and dating of ultra-distal tephra: forcing of Quaternary climate and implications for hominin occupation of India. *Quaternary Geochronology*, http://dx.doi.org/10.1016/j.quageo.2012.12.004

Matthews, N. E., Smith, V. C., Costa, A., Durant, A. J., Pyle, D. M. & Pearce, N. J. G. 2011. Ultra-distal tephra deposits from super-eruptions: examples from Toba, Indonesia and Taupo Volcanic Zone, New Zealand. *Quaternary International*, **258**, 54–79, http://dx.doi.org/10.1016/j.quaint.2011.07.010

McCormick, M. P., Thomason, L. W. & Trepte, C. R. 1995. Atmospheric effects of the Mt Pinatubo eruption. *Nature*, **373**, 399–404, http://dx.doi.org/10.1038/373399a0

McKee, C. O., Johnson, R. W. & Rogerson, R. 1990. Explosive volcanism on Bougainville Island: ignimbrites, calderas, and volcanic hazards. *Proceedings of the Pacific Rim Congress 1990*, **2**, 237–245.

Mehringer, P. J., Jr, Sarna-Wojcicki, A. M., Wollwage, L. K. & Sheets, P. 2005. Age and extent of the Ilopango TBJ Tephra inferred from a Holocene chronostratigraphic reference section, Lago De Yojoa, Honduras. *Quaternary Research*, **63**, 199–205, http://dx.doi.org/10.1016/j.yqres.2004.09.011

Merkt, J., Müller, H., Knabe, W., Müller, P. & Weiser, T. 1993. The early Holocene Saksunarvatn tephra found in lake sediments in NW Germany. *Boreas*, **22**, 93–100, http://dx.doi.org/10.1111/j.1502-3885.1993.tb00168.x

Miotello, A. & Mazzoldi, P. 1982. Numerical analysis of field-assisted sodium migration in electron-irradiated glasses. *Journal of Physics C: Solid State Physics*, **15**, 5615–5621, http://dx.doi.org/10.1088/0022-3719/15/27/017

Monzier, M., Robin, C. & Eissen, J. P. 1994. Kuwae ($\approx$ 1425 AD): the forgotten caldera. *Journal of Volcanology and Geothermal Research*, **59**, 207–218, http://dx.doi.org/10.1016/0377-0273(94)90091-4

MOORE, J. C., BEAUDON, E., KANG, S., DIVINE, D., ISAKSSON, E., POHJOLA, V. A. & VAN DE WAL, R. S. W. 2012. Statistical extraction of volcanic sulphate from nonpolar ice cores. *Journal of Geophysical Research*, **117**, D03306, http://dx.doi.org/10.1029/2011JD016592

MØRK, M. B. E. 1984. Magma mixing in the post-glacial Veidivötn fissure eruption, southeast Iceland: a microprobe study of mineral and glass variations. *Lithos*, **17**, 55–75, http://dx.doi.org/10.1016/0024-4937(84)90006-9

MOTHES, P. A. & HALL, M. L. 2008. The plinian fallout associated with Quilotoa's 800 yr BP eruption, Ecuadorian Andes. *Journal of Volcanology and Geothermal Research*, **176**, 56–69, http://dx.doi.org/10.1016/j.jvolgeores.2008.05.018

NELSON, S. A. 1980. Geology and petrology of Volcan Ceboruco, Nayarit, Mexico. *Geological Society of America Bulletin*, **91**, 2290–2431, http://dx.doi.org/10.1130/GSAB-P2-91-2290

NEWHALL, C. G. 1979. Temporal variation in the lavas of Mayon Volcano, Philippines. *Journal of Volcanology and Geothermal Research*, **6**, 61–83, http://dx.doi.org/10.1016/0377-0273(79)90047-7

NEWHALL, C. & SELF, S. 1982. The volcanic explosivity index (VEI) – an estimate of explosive magnitude for historical volcanism. *Journal of Geophysical Research*, **87**, 1231–1238, http://dx.doi.org/10.1029/JC087iC02p01231

OPPENHEIMER, C. 2003. Climatic, environmental and human consequences of the largest known historic eruption: Tambora volcano (Indonesia) 1815. *Progress in Physical Geography*, **27**, 230–259, http://dx.doi.org/10.1191/0309133303pp379ra

PAIN, C. F., BLONG, R. J. & MCKEE, C. O. 1981. Pyroclastic deposits and eruptive sequences of Long Island. Part 1: Lithology, stratigraphy, and volcanology. *Geological Survey of Papua New Guinea Memoir*, **10**, 101–107.

PÁLFY, J., SMITH, P. L. & MORTENSEN, J. K. 2000. A U–Pb and $^{40}Ar/^{39}Ar$ time scale for the Jurassic. *Canadian Journal of Earth Sciences*, **37**, 923–944, http://dx.doi.org/10.1139/e00-002

PALMER, A. S., VAN OMMEN, T. D., CURRAN, M. A., MORGAN, V., SOUNEY, J. M. & MAYEWSKI, P. A. 2001. High-precision dating of volcanic events (AD 1301–1995) using ice cores from Law Dome, Antarctica. *Journal of Geophysical Research*, **106**, 28 089–28 095, http://dx.doi.org/10.1029/2001JD000330/

PEARCE, N. J. G., PERKINS, W. T., WESTGATE, J. A. & WADE, S. C. 2011. Trace-element microanalysis by LA-ICP-MS: The quest for comprehensive chemical characterisation of single, sub-10 μm volcanic glass shards. *Quaternary International*, **246**, 57–81, http://dx.doi.org/10.1016/j.quaint.2011.07.012

RAMIREZ, E., HOFFMANN, G. ET AL. 2003. A new Andean deep ice core from Nevado Illimani (6350 m), Bolivia. *Earth and Planetary Science Letters*, **212**, 337–350, http://dx.doi.org/10.1016/S0012-821X(03)00240-1

REIMER, P. J., BAILLIE, M. G. L. ET AL. 2009. IntCal09 and Marine09 radiocarbon age calibration curves, 0–50,000 years cal BP. *Radiocarbon*, **51**, 1111–1150, http://hdl.handle.net/10289/3622

ROBINSON, S. D. 2001. Extending the Late Holocene White River Ash distribution, northwestern Canada. *Arctic*, **54**, 157–161, http://www.jstor.org/stable/40512371

ROBOCK, A. 2000. Volcanic eruptions and climate. *Reviews in Geophysics*, **38**, 191–219, http://dx.doi.org/10.1029/1998RG000054

ROSE, W. I. 1972. Notes on the 1902 eruption of Santa Maria volcano, Guatemala. *Bulletin Volcanologique*, **36**, 29–45, http://dx.doi.org/10.1007/BF02596981

ROSE, W. I., CONWAY, F. M., PULLINGER, C. R., DEINO, A. & MCINTOSH, W. C. 1999. An improved age framework for late Quaternary silicic eruptions in northern Central America. *Bulletin of Volcanology*, **61**, 106–120, http://dx.doi.org/10.1007/s004450050266

SARNA-WOJCICKI, A. M., SHIPLEY, S., WAITT, R. B., DZURISIN, D. & WOOD, S. H. 1981. Areal distribution, thickness, mass, volume, and grain size of air-fall ash from six major eruptions of 1980. *In*: LIPMAN, P. W. & MULLINEAUX, D. R. (eds) *The 1980 eruptions of Mount St. Helens, Washington*. US Geological Survey Professional Paper 1250, 577–600, United States Government Printing Office, Washington, DC.

SARNA-WOJCICKI, A. M., MORRISON, S. D., MEYER, C. E. & HILLHOUSE, J. W. 1987. Correlation of upper Cenozoic tephra layers between sediments of the western United States and eastern Pacific Ocean and comparison with biostratigraphic and magnetostratigraphic age data. *Geological Society of America Bulletin*, **98**, 207–223, http://dx.doi.org/10.1130/0016-7606(1987)98<207:COUCTL>2.0.CO;2

SCHUSTER, P. F., WHITE, D. E., NAFTZ, D. L. & CECIL, L. D. 2000. Chronological refinement of an ice core record at Upper Fremont Glacier in south central North America. *Journal of Geophysical Research*, **105**, 4657–4666, http://dx.doi.org/10.1029/1999JD901095

SCHUSTER, P. F., KRABBENHOFT, D. P. ET AL. 2002. Atmospheric mercury deposition during the last 270 years: a glacial ice core record of natural and anthropogenic sources. *Environment Science and Technology*, **36**, 2303–2310, http://dx.doi.org/10.1021/es0157503

SELF, S. 2004. Magma volume, volatile emissions, and stratospheric aerosols from the 1815 eruption of Tambora. *Geophysical Research Letters*, **31**, L20608, http://dx.doi.org/10.1029/2004GL020925

SELF, S. & RAMPINO, M. R. 1981. The 1883 eruption of Krakatau. *Nature*, **294**, 699–704, http://dx.doi.org/10.1038/294699a0

SELF, S., RAMPINO, M. R., NEWTON, M. S. & WOLFF, J. A. 1984. Volcanological study of the great Tambora eruption of 1815. *Geology*, **12**, 659–663, http://dx.doi.org/10.1130/0091-7613(1984)12<659:VSOTGT>2.0.CO;2

SEVERI, M., BECAGLI, S. & CASTELLANO, E. 2007. Synchronisation of the EDML and EDC ice cores for the last 52 kyr by volcanic signature matching. *Climate of the Past*, **3**, 367–374, http://www.clim-past.net/3/367/2007/

SHANE, P. A. R. 1998. Correlation of rhyolitic pyroclastic eruptive units for the Taupo volcanic zone by Fe–Ti oxide compositional data. *Bulletin of Volcanology*, **60**, 224–238, http://dx.doi.org/10.1007/s004450050229

SHANE, P. A. R. 2000. Tephrochronology: a New Zealand case study. *Earth Science Reviews*, **49**, 223–259, http://dx.doi.org/10.1016/S0012-8252(99)00058-6

SHANE, P. A. R., NAIRN, I., MARTIN, S. & SMITH, V. C. 2008. Compositional heterogeneity in tephra deposits resulting from the eruption of multiple magma bodies: implications for tephrochronology. *Quaternary International*, **178**, 44–53, http://dx.doi.org/10.1016/j.quaint.2006.11.014

SIGL, M., MCCONNELL, J. R. *ET AL.* 2013. A new bipolar ice core record of volcanism from WAIS Divide and NEEM and implications for climate forcing of the last 2000 years. *Journal of Geophysical Research*, **118**, 1151–1169, http://dx.doi.org/10.1029/2012JD018603

SIGURDSSON, H. & CAREY, S. 1989. Plinian and co-ignimbrite tephra fall from the 1815 eruption of Tambora volcano. *Bulletin of Volcanology*, **51**, 243–270, http://dx.doi.org/10.1007/BF01073515

SMITH, V. C., SHANE, P. A. R. & NAIRN, I. 2005. Trends in rhyolite geochemistry, mineralogy, and magma storage during the last 50 kyr at Okataina and Taupo volcanic centres, Taupo Volcanic Zone, New Zealand. *Journal of Volcanology and Geothermal Research*, **148**, 372–406, http://dx.doi.org/10.1016/j.jvolgeores.2005.05.005

SMITH, V. C., ISAIA, R. & PEARCE, N. J. G. 2011*a*. Tephrostratigraphy and glass compositions of post-15 kyr Campi Flegrei eruptions: implications for eruption history and chronostratigraphic markers. *Quaternary Science Reviews*, **30**, 3638–3660, http://dx.doi.org/10.1016/j.quascirev.2011.07.012

SMITH, V. C., PEARCE, N. J. G. *ET AL.* 2011*b*. Geochemical fingerprinting of the widespread Toba tephra using biotite compositions. *Quaternary International*, **246**, 97–104, http://dx.doi.org/10.1016/j.quaint.2011. 05.012

SMITH, V. C., MARK, D., STAFF, R. & BLOCKLEY, S. P. E. 2011*c*. Toward establishing precise $^{40}Ar/^{39}Ar$ chronologies for Late Pleistocene palaeoclimate archives: an example from Suigetsu SG06, Japan. *Quaternary Science Reviews*, **30**, 2845–2850, http://dx.doi.org/10.1016/j.quascirev.2011.06.020

SMITH, V. C., STAFF, R. A. *ET AL.* & SUIGETSU PROJECT MEMBERS 2013. Identification and correlation of visible tephras in the Lake Suigetsu SG06 sedimentary archive, Japan: chronostratigraphic markers for synchronising of east Asian/west Pacific palaeoclimatic records across the last 150 ka. *Quaternary Science Reviews*, **67**, 121–137, http://dx.doi.org/10.1016/j.quascirev.2013.01.026

SOWERS, T., RODEBAUGH, A., YOSHIDA, N. & TOYODA, S. 2002. Extending records of the isotopic composition of atmospheric N_2O back to 1800 A.D. from air trapped in snow at the South Pole and the Greenland Ice Sheet Project II ice core. *Global Biogeochemical Cycles*, **16**, 1126–1136, http://dx.doi.org/10.1029/2002GB001911

THOMPSON, L. G. 2000. Ice core evidence for climate change in the Tropics: implications for our future. *Quaternary Science Reviews*, **19**, 19–35, http://dx.doi.org/10.1016/S0277-3791(99)00052-9

THORARINSSON, S. 1954. *The Eruptions of Hekla 1947–1948. II, 3. The Tephra Fall from Hekla.* Vis Islendinga, Reykjavik, **68**.

THORDARSON, T. & SELF, S. 1993. The Laki (Skaftár Fires) and Grimsvotn eruptions in 1783–1785. *Bulletin of Volcanology*, **55**, 233–263, http://dx.doi.org/10.1007/BF00624353

THORDARSON, T., SELF, S., OSKARSSON, N. & HULSEBOSCH, T. 1996. Sulfur, chlorine, and fluorine degassing and atmospheric loading by the 1783–1784 AD Laki (Skaftár Fires) eruption in Iceland. *Bulletin of Volcanology*, **58**, 205–225, http://dx.doi.org/10.1007/s004450050136

TOMLINSON, E., THORDARSON, T., MÜLLER, W. & THIRLWALL, M. 2010. Microanalysis of tephra by LA-ICP-MS-strategies, advantages and limitations assessed using the Thorsmörk Ignimbrite (Southern Iceland). *Chemical Geology*, **279**, 73–89, http://dx.doi.org/10.1016/j.chemgeo.2010.09.013

TOMLINSON, E. L., ARIENZO, I. *ET AL.* 2012*a*. Geochemistry of the Phlegraean Fields (Italy) proximal sources for major Mediterranean tephras: Implications for the dispersal of Plinian and co-ignimbritic components of explosive eruptions. *Geochimica et Cosmochima Acta*, **93**, 102–128, http://dx.doi.org/10.1016/j.gca.2012.05.043

TOMLINSON, E. L., KINVIG, H. S., SMITH, V. C., BLUNDY, J. D., GOTTSMANN, J., MULLER, W. & MENZIES, M. A. 2012*b*. The Upper and Lower Nisyros Pumices: revisions to the Mediterranean tephrostratigraphic record based on micron-beam glass geochemistry. *Journal of Volcanology and Geothermal Research*, **243–244**, 69–80, http://dx.doi.org/10.1016/j.jvolgeores.2012.07.004

TURNEY, C. S. 1998. Extraction of rhyolitic component of Vedde microtephra from minerogenic lake sediments. *Journal of Paleolimnology*, **19**, 199–206, http://dx.doi.org/10.1023/A:1007926322026

WAHLEN, M., ALLEN, D., DECK, B. & HERCHENRODER, A. 1991. Initial measurements of CO_2 concentrations (1530 to 1940 AD) in air occluded in the GISP 2 ice core from central Greenland. *Geophysical Research Letters*, **18**, 1457–1460, http://dx.doi.org/10.1029/91GL01724

WASTEGÅRD, S. & DAVIES, S. M. 2009. An overview of distal tephrochronology in northern Europe during the last 1000 years. *Journal of Quaternary Science*, **24**, 500–512, http://dx.doi.org/10.1002/jqs.1269

WESTRICH, H. R., EICHELBERGER, J. C. & HERVIG, R. L. 1991. Degassing of the 1912 Katmai magmas. *Geophysical Research Letters*, **18**, 1561–1564, http://dx.doi.org/10.1029/91GL01667

WILSON, C. & WALKER, G. 1985. The Taupo Eruption, New Zealand I. General Aspects. *Philosophical Transactions of the Royal Society of London A – Mathematical and Physical Sciences*, **314**, 199–228, http://www.jstor.org/stable/37662

WITTER, J. B. & SELF, S. 2006. The Kuwae (Vanuatu) eruption of AD 1452: potential magnitude and volatile release. *Bulletin of Volcanology*, **69**, 301–318, http://dx.doi.org/10.1007/s00445-006-0075-4

Wolff, E. W. 2011. Greenhouse gases in the Earth system: a palaeoclimate perspective. *Philosophical Transactions of the Royal Society A – Mathematical, Physical and Engineering Sciences*, **369**, 2133–2147, http://dx.doi.org/10.1098/rsta.2010.0225

Yin, J., Jull, A. J. T., Burr, G. S. & Zheng, Y. 2012. A wiggle-match age for the Millennium eruption of Tianchi Volcano at Changbaishan, Northeastern China. *Quaternary Science Reviews*, **47**, 150–159, http://dx.doi.org/10.1016/j.quascirev.2012.05.015

Zielinski, G. A. 1995. Stratospheric loading and optical depth estimates of explosive volcanism over the last 2100 years derived from the Greenland Ice Sheet Project 2 ice core. *Journal of Geophysical Research*, **100**, 20 937–20 955, http://dx.doi.org/10.1029/95JD01751

Technology as a geological phenomenon: implications for human well-being

P. K. HAFF

Division of Earth and Ocean Sciences, Nicholas School of the Environment, Duke University, Box 90233, Durham, North Carolina 27516, USA (e-mail: haff@duke.edu)

Abstract: The technosphere, the interlinked set of communication, transportation, bureaucratic and other systems that act to metabolize fossil fuels and other energy resources, is considered to be an emerging global paradigm, with similarities to the lithosphere, atmosphere, hydrosphere and biosphere. The technosphere is of global extent, exhibits large-scale appropriation of mass and energy resources, shows a tendency to co-opt for its own use information produced by the environment, and is autonomous. Unlike the older paradigms, the technosphere has not yet evolved the ability to recycle its own waste stream. Unless or until it does so, its status as a paradigm remains provisional. Humans are 'parts' of the technosphere – subcomponents essential for system function. Viewed from the inside by its human parts, the technosphere is perceived as a derived and controlled construct. Viewed from outside as a geological phenomenon, the technosphere appears as a quasi-autonomous system whose dynamics constrains the behaviour of its human parts. A geological perspective on technology suggests why strategies to limit environmental damage that consider only the needs of people are likely to fail without parallel consideration of the requirements of technology, especially its need for an abundant supply of energy.

The case for a new geological epoch has been articulated by Crutzen and others (e.g. Crutzen & Stoermer 2000; Crutzen 2002; Steffen *et al.* 2011) with the argument that the impact on and co-option of natural processes by human actions has become sufficiently intense that the characteristics of the recent but mostly natural Holocene world no longer adequately correspond to the suite of Earth surface processes operating today. Among the consequences of human actions that have or are threatening to significantly alter natural Earth function are the exploitation of a large fraction of the Earth's land surface for human purposes, 38% of which was agricultural land in 2009 (FAO 2011), fixation of more atmospheric nitrogen than fixed by all natural terrestrial processes (Vitousek *et al.* 1997), appropriation of terrestrial organic material equivalent to 40% of net primary production (Vitousek *et al.* 1986), extinction of biological species, which, at the current rate, exceeds that leading up to the 'Big Five' mass extinction events of geological history (occurring at the end of the Ordovician, Devonian, Permian, Triassic and Cretaceous periods) (Barnosky *et al.* 2011), interception by dams of 25–30% of global sediment load (Vorosmarty *et al.* 2003), impoundment of up to 15% of global runoff (Nilsson *et al.* 2005) and consumption of energy at *c.*20% of the rate of photosynthetic energy flow through the biosphere (Sassoon *et al.* 2009). Moreover there is an expectation of continued change in Earth function as a consequence of human activity (IPCC 2007), including further warming of the atmosphere, shifts in the geographical distribution of species, continued recession of glaciers, melting of permafrost, vanishing of late-summer Arctic sea ice, and continued sea-level rise. To emphasize the importance of human actions on Earth function, a new name, 'Anthropocene', has been proposed for the modern geological epoch (Crutzen & Stoermer 2000).

Technology from the outside

The logic of inaugurating a new nomenclature to reflect the changed and changing conditions of the globe over the past few centuries is convincing. Here, we offer a slightly different emphasis on the nature of this recent transition out of the Holocene by highlighting the role that technology plays in defining the world we now inhabit. The proliferation of technology across the globe defines the technosphere – the set of large-scale networked technologies that underlie and make possible rapid extraction from the Earth of large quantities of free energy and subsequent power generation, long-distance, nearly instantaneous communication, rapid long-distance energy and mass transport, the existence and operation of modern governmental and other bureaucracies, high-intensity industrial and manufacturing operations including regional, continental and global distribution of food and other goods, and a myriad additional 'artificial' or 'non-natural' processes without which modern

From: Waters, C. N., Zalasiewicz, J. A., Williams, M., Ellis, M. A. & Snelling, A. M. (eds) 2014. *A Stratigraphical Basis for the Anthropocene*. Geological Society, London, Special Publications, **395**, 301–309.
First published online October 24, 2013, http://dx.doi.org/10.1144/SP395.4

civilization and its present 7×10^9 human constituents could not exist. If the term 'anthroposphere' is meant to emphasize the role of human beings as causative agents responsible for Earth transformations that define the Anthropocene, the use of 'technosphere' suggests a more detached view of an emerging geological process that has entrained humans as essential components that support its dynamics. (To aid readability, and where confusion will not result, the word 'technology' is sometimes used below as a synonym for 'technosphere'.)

Because we design, manufacture, deploy and maintain many of the parts, or 'artefacts', of which technology is composed, and then network them together to obtain a desired function, it is natural for humans to see the technosphere from the 'inside' and to think of it as a purely derivative phenomenon, dependent entirely on humans for its creation and continued existence. However, that is only half the story. The human population, at anything like its current size, is deeply dependent on the existence of the technosphere. Without the support structure and services provided by technology, the human population would quickly decline towards its Stone Age base of no more than ten million (US Census Bureau 2012) individuals. The technosphere is not 'just' a human-created phenomenon, because, except for simple artefacts like stone tools, humans did not create technology independently, but only in the context of existing technological systems. From the outside, that is, from its own vantage point, notwithstanding that its human parts are essential, technology appears to have bootstrapped itself into its present state. This is the same process that characterizes all emergent complex systems vis-à-vis their small-scale components; that is, large-scale dynamics appears spontaneously (e.g. Laughlin & Pines 2000; Werner 2003) and defines an environment within which small system components must operate.

Moreover, small components (which may themselves be systems) that are essential to overall system function often do not operate independently, but are constrained by and often depend for their existence on the emergent properties for which their own actions provide necessary support. If the parts are simple enough, they can survive the demise of the emergent system they once helped define, as sand grains remain when the dune disperses. However, more complex parts, like the stratigraphic layers inside a dune, often can maintain their existence only in the environment they help create (signature aeolian stratigraphy vanishes with the dune). Similarly, in the ruins of collapsed civilizations, surviving parts tend to be simple and inert, like the scattered stone blocks that once cooperated to define a building or monument. The more complex, dynamic parts of the civilization – buildings, institutions and cities – often cannot maintain their existence if the larger system they helped define ceases to exist. In a similar way, humans have become entrained within the matrix of technology and are now borne along by a supervening dynamics from which they cannot simultaneously escape and survive.

In this paper we adopt the non-anthropocentric view that technology is a global phenomenon that follows its own dynamics, representing something truly new in the world – the opening phase of a new paradigm of Earth history. In this sense one might say that technology is the next biology.

Commonalities and contrasts between technology and earlier paradigms

Global extent

Identifying commonalities between the technosphere and older geological paradigms helps paint a clearer picture of the potential role of the technosphere as a defining and potentially enduring mode of Earth organization. Identifying contrasts on the other hand helps to show how the technosphere must evolve if it is to finally attain a true paradigmatic status.

Geological paradigms are recognized as such partly because their dynamics are far-reaching – indeed global – in extent, like the paradigms expressed by the hydrological cycle, atmospheric circulation, plate tectonics and biological processes. In recognition of the fact of global coverage we refer to the spheres: the hydrosphere, the atmosphere, the lithosphere and the biosphere. In a similar way technology penetrates to nearly every part of the globe through a web of communication and transportation networks. On land the technosphere transports large quantities of solids further and faster than any natural process except sediment transport by rivers (Haff 2010). The technosphere is also manifest in the wide distribution of myriad artefacts such as needles, motors and medicines, and by technological or technologically assisted processes like pumping and harvesting, as well as by nominally human activities that are closely tied to technological processes, such as watching television or filling out tax forms. Most such localized systems, processes and artefacts derive from, or are connected either directly or indirectly to, the globe-spanning networks of the technosphere.

If we take the State (capitalized to indicate a political entity rather than a generic state of a system) as an example of a large coherent component of the technosphere, then almost all the world's people are parts of (subject to) States and thus are parts of the modern technosphere. The principal exceptions are certain populations that occupy mountainous or other difficult-to-access terrains, such as swamplands. One of the best studied (Scott 2010) examples of these remote, outside-the-modern-technosphere populations is Zomia (population on the order of 100 million), a rugged, mountainous region in SE Asia that is congruent with no State but laps across the boundaries of China, Vietnam, Thailand and other internationally recognized polities. Here, the early technosphere appeared in the last millennium or two in the form of States occupying valley bottoms and depending for their existence on wet-rice cultivation and the presence of large, sedentary human populations. The mountainous hinterlands of Zomia provided protection over many centuries for tribal peoples who, surviving through swidden agriculture and non-sedentary ('nomadic') lifestyles, rejected the hegemony of the valley States, which were unable to easily project power beyond the smooth flatlands that sustained the valley padis. Continued evolution of technology with instantaneous communications over great distances, domination of the third dimension by aircraft, and introduction of heavy machinery to construct roads and railways through difficult terrain had, by the mid-twentieth century, threatened the existence of humanity's largest, and one of its last, refugia from the technosphere. In subduing the last vestiges of mass resistance to assimilation, the technosphere seems to be approaching, as if towards a mathematical limit, domination of 100% of the world's people. The technosphere is in any practical sense a global phenomenon, spanning the planet and absorbing into itself almost all of the world's human population.

Appropriation of resources

The physical components of a geological paradigm are constructed from parts cadged from the Earth's supply of resources, including from resources used by older paradigms. For example, the biosphere is built up in large part of carbon, oxygen and hydrogen (water) abstracted from the atmosphere and the hydrosphere, and incorporates inorganic atoms derived from mineral grains in the lithosphere. The technosphere abstracts water from the hydrosphere for urban consumption, irrigation and to supply industry, appropriates organic material from the biosphere (such as wood and agricultural products), uses material from the lithosphere for building materials and foundations and as a medium to support agriculture, and captures oxygen from the atmosphere to support the combustion of its main energy source, fossil fuels derived ultimately from the biosphere, as well as for the burning of food calories, which power its human and other animal parts.

Redesigned organisms, such as those that comprise livestock and cereal grains or that produce fibres for textiles, can be interpreted as newly constructed parts of the technosphere rather than as appropriated parts of the biosphere. Although the processes that govern domesticated organismal metabolism are essentially the same as those of related organisms in the natural biosphere, the domesticated organisms themselves exist with their present large populations only because of technological processes that support them and which they in turn support. Most of the physical organisms themselves were never part of the natural biosphere but were born, so to speak, directly into the technosphere. Only the information on how to construct them was appropriated. In the same way, human organisms, whose function is essential to the existence of the technosphere, can be viewed as newly constructed technological parts based on old design information (DNA) captured from the biosphere.

Besides appropriating mass and information, a geological paradigm must capture for its own use some fraction of existing energy stores and flows. Because geological paradigms are dynamical systems that help to define the global environment, their power requirements are large (data in the following from Sassoon *et al.* 2009; see also Hermann 2006). Units of power are terawatts, TW (10^{12} W). From an incident solar flux of *c.* 162 000 TW, the atmosphere absorbs energy (or destroys or converts exergy, or usable energy) at a rate of 31 000 TW, of which *c.* 870 TW appears as kinetic energy of the winds. The hydrosphere absorbs another 41 000 TW in evaporating water. Absorption by the biosphere of 15 000 TW leads via photosynthesis to the generation of chemical energy in plant matter at a rate of *c.* 90 TW, of which the technosphere appropriates almost 10 TW (agriculture and forestry). Fossil fuels, uranium and renewable energy sources provide energy to the technosphere at a rate of *c.* 17 TW (IEA 2012). This is an appreciable fraction of the geothermal energy flux (32 TW), the biochemical energy flux (90 TW) and the gravitational power load of the world's rivers (7 TW), suggesting the susceptibility of parts of the land surface, the biosphere and the fluvial portion of the hydrosphere to disruption or appropriation by the technosphere.

Geological paradigms emerge when a large energy source is available and the environment contains many similar parts whose individual properties provide a basis for collective use of available energy to perform work. Molecules of oxygen, nitrogen and water are abundant; they form fluids under a wide range of environmental conditions

(conditions of temperature and pressure that their own collective properties help create), and, under forcing powered ultimately by solar radiation, they constitute large aggregated volumes of water and air that flow across the planet. Because each molecule of a given type has identical properties, essentially all molecules that can potentially participate in their respective cycles actually will participate if the energy supply is large enough. In a similar way, for humans, the distribution over a large number of individuals of characteristics such as acquisitiveness that result in demand for goods and services requiring energy to manufacture and deliver, renders human parts susceptible to entrainment into large-scale collective behaviour in the presence of a suitably large accessible energy source, such as fossil fuels. As with the molecules that support the atmospheric and hydrological paradigms, similarities between parts make it probable that each human participates in the dynamics of the emerging technological paradigm.

If the inclination to acquire is not saturated at a given level of average per capita energy consumption, there is scope for technological energy consumption to increase. The 'principle of maximum entropy production' (PMEP) asserts that sufficiently complex dynamic systems will evolve to a state in which usable energy is consumed as fast as possible, consistent with extant constraints (Kleidon & Lorenz 2005). Although not proven (but see Niven 2009), PMEP appears to have had some success in application to energy consumption by the Earth's atmosphere in predicting the distribution of average meridional temperature and cloudiness (Paltridge 1975). PMEP applies to steady-state systems and thus is not expected to describe the present state of rapidly accelerating technological energy use. However, if a state of higher energy consumption can potentially be realized, that is, if there are no constraints that prohibit a faster rate of energy consumption, then PMEP suggests that the technosphere will tend to evolve towards increased appropriation of usable energy (Haff 2013), bearing its human parts along in the process.

Conservative nature of geological paradigms

Geological paradigms are conservative. The elements and functionality of pre-existing paradigms can survive the emergence of a new geological paradigm, even in the face of capture of some of their structural and metabolic resources. In a thermodynamic transition occurring less than 200 hundred million years after the formation of the molten planet, the Earth's surface solidified (Wilde *et al.* 2001), representing an early crustal paradigm; the pre-existing magma ocean did not disappear, however, but lived on just beneath. Later in Earth history, the biosphere arose and modified the atmosphere (e.g. by oxygenation), the hydrosphere (by influencing precipitation patterns, groundwater levels and river flow) and the solid Earth surface (by enhancing weathering and soil formation and by impeding erosion), but the earlier paradigms continued to function, if in modified ways. The winds still blew, the rivers still flowed, and mountain ranges still rose and fell. More recently, technology has appropriated large quantities of 'natural resources' from the biosphere, hydrosphere and lithosphere, but, at least for the time being, the ancient paradigms continue as globally organized systems. One reason for the conservatism of paradigms is that the conditions required for their emergence and function were defined to a large extent by the pre-existing environment, which provided a framework conducive to the gestation of the new paradigm and without which ramp-up of the emergent dynamics would not be possible. For example, plate tectonics required the continued existence of the fluid Earth – if in modified form – in order for plates to move across the Earth's surface, to subduct and for new plates to form, while biological systems required the continuing function of the hydrological cycle, and so on.

Technology emerged on an Earth filled with materials, energy stores and flows, and functioning dynamical systems that were products or components of these earlier natural paradigms. Processes, systems and stores that today remain essential to human well-being (and thus to technological function) comprise the elements of what is called natural capital (e.g. Daily 1997). Natural capital includes the Earth's thick, fertile soils, mineral resources, bacterial and chemical populations that breakdown or recycle wastes, sources of fresh water, soil mechanisms that filter or detoxify contaminants, a reasonably stable and equable climate, and biological diversity, among many other examples. A principal criticism of the rapidly expanding growth of technology centres on its non-conservative function, that is, on the effects of the destruction of natural capital as measured by (high) rates of, for example, urban growth, land-use change, destruction of rain forests, extinction of biological populations, and consumption of finite stores of oil, gas, coal and other Earth materials on which our well-being depends. Their conservative nature still serves earlier paradigms, but technology seems to be on a course to abandon the conservative dynamics that has been essential for its emergence and function. Technology is the first geological paradigm complex enough to become aware, through its human components, of the essential contribution to its own existence of the support provided by established paradigms. Whether this awareness will lead to the conservation of a sufficient quantity of natural capital to maintain technological function (and thus the well-being of humans) is the basic question of environmental science. The answer to this question may also determine whether or not the technosphere will, in the long run, become an established rather than a failed geological paradigm.

Recycling of mass resources

Recycling of mass resources plays an essential role in the function of established geological paradigms. A water drop falling from a cloud on its way to the ocean must be replaced by an equivalent mass of evaporated (recycled) water or the hydrosphere will soon cease operation for lack of atmospheric moisture. Whether it be the lithosphere, hydrosphere, atmosphere, biosphere or technosphere, over a long enough period of time the (finite) mass resources on which paradigm function depends will be drawn down in the absence of recycling, leading to limitations on paradigmatic activity. In biology, Liebig's law of the minimum (Odum 1971) stipulates that plant population growth is controlled by the availability of the 'scarcest' nutrient. Return of nutrients to the soil by organisms is a recycling process that helps limit the effects of scarcity of essential minerals. The technosphere recycles some kinds of materials. Over much of the history of technology, metals like gold and silver have served an important role in facilitating mass and energy flows through society. Coinage was not discarded after a sale, but instead soon returned to the monetary circulatory system. Many other metals are important nutrients for technological metabolism (e.g. rare earth metals), and a few are used as major structural elements (principally iron and aluminium). As ore deposits are exhausted, metal recycling (e.g. Rauch & Pacyna 2009; Goldstein 2012) will become increasingly necessary if the technosphere is to maintain its high level of metabolic function. However, at the present time, the technosphere is a poor recycler of many of the critical resources that it uses.

In a closed environment like the Earth (essentially no mass input or output), every metabolizing system must eventually recycle its own waste products (or rely on other systems to do so), otherwise accumulation of spent material (i.e. pollutants) will impair system function. If leaf litter produced by a forest were not recycled it would soon build up to a point where the trees that produced it were buried by their own detritus. In the case of the technosphere, the most important example of mass pollution may be the buildup of atmospheric carbon dioxide. Warming of the atmosphere is likely to lead to rapid sea-level rise and submergence of coastal cities and infrastructure, putting as many as 200 million people at risk of becoming environmental refugees (Myers 2002). The ensuing climate disruption could also lead to large-scale crop failures through drought, flood, pest infestations or other possible impacts of climate destabilization (Ehrlich & Ehrlich 2013). Failure of even a modest fraction of the industrial global agriculture enterprise (which in its use of nitrogen fixed through the Haber–Bosch process is estimated to support *c.* 40% of the world's population; Smil 1999) would be a disaster for humanity. If continued carbon emission without recycling should ultimately degrade or eliminate the participation of a large enough number of humans in their active roles as technospheric parts, then the metabolism of the technosphere would decline.

The recycling shortcomings of the technosphere with respect to carbon and other essential inputs may make it appear a poor candidate for a new paradigm, especially when compared to the ability of, say, the biosphere to recycle its own waste. However, this comparison is somewhat misleading. Over geological history the biosphere has also failed to recycle its own wastes, with catastrophic consequences for many species. In the time leading up to the Great Oxidation Event about 2.4 billion years ago (e.g. Lenton & Watson 2011), proto-cyanobacteria evolved that were capable of oxygenic photosynthesis, using sunlight to split water, freeing up electrons for metabolic use. These novel organisms, however, failed to recycle the resulting toxic oxygen waste stream. As a consequence of loading the atmosphere with large quantities of poisonous waste, much of the rest of the biosphere was forced to evolve mechanisms to detoxify or respire oxygen, or to retreat to sequestered anaerobic environments (Sessions *et al.* 2009).

Today, the biosphere is, in the main, an effective recycler of its own wastes, partly as a consequence of the long period of time that evolution has had to select against organisms that degrade their own ability to reproduce by polluting their environment or that are unable to tolerate levels of environmental pollution generated by the rest of the biosphere. That technology exhibits a massive failure to recycle may be a consequence of its status as a new geological phenomenon. Over a long enough period of time, mass flow loops may close. Whether the rapid pace of technological change will generate effective recycling mechanisms for the resulting waste stream soon enough to limit having a large impact on the climate, and hence a braking effect on technology, is unknown. Although technology-driven climate change may exacerbate key vulnerabilities in areas such as food supply, infrastructure and human health (Schneider *et al.* 2007), to the detriment of human as well as technospheric well-being, the ability to recycle at a rate that can maintain current environmental conditions is not necessarily a requirement for the continued emergence of a technological paradigm. The experience of the biosphere in the wake of the Great Oxidation Event shows that even a massive failure to recycle environmental pollutants need not necessarily signal the end of a paradigm. If technological recycling fails to spin up soon enough to avoid catastrophic (as seen by humans) reorganization of global Earth function, that may not be the end of the game but just a change in rules. In the extreme case that *Homo sapiens* became extinct, then presumably the emergence of the technological paradigm would stop, and the evidence of the brief technological excursion by the Earth would be compressed to a thin line in the future sedimentary record. Technology would have been an event, analogous to the Cretaceous–Tertiary impact, not a new paradigm of Earth function. However, if human extinction does not occur, then the technosphere may survive the large changes in environmental variables that it is causing, as has been the case for the biosphere. By adjusting to the new Earth environment, in company with a stressed but surviving human population, the technosphere may evolve in ways we cannot predict or perhaps even imagine.

Autonomy of paradigms

Natural geological paradigms that emerged before humans came on the scene were autonomous. That is, by definition they needed no human deliberation or control in order to function, a condition that might seem to distinguish them from the technological paradigm. We tend to see technology as a human construct under our control. Technology seems to us not autonomous but critically dependent on humans and human actions. Humans design, manufacture, construct, deploy and maintain key elements of the technosphere. Certainly, the technosphere could not exist without its human component. On the other hand, neither can any other system maintain its existence without the participation of its components – the hydrological cycle could not exist without the supporting activity of its water molecules, the rock cycle without its mineral components, and so on. That the technosphere requires for its function the participation of certain critical parts, even if they are people, does not by itself distinguish it from other geological paradigms.

One apparent difference between the technosphere and other paradigms is that for classical geological paradigms like the hydrosphere, the parts seem to come along for the ride: a drop or molecule of water is borne along in the hydrological cycle like an object on a conveyor belt. Technological systems on the other hand are often seen as driven by human decision and human actions (Haff 2012). People are proactive; technology is reactive. The computer responds to human intentions and commands. Not only did people design and manufacture it, but we personally decide how to use it – when to plug it in, turn it on, what keys to hit, and so on. The autonomous nature of technology comes more clearly into view when we move beyond technological artefacts that people interact with directly and consider larger technological systems, which contain people among their parts.

A refrigerator is a technological artefact over which the owner has some control, but the electric power grid to which the refrigerator is connected is not under the owner's control. In fact, the power grid is not really under anyone's control. It is quasi-autonomous in the sense that it cannot be shut down by human decision except for short periods of time, and most of its function occurs without human intervention or even knowledge. However, because modern complex society could not function without an essentially continuous supply of electricity, the grid must be functional most of the time for the present technosphere to exist, independent of the actions of any person or group of persons. Any large-scale attempt to shut down the flow of electricity would meet immediate resistance. Periods of disruption or failure in a tightly coupled system, like a regional component of a national power grid, are possible in the event of sabotage or the chance occurrence of rare but small cascading events, but such failures are more in the nature of glitches from which recovery is eventually possible rather than examples of terminal failure. The makeup of the grid with its rugged components, backup systems, redundancy, reserve capacity, alarms and other security systems make it much more than just a system that satisfies the basic physical requirements for generating and distributing electric power. If this is all it were, then the grid would have the same characteristics as a table-top physics experiment that could be arbitrarily taken down or reconfigured on the fly by human decision. Instead, the grid bristles with protective capabilities that help avoid or defend against challenges, human or otherwise, to its basic function. It is as if the table-top laboratory apparatus were to get a mind of its own and strongly resist any actions by the experimenter that threatened its functionality. In other words, the power grid is autonomous.

Moreover, the property of autonomy is sufficiently well developed that it trickles down even to the technological artefacts that we normally think we control. We open and close the refrigerator on impulse, but it is very difficult for a typical 'owner' to unplug his refrigerator for any extended period of time – say for a few days. This is an objective observation that the reader can test for himself. The refrigerator shares a part of the autonomy of the grid to which it is connected and resists challenges to its functionality, especially attempts to cut off its power source.

The technosphere thus exhibits a number of properties of earlier geological paradigms. It is autonomous. It is a global phenomenon. It appropriates Earth resources, including energy, mass and information, for its own uses on a large scale. However, unlike earlier Earth paradigms, which recycle most of their waste products, the technosphere does little recycling. The future of the technosphere as a paradigm rather than just an episode in Earth history is contingent upon the emergence of effective recycling mechanisms.

Implications of the new paradigm

There is strategic value in looking at technology from the 'outside' as an emerging geological phenomenon. The problem of coping with the consequences of technology and technological change offers different answers, depending on whether we think of technology primarily as a human-generated and controlled phenomenon, or whether we look at it as a quasi-autonomous phenomenon that in effect operates according to its own dynamics. Looking at technology from the inside, we tend to formulate solutions to the problem of natural capital degradation that draw a straight line from the seemingly excessive use of resources like energy to policies that might restrict such use. However, prescriptions such as constricting the resource stream on which the function of technology depends, for example by taxing carbon, tend to encounter resistance. Technology is not passive but has evolved mechanisms for its own defence – a requirement of any dynamic system whose longevity is measured in a large number of internal clock cycles, such as the time between cell phone bills or elections. The most important of these defenses is preemptive in nature and takes advantage of fundamental properties of its human parts, especially the property of acquisitiveness. Technology defends its mode of operation primarily by offering incentives such as abundant food, medicines, instant communication channels and other desiderata that bind, or even addict, humans to the system that produces them, as well as by less subtle mechanisms expressed via legal, judicial, political, military and other elements of the technological armory. The upshot is that attempts to ratchet back the rate of energy use and of consumption of the other resources on which this cornucopia depends, or to interfere with the continuing diversification and penetration of technology as it seeks out new sources of energy and material resources, are automatically resisted by the feedback loops on which technological metabolism is based. This is the nature of a positive feedback loop – it has intrinsic stability against disruption – the twist here being that humans, as sentient components of such a loop, may feel surprise or dismay upon realizing their enforced participation in a dynamics they thought they controlled. Policies that are based only on a consideration of future human well-being and do not take into account the needs of technology, especially the need to continue metabolizing at a high rate – which is the source of the constraints and incentives that channel human behaviour towards technology-friendly activities and is thus the *sine qua non* of technology – are likely to fail or be slow to implement because they consider the implicit two-way compact between system and parts only from the viewpoint of the parts.

With respect to human well-being, a high rate of technological energy consumption is, by itself, not the central problem. Global warming is not a necessary consequence of a high rate of energy use, but of the lack of adequate recycling mechanisms. The hydrosphere consumes energy more than a thousand times faster than the technosphere, but it recycles its own waste (fallen rainwater). The technosphere, in burning fossil fuels, operates without any provision to recycle a major waste product, carbon. From the point of view of an autonomous technosphere, climate change is not a problem to be solved by using less energy, but by using more energy. As seen from the dynamics of the Carnot engine, whatever useful work is done by a system, additional energy is required to power a recycling mechanism.

Efforts to ramp up 'renewable' energy sources such as those based on wind or photovoltaics offer new ways to use energy to do work without the principal drawback of fossil-fuel combustion, that is, without the need to recycle carbon. However, renewables technology is still technology. Trying to fix the climate problem by turning to renewables may therefore not lead where it seems. This will be the case if opportunities offered by renewables appear different to the technosphere than to humans. The emergence of new technological subsystems that can capture abundant but previously unavailable or little-used renewable resources like that provided by sunlight may just as likely be an opening move in the expansion of the technosphere towards massive increases in the use of these new resources than simply a way to substitute new cleaner energy sources for older sources that degrade the environment. For example, with regard to solar energy, there is no reason to expect that the technosphere will limit itself to extracting energy from just the fraction of the solar radiation flux that happens to be incident on the planetary disk. Technology is already extending itself into space and it may not be indifferent to the large fluxes of solar energy available there. Geoengineering proposals to counteract global warming by deploying a cloud of refractive 'flyers' in space to deflect incident solar energy away from the Earth are already on the books (Angel 2006). However, deflecting sunlight is throwing away usable energy. It is a short step to see how this idea might be retuned to do just the opposite – to capture the energy of photons in space that would have missed the Earth and then transmit the energy down to the Earth's surface (e.g. in the form of microwaves; Glaser 1968). This is not a prediction, but illustrates how looking at the technosphere from the outside provides its own perspective on possible Earth futures.

Whatever the future of particular renewable energy sources, the driving forces are already in place for transition to rates of energy consumption that are larger than, and perhaps much larger than, the current power level of fossil-fuel use. Two such drivers are the unsatisfied aspirations of the world's energy-poor humans and the need to recycle wastes. Average global per capita energy use (2010) is about 40% of the rate of per capita energy consumption in Organisation for Economic Co-operation and Development (OECD) countries (IEA 2012). Closing this gap would require an increase in the power level of the technosphere by a factor of about 2.5, from 17 TW to 42 TW. To bring world per capita energy use to the level of North American consumption would require an increase by a factor of 4 to a total of 68 TW. The need to recycle metabolic waste products generated by an increasingly energy-consumptive population would further increase power demand. With the appearance of effective technospheric recycling mechanisms, a third easily imagined driver of future energy consumption would be the realization that the current level of per capita energy use in rich societies does not represent any kind of natural or intrinsic or perhaps even desirable limit to the rate at which one might use energy. If the lives of people in rich societies today are supported by energy use at an average per capita rate of 10 kW (US in 2011; EIA 2012), the question may arise of why not 20 kW in the future, for everyone? Real-time decisions and choices by humans within the context of their cultural and political institutions will help determine the details of how potential drivers translate into future energy use. However, looking at the world from the point of view of a still emerging geological paradigm we should expect that the framework for our decisions and choices will be shaped and guided by the needs of the technosphere as well as by human needs.

I would like to thank N. Cassar and E. Goldstein for critical commentary and for suggestions for improving the manuscript.

References

ANGEL, R. 2006. Feasibility of cooling the Earth with a cloud of small spacecraft near the inner Lagrange point (L1). *Proceedings of the National Academy of Science of the USA*, **103**, 17 184–17 189. http://dx.doi.org/10.1073/pnas.0608163103

BARNOSKY, A. D., MATZKE, N. *ET AL.* 2011. Has the Earth's sixth mass extinction already arrived? *Nature*, **471**, 51–57. http://dx.doi.org/10.1038/nature09678

CRUTZEN, P. J. 2002. Geology of mankind. *Nature*, **415**, 23.

CRUTZEN, P. J. & STOERMER, E. F. 2000. The Anthropocene. *Global Change Newsletter*, **41**, 17–18.

DAILY, G. C. (ed.) 1997. *Nature's Services: Societal Dependence on Natural Ecosystems*. Island Press, Covelo.

EHRLICH, P. R. & EHRLICH, A. H. 2013. Can a collapse of global civilization be avoided? *Proceedings of the Royal Society B*, **280**, 20122845. http://dx.doi.org/10.1098/rspb.2012.2845

EIA 2012. *Annual Energy Review 2011*. US Energy Information Agency, DOE/EIA-0384(2011). http://www.eia.gov/totalenergy/data/annual/pdf/aer.pdf

FAO 2011. *FAOSTAT*. Food and Agricultural Organization of the United Nations. http://faostat.fao.org/site/377/DesktopDefault.aspx?PageID=377#ancor

GLASER, P. E. 1968. Power from the Sun: its future. *Science*, **162**, 857–861.

GOLDSTEIN, E. 2012. *Possible dynamics of technological metals in the Anthropocene. 2012 Fall Meeting of the American Geophysical Union.* Abstract GC53C-1292. http://fallmeeting.agu.org/2012/eposters/eposter/gc53c-1292/

HAFF, P. K. 2010. Hillslopes, rivers, plows, and trucks: mass transport on Earth's surface by natural and technological processes. *Earth Surface Processes and Landforms*, **35**, 1157–1166. http://dx.doi.org/10.1002/esp.1902

HAFF, P. K. 2012. Technology and human purpose: the problem of solids transport on the Earth's surface. *Earth System Dynamics*, **3**, 417–431. http://dx.doi.org/10.5194/esd-3-149-2012

HAFF, P. K. 2013. Maximum entropy production by technology. Accepted for publication. *In*: DEWAR, R. C., LINEWEAVER, C., NIVEN, R. & REGENAUER-LIEB, K. (eds) *Beyond the Second Law: Entropy Production and Non-Equilibrium Systems*. Springer, Berlin, in press

HERMANN, W. A. 2006. Quantifying global exergy resources. *Energy*, **31**, 1685–1702.

IEA 2012. *Key World Energy Statistics*. International Energy Agency. http://www.iea.org/publications/freepublications/publication/kwes.pdf

IPCC 2007. *IPCC Fourth Assessment Report: Climate Change 2007. Intergovernmental Panel on Climate Change*. http://www.ipcc.ch/publications_and_data/publications_and_data_reports.shtml#1

KLEIDON, A. & LORENZ, R. (eds) 2005. *Non-Equilibrium Thermodynamics and the Production of Entropy: Life, Earth, and Beyond*. Springer, Berlin.

LAUGHLIN, R. B. & PINES, D. 2000. The theory of everything. *Proceedings of the National Academy of*

Science of the USA, **97**, 28–31. http://dx.doi.org/10.1073/pnas.97.1.28

Lenton, T. & Watson, A. 2011. *Revolutions That Made the Earth.* Oxford University Press, New York.

Myers, N. 2002. Environmental refugees: a growing phenomenon of the 21st century. *Philosophical Transactions of the Royal Society London B*, **357**, 609–613. http://dx.doi.org/10.1098/rstb.2001.0953

Nilsson, C., Reidy, C. A., Dynesius, M. & Revenga, C. 2005. Fragmentation and flow regulation of the world's large river systems. *Science*, **308**, 405–408.

Niven, R. K. 2009. Steady state of a dissipative flow-controlled system and the maximum entropy production principle. *Physical Review E*, **80**, 021113. http://dx.doi.org/10.1103/PhysRevE.80.021113

Odum, E. P. 1971. *Fundamentals of Ecology.* 3rd edn. Saunders, Philadelphia.

Paltridge, G. W. 1975. Global dynamics and climate – a system of minimum entropy exchange. *Quarterly Journal of the Royal Meteorological Society*, **101**, 475–484.

Rauch, J. N. & Pacyna, J. M. 2009. Earth's global Ag, Al, Cr, Cu, Fe, Ni, Pb, and Zn cycles. *Global Biogeochemical Cycles*, **23**, GB2001. http://dx.doi.org/10.1029/2008GB003376

Sassoon, R. E., Hermann, W. A., Hsiao, I-C., Miljkovic, L., Simon, A. J. & Benson, S. M. 2009. Quantifying the flow of exergy and carbon through the natural and human systems. *In*: Collins, R. T. (ed.) *Materials for Renewable Energy at the Society, Technology Nexus. Materials Research Society Symposium Proceedings* 1170E, 1170-R01-03.

Schneider, S. H., Semenov, S. et al. 2007. Assessing key vulnerabilities and the risk from climate change. *In*: Parry, M. L., Canziani, O. F., Palutikof, J. P., van der Linden, P. J. & Hanson, C. E. (eds) *Climate Change 2007: Impacts, Adaptation and Vulnerability. Contribution of Working Group II to the Fourth Assessment Report of the Intergovernmental Panel on Climate Change.* Cambridge University Press, Cambridge, 779–810.

Scott, J. C. 2010. *The Art of Not Being Governed.* Yale University Press, New Haven.

Sessions, A. L., Doughty, D. M., Welander, P. V., Summons, R. E. & Newman, D. K. 2009. The continuing puzzle of the great oxidation event. *Current Biology*, **19**, R567–R574. http://dx.doi.org/10.1016/j.cub.2009.05.054

Smil, V. 1999. Detonator of the population explosion. *Nature*, **400**, 415.

Steffen, W., Grinevald, J., Crutzen, P. & McNeill, J. 2011. The Anthropocene: conceptual and historical perspectives. *Philosophical Transactions of the Royal Society A*, **369**, 842–867. http://dx.doi.org/10.1098/rsta.2010.0327

US Census Bureau 2012. *International Programs, World Population, Historical Estimates of World Population.* http://www.census.gov/population/international/data/worldpop/table_history.php

Vitousek, P. M., Ehrlich, P. R., Ehrlich, A. H. & Matson, P. A. 1986. Human appropriation of the products of photosynthesis. *BioScience*, **36**, 368–373.

Vitousek, P. M., Mooney, H. A., Lubchenco, J. & Melillo, J. M. 1997. Human domination of Earth's ecosystems. *Science*, **277**, 494–499. http://dx.doi.org/10.1126/science.277.5325.494

Vorosmarty, C. J., Meybeck, M., Balazs, F., Sharma, K., Green, P. & Syvitski, J. P. M. 2003. Anthropogenic sediment retention: major global impact from registered river impoundments. *Global and Planetary Change*, **39**, 169–190.

Werner, B. T. 2003. Modeling landforms as self-organized, hierarchical dynamical systems. *In*: Wilcock, P. R. & Iverson, R. M. (eds) *Prediction in Geomorphology, Geophysical Monograph Series.* American Geophysical Union, Washington, **135**, 133–150. http://dx.doi.org/10.1029/135GM10

Wilde, S. A., Valley, J. W., Peck, W. H. & Graham, C. M. 2001. Evidence from detrital zircons for the existence of continental crust and oceans on the Earth 4.4 Gyr ago. *Nature*, **409**, 175–178. http://dx.doi.org/10.1038/35051550

Index

Page numbers in *italics* refer to Figures. Page numbers in **bold** refer to Tables.